T0203117

Fundamental Solutions of Linear Partial Differential Operators

Norbert Ortner • Peter Wagner

Fundamental Solutions of Linear Partial Differential Operators

Theory and Practice

Springer

Norbert Ortner
Department of Mathematics
University of Innsbruck
Innsbruck, Austria

Peter Wagner
Faculty of Engineering Science
University of Innsbruck
Innsbruck, Austria

ISBN 978-3-319-36799-6   ISBN 978-3-319-20140-5   (eBook)
DOI 10.1007/978-3-319-20140-5

Mathematics Subject Classification (2010): 35E05, 35L25, 35J30, 35K40, 44A10, 46F12, 74H05, 74B05, 78A25

Springer Cham Heidelberg New York Dordrecht London
© Springer International Publishing Switzerland 2015
Softcover reprint of the hardcover 1st edition 2015

Printed on acid-free paper

Springer International Publishing AG Switzerland is part of Springer Science+Business Media (www.springer.com)

We dedicate this book to our wives Maria and Sophia

Introduction

The clarification of the notion of a *fundamental solution* of a linear differential operator

$$P(\partial) = \sum_{|\alpha|\leq m} a_\alpha \partial^\alpha, \quad m \in \mathbf{N}_0, \ a_\alpha \in \mathbf{C}, \ \alpha \in \mathbf{N}_0^n, \ \partial^\alpha = \partial_1^{\alpha_1} \dots \partial_n^{\alpha_n}, \ \partial_j = \frac{\partial}{\partial x_j},$$

is attributed to *Laurent Schwartz* in Malgrange [175, p. 29]. L. Schwartz writes in his seminal treatise "Théorie des distributions":

The usual definition of a fundamental solution as a classical solution of the homogeneous equation with a certain singularity has to be completely rejected. Instead, $E \in \mathcal{D}'(\mathbf{R}^n)$ is a fundamental solution of $P(\partial)$ if and only if the equation $P(\partial)E = \delta$ holds in the sense of distributions, see Schwartz [246, pp. 135, 136].

That distribution theory is necessary not just for the definition of a fundamental solution, but also for its calculation is shown already by the trivial operator $P(\partial) = 1$ (where $E = \delta$), or less trivially, by the simple transport operator $P(\partial) = \partial_1 - \partial_2$, whose fundamental solutions are given by $E = Y(x_1)\delta(x_1 + x_2) + T(x_1 + x_2)$, Y denoting the Heaviside function and T being an arbitrary distribution in $\mathcal{D}'(\mathbf{R}^1)$.

Already in this simple example, two questions immediately arise:

(i) What is $T(x_1 + x_2)$?   (ii) How is $Y(x_1)\delta(x_1 + x_2)$ defined?

Therefore, we collect in **Chap. 1: Distributions and Fundamental Solutions** some facts from distribution theory. E.g., question (i) is answered in Sect. 1.2, where the *composition of distributions with smooth maps* and the *pullback of distributions* are investigated. Similarly, for (ii), one needs pullbacks and multiplication of distributions, and this leads to (an easy case) of a *distribution with support on a hypersurface* (see Example 1.2.14), i.e.,

$$Y(x_1)\delta(x_1 + x_2) : \mathcal{D}(\mathbf{R}^2) \longrightarrow \mathbf{C} : \phi \longmapsto \int_0^\infty \phi(t, -t) \, dt.$$

Our presentation of distribution theory differs from the usual introductions to distributions perhaps mainly by two topics: Definition of *single and double layers* and use of the *jump formula* for the verification of fundamental solutions (see Sect. 1.3), and the treatment of *distribution-valued functions* (see Sect. 1.4).

In this book, only *operators and systems with constant coefficients* are considered. The systems are taken from applications in physics, operators of higher order usually appear as determinants of such systems. The *Malgrange–Ehrenpreis theorem*, see Proposition 2.2.1 in **Chap. 2: General Principles for Fundamental Solutions**, asserts that each not identically vanishing differential operator has a fundamental solution. It can be represented as a (generalized) inverse Fourier-Laplace transform, hence, "in principle" as an n-fold definite integral. Our main goal consists in deriving representations of fundamental solutions which are "as simple as possible". Let us now try to explain this expression by discussing three examples of differential operators of increasing complexity.

In 1788, P.S. de Laplace used the algebraic function $|x|^{-1}$ as a fundamental solution of the operator $\Delta_3 = \partial_1^2 + \partial_2^2 + \partial_3^2$, which justly bears his name. The above definition of a fundamental solution by L. Schwartz requires the additional factor $-1/(4\pi)$, i.e.,

$$E = -\frac{1}{4\pi |x|} = -\frac{1}{4\pi \sqrt{x_1^2 + x_2^2 + x_3^2}} \in L^1_{\mathrm{loc}}(\mathbf{R}^3) \subset \mathcal{D}'(\mathbf{R}^3)$$

fulfills $\Delta_3 E = \delta$, see Example 1.3.14 (a) and the remark in Zeilon [306, p. 2]. Hence, the Laplacean Δ_3 possesses a fundamental solution in the form of an algebraic function. Furthermore, E is uniquely determined by either of the properties of being homogeneous or of vanishing at infinity, respectively (see Sect. 2.4).

In 1959, S.L. Sobolev found the representation

$$E(t, x) = Y(t) \left[\frac{Y(-x_1 x_2 x_3)}{\pi^2 t} K_0 \big(2\sqrt{-x_1 x_2 x_3 / t}\,\big) - \frac{Y(x_1 x_2 x_3)}{2\pi t} N_0 \big(2\sqrt{x_1 x_2 x_3 / t}\,\big) \right]$$

(see Examples 2.3.8 and 2.6.4) of the unique temperate fundamental solution with support in the half-space $\{(t, x) \in \mathbf{R}^4; t \geq 0\}$ (see Proposition 2.4.13 for the uniqueness) of the operator $P(\partial) = \partial_t - \partial_1 \partial_2 \partial_3$. Hence the "simplest" representation of E in this case is by means of the higher transcendental functions K_0, N_0, i.e., Bessel functions which are given by definite integrals over elementary functions. The operator $\partial_t - \partial_1 \partial_2 \partial_3$ is not hyperbolic. We call operators of this type *quasihyperbolic*, see Example 2.2.2, Definition and Proposition 2.4.13, and **Chap. 4: Quasihyperbolic Systems**, in contrast to the literature where such operators are often called "correct in the sense of Petrovsky", cf. Gindikin and Volevich [109, p. 168], or "weakly parabolic", cf. Dautray and Lions [53, p. 222].

Very often, linear differential operators occurring in physics are products of lower order operators or can be reduced to such operators. We show in **Chap. 3: Parameter Integration** how fundamental solutions in such cases can be represented by

integrals with respect to parameters generating the convex combinations of these
lower order operators. As our third example, let us consider Timoshenko's beam
operator, i.e.,

$$\partial_t^2 + \frac{EI}{\rho A}\partial_x^4 + \frac{\rho I}{GA\kappa}\partial_t^4 - \frac{I}{A}\left(1 + \frac{E}{G\kappa}\right)\partial_t^2\partial_x^2,$$

see Examples 2.4.14, 3.5.4, 4.1.6, and 4.3.7. It can be rewritten in the form

$$R(\partial) = (\partial_t^2 - a\partial_x^2 + b)^2 - (c\partial_x^2 - d)^2 - e^2$$

and the forward fundamental solution E of $R(\partial)$ can be represented as a parameter
integral with respect to λ, μ over the forward fundamental solutions $E_{\lambda,\mu}$ of the
operators $[\partial_t^2 - a\partial_x^2 + b + \lambda(c\partial_x^2 - d) + \mu e]^2$. This leads to a representation of E as a
definite integral over Bessel functions, see formulas (3.5.9/3.5.10) in Example 3.5.4.
This representation of E already yields a precise description of the support and of
the singular support, respectively, of E.

Alternatively, the singularities of the forward fundamental solution/matrix of
such a quasihyperbolic operator/system can be deduced a priori from the representation by Laplace inversion (see Sects. 4.1, 4.2, and 4.3). Let us also note that Laplace
inversion in such cases often leads to completely different integral representations
of E, see Example 4.1.6 for Timoshenko's operator. For *homogeneous hyperbolic
operators*, such representation formulas are traditionally known under the name of
"Herglotz–Petrovsky formulas", and we consider these formulas in Sect. 4.4 and the
analogues for *homogeneous elliptic operators* in Sect. 5.2 in **Chap. 5: Fundamental
Matrices of Homogeneous Systems**.

Let us finally summarize that the main goal of this book consists in presenting the
most important procedures for constructing fundamental solutions and in illustrating
them by physically relevant examples. Clearly, we could not include each and every
method and/or operator appearing in the literature. Furthermore, we emphasize
that the many applications of fundamental solutions for theoretical purposes (e.g.,
hypoellipticity, surjectivity of operators, etc.) as well as in practical matters (solution
of Cauchy or boundary value or mixed problems in the natural sciences) have barely
been touched upon in this treatise for the lack of space.

Contents

Chapter 1
Distributions and Fundamental Solutions

This chapter is an introduction to distribution theory illustrated by the verification of fundamental solutions of the classical operators Δ_n^k, $(\lambda - \Delta_n)^k$, $(\Delta_n + \lambda)^k$, $\partial_{\bar{z}}$, $(\partial_t^2 - \Delta_n)^k$, $\partial_1 \cdots \partial_k$, $(\partial_t - \lambda \Delta_n)^k$, $(\partial_t \pm i\lambda \Delta_n)^k$ (for $\lambda > 0, k \in \mathbf{N}$), which are listed in Laurent Schwartz' famous book "Théorie des distributions," see Schwartz [246]. The theory of distributions was developed by L. Schwartz in order to provide a suitable tool for solving problems in the analysis of *several variables*, e.g., in the theory of partial differential equations or in many-dimensional harmonic analysis. Taking into account this emphasis on many dimensions, we present mainly examples in $\mathbf{R}^n$ instead of $\mathbf{R}^1$. For example, in Example 1.4.10 we derive the distributional differentiation formula

$$\partial_j \partial_k \left(\frac{|x|^{2-n} - 1}{2 - n} \right) = \mathrm{vp}\left(\frac{|x|^2 \delta_{jk} - nx_j x_k}{|x|^{n+2}} \right) + \frac{|\mathbf{S}^{n-1}|}{n} \delta_{jk} \delta, \quad 1 \le j, k \le n, \ n \ne 2.$$

In contrast to the classical textbooks on distribution theory (by Barros-Neto [8]; Blanchard and Brüning [16]; Donoghue [61]; Duistermaat and Kolk [65]; Friedlander and Joshi [84]; Gel'fand and Shilov [104]; Grubb [118]; Hervé [129]; Hirsch and Lacombe [131]; Horváth [141]; Petersen [228]; Strichartz [266]; Vladimirov [280]; Zuily [309], etc.), we use the distributional *jump formula* for differentiation in order to verify the formula above and many of the fundamental solutions to the above list of classical operators, see Examples 1.3.10 and 1.3.11.

In Sect. 1.2 we define the *pullback of distributions by mappings*. Due to its importance in Sect. 4.4, we give explicit expressions for $h_1^*(\delta_{a_1}) h_2^*(\delta_{a_2}) = \delta_{a_1}(h_1)\delta_{a_2}(h_2) = \delta(h_1(x) - a_1)\delta(h_2(x) - a_2)$. In Sect. 1.4, we introduce *distribution-valued functions*. Apart from the ubiquitous use of analytic continuation for the construction of fundamental solutions, this is also motivated by providing a theoretical foundation for the partial Fourier transform (in contrast to ad hoc constructions as, e.g., in Treves [274]).

© Springer International Publishing Switzerland 2015
N. Ortner, P. Wagner, *Fundamental Solutions of Linear Partial Differential Operators*, DOI 10.1007/978-3-319-20140-5_1

The *convolvability* and the *convolution* of distributions are investigated in Sect. 1.5 using L. Schwartz' general definition: $S, T \in \mathcal{D}'(\mathbf{R}^n)$ are convolvable if and only if $S(x - y)T(y) \in \mathcal{D}'(\mathbf{R}_x^n) \hat{\otimes} \mathcal{D}'_{L^1}(\mathbf{R}_y^n)$. Their convolution is then defined as

$$S * T = \langle 1_y, S(x - y)T(y) \rangle = \int_{\mathbf{R}^n} S(x - y)T(y) \, dy.$$

These definitions require a study of the space $\mathcal{D}'_{L^1}$ of integrable distributions which is a special case of the spaces $\mathcal{D}'_{L^p}$, $1 \leq p \leq \infty$. As an application of this general definition of convolution for distributions, we derive the Liénard–Wiechert fields of a moving charged particle in electrodynamics.

In Sect. 1.6, we treat the Fourier transform in the space $\mathcal{S}'$ of spherical distributions (omitting the more general Gel'fand–Shilov Fourier transformation $\mathcal{F} : \mathcal{D}' \to \mathcal{Z}'$.) A proof for the injectivity of $\mathcal{F} : L^1 \to \mathcal{BC}$ employing the Carleman transform is already given in Proposition 1.1.8. The non-surjectivity of $\mathcal{F} : L^1 \to \mathcal{C}_0$ is shown by means of new explicit examples in Example 1.6.8. Several methods for calculating the Fourier transform of integrable distributions are compared in Example 1.6.9. In Proposition 1.6.21, we investigate Poisson's summation formula in $\mathcal{D}'_{L^\infty}(\mathbf{R}^n)$.

Let us finally mention that this chapter is an elaboration and extension of a one-semester course on distribution theory given several times by the authors.

1.1 Definition of Test Functions and Distributions: $\mathcal{D}(\Omega)$, $\mathcal{E}(\Omega)$, $\mathcal{D}'(\Omega)$

Let us introduce some basic notation and then repeat the essential definitions of distribution theory, cf. Schwartz [246], Horváth [141], Treves [273], Strichartz [266], Hörmander [139].

Throughout, we denote by Ω a non-empty open subset of $\mathbf{R}^n$ and write $\partial_i = \frac{\partial}{\partial x_i}$ for the partial derivatives. For $x \in \mathbf{R}^n$, $\alpha, \beta \in \mathbf{N}_0^n$, we use multi-index notation in the following form:

$$x^\alpha = x_1^{\alpha_1} \cdots x_n^{\alpha_n}, \quad |\alpha| = \alpha_1 + \cdots + \alpha_n, \quad \alpha! = \alpha_1! \cdots \alpha_n!,$$

$$\alpha \geq \beta \quad \text{if and only if} \quad \forall j : \alpha_j \geq \beta_j,$$

$$\binom{\alpha}{\beta} = \frac{\alpha!}{\beta!(\alpha - \beta)!} \text{ for } \alpha \geq \beta, \quad \partial^\alpha = \partial_1^{\alpha_1} \cdots \partial_n^{\alpha_n} = \frac{\partial^{|\alpha|}}{\partial x_1^{\alpha_1} \cdots \partial x_n^{\alpha_n}}.$$

Generally,

$$P(\partial) = P(\partial_1, \dots, \partial_n) = \sum_{|\alpha| \leq m} a_\alpha \partial^\alpha, \quad a_\alpha \in \mathbf{C}, \; m \in \mathbf{N}_0,$$

denotes a linear differential operator of order at most m with constant coefficients.

Let us next define the two basic locally convex topological vector spaces $\mathcal{E}(\Omega)$ and $\mathcal{D}(\Omega)$. In order to avoid technical complications, the topologies on these spaces are defined through the convergence of sequences instead of by means of seminorms.

Definition 1.1.1

(1) For $\emptyset \neq \Omega \subset \mathbf{R}^n$ open, the **C**-vector space of infinitely often differentiable functions is denoted by

$$\mathcal{E}(\Omega) = C^\infty(\Omega) = \{\phi : \Omega \longrightarrow \mathbf{C}; \phi \text{ is } C^\infty\}.$$

For $\phi \in \mathcal{E}(\Omega)$, its *support* $\operatorname{supp} \phi$ is the closure in Ω of the set $\{x \in \Omega; \phi(x) \neq 0\}$ wherein ϕ does not vanish.
The space of *test functions* on Ω is

$$\mathcal{D}(\Omega) = \{\phi \in \mathcal{E}(\Omega); \operatorname{supp} \phi \text{ is compact}\}.$$

We shall often write $\mathcal{E}$ and $\mathcal{D}$ instead of $\mathcal{E}(\mathbf{R}^n)$ and $\mathcal{D}(\mathbf{R}^n)$, respectively.
(2) The sequence $(\phi_k)_{k\in\mathbf{N}} \in \mathcal{E}(\Omega)^{\mathbf{N}}$ is called *convergent to* ϕ *in* $\mathcal{E}(\Omega)$ iff ϕ_k and the derivatives of ϕ_k converge uniformly to ϕ and the corresponding derivatives of ϕ, respectively, on each compact subset of Ω, i.e.,

$$\lim_{k\to\infty} \phi_k = \phi \text{ in } \mathcal{E}(\Omega) \text{ (or } \phi_k \to \phi \text{ in } \mathcal{E}(\Omega)) \Longleftrightarrow$$

$$\forall \alpha \in \mathbf{N}_0^n : \forall K \subset \Omega \text{ compact} : \forall \epsilon > 0 : \exists N \in \mathbf{N} : \forall k \geq N : \|\partial^\alpha \phi_k - \partial^\alpha \phi\|_{K,\infty} \leq \epsilon$$

wherein $\|\phi\|_{K,\infty} = \sup_{x\in K} |\phi(x)|$. The sequence $(\phi_k)_{k\in\mathbf{N}} \in \mathcal{D}(\Omega)^{\mathbf{N}}$ *converges to* ϕ *in* $\mathcal{D}(\Omega)$ if, in addition to being convergent in $\mathcal{E}(\Omega)$, the supports of all ϕ_k are contained in a fixed compact subset of Ω, i.e.,

$$\lim_{k\to\infty} \phi_k = \phi \text{ in } \mathcal{D}(\Omega) \Longleftrightarrow$$

(i) $\phi_k \to \phi$ in $\mathcal{E}(\Omega)$ (ii) $\exists K \subset \Omega$ compact : $\forall k \in \mathbf{N} : \operatorname{supp} \phi_k \subset K$.

Addition, multiplication with scalars, and point-wise multiplication render $\mathcal{E}(\Omega)$ and $\mathcal{D}(\Omega)$ **C**-algebras.

L.A. Cauchy's celebrated example of a non-analytic C^∞ function shows that $\mathcal{D}(\Omega)$ is non-trivial:

Example 1.1.2 Let $\chi : \mathbf{R} \longrightarrow \mathbf{R} : x \longmapsto \left\{ \begin{array}{ll} 0, & \text{if } x \leq 0, \\ e^{-1/x}, & \text{if } x > 0 \end{array} \right\}$. Then $\chi \in C^1(\mathbf{R})$ since

$$\lim_{x \searrow 0} \frac{\chi(x) - \chi(0)}{x} = \lim_{x \searrow 0} \frac{e^{-1/x}}{x} = 0 = \lim_{x \searrow 0} \frac{e^{-1/x}}{x^2}$$

(by de l'Hôpital's rule), and hence $\chi'(x) = \begin{cases} 0, & \text{if } x \leq 0, \\ x^{-2}e^{-1/x}, & \text{if } x > 0 \end{cases}$ is again

continuous. By induction, we infer that

$$\chi^{(k)}(x) = \begin{cases} 0, & \text{if } x \leq 0, \\ P_k(x)x^{-2k}e^{-1/x}, & \text{if } x > 0 \end{cases}$$

for $k \in \mathbf{N}$ and certain polynomials P_k, and thus $\chi \in C^k(\mathbf{R})$. This implies $\chi \in \mathcal{E} = \bigcap_{k \in \mathbf{N}} C^k(\mathbf{R})$.

For $x_0 \in \mathbf{R}^n$ and $\epsilon > 0$ such that the closed ball $B_\epsilon(x_0) = \{x \in \mathbf{R}^n; |x - x_0| \leq \epsilon\}$ is contained in Ω, we have $\chi(1 - |x - x_0|^2/\epsilon^2) \in \mathcal{D}(\Omega)$ and hence $\mathcal{D}(\Omega) \neq \{0\}$.

□

Example 1.1.3 For $\phi \in \mathcal{D} \setminus \{0\}$, we can consider the sequences

$$\phi_k = \frac{1}{k}\phi \quad \text{and} \quad \psi_k(x) = \frac{1}{k}\phi\Big(\frac{x}{k}\Big).$$

Whereas ϕ_k converges to 0 in $\mathcal{E}$ *and* in $\mathcal{D}$, ψ_k converges (to 0) in $\mathcal{E}$, but not in $\mathcal{D}$, since the supports of ψ_k are not uniformly bounded.   □

The *space of distributions* $\mathcal{D}'(\Omega)$ is the dual space of $\mathcal{D}(\Omega)$ in the sense of locally convex vector spaces. We define $\mathcal{D}'(\Omega)$ directly, thereby circumventing the non-metrizability of $\mathcal{D}(\Omega)$.

Definition 1.1.4

(1) For $\emptyset \neq \Omega \subset \mathbf{R}^n$ open,

$$\mathcal{D}'(\Omega) = \{T : \mathcal{D}(\Omega) \longrightarrow \mathbf{C}; \ T \ \mathbf{C}\text{-linear and } T(\phi_k) \to 0 \text{ if } \phi_k \to 0 \text{ in } \mathcal{D}(\Omega)\}.$$

(2) The sequence $(T_k)_{k \in \mathbf{N}} \in \mathcal{D}'(\Omega)^{\mathbf{N}}$ is called *convergent to* T *in* $\mathcal{D}'(\Omega)$ (and we write $\lim_{k \to \infty} T_k = T$) iff $T \in \mathcal{D}'(\Omega)$ and the sequences $T_k(\phi)$ converge to $T(\phi)$ for each $\phi \in \mathcal{D}(\Omega)$. Analogously, we define $\lim_{\lambda \to \lambda_0} T_\lambda = T$ for a family T_λ of distributions depending on a parameter λ in $\mathbf{C}^k$ (or any metrizable topological space).

The convergence of sequences in $\mathcal{D}'(\Omega)$ is often called *weak* convergence, since it refers to the "weak" topology on $\mathcal{D}'(\Omega)$. The evaluation $T(\phi)$ of a distribution T on a test function ϕ is often written as $\langle \phi, T \rangle$, which is a hint at the bilinearity of the evaluation mapping

$$\mathcal{D}(\Omega) \times \mathcal{D}'(\Omega) \longrightarrow \mathbf{C} : (\phi, T) \longmapsto T(\phi) = \langle \phi, T \rangle.$$

In order to indicate the "active" variable in a distribution T, it is often convenient to write $T \in \mathcal{D}'(\mathbf{R}_x^n)$ or T_x or $T(x)$; e.g., this is necessary in $\langle \phi(x,y), T(x) \rangle$ and in $\langle \phi(x,y), T(y) \rangle$, $\phi \in \mathcal{D}(\mathbf{R}_{x,y}^{2n})$, which numbers are, in general, different.

Proposition 1.1.5 $\mathcal{D}'(\Omega)$ *is sequentially complete with respect to the weak topology, i.e., if* $(T_k)_{k \in \mathbf{N}} \in \mathcal{D}'(\Omega)^{\mathbf{N}}$ *and* $\lim_{k \to \infty} \langle \phi, T_k \rangle$ *exists for each* $\phi \in \mathcal{D}(\Omega)$*, then*

$$\langle \phi, T \rangle := \lim_{k \to \infty} \langle \phi, T_k \rangle$$

defines a distribution $T \in \mathcal{D}'(\Omega)$.

For the proof, which relies essentially on the Banach–Steinhaus theorem for the barrelled space $\mathcal{D}(\Omega)$, we refer to Robertson and Robertson [236], Ch. IV, Cor. 1 to Thm. 3, p. 69; Donoghue [61], Sect. 20, p. 100; Vladimirov [280], 1.4, pp. 14, 15; Gel'fand and Shilov [104], App. A, pp. 368, 369.

Distributions generalize Lebesgue (locally) integrable functions and Radon measures.

Definition 1.1.6 A Lebesgue measurable function $f : \Omega \longrightarrow \mathbf{C}$ is called *locally integrable* on Ω iff $\int_K |f(x)|\,dx$ is finite for all compact $K \subset \Omega$.

For example, every continuous function on Ω is locally integrable, $\frac{1}{x}$ is locally integrable on $\mathbf{R} \setminus \{0\}$, but not on $\mathbf{R}$, whereas $\frac{1}{\sqrt{|x|}}$ is locally integrable also on $\mathbf{R}$.

Proposition 1.1.7

(1) *If* $f : \Omega \longrightarrow \mathbf{C}$ *is locally integrable, then the associated linear functional*

$$T_f : \mathcal{D}(\Omega) \longrightarrow \mathbf{C} : \phi \longmapsto \int_\Omega \phi(x) f(x)\,dx$$

is a distribution.

(2) *For* $f, g : \Omega \longrightarrow \mathbf{C}$ *locally integrable, we have* $T_f = T_g$ *if and only if* $f = g$ *almost everywhere, i.e.,* $\{x \in \Omega; f(x) \neq g(x)\}$ *is a set of Lebesgue measure 0.*

We can therefore identify the equivalence classes of locally integrable functions which are equal almost everywhere with their associated distribution. For shortness, we shall often write just f instead of T_f.

Proof

(1) Obviously, T_f is linear. On the other hand, if $\phi_k \to \phi$ in $\mathcal{D}(\Omega)$, then

$$\left| \int_\Omega \left(\phi_k(x) - \phi(x) \right) f(x)\,dx \right| \leq \|\phi_k - \phi\|_{K,\infty} \cdot \int_K |f(x)|\,dx$$

for compact $K \subset \Omega$ such that supp $\phi_k \subset K$ for all $k \in \mathbf{N}$. Hence $T_f(\phi_k) \to T_f(\phi)$ for $k \to \infty$, i.e., $T_f \in \mathcal{D}'(\Omega)$.

(2) $T_f = T_g$ means that $\int_\Omega (f(x) - g(x))\phi(x)\,\mathrm{d}x = 0$ for each $\phi \in \mathcal{D}(\Omega)$ and this
implies

$$\int_{\mathbf{R}^n} (f(x) - g(x))\phi(x)\mathrm{e}^{-\mathrm{i}\xi x}\,\mathrm{d}x = 0$$

for each $\phi \in \mathcal{D}(\Omega)$ and each $\xi \in \mathbf{R}^n$. By the injectivity of the (classical)
Fourier transform (see Proposition 1.1.8), we conclude that $(f(x)-g(x))\phi(x) = 0$ almost everywhere for $\phi \in \mathcal{D}(\Omega)$ and hence $f = g$ almost everywhere.   □

Proposition 1.1.8 *Let*

$$\mathcal{BC}(\mathbf{R}^n) = \{f : \mathbf{R}^n \longrightarrow \mathbf{C}; f \text{ is continuous and bounded}\}.$$

Then

$$\mathcal{F} : L^1(\mathbf{R}^n) \longrightarrow \mathcal{BC}(\mathbf{R}^n) : f \longmapsto \left(\xi \mapsto \int_{\mathbf{R}^n} f(x)\mathrm{e}^{-\mathrm{i}\xi x}\,\mathrm{d}x\right)$$

is well defined, linear, continuous, and injective.

Proof

(a) The continuity of $\mathcal{F}f$ follows from Lebesgue's theorem on dominated convergence, and we obviously have $\|\mathcal{F}f\|_\infty \le \|f\|_1$. This implies $\mathcal{F}f \in \mathcal{BC}(\mathbf{R}^n)$ and the continuity of $\mathcal{F}$.
(b) Let us show that the mapping $\mathcal{F}$ is injective by using the Carleman transform, see Carleman [42], (44), (45), p. 27; Gurarii [121], Sect. 12, pp. 147–158; Hörmander [137], Example 1.4.12, p. 1.19; Newman [189]. We first assume $n = 1$ and take $f \in L^1(\mathbf{R}^1)$ with $\mathcal{F}f = 0$. The function

$$g(z) = \begin{cases} \displaystyle\int_{-\infty}^0 f(x)\mathrm{e}^{-\mathrm{i}xz}\,\mathrm{d}x, & \text{if } \mathrm{Im}\, z \ge 0, \\ \displaystyle-\int_0^\infty f(x)\mathrm{e}^{-\mathrm{i}xz}\,\mathrm{d}x, & \text{if } \mathrm{Im}\, z \le 0 \end{cases}$$

is well defined on $\mathbf{R}$ due to $\mathcal{F}f = 0$, continuous on $\mathbf{C}$, and obviously analytic on $\mathbf{C} \setminus \mathbf{R}$. Morera's theorem (or Rudin [238], Thm., p. 4) implies that g is entire. Since g is bounded (due to $\|g\|_\infty \le \|f\|_1$), and $\lim_{y\to\infty} g(\mathrm{i}y) = 0$, we conclude from Liouville's theorem that g vanishes identically. Hence $g(0) = \int_{-\infty}^0 f(x)\,\mathrm{d}x = 0$, and use of the shifted functions $x \mapsto f(x + \xi)$ yields $\int_{-\infty}^\xi f(x)\,\mathrm{d}x = 0$ for all $\xi \in \mathbf{R}$ and hence $f = 0$ almost everywhere.
(c) Finally, we use induction on the dimension n. If $f \in L^1(\mathbf{R}^n)$, then

$$f_{x_1} : \mathbf{R}^{n-1} \longrightarrow \mathbf{C} : x' \longmapsto f(x_1, x')$$

belongs, by Fubini's theorem, to $L^1(\mathbf{R}^{n-1})$ for $x_1 \in \mathbf{R} \setminus N$, where N is a null-set in $\mathbf{R}$. Furthermore, for $\xi' \in \mathbf{R}^{n-1}$ fixed,

$$g_{\xi'}(x_1) := (\mathcal{F}f_{x_1})(\xi') = \int_{\mathbf{R}^{n-1}} f(x_1, x')e^{-i\xi' x'}\, dx'$$

is integrable for all $\xi' \in \mathbf{R}^{n-1}$, and $\mathcal{F}g_{\xi'}(\xi_1) = \mathcal{F}f(\xi) = 0$ for $\xi_1 \in \mathbf{R}$. Hence, by part (b) above, $g_{\xi'}$ vanishes almost everywhere. Therefore, Fubini's theorem implies

$$0 = \int_{\mathbf{R}^{n-1}} \int_{\mathbf{R}\setminus N} |g_{\xi'}(x_1)|\, dx_1 d\xi' = \int_{\mathbf{R}\setminus N} \int_{\mathbf{R}^{n-1}} |g_{\xi'}(x_1)|\, d\xi' dx_1.$$

Thus $\int_{\mathbf{R}^{n-1}} |g_{\xi'}(x_1)|\, d\xi' = 0$ for $x_1 \in \mathbf{R}\setminus N_1$ where $N \subset N_1$ is another null-set in $\mathbf{R}$. By the continuity of $\xi' \mapsto g_{\xi'}(x_1)$, this implies $g_{\xi'}(x_1) = 0$ for all $\xi' \in \mathbf{R}^{n-1}$ and $x_1 \in \mathbf{R} \setminus N_1$. The induction hypothesis then furnishes $\int_{\mathbf{R}^{n-1}} |f(x_1, x')|\, dx' = 0$ for $x_1 \in \mathbf{R} \setminus N_1$ and thus $f = 0$ almost everywhere. $\square$

The Riemann–Lebesgue lemma states that the range of the Fourier transform in Proposition 1.1.8 is actually contained in the space $\mathcal{C}_0(\mathbf{R}^n)$ of continuous functions vanishing at infinity.

Definition 1.1.9 Assume $\emptyset \neq \Omega \subset \mathbf{R}^n$ open.

(1) We define the space $L^1_{\mathrm{loc}}(\Omega)$ of *locally integrable functions* as a subspace of $\mathcal{D}'(\Omega)$:

$$L^1_{\mathrm{loc}}(\Omega) = \{T_f \in \mathcal{D}'(\Omega); f : \Omega \longrightarrow \mathbf{C} \text{ locally integrable}\}.$$

(2) Similarly, for $1 \leq p < \infty$, we define

$$L^p_{\mathrm{loc}}(\Omega) = \{T_f \in \mathcal{D}'(\Omega); |f|^p \text{ is locally integrable}\},$$

$$L^p_{\mathrm{c}}(\Omega) = \{T_f \in L^p_{\mathrm{loc}}(\Omega); f = 0 \text{ outside a compact subset of } \Omega\},$$

$$L^\infty_{\mathrm{loc}}(\Omega) = \{T_f \in \mathcal{D}'(\Omega); |f| \text{ is locally bounded}\}, \quad L^\infty_{\mathrm{c}}(\Omega) = L^\infty_{\mathrm{loc}}(\Omega) \cap L^1_{\mathrm{c}}(\Omega).$$

(3) Finally, the spaces of *Radon measures* $\mathcal{M}(\Omega)$, of *integrable (or bounded) measures* $\mathcal{M}^1(\Omega)$, and of *measures with compact support* $\mathcal{M}_{\mathrm{c}}(\Omega)$, respectively, are defined as the subspaces of $\mathcal{D}'(\Omega)$ arising by the application of the corresponding class of measures to test functions. (In a similar vein as in Proposition 1.1.7, the Riesz–Markov theorem implies that two complex measures μ_1, μ_2 fulfilling $|\mu_j|(K) < \infty$ for all compacts sets $K \subset \Omega$ coincide as measures if and only if they coincide as distributions.)

In order to convince the reader that distributions generalize measures and functions, let us present the following table, cf. Horváth [145], p. 10; Vo-Khac Koan [282], p. 168 and p. 175.

Table 1.1.10 If $\emptyset \neq \Omega \subset \mathbf{R}^n$ open and $1 < p \leq q < \infty$, then the following inclusions hold:

$$L_c^\infty(\Omega) \subset L_c^q(\Omega) \subset L_c^p(\Omega) \subset L_c^1(\Omega) \subset \mathcal{M}_c(\Omega)$$

$$\cap \qquad \cap \qquad \cap \qquad \cap \qquad \cap$$

$$L^\infty(\Omega) \quad L^q(\Omega) \quad L^p(\Omega) \quad L^1(\Omega) \subset \mathcal{M}^1(\Omega)$$

$$\cap \qquad \cap \qquad \cap \qquad \cap \qquad \cap$$

$$L_{\text{loc}}^\infty(\Omega) \subset L_{\text{loc}}^q(\Omega) \subset L_{\text{loc}}^p(\Omega) \subset L_{\text{loc}}^1(\Omega) \subset \mathcal{M}(\Omega) \subset \mathcal{D}'(\Omega)$$

Example 1.1.11 A particularly important example of a measure with compact support is the *Dirac measure*. For $a \in \Omega \subset \mathbf{R}^n$ open, we set

$$\delta_a : \mathcal{D}(\Omega) \longrightarrow \mathbf{C} : \phi \longmapsto \phi(a).$$

In particular δ is short for $\delta_0 \in \mathcal{D}'(\mathbf{R}^n)$.

Then $\delta_a \in \mathcal{M}_c(\Omega) \setminus L_{\text{loc}}^1(\Omega)$, since, by Lebesgue's theorem on dominated convergence,

$$\lim_{j \to \infty} \langle \rho_j, T \rangle = 0 \quad \text{if} \quad \rho_j(x) = \chi(1 - j|x - a|^2)$$

and χ is as in Example 1.1.2 and $T = T_f \in L_{\text{loc}}^1(\Omega)$; on the other hand, $\langle \rho_j, \delta_a \rangle = \chi(1) \neq 0$.

Although δ cannot be represented by locally integrable functions, it is the (weak) limit of such functions. In fact, if $k \in L^1(\mathbf{R}^n)$ with $\int k(x)\,dx = 1$, then

$$\lim_{\epsilon \searrow 0} \epsilon^{-n} k\left(\frac{x}{\epsilon}\right) = \delta \quad \text{in} \quad \mathcal{D}'(\mathbf{R}^n),$$

since, for $\phi \in \mathcal{D}$, Lebesgue's theorem implies

$$\langle \phi, \epsilon^{-n} k\left(\frac{x}{\epsilon}\right) \rangle = \epsilon^{-n} \int_{\mathbf{R}^n} \phi(x) k\left(\frac{x}{\epsilon}\right) dx = \int_{\mathbf{R}^n} \phi(\epsilon y) k(y)\, dy \to$$

$$\to \int_{\mathbf{R}^n} \phi(0) k(y)\, dy = \phi(0) = \langle \phi, \delta \rangle.$$

Well-known special cases of this are the following (Y denotes the Heaviside function, i.e., $Y(t) = 1$ for $t > 0$ and $Y(t) = 0$ for $t \leq 0$) :

(i) $\lim_{\epsilon \searrow 0} C \epsilon^{-n} Y(\epsilon - |x|) = \delta$ with $C = \pi^{-n/2} \Gamma(\frac{n}{2} + 1)$;

(ii) $\lim_{\epsilon \searrow 0} \dfrac{C\epsilon}{(|x|^2 + \epsilon^2)^{(n+1)/2}} = \delta$ with $C = \pi^{-(n+1)/2} \Gamma(\frac{n+1}{2})$,

   cf. Example 1.6.12 below and Duoandikoetxea [67], (1.30), p. 19;

(iii) $\lim_{\epsilon \searrow 0} C \epsilon^{-n/2} e^{-|x|^2/\epsilon} = \delta$ with $C = \pi^{-n/2}$.

The examples (ii) and (iii) correspond to the classical summability methods named after Abel–Poisson and Gauß–Weierstraß, respectively. $\qquad\square$

Example 1.1.12 Let us next consider distributions that cannot be represented by measures. For $f \in L^1(\mathbf{S}^{n-1})$ (where $\mathbf{S}^{n-1}$ is equipped with the usual surface measure $d\sigma$) satisfying

$$\int_{\mathbf{S}^{n-1}} f(\omega)\,d\sigma(\omega) = 0 \quad \text{(the so-called "mean-value zero condition"),} \qquad (1.1.1)$$

we define the *principal value* (in French "valeur principale") $\mathrm{vp}\big(|x|^{-n}f(\tfrac{x}{|x|})\big) \in \mathcal{D}'(\mathbf{R}^n)$ by the limit

$$\mathrm{vp}\big(|x|^{-n}f(\tfrac{x}{|x|})\big) := \lim_{\epsilon \searrow 0} Y(|x| - \epsilon)|x|^{-n}f\left(\frac{x}{|x|}\right),$$

i.e.,

$$\big\langle \phi, \mathrm{vp}\big(|x|^{-n}f(\tfrac{x}{|x|})\big)\big\rangle = \lim_{\epsilon \searrow 0} \int_{|x| \geq \epsilon} \frac{\phi(x)}{|x|^n} f\left(\frac{x}{|x|}\right) dx, \qquad \phi \in \mathcal{D}.$$

In order to show that this limit exists and yields a distribution, let us give yet another representation of it. If $\phi \in \mathcal{D}$ and $R > 0$ is such that $\mathrm{supp}\,\phi \subset \{x \in \mathbf{R}^n; |x| \leq R\}$, then Eq. (1.1.1) implies (cf. Duoandikoetxea [67], (4.2), p. 69)

$$\big\langle \phi, \mathrm{vp}\big(|x|^{-n}f(\tfrac{x}{|x|})\big)\big\rangle = \lim_{\epsilon \searrow 0} \int_{\epsilon \leq |x| \leq R} \frac{\phi(x)}{|x|^n} f\left(\frac{x}{|x|}\right) dx$$

$$= \lim_{\epsilon \searrow 0} \int_{\epsilon \leq |x| \leq R} \frac{\phi(x) - \phi(0)}{|x|^n} f\left(\frac{x}{|x|}\right) dx$$

$$= \int_{|x| \leq R} \frac{\phi(x) - \phi(0)}{|x|^n} f\left(\frac{x}{|x|}\right) dx,$$

where the last integral is convergent due to $|\phi(x) - \phi(0)| \leq \|\nabla\phi\|_\infty \cdot |x|$. This inequality also furnishes that

$$\big\langle \phi_k, \mathrm{vp}\big(|x|^{-n}f(\tfrac{x}{|x|})\big)\big\rangle \to 0 \quad \text{if} \quad \phi_k \to 0 \text{ in } \mathcal{D},$$

and hence that $\mathrm{vp}\big(|x|^{-n}f(\tfrac{x}{|x|})\big)$ is a distribution. (Alternatively, this is also implied by Proposition 1.1.5.)

The distributions $\mathrm{vp}\big(|x|^{-n}f(\tfrac{x}{|x|})\big)$ are the kernels of the classical singular integral operators of the "first generation," cf. Meyer and Coifman [179], pp. 2, 3. Particular cases are $\mathrm{vp}\,\tfrac{1}{x} = \mathrm{vp}\big(|x|^{-1}\,\mathrm{sign}(\tfrac{x}{|x|})\big) \in \mathcal{D}'(\mathbf{R}^1)$, which is the kernel of the Hilbert transform on $\mathbf{R}^1$, and the distributions

$$T_{jk} = \mathrm{vp}\Big(\frac{\delta_{jk}|x|^2 - nx_jx_k}{|x|^{n+2}}\Big) \in \mathcal{D}'(\mathbf{R}^n), \qquad 1 \le j, k \le n, \qquad \delta_{jk} = \begin{cases} 1 \text{ if } j = k, \\ 0 \text{ else,} \end{cases}$$

which originate as second derivatives of the kernel $|x|^{2-n}$, if $n \ne 2$, or $\log|x|$, if $n = 2$, respectively, i.e., of the Newtonian potential. Here $f_{jk}(\omega) = \delta_{jk} - n\omega_j\omega_k$ obviously fulfills the mean-value zero condition (1.1.1).

A complex approximation of $\mathrm{vp}\,\frac{1}{x} \in \mathcal{D}'(\mathbf{R}^1)$ follows from Sokhotski's formula

$$\lim_{\epsilon \searrow 0} \frac{1}{x \pm i\epsilon} = \mathrm{vp}\,\frac{1}{x} \mp i\pi\delta, \tag{1.1.2}$$

see Sokhotski [256]. In fact, by Example 1.1.11, (ii),

$$\lim_{\epsilon \searrow 0} \frac{1}{x \pm i\epsilon} = \lim_{\epsilon \searrow 0} \frac{x \mp i\epsilon}{x^2 + \epsilon^2} = \lim_{\epsilon \searrow 0} \frac{x}{x^2 + \epsilon^2} \mp i\pi\delta.$$

On the other hand, for $\phi \in \mathcal{D}(\mathbf{R}^1)$ with $\mathrm{supp}\,\phi \subset [-R, R]$ and $\epsilon \searrow 0$, we obtain

$$\langle \phi, \frac{x}{x^2 + \epsilon^2} \rangle = \int_{-R}^{R} \frac{(\phi(x) - \phi(0))x}{x^2 + \epsilon^2}\,\mathrm{d}x \to \int_{-R}^{R} \frac{\phi(x) - \phi(0)}{x}\,\mathrm{d}x = \langle \phi, \mathrm{vp}\,\frac{1}{x} \rangle.$$

Let us generalize Sokhotski's formula to the distributions $\mathrm{vp}\big(|x|^{-n}f(\frac{x}{|x|})\big)$.

Whereas a straight-forward calculation yields the complex limit representation

$$\lim_{\epsilon \searrow 0} \frac{f(\frac{x}{|x|})}{|x|^n \pm i\epsilon} = \mathrm{vp}\big(|x|^{-n}f(\frac{x}{|x|})\big),$$

Sokhotski's formula corresponds to a different kind. Let us assume that $f \in L^1(\mathbf{S}^{n-1})$ is real-valued, fulfills (1.1.1), and $f\log|f| \in L^1(\mathbf{S}^{n-1})$ (where we formally set $0\log 0 = 0$.) Then

$$\lim_{\epsilon \searrow 0} \frac{f(\frac{x}{|x|})}{|x|^n \pm i\epsilon f(\frac{x}{|x|})} = \mathrm{vp}\big(|x|^{-n}f(\frac{x}{|x|})\big) + C_\pm\delta, \tag{1.1.3}$$

where

$$C_\pm = -\frac{1}{n}\int_{\mathbf{S}^{n-1}} f(\omega)\log|f(\omega)|\,\mathrm{d}\sigma(\omega) \mp \frac{i\pi}{2n}\int_{\mathbf{S}^{n-1}} |f(\omega)|\,\mathrm{d}\sigma(\omega).$$

Indeed, under the above conditions on f, the functions $f(\frac{x}{|x|})/(|x|^n \pm i\epsilon f(\frac{x}{|x|}))$ are locally integrable for $\epsilon > 0$. Furthermore, for $\phi \in \mathcal{D}(\mathbf{R}^n)$ with $|x| \le R$ for all

$x \in \operatorname{supp} \phi$, we have

$$\lim_{\epsilon \searrow 0} \int \phi(x) \frac{f(\frac{x}{|x|})}{|x|^n \pm i\epsilon f(\frac{x}{|x|})}\, dx = \lim_{\epsilon \searrow 0} \int_{|x| \leq R} \frac{\phi(x) - \phi(0)}{|x|^n \pm i\epsilon f(\frac{x}{|x|})} f(\frac{x}{|x|})\, dx$$

$$+ \phi(0) \lim_{\epsilon \searrow 0} \int_0^R r^{n-1} dr \int_{S^{n-1}} \frac{f(\omega)\, d\sigma(\omega)}{r^n \pm i\epsilon f(\omega)}$$

$$= \langle \phi, \operatorname{vp}(|x|^{-n} f(\tfrac{x}{|x|})) + C_\pm \delta \rangle,$$

where

$$C_\pm = \lim_{\epsilon \searrow 0} \frac{1}{n} \int_{S^{n-1}} f(\omega) \log(r^n \pm i\epsilon f(\omega))|_{r=0}^R\, d\sigma(\omega)$$

$$= \lim_{\epsilon \searrow 0} \frac{1}{n} \int_{S^{n-1}} f(\omega) \left[\log R^n - \log(\epsilon|f(\omega)|) \mp \frac{i\pi}{2} \operatorname{sign} f(\omega) \right] d\sigma(\omega)$$

$$= -\frac{1}{n} \int_{S^{n-1}} f(\omega) \log |f(\omega)|\, d\sigma(\omega) \mp \frac{i\pi}{2n} \int_{S^{n-1}} |f(\omega)|\, d\sigma(\omega).$$

In particular, Heisenberg's formula

$$\lim_{\epsilon \searrow 0} \left(\frac{1}{x + i\epsilon} - \frac{1}{x - i\epsilon} \right) = -2\pi i \delta$$

can be generalized in the following way to n dimensions:

$$\lim_{\epsilon \searrow 0} \left(\frac{f(\frac{x}{|x|})}{|x|^n + i\epsilon f(\frac{x}{|x|})} - \frac{f(\frac{x}{|x|})}{|x|^n - i\epsilon f(\frac{x}{|x|})} \right) = -\frac{i\pi}{n} \|f\|_1 \delta$$

if $f \in L^1(S^{n-1})$ is real-valued.

To give a concrete example, let us consider $f(\omega) = \omega_j \omega_k$ for $1 \leq j < k \leq n$ (comp. T_{jk} above). Then

$$\int_{S^{n-1}} f(\omega) \log |f(\omega)|\, d\sigma(\omega) = 0$$

and

$$\int_{S^{n-1}} |f(\omega)|\, d\sigma(\omega) = 4 \int_{\substack{\omega \in S^{n-1} \\ \omega_1 \geq 0, \omega_n \geq 0}} \omega_1 \omega_n\, d\sigma(\omega) = 4 \int_{\substack{\omega \in S^{n-1} \\ \omega_1 \geq 0, \omega_n \geq 0}} \omega^T \cdot v(\omega)\, d\sigma(\omega),$$

where $v(x) = (0, \ldots, 0, x_1)^T \in \mathbf{R}^n$. Since $\operatorname{div} v = 0$, Gauß' divergence theorem implies

$$\int_{\mathbf{S}^{n-1}} |f(\omega)| \, d\sigma(\omega) = 4 \int_{\substack{x' \in \mathbf{R}^{n-1}, |x'| \leq 1 \\ x_1 \geq 0}} x_1 \, dx' = \frac{2\pi^{\frac{n}{2}-1}}{\Gamma(\frac{n}{2}+1)}.$$

The last equation follows from Guldin's rule. For $1 \leq j < k \leq n$, we thus obtain

$$\lim_{\epsilon \searrow 0} \frac{\omega_j \omega_k}{|x|^n \pm i\epsilon\omega_j\omega_k} = \operatorname{vp}\left(\frac{x_j x_k}{|x|^{n+2}}\right) \mp \frac{i\pi^{n/2}}{n\Gamma(\frac{n}{2}+1)} \delta.$$

$\square$

For a further generalization of Sokhotski's formula, see part (b) of the proof of Proposition 4.4.1.

The two equations in formulae (1.1.2) and (1.1.3) actually follow one from the other by complex conjugation if we take into account the following definition.

Definition 1.1.13 For $T \in \mathcal{D}'(\Omega)$, the *complex conjugate* $\overline{T}$, the *real part* $\operatorname{Re} T$, and the *imaginary part* $\operatorname{Im} T$ are defined by

$$\langle \phi, \overline{T} \rangle = \overline{\langle \overline{\phi}, T \rangle}, \quad \operatorname{Re} T = \frac{1}{2}(T + \overline{T}), \quad \operatorname{Im} T = \frac{1}{2i}(T - \overline{T}),$$

respectively.

Example 1.1.14 Let us now generalize the setting of Example 1.1.12. We take again $f \in L^1(\mathbf{S}^{n-1})$, and we assume all moments up to the order $l \in \mathbf{N}_0$ vanish, i.e.,

$$\forall \alpha \in \mathbf{N}_0^n \text{ with } |\alpha| \leq l : \int_{\mathbf{S}^{n-1}} \omega^\alpha f(\omega) \, d\sigma(\omega) = 0.$$

If, furthermore, $g : (0, \infty) \longrightarrow \mathbf{C}$ is measurable and fulfills $\int_0^1 |g(r)| r^{l+n} \, dr < \infty$, then again the distribution

$$\operatorname{vp}\big(g(|x|)f(\tfrac{x}{|x|})\big) = \lim_{\epsilon \searrow 0} Y(|x| - \epsilon)g(|x|)f(\tfrac{x}{|x|}) \in \mathcal{D}'(\mathbf{R}^n)$$

is well defined. In fact, if $\phi(x) = 0$ for $|x| \geq R$, then an appeal to Taylor's theorem furnishes, similarly as in Example 1.1.12,

$$\langle \phi, \operatorname{vp}(g(|x|)f(\tfrac{x}{|x|})) \rangle = \int_{|x| \leq R} \left(\phi(x) - \sum_{|\alpha| \leq l} \frac{\partial^\alpha \phi(0)}{\alpha!} x^\alpha\right) g(|x|)f(\tfrac{x}{|x|}) \, dx.$$

For example, if $n = 3$, $k > 0$, and $Y_{lm}(\omega)$, $l \in \mathbf{N}_0$, $m \in \mathbf{Z}$, $|m| \leq l$, are the usual spherical harmonics, then

$$T_{lm} = \mathrm{vp}\big(|x|^{-1/2}J_{-l-1/2}(k|x|)Y_{lm}(\tfrac{x}{|x|})\big) \in \mathcal{D}'(\mathbf{R}^3)$$

continues the respective $\mathcal{C}^\infty$ function on $\mathbf{R}^3 \setminus \{0\}$, which originates as a solution of the Helmholtz equation $(\Delta_3 + k^2)u = 0$ by separation of variables. These distributions are important in the application of the pseudopotential method, see Stampfer and Wagner [260, 261]. □

1.2 Multiplication, Support, Composition

In order to extend operations defined on functions to distributions, we use the *method of transposition* (similarly as in Definition 1.1.13): The evaluation of the transformed distribution AT on test functions ϕ is expressed by evaluation of T on $A_1\phi$, where A_1 is a transposed operator constructed such that AT_f coincides with T_{Af}, i.e., the equation

$$\langle \phi, T_{Af} \rangle = \langle A_1\phi, T_f \rangle, \qquad \phi \in \mathcal{D}(\Omega),$$

is, as a definition, extended to general $T \in \mathcal{D}'(\Omega)$.

Definition 1.2.1 For $g \in \mathcal{E}(\Omega)$ and $T \in \mathcal{D}'(\Omega)$, the *multiplication* $g \cdot T$ is defined by $\langle \phi, g \cdot T \rangle = \langle \phi \cdot g, T \rangle$, $\phi \in \mathcal{D}(\Omega)$.

Obviously $g \cdot T \in \mathcal{D}'(\Omega)$, since $\phi_k \to \phi$ in $\mathcal{D}(\Omega)$ implies $g\phi_k \to g\phi$ in $\mathcal{D}(\Omega)$. More abstractly, the mapping $\mathcal{D}'(\Omega) \longrightarrow \mathcal{D}'(\Omega) : T \longmapsto g \cdot T$ is the transpose of the linear continuous mapping $\mathcal{D}(\Omega) \longrightarrow \mathcal{D}(\Omega) : \phi \longmapsto g \cdot \phi$. Note that this multiplication is consistent with that for locally integrable functions, i.e., $g \cdot T_f = T_{g \cdot f}$, since

$$\langle \phi, T_{g \cdot f} \rangle = \int_\Omega \phi(x)g(x)f(x)\,\mathrm{d}x = \langle \phi \cdot g, T_f \rangle.$$

Similarly, we define next the restriction of distributions in $\mathcal{D}'(\Omega)$ to an open subset $\Omega_1 \subset \Omega$ as the transpose of the imbedding $\mathcal{D}(\Omega_1) \hookrightarrow \mathcal{D}(\Omega)$. This also furnishes the concept of support for distributions.

Definition 1.2.2

(1) If $\emptyset \neq \Omega_1 \subset \Omega \subset \mathbf{R}^n$ are open sets, then the *restriction* is defined by
$\mathcal{D}'(\Omega) \longrightarrow \mathcal{D}'(\Omega_1) : T \longmapsto T|_{\Omega_1} : \phi \mapsto T(\hat{\phi})$,

$$\text{where } \hat{\phi}(x) = \left\{ \begin{array}{ll} \phi(x), & \text{if } x \in \Omega_1, \\ 0, & \text{if } x \in \Omega \setminus \Omega_1 \end{array} \right\} \quad \text{for } \phi \in \mathcal{D}(\Omega_1).$$

(2) For $T \in \mathcal{D}'(\Omega)$, we define the *support* by

$$\operatorname{supp} T := \Omega \setminus \cup\{\Omega_1;\ \Omega_1 \subset \Omega \text{ open}, T|_{\Omega_1} = 0\}.$$

Apparently, for $f \in \mathcal{E}(\Omega)$, $\operatorname{supp} T_f = \operatorname{supp} f$. In contrast, for $T_f \in L^1_{\text{loc}}(\Omega)$, $U = \Omega \setminus \operatorname{supp} T_f$ is the largest open set such that $\int_U |f(x)|\,dx = 0$.

The next proposition shows that T actually vanishes outside its support. This also implies that distributions are "local objects," or, in other words, that the spaces $\mathcal{D}'(\Omega)$ constitute a sheaf.

Proposition 1.2.3

(1) *For $T \in \mathcal{D}'(\Omega)$ holds $T|_{\Omega \setminus \operatorname{supp} T} = 0$.*
(2) *Let $\Omega = \cup_{i \in I} \Omega_i$ with $\emptyset \neq \Omega, \Omega_i \subset \mathbf{R}^n$ open for $i \in I$, I being an arbitrary index set. Then the following holds:*

    (a) *If $S, T \in \mathcal{D}'(\Omega)$ fulfill $S|_{\Omega_i} = T|_{\Omega_i}$ for all $i \in I$, then $S = T$.*
    (b) *If $T_i \in \mathcal{D}'(\Omega_i)$ fulfill $T_i|_{\Omega_i \cap \Omega_j} = T_j|_{\Omega_i \cap \Omega_j}$ for all $i, j \in I$ with $\Omega_i \cap \Omega_j \neq \emptyset$, then $\exists_1 T \in \mathcal{D}'(\Omega): \forall i \in I : T|_{\Omega_i} = T_i$.*

Proof

(1) If $T \in \mathcal{D}'(\Omega)$ and $\phi \in \mathcal{D}(\Omega)$ with $\operatorname{supp} \phi \cap \operatorname{supp} T = \emptyset$, then there exist $0 < \epsilon_1 < \epsilon_2$ and open balls

$$B_{ij} := \{x \in \mathbf{R}^n;\ |x - m_i| < \epsilon_j\}, \quad i = 1, \dots, l, \quad j = 1, 2,$$

such that $\operatorname{supp} \phi \subset \cup_{i=1}^l B_{i1}$ and $\forall i : B_{i2} \subset \Omega$ and $T|_{B_{i2}} = 0$.

Let $\chi(t) = Y(t)e^{-1/t}$, $t \in \mathbf{R}$, be as in Example 1.1.2, and set $\rho_i(x) = \chi(1 - |x - m_i|^2/\delta^2)$, $x \in \mathbf{R}^n$, for some $\delta \in (\epsilon_1, \epsilon_2)$. Then $\psi_i(x) = \rho_i(x)/[\sum_{j=1}^l \rho_j(x)]$ is a partition of unity on $\cup_{i=1}^l B_{i1}$ i.e., $\sum_{i=1}^l \psi_i(x) = 1$ in $\cup_{i=1}^l B_{i1}$, and such that $\psi_i \cdot \phi \in \mathcal{D}(B_{i2})$. Therefore,

$$\langle \phi, T \rangle = \langle \sum_{i=1}^l \psi_i \cdot \phi, T \rangle = \sum_{i=1}^l \langle \psi_i \cdot \phi, T|_{B_{i2}} \rangle = 0.$$

(2) Note that condition (a) follows from (1): $(S - T)|_{\Omega_i} = 0$ implies $S = T$ in $\mathcal{D}'(\Omega)$.

For (b), we can construct—similarly as in (1)—a partition of unity subordinate to the covering Ω_i, $i \in I$, i.e., $\psi_i \in \mathcal{D}(\Omega_i)$ satisfying

    (i) $\forall K \subset \Omega$ compact: $\operatorname{supp} \psi_i \cap K = \emptyset$ for all but finitely many $i \in I$;
    (ii) $\forall x \in \Omega : \sum_{i \in I} \psi_i(x) = 1$,

cf. Schwartz [246], Ch. I, Sect. 2, Thm. II, p. 22.

We then define T by $\langle \phi, T \rangle = \sum_{i \in I} \langle \phi \psi_i, T_i \rangle$. Then $T \in \mathcal{D}'(\Omega)$, and for $\phi \in \mathcal{D}(\Omega_j)$ it follows

$$\langle \phi, T \rangle = \sum_{i \in I} \langle \phi \psi_i, T_i \rangle = \sum_{i \in I} \langle \phi \psi_i, T_i |_{\Omega_i \cap \Omega_j} \rangle = \sum_{i \in I} \langle \phi \psi_i, T_j \rangle = \langle \phi, T_j \rangle. \qquad \square$$

Definition 1.2.4 For $\emptyset \neq \Omega \subset \mathbf{R}^n$ open, let the space $\mathcal{E}'(\Omega)$ of *distributions on Ω with compact support* be defined as

$$\mathcal{E}'(\Omega) = \{T \in \mathcal{D}'(\Omega); \text{ supp } T \text{ is compact}\}.$$

Example 1.2.5 For $1 \leq p \leq \infty$, we obviously have $L_c^p(\Omega) \subset \mathcal{E}'(\Omega)$.

We also observe that distributions in $\mathcal{E}'(\Omega)$ can be continued by 0 to yield distributions in $\mathcal{E}'(\mathbf{R}^n)$:

$$\mathcal{E}'(\Omega) \hookrightarrow \mathcal{E}'(\mathbf{R}^n) : T \longmapsto (\phi \mapsto \langle \phi \psi, T \rangle), \qquad \phi \in \mathcal{D}(\mathbf{R}^n),$$

where $\psi \in \mathcal{D}(\Omega)$ is such that $\psi = 1$ on supp T.

We also mention that, by the same token, $\mathcal{E}'(\Omega)$ coincides with the dual of $\mathcal{E}(\Omega)$: For $T \in \mathcal{E}'(\Omega)$, the mapping

$$\hat{T} : \mathcal{E}(\Omega) \longrightarrow \mathbf{C} : \phi \longmapsto \langle \phi \psi, T \rangle$$

(with ψ as above) is well defined, linear, and continuous, i.e., $\phi_k \to \phi$ in $\mathcal{E}(\Omega)$ implies $\hat{T}(\phi_k) \to \hat{T}(\phi)$.

Finally, the equation supp $\delta_a = \{a\}$ for $a \in \Omega$ shows once more that $\delta_a \in \mathcal{E}'(\Omega) \setminus L_c^1(\Omega)$. $\qquad \square$

Example 1.2.6 Whereas distributions in Ω with compact support can always be extended to $\mathbf{R}^n$ (see Example 1.2.5), this is not necessarily the case for $S \in \mathcal{D}'(\Omega)$.

(a) We first consider a situation in which extension is possible. If $f \in L^1(\mathbf{S}^{n-1})$ does not satisfy the mean-value zero condition (1.1.1), then $\lim_{\epsilon \searrow 0} Y(|x| - \epsilon)|x|^{-n}f\left(\frac{x}{|x|}\right)$ does not exist in $\mathcal{D}'(\mathbf{R}^n)$. Nevertheless, we can continue

$$S = |x|^{-n}f\left(\frac{x}{|x|}\right) \in L_{\text{loc}}^1(\mathbf{R}^n \setminus \{0\}) \subset \mathcal{D}'(\mathbf{R}^n \setminus \{0\})$$

to a distribution T in $\mathcal{D}'(\mathbf{R}^n)$. In fact,

$$\langle \phi, T \rangle = \int_{|x| \leq 1} \frac{\phi(x) - \phi(0)}{|x|^n} f\left(\frac{x}{|x|}\right) dx + \int_{|x| > 1} \frac{\phi(x)}{|x|^n} f\left(\frac{x}{|x|}\right) dx, \qquad \phi \in \mathcal{D}(\mathbf{R}^n),$$

yields a distribution in $\mathcal{D}'(\mathbf{R}^n)$ such that $T|_{\mathbf{R}^n \setminus \{0\}} = S$.

Note that now—in contrast to Example 1.1.12—there does not exist any "canonical" extension of S in $\mathcal{D}'(\mathbf{R}^n)$: If the above partition of the integration domain into $|x| \leq 1$ and $|x| > 1$ is changed, this changes T by an additional δ-term. More generally, the condition $T|_{\mathbf{R}^n \setminus \{0\}} = S$ makes T unique up to a distribution with support in $\{0\}$.

However, as in Example 1.1.12, T can be represented by a limit in $\mathcal{D}'(\mathbf{R}^n)$:

$$\langle \phi, T \rangle = \lim_{\epsilon \searrow 0} \left[\int_{\epsilon \leq |x| \leq 1} \frac{\phi(x) - \phi(0)}{|x|^n} f\left(\frac{x}{|x|}\right) dx + \int_{|x|>1} \frac{\phi(x)}{|x|^n} f\left(\frac{x}{|x|}\right) dx \right]$$

$$= \lim_{\epsilon \searrow 0} \left[\langle \phi, Y(|x| - \epsilon)|x|^{-n} f\left(\frac{x}{|x|}\right) \rangle - \phi(0) \int_\epsilon^1 \frac{dr}{r} \int_{S^{n-1}} f(\omega) \, d\sigma(\omega) \right]$$

and hence

$$T = \lim_{\epsilon \searrow 0} \left[Y(|x| - \epsilon)|x|^{-n} f\left(\frac{x}{|x|}\right) + \log \epsilon \int_{S^{n-1}} f(\omega) \, d\sigma(\omega) \, \delta \right].$$

In particular, if $\int_{S^{n-1}} f(\omega) \, d\sigma(\omega) = 0$, then $T = \mathrm{vp}\left(|x|^{-n} f\left(\frac{x}{|x|}\right)\right)$.
If $n = 1$ and $f(\omega) = Y(\omega)$, then

$$x_+^{-1} = \lim_{\epsilon \searrow 0} \left[Y(x - \epsilon)x^{-1} + (\log \epsilon)\delta \right]$$

is an extension of $Y(x-\epsilon)x^{-1} \in L^1_{\mathrm{loc}}(\mathbf{R} \setminus \{0\})$ in $\mathcal{D}'(\mathbf{R}^1)$. Note that $x_+^{-1} \notin L^1_{\mathrm{loc}}(\mathbf{R})$, since $Y(x)x^{-1}$ is not locally integrable on $\mathbf{R}$.

Similarly as in Example 1.1.14, $g(|x|)f\left(\frac{x}{|x|}\right)$ can always be extended to yield a distribution in $\mathcal{D}'(\mathbf{R}^n)$ if $f \in L^1(S^{n-1})$ and $g : (0, \infty) \longrightarrow \mathbf{C}$ is measurable and fulfills $\int_0^1 |g(r)| r^{l+n} \, dr < \infty$ for some $l \in \mathbf{N}_0$. In fact, for $\phi \in \mathcal{D}(\mathbf{R}^n)$,

$$\langle \phi, T \rangle = \int_{|x|<1} \left(\phi(x) - \sum_{|\alpha| \leq l} \frac{\partial^\alpha \phi(0)}{\alpha!} x^\alpha \right) g(|x|) f\left(\frac{x}{|x|}\right) dx$$

$$+ \int_{|x|>1} \phi(x) g(|x|) f\left(\frac{x}{|x|}\right) dx,$$

furnishes a distribution T with $T|_{\mathbf{R}^n \setminus \{0\}} = g(|x|)f\left(\frac{x}{|x|}\right)$. Again, there is no "canonical" extension.

Seen from a more general point of view, the above continuation process can be subsumed under the title "regularization of algebraic singularities," cf. Gel'fand and Shilov [104], Ch. I, 1., pp. 45–81, and Ch. III, 4., pp. 313–329; Komech [154], Ch. 3, Sect. 1, 2, pp. 164–172, and Ch. 4, Sect. 1, 2, pp. 186–195; Palamodov [224], Ch. 1, Sect. 3, pp. 11–14, and Ch. 4, Sect. 2, pp. 67–70.

(b) Let us now treat a case where extension is impossible. We take $S(x) = Y(x)e^{1/x} \in L^1_{\mathrm{loc}}(\mathbf{R} \setminus \{0\})$ and we will show that there does not exist any $T \in \mathcal{D}'(\mathbf{R})$ with $T|_{\mathbf{R} \setminus \{0\}} = S$. In fact, let $\phi \in \mathcal{D}(\mathbf{R})$ with

(i) $\mathrm{supp}\, \phi \subset (0, \infty)$,   (ii) $\forall x \in \mathbf{R} : \phi(x) \geq 0$,   (iii) $\forall x \in [1, 2] : \phi(x) \geq 1$, and set $\phi_k(x) = e^{-k}\phi(k^2 x)$. Then $\phi_k \to 0$ in $\mathcal{D}(\mathbf{R})$ (but not in $\mathcal{D}(\mathbf{R} \setminus \{0\})$ since the supports of ϕ_k are not uniformly bounded in $\mathbf{R} \setminus \{0\}$.) On the other hand,

$$\langle \phi_k, S \rangle = e^{-k} \int_0^\infty e^{1/x} \phi(k^2 x)\, dx \geq e^{-k} \int_{k^{-2}}^{2k^{-2}} e^{1/x}\, dx \geq \frac{1}{k^2} \exp(-k + \tfrac{1}{2}k^2) \to \infty,$$

in contradiction to $\langle \phi_k, S \rangle = \langle \phi_k, T \rangle \to 0$ if an extension $T \in \mathcal{D}'(\mathbf{R})$ did exist.

More generally, one sees in the same way that $f \in C(\mathbf{R}^n \setminus \{0\})$ has no distributional extension in $\mathcal{D}'(\mathbf{R}^n)$ if f is positive and $\forall k \in \mathbf{N} : \lim_{|x| \to 0} |x|^k f(x) = \infty$, cf. Hirsch and Lacombe [131], Ch. III, 7, 2F, Ex. 4b, p. 275; Zuily [309], Exercise 14, p. 31, and Sol. Exercise 14, pp. 41, 42. □

Let us next treat the *composition of distributions with diffeomorphisms*. If $\emptyset \neq \Omega_1, \Omega_2 \subset \mathbf{R}^n$ and $h : \Omega_1 \longrightarrow \Omega_2$ is a diffeomorphism and $f : \Omega_2 \longrightarrow \mathbf{C}$ is locally integrable, then $T_{f \circ h}$ is given by the following formula:

$$\langle \phi, T_{f \circ h} \rangle = \int_{\Omega_1} \phi(x) f(h(x))\, dx$$

$$= \int_{\Omega_2} \phi(h^{-1}(y)) f(y) |\det(h^{-1})'(y)|\, dy = \langle \frac{\phi}{|\det h'|} \circ h^{-1}, T_f \rangle.$$

This shows that the following definition of composition for distributions is consistent with that for functions.

Definition 1.2.7 For a diffeomorphism $h : \Omega_1 \longrightarrow \Omega_2$ of open sets in $\mathbf{R}^n$ and $T \in \mathcal{D}'(\Omega_2)$, we define the *composition* $T \circ h = T(h(x)) \in \mathcal{D}'(\Omega_1)$ *of T with h* by

$$\langle \phi, T \circ h \rangle = \langle \frac{\phi}{|\det h'|} \circ h^{-1}, T \rangle, \qquad \phi \in \mathcal{D}(\Omega_1).$$

Example 1.2.8 If $a \in \Omega_1$ and $b = h(a) \in \Omega_2$, then $\delta_b \circ h = |\det h'(a)|^{-1} \delta_a$, cf. Hörmander [139], Ch. VI, Ex. 6.1.3, p. 136. In particular, if $A \in \mathrm{Gl}_n(\mathbf{R})$, then $\delta \circ A = |\det A|^{-1} \delta$, and, if $A = c I_n, c > 0$, this equation implies that δ is, according to the following definition, homogeneous of degree $-n$, i.e., $\delta(cx) = c^{-n} \delta$. □

Definition 1.2.9 Let $\emptyset \neq \Omega \subset \mathbf{R}^n$ be an open cone, i.e. an open subset fulfilling $c\Omega = \Omega$ for all $c > 0$. Then $T \in \mathcal{D}'(\Omega)$ is called *homogeneous of degree* $\lambda \in \mathbf{C}$ iff

$$\forall c > 0 : T(cx) = T \circ c I_n = c^\lambda T.$$

Example 1.2.10 If $f \in L^1(\mathbf{S}^{n-1})$ fulfills the mean-value zero condition (1.1.1), then we easily see that $T = \mathrm{vp}(|x|^{-n} f(\frac{x}{|x|})) \in \mathcal{D}'(\mathbf{R}^n)$ is homogeneous of degree $-n$. In contrast, if $\int_{\mathbf{S}^{n-1}} f(\omega)\, d\sigma(\omega) \neq 0$, then $|x|^{-n} f(\frac{x}{|x|}) \in L^1_{\mathrm{loc}}(\mathbf{R}^n \setminus \{0\}) \subset \mathcal{D}'(\mathbf{R}^n \setminus \{0\})$

is still homogeneous of degree $-n$, but the extension constructed in Example 1.2.6
(a) ceases to be homogeneous in $\mathcal{D}'(\mathbf{R}^n)$. If $c > 0$, then

$$
T(cx) = \lim_{\epsilon \searrow 0}\left[c^{-n}Y(c|x| - \epsilon)|x|^{-n}f\left(\frac{x}{|x|}\right) + c^{-n}\log\epsilon \int_{S^{n-1}} f(\omega)\, d\sigma(\omega)\, \delta \right]
$$

$$
= c^{-n}\lim_{\epsilon \searrow 0}\left[Y\left(|x| - \frac{\epsilon}{c}\right)|x|^{-n}f\left(\frac{x}{|x|}\right) + \log\frac{\epsilon}{c}\int_{S^{n-1}} f(\omega)\, d\sigma(\omega)\, \delta \right]
$$

$$
+ c^{-n}(\log c)\int_{S^{n-1}} f(\omega)\, d\sigma(\omega)\, \delta,
$$

and hence we conclude that

$$
T(cx) = c^{-n}T + c^{-n}(\log c)\int_{S^{n-1}} f(\omega)\, d\sigma(\omega)\, \delta. \qquad\qquad \square
$$

Example 1.2.11 Let us consider next the composition of principal values with linear
mappings.

Obviously, if $T \in \mathcal{D}'(\mathbf{R}^n)$ is homogeneous of degree λ, then the same holds for
$T \circ A$ if $A \in \mathrm{Gl}_n(\mathbf{R})$. In particular, if $f \in L^1(S^{n-1})$ fulfills the mean-value zero
condition (1.1.1), then $T = \mathrm{vp}\left(|x|^{-n}f(\frac{x}{|x|})\right) \in \mathcal{D}'(\mathbf{R}^n)$ is homogeneous of degree $-n$
and the same holds for $T \circ A$. We evidently have

$$
T \circ A|_{\mathbf{R}^n \setminus \{0\}} = |Ax|^{-n}f\left(\frac{Ax}{|Ax|}\right) \in L^1_{\mathrm{loc}}(\mathbf{R}^n \setminus \{0\}),
$$

and hence $T \circ A$ coincides with $S = \mathrm{vp}\left(|Ax|^{-n}f(\frac{Ax}{|Ax|})\right)$ outside the origin.

Let us determine the difference $T \circ A - S \in \mathcal{D}'(\mathbf{R}^n)$. For $\phi \in \mathcal{D}(\mathbf{R}^n)$, we have

$$
\langle \phi, T \circ A - S \rangle = \lim_{\epsilon \searrow 0}\langle \phi, \left(Y(|x| - \epsilon)|x|^{-n}f\left(\frac{x}{|x|}\right)\right) \circ A - Y(|x| - \epsilon)|Ax|^{-n}f\left(\frac{Ax}{|Ax|}\right)\rangle
$$

$$
= \lim_{\epsilon \searrow 0}\int_{\mathbf{R}^n} \phi(x)|Ax|^{-n}f\left(\frac{Ax}{|Ax|}\right)\left[Y(|Ax| - \epsilon) - Y(|x| - \epsilon)\right] dx
$$

$$
= \lim_{\epsilon \searrow 0} |\det A|^{-1}\times
$$

$$
\times \int_{\mathbf{R}^n} \phi(A^{-1}y)|y|^{-n}f\left(\frac{y}{|y|}\right)\left[Y(|y| - \epsilon) - Y(|A^{-1}y| - \epsilon)\right] dy
$$

$$
= \lim_{\epsilon \searrow 0} |\det A|^{-1}\times
$$

$$
\times \int_{S^{n-1}} f(\omega)\int_0^\infty \phi(rA^{-1}\omega)\left[Y(r - \epsilon) - Y(r|A^{-1}\omega| - \epsilon)\right]\frac{dr}{r}\, d\sigma(\omega).
$$

Since

$$\int_0^\infty \phi(rA^{-1}\omega)\big[Y(r-\epsilon) - Y(r|A^{-1}\omega| - \epsilon)\big] \frac{dr}{r} = -\phi(0)\log|A^{-1}\omega| + O(\epsilon)$$

for $\epsilon \searrow 0$, we conclude that

$$T \circ A = \mathrm{vp}\Big(|Ax|^{-n}f\Big(\frac{Ax}{|Ax|}\Big)\Big) - \frac{\delta}{|\det A|}\int_{S^{n-1}} f(\omega)\log|A^{-1}\omega|\, d\sigma(\omega).$$

(For the particular case of $A = cI_n$, we recover the result of Example 1.2.10.)

Note that, more generally, if $h: \mathbf{R}^n \longrightarrow \mathbf{R}^n$ is a diffeomorphism fulfilling $h(0) = 0$, then an analogous calculation yields the formula

$$\Big(\mathrm{vp}(|x|^{-n}f(\tfrac{x}{|x|}))\Big) \circ h = \mathrm{vp}\Big((|x|^{-n}f(\tfrac{x}{|x|})) \circ h\Big)$$

$$- \frac{\delta}{|\det h'(0)|}\int_{S^{n-1}} f(\omega)\log|h'(0)^{-1}\omega|\, d\sigma(\omega).$$

(Here, as above,

$$\mathrm{vp}\Big((|x|^{-n}f(\tfrac{x}{|x|})) \circ h\Big) = \lim_{\epsilon \searrow 0} Y(|x| - \epsilon)|h(x)|^{-n}f\Big(\frac{h(x)}{|h(x)|}\Big)$$

in $\mathcal{D}'(\mathbf{R}^n)$.) Using a partition of unity we see that the last formula also holds if h is continuous, $h^{-1}(0) = \{0\}$, and h is a C^1 bijection near 0. This is already an instance where we use the composition of distributions with more general, not necessarily diffeomorphic, mappings.   □

Definition 1.2.12 Let $\emptyset \ne \Omega \subset \mathbf{R}^n$ be open, $h : \Omega \longrightarrow \mathbf{R}$ be C^∞ and submersive, i.e., $\forall x \in \Omega : \nabla h(x) \ne 0$, and $T \in \mathcal{D}'(\mathbf{R})$. Then the *pullback* h^*T of T by h is defined by

$$\langle \phi, h^*T \rangle = \langle \frac{\mathrm{d}}{\mathrm{d}s}\int_\Omega \phi(x) Y(s - h(x))\, dx, T_s \rangle.$$

(Cf. Friedlander and Joshi [84], (7.2.4/5), p. 82.) We shall also often write $T \circ h$ or simply $T(h(x))$ instead of h^*T.

Proposition 1.2.13 *The mapping $h^* : \mathcal{D}'(\mathbf{R}) \longrightarrow \mathcal{D}'(\Omega) : T \longmapsto h^*T = T \circ h$ is well defined and sequentially continuous. Furthermore, if $T = T_f \in L^1_{\mathrm{loc}}(\mathbf{R})$, then $T \circ h$ is locally integrable and it coincides with the classical composition $T_{f \circ h}$ of functions.*

Proof

(1) Let $\phi \in \mathcal{D}(\Omega)$. Using a partition of unity and appropriate coordinates, we can assume that $\frac{\partial h}{\partial x_1} \ne 0$ and that $\xi_1 = h$, $\xi_2 = x_2, \ldots, \xi_n = x_n$ are coordinates in a

neighborhood of supp ϕ. Substituting the new variables we obtain

$$\frac{d}{ds}\int_\Omega \phi(x)Y(s-h(x))\,dx = \frac{d}{ds}\int \psi(\xi)Y(s-\xi_1)\,d\xi = \int \psi(s,\xi_2,\ldots,\xi_n)\,d\xi' =: \chi(s),$$

where

$$\psi(\xi) := \phi(x(\xi))\left|\det\left(\frac{\partial x_i}{\partial \xi_j}\right)\right| = \frac{\phi(x(\xi))}{|(\partial_1 h)(x(\xi))|}, \qquad \xi' = (\xi_2,\ldots,\xi_n),$$

and $\chi \in \mathcal{D}(\mathbf{R}^1)$. Hence h^*T is defined by the formula $\langle \phi, h^*T \rangle = \langle \chi, T \rangle$, and h^*T is a distribution because the mapping $\mathcal{D}(\Omega) \longrightarrow \mathcal{D}(\mathbf{R}) : \phi \longmapsto \chi$ is obviously linear and, by Lebesgue's theorem on dominated convergence, sequentially continuous. Furthermore, $T_k \to T$ implies $T_k \circ h \to T \circ h$.

(2) If $T = T_f \in L^1_{loc}(\mathbf{R})$ and ϕ, ξ, χ are as above, then

$$\langle \phi, T_f \circ h \rangle = \langle \chi, T_f \rangle = \int \chi(s)f(s)\,ds$$

$$= \int f(s)\int \psi(s,\xi_2,\ldots,\xi_n)\,d\xi_2\ldots d\xi_n ds$$

$$= \int f(\xi_1)\phi(x(\xi))\left|\det\left(\frac{\partial x_i}{\partial \xi_j}\right)\right|\,d\xi$$

$$= \int_\Omega f(h(x))\phi(x)\,dx = \langle \phi, T_{f\circ h} \rangle$$

by Fubini's theorem and the substitution formula for multiple integrals. $\square$

Let us remark that the pullback $h^* : \mathcal{D}'(\Omega_2) \longrightarrow \mathcal{D}'(\Omega_1)$ can be defined similarly if $\Omega_1 \subset \mathbf{R}^n$ and $\Omega_2 \subset \mathbf{R}^m$ are open subsets and $h : \Omega_1 \longrightarrow \Omega_2$ is submersive, i.e., the rank of $h'(x)$ is m for all $x \in \Omega_1$, cf. Friedlander and Joshi [84], Thm. 7.2.2, p. 84; Hörmander [139], Thm. 6.1.2, p. 134. Furthermore, $h^*T = T \circ h$ is also well defined if $T \in \mathcal{D}'(\Omega_2)$ is a continuous function in a neighborhood of the set where h is not submersive. This follows easily by a partition of unity argument, cf. Komech [154], Ch. I, Sect. 1, Rem. 1.1, p. 132.

Example 1.2.14 Let us specialize now Definition 1.2.12 to the case of $T = \delta_a \in \mathcal{D}'(\mathbf{R}^1)$. We assume that $\emptyset \neq \Omega \subset \mathbf{R}^n$ is open, $h : \Omega \longrightarrow \mathbf{R}$ C^∞, $a \in h(\Omega)$, and h is submersive in a neighborhood of $M = h^{-1}(a)$. Then M is a C^∞ hypersurface of $\mathbf{R}^n$, and we equip it with the Riemannian metric g induced by the standard metric $\sum_{i=1}^n dx_i \otimes dx_i$ on $\mathbf{R}^n$. The metric g generates the surface measure $d\sigma$, i.e.,

$$\int_M \phi(x)\,d\sigma(x) = \int_U \phi(x(u))\sqrt{\det(g_{jk}(u))}\,du$$

if $\phi \in \mathcal{C}(M)$ such that $\operatorname{supp} \phi$ lies in a coordinate patch parameterized by $x = x(u)$, $u \in U \subset \mathbf{R}^{n-1}$, and $g = \sum_{j,k=1}^{n-1} g_{jk}(u) \, du_j \otimes du_k$.

If $f : M \longrightarrow \mathbf{C}$ is locally integrable (with respect to $d\sigma$), then the *single layer distribution* $S_M(f) \in \mathcal{D}'(\Omega)$ *with density* f is defined by

$$\langle \phi, S_M(f) \rangle = \int_M \phi(x) f(x) \, d\sigma(x), \qquad \phi \in \mathcal{D}(\Omega). \tag{1.2.1}$$

(Note that $S_M(f) \in \mathcal{D}'(\Omega)$ is well defined, since $M \cap \operatorname{supp} \phi$ is a closed subset of $\operatorname{supp} \phi$ and hence is compact for $\phi \in \mathcal{D}(\Omega)$.) In a physical context, $S_M(\rho)$ describes a mass or charge distribution of density ρ on the submanifold M.

Since $\delta_a = \lim_{\epsilon \searrow 0} \frac{1}{2\epsilon} Y(\epsilon - |s - a|)$ in $\mathcal{D}'(\mathbf{R}_s^1)$ (see Example 1.1.11), we infer from Proposition 1.2.13 that $\delta_a \circ h = \lim_{\epsilon \searrow 0} \frac{1}{2\epsilon} Y(\epsilon - |h(x) - a|)$ in $\mathcal{D}'(\Omega)$ is the limit of constant mass densities on the layers $\{x \in \Omega; \, a - \epsilon < h(x) < a + \epsilon\}$. These layers have the approximate width $\frac{2\epsilon}{|\nabla h(x)|}$ and this gives intuitive understanding to the formula

$$\langle \phi, \delta_a \circ h \rangle = \int_M \frac{\phi(x)}{|\nabla h(x)|} \, d\sigma(x), \qquad \phi \in \mathcal{D}(\Omega), \tag{1.2.2}$$

cf. Friedlander and Joshi [84], (7.2.10), p. 83; Hörmander [139], Thm. 6.1.5, p. 136.

In order to verify formula (1.2.2), we introduce coordinates $\xi_1 = h$, $\xi_2 = x_2, \ldots, \xi_n = x_n$ as in the proof of Proposition 1.2.13. Then $x_2, \ldots, x_n$ are local coordinates on M, and employing

$$0 = dh = \sum_{j=1}^n \partial_j h \, dx_j \implies dx_1 = - \sum_{j=2}^n \frac{\partial_j h}{\partial_1 h} \, dx_j \quad \text{on } M$$

we can express the metric g on M in the following form:

$$g = \sum_{j,k=2}^n \left(\delta_{jk} + \frac{\partial_j h \cdot \partial_k h}{(\partial_1 h)^2} \right) dx_j \otimes dx_k = \sum_{j,k=2}^n g_{jk} dx_j \otimes dx_k.$$

Hence

$$(g_{jk})_{j,k=2,\ldots,n} = I_{n-1} + v \cdot v^T \qquad \text{with } v = \frac{1}{\partial_1 h} \begin{pmatrix} \partial_2 h \\ \vdots \\ \partial_n h \end{pmatrix},$$

and Schur's formula yields

$$\det\left((g_{jk})_{j,k=2,\ldots,n} \right) = 1 + |v|^2 = \frac{|\nabla h|^2}{(\partial_1 h)^2}$$

and

$$\langle \phi, \delta_a \circ h \rangle = \langle \chi, \delta_a \rangle = \chi(a) = \int \frac{\phi(x(a, \xi'))}{|\partial_1 h(x(a, \xi'))|} \, d\xi'$$

$$= \int \frac{\phi}{|\nabla h|}(x(a, \xi')) \cdot \sqrt{\det(g_{jk})} \, d\xi' = \int_M \frac{\phi(x)}{|\nabla h(x)|} \, d\sigma(x),$$

where $\xi' = (\xi_2, \ldots, \xi_n)$ and $\phi \in \mathcal{D}(\Omega), \chi \in \mathcal{D}(\mathbf{R}^1)$ are as in the proof of Proposition 1.2.13. Thus, in short-hand, formula (1.2.2) can be written as

$$\delta_a \circ h = \delta(h(x) - a) = S_M(|\nabla h|^{-1}). \tag{1.2.2'}$$

Note that $\mathrm{supp}(\delta_a \circ h) = M$.

Let us next specialize formula (1.2.2) to some particular cases.

(a) If $n = 1$, then $h : \Omega \longrightarrow \mathbf{R}$ (with $\Omega \subset \mathbf{R}$ open) fulfills the submersion condition iff $h'(x) \neq 0$ for all $x \in M = h^{-1}(a)$, which must be a discrete set in Ω. Then we obtain

$$h^* \delta_a = \delta_a \circ h = \sum_{x \in h^{-1}(a)} \frac{1}{|h'(x)|} \delta_x \in \mathcal{D}'(\Omega).$$

For example,

$$\delta_a(x^2) = \delta(x^2 - a) = \frac{1}{2\sqrt{a}} \left(\delta_{\sqrt{a}} + \delta_{-\sqrt{a}} \right) \in \mathcal{D}'(\mathbf{R}), \qquad a > 0,$$

$$\delta \circ \sin = \delta(\sin x) = \sum_{k \in \mathbf{Z}} \delta_{k\pi} \in \mathcal{D}'(\mathbf{R}),$$

$$\delta(\sin \tfrac{1}{x}) = \frac{1}{\pi^2} \sum_{k \in \mathbf{Z} \setminus \{0\}} k^{-2} \delta_{1/(k\pi)} \in \mathcal{D}'(\mathbf{R} \setminus \{0\}).$$

In the last example, we have $h : \Omega = \mathbf{R} \setminus \{0\} \longrightarrow \mathbf{R} : x \longmapsto \sin \tfrac{1}{x}$ and $M = \{(k\pi)^{-1}; k \in \mathbf{Z} \setminus \{0\}\}$. Note that the sum for $\delta(\sin \tfrac{1}{x})$ converges also in $\mathcal{D}'(\mathbf{R})$, since $\sum_{k=1}^{\infty} k^{-2}$ converges; in contrast, $\delta(\exp(-\tfrac{1}{|x|}) \sin \tfrac{1}{x}) \in \mathcal{D}'(\mathbf{R} \setminus \{0\})$ does not even have an extension in $\mathcal{D}'(\mathbf{R})$, cf. Example 1.2.6 (b).

(b) We suppose next that h is a positive definite quadratic form, i.e., $h(x) = x^T C x$ with $C = C^T \in \mathrm{Gl}_n(\mathbf{R})$ positive definite. Then there exists a linear map $A \in \mathrm{Gl}_n(\mathbf{R})$ such that $h(Ay) = y^T A^T C A y = |y|^2$, and hence

$$\langle \phi, \delta(x^T C x - 1) \rangle = \langle \phi, \delta_1 \circ h \rangle = \langle \phi, \delta_1 \circ h \circ A \circ A^{-1} \rangle$$

$$= |\det A| \langle \phi \circ A, \delta_1(|y|^2) \rangle = \frac{|\det A|}{2} \int_{\mathbf{S}^{n-1}} \phi(A\omega) \, d\sigma(\omega),$$

since $\big| \nabla |y|^2 \big| = 2|y| = 2$ for $y = \omega \in \mathbf{S}^{n-1}$. Note that $|\det A| = (\det C)^{-1/2}$.

If, e.g., $h(x) = \sum_{i=1}^{3} x_i^2/a_i^2$, $x \in \mathbf{R}^3$, $a_1, a_2, a_3 \in (0, \infty)$, and $\phi \in \mathcal{D}(\mathbf{R}^3)$, then

$$\langle \phi, \delta_1 \circ h \rangle = \langle \phi, \delta\left(1 - \sum_{i=1}^{3} \frac{x_i^2}{a_i^2}\right)\rangle$$

$$= \frac{a_1 a_2 a_3}{2} \int_0^{\pi} \int_0^{2\pi} \phi(a_1 \cos \varphi \sin \vartheta, a_2 \sin \varphi \sin \vartheta, a_3 \cos \vartheta) \sin \vartheta \, d\varphi d\vartheta.$$

More generally, if $h(x) = x^T C x$ is a definite quadratic form with r positive and $s = n - r$ negative eigenvalues, then there exists $A \in \mathrm{Gl}_n(\mathbf{R})$ such that

$$h(Ay) = |y'|^2 - |y''|^2, \qquad y = (y', y''), \qquad y' \in \mathbf{R}^r, \qquad y'' \in \mathbf{R}^s,$$

and hence

$$\langle \phi, \delta(x^T C x - 1) \rangle = \langle \phi, \delta_1 \circ h \rangle = |\det A| \langle \phi \circ A, \delta_1(|y'|^2 - |y''|^2) \rangle$$

$$= \frac{|\det A|}{2} \int_0^{\infty} \int_{S^{r-1} \times S^{s-1}} \phi\left(A \begin{pmatrix} \omega_1 \, \mathrm{ch}\, t \\ \omega_2 \, \mathrm{sh}\, t \end{pmatrix}\right) \mathrm{ch}^{r-1} t \, \mathrm{sh}^{s-1} t \, d\sigma(\omega_1) d\sigma(\omega_2) dt,$$

since $|\nabla(|y'|^2 - |y''|^2)| = 2|y| = 2\sqrt{\mathrm{ch}^2 t + \mathrm{sh}^2 t}$ for $y = \begin{pmatrix} \omega_1 \, \mathrm{ch}\, t \\ \omega_2 \, \mathrm{sh}\, t \end{pmatrix}$ and $d\sigma(y) = \sqrt{\mathrm{ch}^2 t + \mathrm{sh}^2 t} \, \mathrm{ch}^{r-1} t \, \mathrm{sh}^{s-1} t \, d\sigma(\omega_1) \otimes d\sigma(\omega_2) \otimes dt$.

If also a linear term is present in h, i.e.,

$$h(x) = h(Ay) = |y'|^2 - |y''|^2 - y_n, \, y = (y', y'', y_n), \, y' \in \mathbf{R}^r, \, y'' \in \mathbf{R}^s, \, r+s+1 = n,$$

and $A \in \mathrm{Gl}_n(\mathbf{R})$, then

$$\langle \phi, \delta \circ h \rangle = |\det A| \int_0^{\infty} \int_0^{\infty} \int_{S^{r-1} \times S^{s-1}} \phi\left(A \begin{pmatrix} t_1 \omega_1 \\ t_2 \omega_2 \\ t_1^2 - t_2^2 \end{pmatrix}\right) t_1^{r-1} t_2^{s-1} d\sigma(\omega_1) d\sigma(\omega_2) dt_1 dt_2.$$

Of course, this is trivial, since generally

$$\langle \phi, \delta(x_n - g(x')) \rangle = \int_{\mathbf{R}^{n-1}} \phi(x', g(x')) \, dx', \qquad \phi \in \mathcal{D}(\mathbf{R}^n),$$

for a C^1 function $g : \mathbf{R}^{n-1} \longrightarrow \mathbf{R}$.   □

Example 1.2.15 General formulae for fundamental solutions often contain expressions of the type $\delta_{a_1}(h_1)\delta_{a_2}(h_2)$. Such a product has to be interpreted in the sense of the remark following Proposition 1.2.13, i.e., as

$$h^*\delta_a = \delta_a \circ h, \quad a = \begin{pmatrix} a_1 \\ a_2 \end{pmatrix} \in \mathbf{R}^2, \quad h = \begin{pmatrix} h_1 \\ h_2 \end{pmatrix} : \Omega \longrightarrow \mathbf{R}^2, \quad \delta_a \in \mathcal{D}'(\mathbf{R}^2).$$

Analogously to Example 1.2.14, let us suppose that $h : \Omega \longrightarrow \mathbf{R}^m$ is submersive on $M = h^{-1}(a)$, $a \in \mathbf{R}^m$. Then $\delta_a \circ h$ can be expressed by the surface measure $d\sigma$ on the submanifold $M \subset \Omega$ of codimension m :

$$\langle \phi, \delta_a \circ h \rangle = \int_M \frac{\phi(x)\,d\sigma(x)}{\sqrt{\det(h'(x) \cdot h'(x)^T)}}, \qquad \phi \in \mathcal{D}(\Omega), \tag{1.2.3}$$

or, in short-hand, $\delta_a \circ h = \dfrac{d\sigma}{\sqrt{\det(h' \cdot h'^T)}}$, wherein $h' = \begin{pmatrix} \partial_1 h_1 & \ldots & \partial_n h_1 \\ \vdots & & \vdots \\ \partial_1 h_m & \ldots & \partial_n h_m \end{pmatrix}$ denotes

the Jacobian of h.

Let us apply formula (1.2.3) to two particular cases.

(a) The density pertaining to a point mass or an electrical charge moving on the trajectory $x_1 = u_1(t), \ldots, x_{n-1} = u_{n-1}(t)$, $u\ \mathcal{C}^1$, in space-time $\mathbf{R}^n_{t,x}$ can be defined by the distribution $\delta \circ h \in \mathcal{D}'(\mathbf{R}^n)$ where $h : \mathbf{R}^n \longrightarrow \mathbf{R}^{n-1} : (t, x) \longmapsto x - u(t)$. As is intuitively clear,

$$\langle \phi, \delta(x - u(t)) \rangle = \int_{\mathbf{R}} \phi(t, u(t))\,dt, \qquad \phi \in \mathcal{D}(\mathbf{R}^n).$$

This also results from formula (1.2.3), since

$$h(t, x) = x - u(t) \Longrightarrow h' = (-\dot{u}, I_{n-1}) \text{ and } h' \cdot h'^T = I_{n-1} + \dot{u} \cdot \dot{u}^T$$

$$\Longrightarrow \sqrt{\det(h' \cdot h'^T)} = \sqrt{1 + |\dot{u}|^2} = \frac{d\sigma}{dt}.$$

(b) An elementary example of a distribution with support on a codimension 2 submanifold is provided by

$$\delta \circ h, \quad h = \begin{pmatrix} |x|^2 - R^2 \\ b^T x - 1 \end{pmatrix}, \quad R > 0, \quad b \in \mathbf{R}^n \setminus \{0\}.$$

Then

$$\langle \phi, h^* \delta \rangle = \langle \phi, \delta(|x|^2 - R^2)\delta(b^T x - 1) \rangle$$

$$= \frac{Y(|b|R - 1)}{2|b|^{n-2}} (|b|^2 R^2 - 1)^{(n-3)/2} \times$$

$$\times \int_{\mathbf{S}^{n-2}} \phi\left(\frac{b}{|b|^2} + \sqrt{R^2 - |b|^{-2}} \sum_{i=1}^{n-1} \omega_i v_i \right) d\sigma(\omega)$$

if $v_1, \dots, v_{n-1}, \frac{b}{|b|}$ is an orthonormal basis in $\mathbf{R}^n$. In fact,

$$h' = \begin{pmatrix} 2x^T \\ b^T \end{pmatrix}, \quad h' \cdot h'^T = \begin{pmatrix} 4|x|^2 & 2x^T b \\ 2x^T b & |b|^2 \end{pmatrix}, \quad \det(h' \cdot h'^T) = 4(|b|^2 R^2 - 1) \text{ on } M$$

and $d\sigma(x) = (R^2 - |b|^{-2})^{(n-2)/2} d\sigma(\omega)$. $\qquad\qquad\qquad\qquad\qquad\square$

1.3 Differentiation

If $\emptyset \neq \Omega \subset \mathbf{R}^n$ is open and $f \in \mathcal{C}^1(\Omega)$, $\phi \in \mathcal{D}(\Omega)$, then we can extend the functions $f \cdot \phi$, $\partial_1 f \cdot \phi$, $f \cdot \partial_1 \phi$ to all of $\mathbf{R}^n$ by 0. These extended functions are in $\mathcal{C}^1(\mathbf{R}^n)$ and $C(\mathbf{R}^n)$, respectively, and fulfill (in $\mathbf{R}^n$) $\partial_1(f\phi) = \phi \cdot \partial_1 f + f \cdot \partial_1 \phi$. Therefore, if $\operatorname{supp} \phi$ is contained in the strip

$$\{x \in \mathbf{R}^n; |x_1| \leq R\} = [-R, R] \times \mathbf{R}^{n-1},$$

then Fubini's theorem yields

$$\langle \phi, T_{\partial_1 f} \rangle = \int_\Omega \phi \cdot \partial_1 f \, dx = \int_{\mathbf{R}^n} \phi \cdot \partial_1 f \, dx = \int_{\mathbf{R}^n} (-f \cdot \partial_1 \phi + \partial_1 (f \cdot \phi)) \, dx$$

$$= -\int_\Omega f \cdot \partial_1 \phi \, dx + \int_{\mathbf{R}^{n-1}} [(f\phi)(R, x') - (f\phi)(-R, x')] \, dx' = -\langle \partial_1 \phi, T_f \rangle.$$

This shows that the following definition of $\partial_j T$ is consistent with the usual differentiation in case T is $\mathcal{C}^1$.

Definition 1.3.1 For $\emptyset \neq \Omega \subset \mathbf{R}^n$ open, $T \in \mathcal{D}'(\Omega)$, and $\alpha \in \mathbf{N}_0^n$, we define the *(higher) partial derivatives* of T by

$$\partial^\alpha T : \mathcal{D}(\Omega) \longrightarrow \mathbf{C} : \phi \longmapsto (-1)^{|\alpha|} \langle \partial^\alpha \phi, T \rangle.$$

Proposition 1.3.2 $\partial^\alpha : \mathcal{D}'(\Omega) \longrightarrow \mathcal{D}'(\Omega)$ *is well defined, linear, and sequentially continuous. If* $f \in C^m(\Omega)$ *and* $|\alpha| \leq m$, *then* $\partial^\alpha T_f = T_{\partial^\alpha f}$. *Furthermore,* $\partial_1 T = \lim_{\epsilon \to 0} \frac{1}{\epsilon}(T(x_1 + \epsilon, x') - T)$, *if* $T(x_1 + \epsilon, x') = T \circ h$ *with* $h(x) = (x_1 + \epsilon, x_2, \ldots, x_n)$ *(cf. Definition 1.2.7), and similarly for the other derivatives.*

Proof The first part is implied by the linearity and continuity of the mapping

$$\partial^\alpha : \mathcal{D}(\Omega) \longrightarrow \mathcal{D}(\Omega) : \phi \longmapsto (-1)^{|\alpha|} \partial^\alpha \phi.$$

The equation $\partial^\alpha T_f = T_{\partial^\alpha f}$ follows by induction from the introduction above.
Finally, for $\phi \in \mathcal{D}(\Omega)$,

$$\lim_{\epsilon \to 0} \frac{1}{\epsilon} \langle \phi, T(x_1 + \epsilon, x') - T \rangle = \lim_{\epsilon \to 0} \left\langle \frac{\phi(x_1 - \epsilon, x') - \phi}{\epsilon}, T \right\rangle = -\langle \partial_1 \phi, T \rangle,$$

since

$$\lim_{\epsilon \to 0} \frac{\phi(x_1 - \epsilon, x') - \phi}{\epsilon} = -\lim_{\epsilon \to 0} \int_0^1 (\partial_1 \phi)(x_1 + \epsilon t, x') \, dt = -\partial_1 \phi$$

holds in $\mathcal{D}(\Omega)$. $\square$

Proposition 1.3.3 *For* $\emptyset \neq \Omega \subset \mathbf{R}^n$ *open,* $f \in C^\infty(\Omega)$, $T \in \mathcal{D}'(\Omega)$, $h : \Omega \longrightarrow \mathbf{R}$ *submersive,* $\alpha, \beta \in \mathbf{N}_0^n$, $S \in \mathcal{D}'(\mathbf{R})$, *we have*

(a) $\partial^\alpha \partial^\beta T = \partial^\beta \partial^\alpha T$ *(commutativity);*
(b) $\partial^\alpha (f \cdot T) = \sum_{\beta \leq \alpha} \binom{\alpha}{\beta} \partial^\beta f \cdot \partial^{\alpha-\beta} T$ *(Leibniz formula);*
(c) $\partial_j (S \circ h) = \frac{\partial h}{\partial x_j} \cdot (S' \circ h)$ *(chain rule).*

Proof This follows by transposition from the corresponding rules in $\mathcal{D}(\Omega)$. $\square$

Example 1.3.4 Whereas densities of point charges as, e.g., the Dirac measure (cf. Example 1.1.11) can be described in the framework of measure theory, this is no longer the case for dipoles (cf. Schwartz [246], Ch. I, Sect. 2, p. 20) and for "double layers" on surfaces, which are both genuine distributions.

(a) If $\omega \in \mathbf{S}^{n-1}$, $l \in \mathbf{R}$, and $a \in \mathbf{R}^n$, then Proposition 1.3.2 implies that

$$\lim_{\epsilon \to 0} \frac{l}{\epsilon}(\delta_{a+\epsilon\omega} - \delta_a) = \lim_{\epsilon \to 0} \frac{l}{\epsilon}(\delta_a(x - \epsilon\omega) - \delta_a) = -l \sum_{j=1}^n \omega_j \partial_j \delta_a$$

$$= -l \omega^T \cdot \nabla \delta_a \in \mathcal{D}'(\mathbf{R}^n),$$

and hence $-l\omega^T \cdot \nabla \delta_a$ corresponds to a dipole in direction ω located at a and with strength l. For $\phi \in \mathcal{D}$, we have

$$\langle \phi, -l\omega^T \cdot \nabla \delta_a \rangle = l\omega^T \cdot \nabla \phi(a).$$

(b) Let us next assume that M is a C^2 hypersurface in $\mathbf{R}^n$ given by $M = h^{-1}(a)$ with $h : \Omega \longrightarrow \mathbf{R}$ C^2 and submersive near M, cf. Example 1.2.14. M is orientable and we shall orient it by the unit normal $\nu = \frac{\nabla h}{|\nabla h|}$. For $f : M \longrightarrow \mathbf{C}$ locally integrable, we have defined the single layer distribution $S_M(f) \in \mathcal{D}'(\Omega)$ in (1.2.1). Similarly, we now define the *double layer distribution* $D_M(f) \in \mathcal{D}'(\Omega)$ *with density* f by

$$\langle \phi, D_M(f) \rangle = -\int_M f(x) \cdot \partial_\nu \phi(x) \, d\sigma(x), \qquad \phi \in \mathcal{D}(\Omega), \tag{1.3.1}$$

where $\partial_\nu \phi(x) = \nu(x)^T \cdot \nabla \phi(x) = \sum_{j=1}^n \nu_j(x) \partial_j \phi(x)$, $x \in M$, is often called the *normal derivative* of ϕ on M. According to (a), $D_M(f)$ corresponds to a distribution of dipoles in direction ν spread out on M with density $-f$. Note that, similarly to the equation $g \cdot \delta' = g(0)\delta' - g'(0)\delta \in \mathcal{D}'(\mathbf{R})$, one readily obtains

$$g \cdot D_M(f) = D_M(f \cdot g|_M) - S_M(f \cdot \partial_\nu g), \qquad g \in \mathcal{E}(\Omega), \tag{1.3.2}$$

by evaluation on test functions.

Let us express next $\delta_a' \circ h$ by single and double layer distributions. Employing Proposition 1.3.3 (c) and (1.2.2) we find (for $\phi \in \mathcal{D}(\Omega)$)

$$\langle \phi, \delta_a' \circ h \rangle = \sum_{j=1}^n \langle \phi, \frac{\partial_j h}{|\nabla h|^2} \partial_j(\delta_a \circ h) \rangle = -\sum_{j=1}^n \langle \partial_j \left(\frac{\phi \, \partial_j h}{|\nabla h|^2} \right), \delta_a \circ h \rangle$$

$$= -\sum_{j=1}^n \int_M \partial_j \left(\frac{\phi \, \partial_j h}{|\nabla h|^2} \right) \frac{d\sigma}{|\nabla h|}$$

$$= -\int_M \partial_\nu \phi \, \frac{d\sigma}{|\nabla h|^2} - \int_M \phi \sum_{j=1}^n \partial_j \left(\frac{\partial_j h}{|\nabla h|^2} \right) \frac{d\sigma}{|\nabla h|},$$

and hence

$$\delta_a' \circ h = D_M(|\nabla h|^{-2}) - S_M\left(|\nabla h|^{-1} \cdot \nabla^T (\tfrac{\nu}{|\nabla h|})\right) \text{ in } \mathcal{D}'(\Omega). \tag{1.3.3}$$

(c) For illustration, let us apply the above in two easy cases where M is a sphere. If $h(x) = |x|$ and $a > 0$, then $M = h^{-1}(a) = aS^{n-1}$, and $|\nabla h| = 1$ yields $\delta_a \circ h = \delta(|x| - a) = S_M(1)$, i.e.,

$$\langle \phi, \delta_a \circ h \rangle = \int_{aS^{n-1}} \phi(x) \, d\sigma(x) = a^{n-1} \int_{S^{n-1}} \phi(a\omega) \, d\sigma(\omega), \qquad \phi \in \mathcal{D}(\mathbf{R}^n).$$

Similarly, from $\sum_{j=1}^{n} \partial_j \left(\frac{\partial_j h}{|\nabla h|^2} \right) = \frac{n-1}{|x|}$ we obtain $\delta'_a \circ h = \delta'(|x| - a) = D_M - \frac{n-1}{a} S_M$, where $S_M := S_M(1)$, $D_M := D_M(1)$. Evaluation on a test function yields

$$\langle \phi, \delta'_a \circ h \rangle = -a^{n-1} \int_{S^{n-1}} \omega^T \cdot \nabla \phi(a\omega) \, d\sigma(\omega) - (n-1)a^{n-2} \int_{S^{n-1}} \phi(a\omega) \, d\sigma(\omega)$$

$$= -\frac{\partial}{\partial a} \int_{S^{n-1}} a^{n-1} \phi(a\omega) \, d\sigma(\omega) = -\frac{\partial}{\partial a} \langle \phi, \delta_a \circ h \rangle$$

in consistence with Proposition 1.3.2, which implies

$$\delta'_a = \lim_{\epsilon \to 0} \frac{1}{\epsilon}(\delta_a(x + \epsilon) - \delta_a) = \lim_{\epsilon \to 0} \frac{1}{\epsilon}(\delta_{a-\epsilon} - \delta_a) = -\frac{\partial}{\partial a} \delta_a \text{ in } \mathcal{D}'(\mathbf{R}),$$

cf. also Seeley [249], p. 3.

Analogously, if $P(x) = |x|^2$ and $a > 0$, then

$$\delta_{a^2} \circ P = S_M(|\nabla P|^{-1}) = \frac{1}{2a} S_M = \frac{1}{2a} \delta_a \circ h$$

and $|\nabla P|^{-1} \cdot \nabla^T (\frac{v}{|\nabla P|}) = \frac{n-2}{4|x|^3}$ yields

$$\delta'_{a^2} \circ P = \frac{1}{4a^2} D_M - \frac{n-2}{4a^3} S_M,$$

or

$$\langle \phi, \delta'_{a^2} \circ P \rangle = -\frac{a^{n-3}}{4} \int_{S^{n-1}} \omega^T \cdot \nabla \phi(a\omega) \, d\sigma(\omega) - \frac{(n-2)a^{n-4}}{4} \int_{S^{n-1}} \phi(a\omega) \, d\sigma(\omega),$$

cf. Seeley [249], p. 3, where a missing sign should be inserted. Again, this follows also from

$$\delta'_{a^2} \circ P = -\frac{1}{2a} \frac{\partial}{\partial a} (\delta_{a^2} \circ P) = -\frac{1}{2a} \frac{\partial}{\partial a} (\frac{1}{2a} \delta_a \circ h) = \frac{1}{4a^2} \delta'_a \circ h + \frac{1}{4a^3} \delta_a \circ h.$$

$\square$

Definition 1.3.5 Let $\emptyset \neq \Omega \subset \mathbf{R}^n$ open and $P(x, \partial) = \sum_{|\alpha| \leq m} a_\alpha(x) \partial^\alpha$ be a linear differential operator with coefficients $a_\alpha \in C^\infty(\Omega)$. A distribution $E \in \mathcal{D}'(\Omega)$ is called a *fundamental solution* of $P(x, \partial)$ at $\xi \in \Omega$ iff $P(x, \partial)E = \delta_\xi$ holds in $\mathcal{D}'(\Omega)$.

If $P(\partial) = \sum_{|\alpha| \leq m} a_\alpha \partial^\alpha$ has constant coefficients, then $E \in \mathcal{D}'(\mathbf{R}^n)$ is called a *fundamental (or elementary) solution* of $P(\partial)$ iff $P(\partial)E = \delta$ holds in $\mathcal{D}'(\mathbf{R}^n)$.

(Cf. Schwartz [246], Ch. V, Sect. 6, Eq. (V, 6; 24), p. 136; Zeilon [306], pp. 1, 2.)

Example 1.3.6 The Heaviside function Y is a fundamental solution of the differentiation operator $\frac{d}{dx}$ in a single variable, i.e., $\frac{d}{dx} Y = \delta$ in $\mathcal{D}'(\mathbf{R})$, since

$$\langle \phi, \frac{d}{dx} Y \rangle = -\langle \phi', Y \rangle = -\int_0^\infty \phi'(x)\, dx = \phi(0) = \langle \phi, \delta \rangle, \qquad \phi \in \mathcal{D}(\mathbf{R}).$$

More generally, this reasoning applies if $f : \mathbf{R} \longrightarrow \mathbf{C}$ is continuously differentiable outside a discrete set D (i.e., $D \cap K$ is finite for each compact set $K \subset \mathbf{R}$), if f has right and left limits in all points of D (and hence $T_f \in L^\infty_{\text{loc}}(\mathbf{R})$), and if f', which is defined on $\mathbf{R} \setminus D$, is locally integrable on $\mathbf{R}$. We then have

$$(T_f)' = T_{f'} + \sum_{a \in D} s(f, a)\delta_a, \tag{1.3.4}$$

where $s(f, a) = \lim_{\epsilon \searrow 0}[f(a + \epsilon) - f(a - \epsilon)]$ is the jump of f at a. Formula (1.3.4) is often called the *distributional jump formula*, see Schwartz [246], Eq. (II, 2; 7), p. 37 (where [f'] stands for $T_{f'}$); Vo-Khac Koan [282], BC, IV, p. 186, Prop.; Hirsch and Lacombe [131], Ch. III, Thm. 2.10, p. 300.

   Formula (1.3.4) can easily be generalized to higher derivatives by induction. This gives the following: Let $f : \mathbf{R} \longrightarrow \mathbf{C}$ be m times continuously differentiable outside the discrete set D, and such that $f^{(k)}$, $0 \le k < m$, have limits from the left and from the right and $f^{(m)}$, defined on $\mathbf{R} \setminus D$, is locally integrable on $\mathbf{R}$. Then

$$\frac{d^m}{dx^m}(T_f) = T_{f^{(m)}} + \sum_{k=0}^{m-1}\sum_{a \in D} s(f^{(m-k-1)}, a)\delta_a^{(k)}, \tag{1.3.5}$$

cf. Schwartz [246], (II, 2; 8), p. 37.

   Applied to the function $f(x) = Y(x)e^{\lambda x}$, $\lambda \in \mathbf{C}$, formula (1.3.4) yields

$$T_f' = \lambda Y \cdot e^{\lambda x} + \delta = \lambda f + \delta, \qquad \left(\frac{d}{dx} - \lambda\right) f = \delta,$$

i.e., $f = Y \cdot e^{\lambda x} \in L^\infty_{\text{loc}}(\mathbf{R})$ is a fundamental solution of $\frac{d}{dx} - \lambda$. In the next proposition, we similarly derive from (1.3.5) a fundamental solution of $\prod_{j=1}^m (\frac{d}{dx} - \lambda_j)^{\alpha_j + 1}$, $\alpha \in \mathbf{N}_0^m$. $\qquad \square$

Proposition 1.3.7 *Let $m \in \mathbf{N}$, $\alpha \in \mathbf{N}_0^m$, and $\lambda_1, \ldots, \lambda_m \in \mathbf{C}$ be pairwise different. Then the ordinary differential operator*

$$P_{\lambda, \alpha}(\tfrac{d}{dx}) = \prod_{j=1}^m \left(\frac{d}{dx} - \lambda_j\right)^{\alpha_j + 1}$$

has as fundamental solution the L^∞_{loc} function $E_{\lambda, \alpha}$ given by

$$E_{\lambda,\alpha}(x) = \frac{Y(x)}{\alpha!}\left(\frac{\partial}{\partial\lambda}\right)^{\alpha}\sum_{j=1}^{m}e^{\lambda_j x}\prod_{k\neq j}(\lambda_j - \lambda_k)^{-1} \qquad (1.3.6)$$

$$= Y(x)\sum_{j=1}^{m}\frac{1}{\alpha_j!}\left(\frac{\partial}{\partial\lambda_j}\right)^{\alpha_j}\left(e^{\lambda_j x}\prod_{k\neq j}(\lambda_j - \lambda_k)^{-\alpha_k-1}\right).$$

$E_{\lambda,\alpha}$ *is the only fundamental solution of* $P_{\lambda,\alpha}(\frac{d}{dx})$ *with support in the interval* $[0,\infty)$.

Proof

(1) We consider first the case $\alpha = 0$ and set up $E = E_{\lambda,0}$ in the form

$$E = Y\cdot\sum_{j=1}^{m}a_j e^{\lambda_j x}, \qquad a_j \in \mathbf{C}.$$

Then we obviously have $P_{\lambda,0}(\frac{d}{dx})E|_{\mathbf{R}\setminus\{0\}} = 0$. If the coefficients a_j are chosen such that E is $\mathcal{C}^{m-2}$ and $E^{(m-1)}$ has a jump of height one at 0, i.e., if

$$0 = s(E,0) = \sum_{j=1}^{m}a_j, \quad 0 = s(E',0) = \sum_{j=1}^{m}\lambda_j a_j, \quad \dots,$$

$$0 = s(E^{(m-2)},0) = \sum_{j=1}^{m}\lambda_j^{m-2}a_j, \quad 1 = s(E^{(m-1)},0) = \sum_{j=1}^{m}\lambda_j^{m-1}a_j,$$

then (1.3.5) implies that $P_{\lambda,0}(\frac{d}{dx})E = \delta$. The coefficients a_j are the solution of Vandermonde's system of linear equations

$$\begin{pmatrix} 1 & 1 & \dots & 1 \\ \lambda_1 & \lambda_2 & \dots & \lambda_m \\ \vdots & \vdots & \vdots & \vdots \\ \lambda_1^{m-1} & \lambda_2^{m-1} & \dots & \lambda_m^{m-1} \end{pmatrix}\begin{pmatrix} a_1 \\ a_2 \\ \vdots \\ a_m \end{pmatrix} = \begin{pmatrix} 0 \\ 0 \\ \vdots \\ 1 \end{pmatrix}.$$

The solution of this system is given by

$$a_j = \prod_{k\neq j}(\lambda_j - \lambda_k)^{-1} = P'_{\lambda,0}(\lambda_j)^{-1}.$$

(2) $E_{\lambda,0}$ depends holomorphically on λ, i.e., the mappings

$$\{\lambda \in \mathbf{C}^m;\, \lambda_1,\dots,\lambda_m \text{ are pairwise different}\} \longrightarrow \mathbf{C} : \lambda \longmapsto \langle\phi, E_{\lambda,0}\rangle$$

are holomorphic for each $\phi \in \mathcal{D}(\mathbf{R})$. In accordance with (1.3.6), the α-th derivative with respect to λ of $E_{\lambda,0}$ yields the distribution $\alpha! E_{\lambda,\alpha}$. Let us check that $E_{\lambda,\alpha}$ is a fundamental solution of $P_{\lambda,\alpha}(\frac{d}{dx})$ by induction with respect to $|\alpha|$. If $\phi \in \mathcal{D}$, $\alpha = (\alpha_1, \alpha') \in \mathbf{N}_0^m$, and $P_{\lambda,\alpha}(\frac{d}{dx}) E_{\lambda,\alpha} = \delta$, then we infer that

$$0 = \frac{\partial}{\partial \lambda_1} \phi(0) = \frac{\partial}{\partial \lambda_1} \langle \phi, P_{\lambda,\alpha}(\tfrac{d}{dx}) E_{\lambda,\alpha} \rangle = \frac{\partial}{\partial \lambda_1} \langle P_{\lambda,\alpha}(-\tfrac{d}{dx}) \phi, E_{\lambda,\alpha} \rangle$$

$$= -(1 + \alpha_1) \langle P_{\lambda,(\alpha_1 - 1, \alpha')}(-\tfrac{d}{dx}) \phi, E_{\lambda,\alpha} \rangle + \langle P_{\lambda,\alpha}(-\tfrac{d}{dx}) \phi, \frac{\partial E_{\lambda,\alpha}}{\partial \lambda_1} \rangle;$$

if $\phi = (-\frac{d}{dx} - \lambda_1) \psi$ with $\psi \in \mathcal{D}(\mathbf{R})$, this implies

$$\langle P_{\lambda,(\alpha_1 + 1, \alpha')}(-\tfrac{d}{dx}) \psi, \frac{\partial E_{\lambda,\alpha}}{\partial \lambda_1} \rangle = (1 + \alpha_1) \langle P_{\lambda,\alpha}(-\tfrac{d}{dx}) \psi, E_{\lambda,\alpha} \rangle = (1 + \alpha_1) \psi(0),$$

and hence $\frac{1}{1+\alpha_1} \frac{\partial E_{\lambda,\alpha}}{\partial \lambda_1}$ is a fundamental solution of $P_{\lambda,(\alpha_1 + 1, \alpha')}(\frac{d}{dx})$.

(3) The uniqueness of $E_{\lambda,\alpha}$ follows from the fact that $P_{\lambda,\alpha}(\frac{d}{dx})$ is a hyperbolic operator, see Definition 2.4.10 and Proposition 2.4.11 below or Hörmander [138], Def. 12.3.3, p. 112, and Thm. 12.5.1, p. 120. Alternatively, this is implied by the fact that the distributional solutions T of the homogeneous equation $P_{\lambda,\alpha}(\frac{d}{dx}) T = 0$ are classical solutions and hence are real-analytic, cf. Proposition 1.3.18 below. $\qquad \square$

Note that the transition from $E_{\lambda,0}$ to $E_{\lambda,\alpha}$ as in part (2) of the proof above is a special case of Proposition 1.4.2 below.

Example 1.3.8 Let us illustrate formula (1.3.6) by considering some physically relevant specific cases.

(a) If $P(\frac{d}{dx}) = P_{\lambda,0}(\frac{d}{dx}) = \prod_{j=1}^m (\frac{d}{dx} - \lambda_j)$ with pairwise different $\lambda_j \in \mathbf{C}$, then

$$E = E_{\lambda,0} = Y \cdot \sum_{j=1}^m e^{\lambda_j x} \prod_{k \neq j} (\lambda_j - \lambda_k)^{-1}.$$

In particular, for $\lambda_1 = -\lambda_2 = i\omega$, $\omega \in \mathbf{C} \setminus \{0\}$, we obtain the fundamental solution $E(x) = \frac{1}{\omega} Y(x) \sin(\omega x)$ to the operator $\frac{d^2}{dx^2} + \omega^2$, which will frequently appear below.

(b) If $P(\frac{d}{dx}) = (\frac{d}{dx} - \lambda)^{r+1}$, then we obtain $E = \frac{1}{r!} Y(x) x^r e^{\lambda x}$. More generally, for $P(\frac{d}{dx}) = (\frac{d}{dx} - \lambda)^{r+1} (\frac{d}{dx} - \mu)^{s+1}$, $\lambda \neq \mu \in \mathbf{C}$, $r, s \in \mathbf{N}_0$, we have

$$E = -\frac{1}{r! s!} Y(x) \left(e^{\lambda x} (-1)^s \sum_{j=0}^r \frac{\binom{r}{j}(r + s - j)! \, x^j}{(\mu - \lambda)^{r+s-j+1}} \right.$$

$$\left. + e^{\mu x} (-1)^r \sum_{j=0}^s \frac{\binom{s}{j}(r + s - j)! \, x^j}{(\lambda - \mu)^{r+s-j+1}} \right).$$

In particular, if $r = s$ and $\mu = -\lambda \neq 0$, then we obtain the following fundamental solution of $(\frac{d^2}{dx^2} - \lambda^2)^{r+1}$:

$$E = \frac{2(-1)^r}{r!^2} Y(x) \sum_{j=0}^{r} \binom{r}{j} \frac{(2r-j)!\, x^j}{(2\lambda)^{2r-j+1}} \cdot \begin{cases} \sinh(\lambda x), & \text{if } j \text{ is even,} \\ -\cosh(\lambda x), & \text{if } j \text{ is odd} \end{cases}$$

$$= \frac{1}{r!} Y(x) \left(\frac{1}{2\lambda} \frac{\partial}{\partial \lambda} \right)^r \frac{\sinh(\lambda x)}{\lambda}.$$

The last equation can be checked, quite laboriously, by induction. It also follows, more easily, from Proposition 1.4.2 below.   □

Example 1.3.9 More generally as in Example 1.2.6 (a), where the distribution x_+^{-1} is defined such as to yield an extension in $\mathcal{D}'(\mathbf{R})$ of $Y(x)x^{-1} \in L^1_{\text{loc}}(\mathbf{R} \setminus \{0\})$, we will now extend $Y(x)x^\lambda \in L^1_{\text{loc}}(\mathbf{R} \setminus \{0\})$, $\lambda \in \mathbf{C}$, by differentiation.

For $\operatorname{Re} \lambda > -1$, we set $x_+^\lambda = Y(x)x^\lambda \in L^1_{\text{loc}}(\mathbf{R}) \subset \mathcal{D}'(\mathbf{R})$. If $\operatorname{Re} \lambda > 0$, then the distributional jump formula (1.3.4) implies

$$(x_+^\lambda)' = \lim_{\epsilon \searrow 0} \frac{d}{dx} \left(Y(x-\epsilon)x^\lambda \right) = \lim_{\epsilon \searrow 0} \left((Y(x-\epsilon)\lambda x^{\lambda-1} + \epsilon^\lambda \delta_\epsilon) \right) = \lambda x_+^{\lambda-1}.$$

Therefore, the following definition of x_+^λ for $\lambda \in \mathbf{C} \setminus -\mathbf{N}$ by

$$x_+^\lambda = \frac{1}{(\lambda+1)\cdots(\lambda+m)} \frac{d^m}{dx^m}(x_+^{\lambda+m}), \qquad \operatorname{Re} \lambda > -m-1,$$

is unambiguous, i.e., it does not depend on the choice of $m \in \mathbf{N}_0$ satisfying $m > -\operatorname{Re} \lambda - 1$; cf. Hörmander [139], (3.2.3), p. 68.

Similarly, we obtain $x_+^{-1} = \frac{d}{dx}(Y(x) \log x)$, since

$$(Y(x) \log x)' = \lim_{\epsilon \searrow 0} \frac{d}{dx} \left(Y(x-\epsilon) \log x \right) = \lim_{\epsilon \searrow 0} \left(Y(x-\epsilon)x^{-1} + \delta_\epsilon \cdot \log \epsilon \right)$$

$$= \lim_{\epsilon \searrow 0} \left(Y(x-\epsilon)x^{-1} + \delta \cdot \log \epsilon \right) = x_+^{-1},$$

cf. Example 1.2.6 (a).

Note, however, that x_+^{-k}, $k = 2, 3, \ldots$, is defined as the finite part at $\lambda = -k$ of the meromorphic distribution-valued function $\lambda \longmapsto x_+^\lambda$, and it coincides with $\frac{(-1)^{k-1}}{(k-1)!}(Y(x) \log x)^{(k)}$ in $\mathcal{C}^\infty(\mathbf{R} \setminus \{0\})$ but not in $\mathcal{D}'(\mathbf{R})$, cf. Example 1.4.8 below.

Let us finally represent x_+^λ, $\lambda \in \mathbf{C} \setminus -\mathbf{N}$, as a distributional limit. For $m \in \mathbf{N}_0$ with $\operatorname{Re} \lambda > -m - 1$, we infer from the jump formula (1.3.5) that

$$x_+^\lambda = \frac{1}{(\lambda + 1)\cdots(\lambda + m)} \lim_{\epsilon \searrow 0} \frac{d^m}{dx^m}\left(Y(x - \epsilon)x^{\lambda + m}\right)$$

$$= \lim_{\epsilon \searrow 0}\left(Y(x - \epsilon)x^\lambda + \sum_{j=0}^{m-1} \frac{\epsilon^{\lambda + j + 1}}{(\lambda + 1)\cdots(\lambda + j + 1)} \delta_\epsilon^{(j)}\right).$$

Since

$$\delta_\epsilon^{(j)} = \sum_{k=0}^{N} \frac{(-\epsilon)^k}{k!} \delta^{(j+k)} + O(\epsilon^{N+1}) \qquad \text{for } \epsilon \to 0,$$

we obtain

$$x_+^\lambda = \lim_{\epsilon \searrow 0}\left(Y(x - \epsilon)x^\lambda + \sum_{i=0}^{m-1} A_{\lambda,i}\, \epsilon^{\lambda + i + 1} \delta^{(i)}\right),$$

$$\text{where } A_{\lambda,i} = \sum_{j=0}^{i} \frac{(-1)^{i+j}}{(i-j)!(\lambda + 1)\cdots(\lambda + j + 1)}.$$

The polynomial

$$P_i(\lambda) = (\lambda + 1)\cdots(\lambda + i + 1)A_{\lambda,i} = \sum_{k=0}^{i} (-1)^k \binom{\lambda + i + 1}{k}$$

has the degree i and the zeros $-1, -2, \ldots, -i$, and it fulfills $P_i(-i - 1) = 1$. Thus we conclude that

$$P_i(\lambda) = \frac{(-1)^i}{i!}(\lambda + 1)\cdots(\lambda + i), \qquad A_{\lambda,i} = \frac{(-1)^i}{i!(\lambda + i + 1)},$$

and hence

$$x_+^\lambda = \lim_{\epsilon \searrow 0}\left(Y(x - \epsilon)x^\lambda + \sum_{i=0}^{m-1} \frac{(-1)^i \epsilon^{\lambda + i + 1}}{i!(\lambda + i + 1)} \delta^{(i)}\right), \quad \lambda \in \mathbf{C} \setminus -\mathbf{N},\ \operatorname{Re} \lambda > -m - 1,$$

$$(1.3.7)$$

cf. Schwartz [246], (II, 2; 26), p. 42; Horváth [143], 2.2.5.5, p. 87.

In particular, we have

$$x_+^{-3/2} = \lim_{\epsilon \searrow 0} \left(Y(x - \epsilon) x^{-3/2} - \frac{2}{\sqrt{\epsilon}} \delta \right),$$

$$x_+^{-5/2} = \lim_{\epsilon \searrow 0} \left(Y(x - \epsilon) x^{-5/2} - \frac{2}{3\epsilon^{3/2}} \delta + \frac{2}{\sqrt{\epsilon}} \delta' \right). \qquad \square$$

Example 1.3.10 Let us now generalize the one-dimensional distributional jump formula (1.3.4) to several dimensions.

We suppose that $\emptyset \neq \Omega \subset \mathbf{R}^n$ is open, $M \subset \Omega$ is a closed C^1-hypersurface, and $f : \Omega \setminus M \longrightarrow \mathbf{R}$ is C^1. We also assume that f has (in general different) boundary values from both sides of M and that the partial derivatives $\partial_j f$, defined in $\Omega \setminus M$, are locally integrable on Ω.

Let us then define the *jump vector field* $s(f)$ of f along M by the formula

$$s(f) : M \longrightarrow \mathbf{R}^n : x \longmapsto \nu(x) \cdot \lim_{\epsilon \to 0} \big(f(x + \epsilon \nu(x)) - f(x - \epsilon \nu(x)) \big),$$

where $\nu(x)$ is a unit normal of M at x. (Note that $s(f)(x)$ does not depend on the choice of $\nu(x)$.)

By our assumptions, the gradient ∇f is locally integrable, i.e., $T_{\nabla f} \in L^1_{\mathrm{loc}}(\Omega)^n \subset \mathcal{D}'(\Omega)^n$. The *distributional jump formula* corresponding to (1.3.4) can then be stated as

$$\nabla T_f = T_{\nabla f} + S_M(s(f)), \tag{1.3.8}$$

where $S_M(s(f)) \in \mathcal{D}'(\Omega)^n$ is—analogously to Example 1.2.14—given by

$$\langle \phi, S_M(s(f)) \rangle = \int_M \phi(x) s(f)(x) \, d\sigma(x) \in \mathbf{R}^n, \qquad \phi \in \mathcal{D}(\Omega);$$

cf. Schwartz [245], (II, 2; 43), p. 94; Schwartz [246], Eqs. (II, 3; 1), (II, 3; 2), p. 43.

In fact, in order to prove (1.3.8), we can use a partition of unity argument and assume that M is locally given by $h = 0$ for some C^1 function h. Then Gauß' theorem yields for a test function $\phi \in \mathcal{D}(\Omega)$ the following:

$$\langle \phi, \partial_j T_f \rangle = -\int_\Omega (\partial_j \phi) f \, dx = -\int_{h(x) < 0} (\partial_j \phi) f \, dx - \int_{h(x) > 0} (\partial_j \phi) f \, dx$$

$$= \int_{\Omega \setminus M} \phi \, \partial_j f \, dx - \int_M \phi(x) \frac{\partial_j h}{|\nabla h|} \lim_{\epsilon \searrow 0} f(x - \epsilon \nabla h(x)) \, d\sigma(x)$$

$$+ \int_M \phi(x) \frac{\partial_j h}{|\nabla h|} \lim_{\epsilon \searrow 0} f(x + \epsilon \nabla h(x)) \, d\sigma(x)$$

$$= \langle \phi, T_{\partial_j f} \rangle + \langle \phi, S_M(s(f)_j) \rangle.$$

This implies (1.3.8).

In particular, if $f, h \in C^1(\Omega)$ and h is submersive on $M = h^{-1}(0)$, and $v(x) = \frac{\nabla h(x)}{|\nabla h(x)|}$, then

$$\nabla(Y(h) \cdot f) = Y(h) \cdot \nabla f + S_M(v \cdot f), \qquad (1.3.9)$$

since

$$s(f)(x) = \frac{\nabla h(x)}{|\nabla h(x)|} \lim_{\epsilon \searrow 0} f(x + \epsilon \nabla h(x)) = f(x)v(x), \qquad x \in M.$$

Of course, (1.3.9) also follows from (1.2.2), since, by Proposition 1.3.3, $\nabla(Y(h) \cdot f) = Y(h) \cdot \nabla f + f \cdot \nabla h \cdot \delta \circ h$. (Note that the measure $\delta \circ h$ can be multiplied with the continuous function $f \cdot \nabla h$.)

For a vector field $v \in C^1(\Omega)^n$ and h as above, there hold formulas similar to (1.3.9), e.g.,

$$\nabla^T(Y(h) \cdot v) = Y(h) \cdot \nabla^T v + S_M(v^T \cdot v),$$

where the divergence $\nabla^T v$ of v is usually denoted by div v, and, for $n = 3$,

$$\nabla \times (Y(h) \cdot v) = Y(h) \cdot \nabla \times v + S_M(v \times v),$$

where $\nabla \times v$ is the curl of v.

The most general version of the jump formula refers to currents, i.e., differential forms with distributional coefficients, and it can be expressed in the form

$$d(Y(h) \cdot \omega) = Y(h) \cdot d\omega + S_M\left(\tfrac{dh}{|\nabla h|} \wedge \omega\right),$$

cf. Schwartz [246], (IX, 3; 11), p. 346. $\square$

Example 1.3.11 In order to calculate the *second* distributional derivatives of $Y(h) \cdot f$, where $f, h \in C^2(\Omega)$ and h is submersive on $M = h^{-1}(0)$, we use the notation

$$\nabla\nabla^T T = \begin{pmatrix} \partial_1^2 T & \dots & \partial_1 \partial_n T \\ \vdots & \vdots & \vdots \\ \partial_n \partial_1 T & \dots & \partial_n^2 T \end{pmatrix}, \quad \nabla = \begin{pmatrix} \partial_1 \\ \vdots \\ \partial_n \end{pmatrix},$$

for the Hesse matrix of $T \in \mathcal{D}'(\Omega)$. Applying formula (1.3.9) twice we obtain (with $S_M = S_M(1)$ and $v = \frac{\nabla h(x)}{|\nabla h(x)|}$)

$$\nabla\nabla^T(Y(h) \cdot f) = \nabla[Y(h) \cdot \nabla^T f + f v^T \cdot S_M]$$

$$= Y(h) \cdot \nabla\nabla^T f + v \cdot \nabla^T f \cdot S_M + \nabla f \cdot v^T \cdot S_M + f \cdot \nabla(v^T) \cdot S_M$$

$$+ f \cdot (\nabla S_M) \cdot v^T. \qquad (1.3.10)$$

For the calculation of ∇S_M, we refer to Proposition 1.3.12 below, where we obtain

$$\nabla S_M = D_M(v) - S_M(v \cdot \operatorname{tr} W), \tag{1.3.11}$$

where $\operatorname{tr} W = \nabla^T v$ is the trace of the Weingarten mapping, i.e., $(n-1)$ times the mean curvature of M. Of course, Eq. (1.3.11) has to be understood componentwise, i.e., as $\partial_j S_M = D_M(v_j) - S_M(v_j \operatorname{tr} W), j = 1, \ldots, n$.

Upon inserting (1.3.11) into (1.3.10) we conclude that

$$\nabla\nabla^T\big(Y(h) \cdot f\big) = Y(h) \cdot \nabla\nabla^T f + f \cdot D_M(v) \cdot v^T$$
$$+ \big(v \cdot \nabla^T f + \nabla f \cdot v^T + f \cdot \nabla(v^T) - f \operatorname{tr} W \cdot v v^T\big) S_M.$$

When we use formula (1.3.2), we obtain

$$f \cdot D_M(v) \cdot v^T = D_M(f v v^T) - S_M\big(v \cdot \partial_v(f v^T)\big),$$

and we finally arrive at formula (1.3.12) in the next lemma, which resumes our assumptions and the result for the Hesse matrix of the discontinuous function $Y(h) \cdot f$.

Lemma *Let $\emptyset \neq \Omega \subset \mathbf{R}^n$ be open, $f, h \in C^2(\Omega)$ with h submersive on $M :=$ $h^{-1}(0)$. Let $v = \frac{\nabla h}{|\nabla h|}$ and $W = (I_n - v v^T)\nabla v^T$ be the Weingarten map with the trace $\operatorname{tr} W = \nabla^T v$. (Thus $\operatorname{tr} W$ is the sum of the eigenvalues of W, i.e., the principal curvatures of M.) Let the single and double layer distributions $S_M(g), D_M(g), g \in C(M)$, be defined as in* Examples 1.2.14 *and* 1.3.4, *respectively. Then the following holds in $\mathcal{D}'(\Omega)$:*

$$\nabla\nabla^T\big(Y(h) \cdot f\big) = Y(h) \cdot \nabla\nabla^T f + D_M(f v v^T) \tag{1.3.12}$$
$$+ S_M\big(v \cdot \nabla^T f + \nabla f \cdot v^T - \partial_v(f) v v^T\big) + S_M\big(f \cdot (W - v v^T \operatorname{tr} W)\big).$$

$\square$

Let us incidentally observe that the Weingarten map is originally given by the linear mapping

$$\tilde{W} : T_x M \longrightarrow T_x M : w \longmapsto (w^T \nabla)v = (\nabla v^T)^T w,$$

$T_x M$ denoting the tangent space of M at $x \in M$. If we continue $\tilde{W}$ by 0 on the normal space $\mathbf{R} \cdot v$, we obtain the symmetric $n \times n$ matrix

$$W = (\nabla v^T)^T \cdot (I_n - v v^T) = (I_n - v v^T) \cdot \nabla v^T,$$

referred to as Weingarten map in the lemma above.

Let us illustrate formula (1.3.12) in the case of $h(x) = |x| - R$, i.e., $M = RS^{n-1}$, $R > 0$. Then

$$v = \frac{x}{|x|}, \qquad \nabla v^T = \frac{|x|^2 I_n - xx^T}{|x|^3} = W, \qquad \operatorname{tr} W = \frac{n-1}{R},$$

and thus

$$\nabla \nabla^T \big(Y(|x| - R) \cdot f \big) = Y(|x| - R) \cdot \nabla \nabla^T f + D_{RS^{n-1}} \Big(\frac{fxx^T}{R^2} \Big)$$

$$+ S_{RS^{n-1}} \Big(\frac{x}{R} \cdot \nabla^T f + \nabla f \cdot \frac{x^T}{R} - (x^T \nabla f) \cdot \frac{xx^T}{R^3} + f \cdot \frac{R^2 I_n - nxx^T}{R^3} \Big).$$

In particular, for a rotationally symmetric function $f(x) = g(|x|)$, $g \in C^2(\mathbf{R} \setminus \{0\})$, we have

$$\nabla \nabla^T \big(Y(|x| - R) \cdot g(|x|) \big) = Y(|x| - R) \Big(\frac{I_n |x|^2 - xx^T}{|x|^3} g'(|x|) + \frac{xx^T}{|x|^2} g''(|x|) \Big)$$

$$\tag{1.3.13}$$

$$+ \frac{g(R)}{R^2} D_{RS^{n-1}}(xx^T) + \Big(\frac{g'(R)}{R^2} - \frac{ng(R)}{R^3} \Big) S_{RS^{n-1}}(xx^T) + \frac{g(R)}{R} I_n S_{RS^{n-1}}(1).$$

Proposition 1.3.12 *For $\emptyset \neq \Omega \subset \mathbf{R}^n$ open and $M \subset \Omega$ a closed oriented C^2-hypersurface with unit normal v, let $S_M(f), D_M(f)$ be defined as in formulas (1.2.1), (1.3.1), respectively. Then*

$$\nabla S_M(1) = D_M(v) - S_M(v \cdot \nabla^T v).$$

(Here $\nabla^T v$ is defined by extending v arbitrarily as a C^1 unit vector field near M. Then $\operatorname{div} v = \nabla^T v$ coincides with $\operatorname{tr} W$ on M, and hence, on M, $\nabla^T v$ does not depend on the specific extension of v.)

We shall give two different proofs of this equation.

First proof Locally, we can represent M as $h^{-1}(0)$ with $h \, C^2$ and submersive. When we set $v = \nabla h / |\nabla h|$ and $B = |\nabla h|^{-1} \nabla \nabla^T h$, we infer from Proposition 1.3.3 and from the formulas (1.2.2), (1.3.2), (1.3.3) the following:

$$\nabla S_M(1) = \nabla(|\nabla h| \cdot \delta \circ h) = \nabla(|\nabla h|) \cdot \delta \circ h + |\nabla h|^2 v \cdot \delta' \circ h$$

$$= B \nabla h \cdot \delta \circ h + |\nabla h|^2 v \cdot \Big(D_M(|\nabla h|^{-2}) - S_M \big(|\nabla h|^{-1} \cdot \nabla^T (|\nabla h|^{-1} v) \big) \Big)$$

$$= S_M(Bv) + D_M(v) - S_M \big(|\nabla h|^{-2} \partial_v (|\nabla h| \nabla h) \big) - S_M \big(\nabla h \cdot \nabla^T (|\nabla h|^{-2} \nabla h) \big).$$

Since

$$|\nabla h|^{-2} \partial_\nu (|\nabla h| \nabla h) = (\nu^T B \nu)\nu + B\nu$$

and

$$\nabla h \cdot \nabla^T (|\nabla h|^{-2} \nabla h) = -2(\nu^T B \nu)\nu + \frac{\Delta h}{|\nabla h|} \cdot \nu,$$

we conclude that

$$\nabla S_M(1) = D_M(\nu) + S_M\left(\left(\nu^T B \nu - \frac{\Delta h}{|\nabla h|}\right)\nu\right) = D_M(\nu) - S_M(\operatorname{tr} W \cdot \nu).$$

The last equation follows from $\nabla \nu^T = B(I_n - \nu\nu^T)$, which implies

$$\operatorname{tr} W = \operatorname{tr}\left((I_n - \nu\nu^T)B(I_n - \nu\nu^T)\right) = \operatorname{tr} B - \nu^T B \nu = \frac{\Delta h}{|\nabla h|} - \nu^T B \nu. \qquad \square$$

Second proof We apply here Gauß' divergence theorem, and for this reason, we have to restrict a little the generality supposing that M is C^3. Let $x \in M$ and $h :$ $U \longrightarrow \mathbf{R} \; C^3$ on a ball U around x such that $M \cap U = h^{-1}(0)$ and $\nabla h \neq 0$ in U. We set again $\nu = \nabla h / |\nabla h|$ on U.

For $\phi \in \mathcal{D}(U)$ and fixed $j \in \{1, \ldots, n\}$, we define the C^1 vector field $w = (w_k)_{k=1,\ldots,n}$ by

$$w_k(x) = \nabla^T(\nu\phi)\delta_{jk} - \partial_j(\nu_k\phi).$$

Then $\operatorname{div} w = 0$ and hence

$$0 = \int_{x \in U; h(x) < 0} \operatorname{div} w \, dx = \int_{M \cap U} \nu^T w \, d\sigma = \int_{M \cap U} [\nu_j \nabla^T(\nu\phi) - \nu^T \partial_j(\nu\phi)] \, d\sigma$$

$$= \int_{M \cap U} [\nu_j \partial_\nu \phi + \nu_j \phi \, \nabla^T \nu - \phi \, \underbrace{\nu^T \partial_j \nu}_{=0} - \partial_j \phi] \, d\sigma$$

$$= \langle \phi, -D_M(\nu_j) + S_M(\nu_j \nabla^T \nu) + \partial_j S_M(1) \rangle,$$

i.e., $\partial_j S_M(1) = D_M(\nu_j) - S_M(\nu_j \operatorname{tr} W)$ taking into account that $\nabla^T \nu = \operatorname{tr}(\nabla \nu^T) = \operatorname{tr}\left(B(I_n - \nu\nu^T)\right) = \operatorname{tr} W$. $\qquad \square$

Let us mention that the above jump formulas are—in different notation—also treated in Gel'fand and Shilov [104], Ch. III, Section 1, p. 209, and in Estrada and Kanwal [71], 2.7, p. 68. An extension to covariant derivatives in general Riemannian spaces is given in Wagner [297].

Let us apply now the many-dimensional jump formula (1.3.12) to *verify* fundamental solutions of some hypoelliptic operators by distributional differentiation.

Definition 1.3.13

(1) For $T \in \mathcal{D}'(\Omega)$, the *singular support* sing supp T is the complement in Ω of the largest open set $U \subset \Omega$ such that $T|_U \in C^\infty(U)$.

(2) The differential operator with constant coefficients $P(\partial) = \sum_{|\alpha| \leq m} a_\alpha \partial^\alpha$, $a_\alpha \in \mathbf{C}$, is called *hypoelliptic* if and only if sing supp$(P(\partial)T) = $ sing supp T for all $T \in \mathcal{D}'$.

To give a trivial example, sing supp $Y^{(m)} = \{0\}$ for all $m \in \mathbf{N}_0$. Clearly, a fundamental solution E of a hypoelliptic operator fulfills sing supp $E = \{0\}$. Conversely, by Schwartz [246], Ch. V, Thm. XII, p. 143, an operator possessing such a fundamental solution is necessarily hypoelliptic. For example, it follows from Proposition 1.3.7 that all linear differential operators with constant coefficients in *one* dimension are hypoelliptic.

Example 1.3.14

(a) Let us employ the jump formula (1.3.12) to show that

$$E = \begin{cases} \dfrac{1}{2\pi} \log |x|, \text{ if } n = 2, \\ c_n |x|^{2-n}, \quad \text{ if } n \neq 2 \end{cases}, \qquad c_n = \frac{\Gamma(\frac{n}{2})}{(2-n)2\pi^{n/2}},$$

is a fundamental solution of the *Laplacean* $\Delta_n = \sum_{i=1}^n \partial_i^2$, cf. Schwartz [246], (II, 3; 10), (II, 3; 14), pp. 45, 46, Ex. 2, p. 288.

Note that E is C^∞ outside the origin, in agreement with the (hypo)ellipticity of Δ_n, and that $\Delta_n E = 0$ holds in $C^\infty(\mathbf{R}^n \setminus \{0\})$. Therefore, formula (1.3.13) yields

$$\Delta_n(Y(|x| - R) \cdot E) = \mathrm{tr}\big(\nabla\nabla^T(Y(|x| - R) \cdot E)\big)$$

$$= g(R)D_{RS^{n-1}}(1) + \big(g'(R) - \tfrac{n}{R}g(R) + \tfrac{n}{R}g(R)\big)S_{RS^{n-1}}(1),$$

if $E(x) = g(|x|)$. Since

$$\langle \phi, g(R)D_{RS^{n-1}}(1) \rangle = -g(R)R^{n-1} \int_{S^{n-1}} \omega^T \nabla\phi(R\omega)\, d\sigma(\omega)$$

converges to 0 for $R \searrow 0$ and fixed $\phi \in \mathcal{D}$, and

$$\langle \phi, g'(R)S_{RS^{n-1}}(1) \rangle = g'(R)R^{n-1} \int_{S^{n-1}} \phi(R\omega)\, d\sigma(\omega)$$

converges to $\phi(0)$ (due to $(2-n)c_n \int_{S^{n-1}} d\sigma = 1$), we conclude that

$$\Delta_n E = \lim_{R \searrow 0} \Delta_n \big(Y(|x| - R) \cdot E \big) = \delta.$$

More generally, the same calculation yields for the Hesse matrix of E the formula

$$\nabla \nabla^T E = \frac{1}{n} I_n \delta + \frac{1}{|S^{n-1}|} \operatorname{vp}\left(\frac{|x|^2 I_n - n x x^T}{|x|^{n+2}} \right)$$

with $|S^{n-1}| = \int_{S^{n-1}} d\sigma = \frac{2\pi^{n/2}}{\Gamma(n/2)}$, cf. also, for $n = 3$, Frahm [80], (3), p. 826; Calderón [41], p. 428.

(b) Exactly in the same way, we obtain that $E = \frac{1}{2\pi(x_1 + i x_2)} = \frac{1}{2\pi z}$ is a fundamental solution of the *Cauchy–Riemann operator* $\partial_1 + i\partial_2$, cf. Schwartz [246], (II, 3; 28), p. 49.

In fact, by means of the jump formula (1.3.9),

$$(\partial_1 + i\partial_2)\big(Y(|x| - R) \cdot E \big) = \frac{1}{2\pi R} S_{RS^1}(1),$$

and hence

$$(\partial_1 + i\partial_2)\frac{1}{2\pi z} = \lim_{R \searrow 0} (\partial_1 + i\partial_2)\big(Y(|x| - R) \cdot E \big) = \delta.$$

(c) Similarly, the *Helmholtz operator* $\Delta_n + \lambda$, $\lambda > 0$, is (hypo)elliptic and has the fundamental solution

$$E = g(|x|) = d_n(\lambda) |x|^{1-n/2} N_{n/2-1}(\sqrt{\lambda}\,|x|), \qquad d_n(\lambda) = \frac{\lambda^{n/4-1/2}}{2^{1+n/2}\pi^{n/2-1}},$$

where N_α denotes the Neumann function of order α (cf. Gradshteyn and Ryzhik [113], 8.403, p. 951; Schwartz [246], (VII, 10; 17), p. 287).

In fact, $(\Delta_n + \lambda)E$ vanishes classically for $x \neq 0$, since

$$\left(\frac{d^2}{dr^2} + \frac{n-1}{r}\frac{d}{dr} + \lambda \right)\left(r^{1-n/2} N_{n/2-1}(\sqrt{\lambda}\,r) \right) = 0$$

for $r > 0$, cf. Gradshteyn and Ryzhik [113], Eq. 8.491.6, p. 971. On the other hand, $\lim_{R \searrow 0} g(R) R^{n-1} = 0$ and

$$\lim_{R \searrow 0} g'(R) R^{n-1} = -d_n(\lambda)\sqrt{\lambda} \lim_{R \searrow 0} R^{n/2} N_{n/2}(\sqrt{\lambda}\,R)$$

$$= d_n(\lambda) \lambda^{1/2-n/4} \frac{2^{n/2}\Gamma(\frac{n}{2})}{\pi} = \frac{\Gamma(\frac{n}{2})}{2\pi^{n/2}} = \frac{1}{|S^{n-1}|},$$

cf. Abramowitz and Stegun [1], (9.19), p. 360, and hence formula (1.3.13) implies

$$(\Delta_n + \lambda)E = \lim_{R\searrow 0}(\Delta_n + \lambda)\big(Y(|x| - R) \cdot E\big)$$

$$= \lim_{R\searrow 0}\big(g(R)D_{RS^{n-1}}(1) + g'(R)S_{RS^{n-1}}(1)\big) = \delta.$$

Note that, in the case of $n \geq 3$, the limit of E for $\lambda \searrow 0$ tends to the fundamental solution of Δ_n in part (a), whereas in the cases $n = 1, 2$, one has to subtract from E suitable constants (depending on λ) in order to obtain convergence of the limit in $\mathcal{D}'$, see Ortner [197], 4.6, pp. 15–18.

(d) Let us finally verify, by the same method, that

$$E(t, x) = \frac{Y(t)}{(4\pi t)^{n/2}}\, e^{-|x|^2/(4t)} \in L^1_{\mathrm{loc}}(\mathbf{R}^{n+1}_{t,x}) \tag{1.3.14}$$

is a fundamental solution of the *heat operator* $\partial_t - \Delta_n$, cf. Schwartz [246], (VII, 10; 26), p. 289.

Similarly to the reasoning above, we set $M_\epsilon = \{(t, x) \in \mathbf{R}^{n+1};\ t = \epsilon\}$ and obtain

$$(\partial_t - \Delta_n)E = \lim_{\epsilon\searrow 0}(\partial_t - \Delta_n)\big(Y(t - \epsilon)\cdot E\big) = \lim_{\epsilon\searrow 0} S_{M_\epsilon}\Big(\frac{e^{-|x|^2/(4\epsilon)}}{(4\pi\epsilon)^{n/2}}\Big),$$

and

$$\langle\phi, S_{M_\epsilon}\Big(\frac{e^{-|x|^2/(4\epsilon)}}{(4\pi\epsilon)^{n/2}}\Big)\rangle = (4\pi\epsilon)^{-n/2}\int_{\mathbf{R}^n}\phi(\epsilon, x)\,e^{-|x|^2/(4\epsilon)}\,dx$$

$$= \pi^{-n/2}\int_{\mathbf{R}^n}\phi(\epsilon, 2\sqrt{\epsilon}\,y)\,e^{-|y|^2}\,dy \to \langle\phi, \delta\rangle$$

for $\epsilon \searrow 0$ and $\phi \in \mathcal{D}(\mathbf{R}^{n+1})$ due to Lebesgue's theorem on dominated convergence, cf. also Hirsch and Lacombe [131], 8.3, Thm. 3.3, p. 310.

Note that the above verifications of fundamental solutions can also be performed without using the jump formula if $\mathcal{C}^\infty$ approximations are employed. For example, for Δ_n, one can utilize

$$E = \lim_{\epsilon\searrow 0} g(\sqrt{|x|^2 + \epsilon^2})\quad \text{or}\quad E = \lim_{k\to\infty}\big(1 - \psi(k|x|)\big)g(|x|),$$

with g as in (a) and $\psi \in \mathcal{D}(\mathbf{R}^1)$, $\psi = 1$ near 0, cf. Folland [76], Thm. 2.16, p. 30; [77], (2.17), p. 75. $\qquad\square$

Proposition 1.3.15 *Let $x_0 \in \Omega \subset \mathbf{R}^n$ open and $T \in \mathcal{D}'(\Omega)$ with $\operatorname{supp} T = \{x_0\}$. Then there exists a linear differential operator $P(\partial) = \sum_{|\alpha| \leq m} a_\alpha \partial^\alpha$, $m \in \mathbf{N}_0$, $a_\alpha \in \mathbf{C}$, such that $T = P(\partial)\delta_{x_0}$.*

For the proof, we refer to Schwartz [246], Ch. III, Thm. XXXV, p. 100; Donoghue [61], Sect. 21, p. 103; Friedlander and Joshi [84], Thm. 3.2.1, p. 36; Hörmander [139], Thm. 2.3.4, p. 46.

Example 1.3.16 Let us verify now the fundamental solution of the Laplacean in Example 1.3.14 (a) by employing Proposition 1.3.15. First we note that a *homogeneous* distribution $T \in \mathcal{D}'(\mathbf{R}^n)$ of degree λ with support in $\{0\}$ is necessarily of the form

$$T = \sum_{|\alpha|=m} a_\alpha \partial^\alpha \delta, \qquad a_\alpha \in \mathbf{C}, \qquad \lambda = -n - m,$$

for some $m \in \mathbf{N}_0$. Hence, if $n \neq 2$ and $E = c_n |x|^{2-n}$ as in Example 1.3.14 (a), then $\Delta_n E$, which is homogeneous of degree $-n$ and has its support confined to 0, must be a multiple of δ, i.e., $\Delta E = a\delta$, $a \in \mathbf{R}$. Finally, by application to the rotationally invariant test function $\chi(|x|) \in \mathcal{D}(\mathbf{R}^n)$, $\chi \in \mathcal{D}(\mathbf{R}^1)$, $\chi = 1$ near 0, we obtain

$$a = \langle \chi(|x|), \Delta_n E \rangle = \langle \Delta_n(\chi(|x|)), E \rangle = c_n |\mathbf{S}^{n-1}| \int_0^\infty \left(\chi''(r) + \tfrac{n-1}{r} \chi'(r) \right) r \, dr$$

$$= c_n |\mathbf{S}^{n-1}| \left(r\chi'(r) + (n-2)\chi(r) \right)\big|_{r=0}^\infty = (2-n)c_n \chi(0)|\mathbf{S}^{n-1}| = 1.$$

The same procedure also works for the Cauchy–Riemann operator $\partial_1 + i\,\partial_2$ and for the heat operator $\partial_t - \Delta_n$. For the latter operator, one uses the "quasihomogeneity" of E, i.e., the property

$$E(c^2 t, cx) = c^{-n} E(t, x), \qquad c > 0. \qquad \square$$

Example 1.3.17 Let us consider here the solution of the equation $h(x)^m \cdot T = 0$ for $T \in \mathcal{D}'(\Omega)$, h submersive on $M = h^{-1}(0) \subset \Omega$, $m \in \mathbf{N}$.

(a) First, if $T \in \mathcal{D}'(\mathbf{R}^1)$, $m \in \mathbf{N}$, and $x^m \cdot T = 0$, then $\operatorname{supp} T \subset \{0\}$ and Proposition 1.3.15 implies that $T = \sum_{j=0}^k a_j \delta^{(j)}$ for some $k \in \mathbf{N}_0$ and $a_j \in \mathbf{C}$, $j = 0, \ldots, k$, $a_k \neq 0$. If k were larger than or equal to m, then $0 = \langle \phi, x^m \cdot T \rangle = (-1)^k k! \, a_k \neq 0$ for $\phi(x) = x^{k-m}\chi(x)$, $\chi \in \mathcal{D}(\mathbf{R})$ with $\chi = 1$ near 0, yields a contradiction. Conversely, $x^m \cdot \delta^{(j)} = 0$ for $j = 0, \ldots, m-1$, and hence

$$\{T \in \mathcal{D}'(\mathbf{R}^1); \, x^m \cdot T = 0\} = \left\{ \sum_{j=0}^{m-1} a_j \delta^{(j)}; \, a_j \in \mathbf{C} \right\}.$$

(b) More generally, by Schwartz [246], Ch. III, Thm. XXXVI, p. 101, the equation $x_1^m T = 0$, $m \in \mathbf{N}$, $T \in \mathcal{D}'(\mathbf{R}^n)$, holds if and only if T can be represented (uniquely) in the following form:

$$T = \sum_{j=0}^{m-1} \delta^{(j)} \otimes T_j, \qquad T_j \in \mathcal{D}'(\mathbf{R}^{n-1}), \quad \text{i.e.,}$$

$$\langle \phi, T \rangle = \sum_{j=0}^{m-1} \langle ((-\partial_1)^j \phi)(0, x'), T_j(x') \rangle, \qquad \phi \in \mathcal{D}(\mathbf{R}^n),$$

see Definition 1.5.1 below.

Still more generally, by Schwartz [246], Ch. III, Thm. XXXVII, p. 102, we have

$$\forall T \in \mathcal{D}'(\Omega) : h(x)^m \cdot T = 0 \iff \exists_1 T_0, \ldots, T_{m-1} \in \mathcal{D}'(M) : T = \sum_{j=0}^{m-1} L_M^{(j)}(T_j)$$

if $h : \Omega \longrightarrow \mathbf{R}$ is submersive on $M = h^{-1}(0) \subset \Omega$, $m \in \mathbf{N}$, $\mathcal{D}'(M) = \{S : \mathcal{D}(M) \longrightarrow \mathbf{C}$ linear and sequentially continuous$\}$, and $L_M^{(j)}(S) \in \mathcal{D}'(\Omega)$ is the *multi-layer potential* on M defined by

$$\langle \phi, L_M^{(j)}(S) \rangle = \langle (-\partial_\nu)^j \phi |_M, S \rangle, \qquad S \in \mathcal{D}'(M), \qquad \phi \in \mathcal{D}(\Omega),$$

where $\partial_\nu \phi = \nu^T \cdot \nabla \phi$, $\nu = \frac{\nabla h}{|\nabla h|}$ in a neighborhood of M. Note that the single and double layer potentials defined in (1.2.1), (1.3.1), respectively, are special cases of this:

$$S_M(f) = L_M^{(0)}(f), \quad D_M(f) = L_M^{(1)}(f),$$

if

$$L_{loc}^1(M) \hookrightarrow \mathcal{D}'(M) : f \longmapsto (\phi \mapsto \int_M f \cdot \phi \, d\sigma). \tag{1.3.15}$$

We also emphasize that $(-\partial_\nu)^j \phi|_M$ and hence $S_M^{(j)}$ does not depend on M only, but also on h for $j \geq 2$.

(c) In order to illustrate these assertions, let us consider $h(x) = |x|^2 - R^2$, $R > 0$, such that $M = R\mathbf{S}^{n-1}$. By the above, the distributional equation $(|x|^2 - R^2) \cdot T = 0$ has the solutions $T = S_M(T_0)$, $T_0 \in \mathcal{D}'(M)$. For $n = 2$, e.g., we have, by expansion with respect to the angle ϑ,

$$h(x) \cdot T = 0 \iff \exists \mu_0, \ldots, \mu_l \in \mathcal{M}^1([0, 2\pi]) : T = \sum_{k=0}^{l} \mu_k(\vartheta) \frac{d^k}{d\vartheta^k} (\delta \circ h),$$

i.e.

$$\langle \phi, T \rangle = \frac{1}{2R} \sum_{k=0}^{l} (-1)^k \langle \frac{d^k \tilde{\phi}}{d\vartheta^k}, \mu_k \rangle,$$

where $\phi \in \mathcal{D}(\mathbf{R}^2)$ and $\tilde{\phi}(\vartheta) = \phi(R \cos \vartheta, R \sin \vartheta) \in C^\infty(\mathbf{R})$. Herein, we used the structure theorem for distributions with compact support, see Schwartz [246], Ch. III, Thm. XXVI, p. 91.   □

Similarly as for the equation $x^m \cdot T = 0$, $T \in \mathcal{D}'(\mathbf{R})$, we can also describe all distributional solutions of the equation $T^{(m)} = 0$, cf. Horváth [141], Ch. 4, Sect. 3, Prop. 3, p. 327; Hörmander [139], Thm. 3.1.4, Cor. 3.1.6.

Proposition 1.3.18 Let $a < b \in \mathbf{R}$ and set $\Omega = (a, b) \subset \mathbf{R}^1$.

(1) $\forall T \in \mathcal{D}'(\Omega) : T' = 0 \Longleftrightarrow T \in \mathbf{C} \subset \mathcal{D}'(\Omega)$.
(2) $\forall m \in \mathbf{N} : \forall T \in \mathcal{D}'(\Omega) : T^{(m)} = 0 \Longleftrightarrow T = \sum_{j=0}^{m-1} c_j x^j, c_j \in \mathbf{C}$.
(3) $\frac{d}{dx} : \mathcal{D}'(\Omega) \longrightarrow \mathcal{D}'(\Omega)$ has a sequentially continuous linear "right-inverse" $R : \mathcal{D}'(\Omega) \longrightarrow \mathcal{D}'(\Omega)$, i.e., R fulfills $\forall T \in \mathcal{D}'(\Omega) : R(T)' = T$.
(4) A linear constant coefficient ordinary differential operator $P(\frac{d}{dx})$ has only classical solutions, i.e., $P(\frac{d}{dx})T = 0$ for $T \in \mathcal{D}'(\Omega)$ implies that $T \in C^\infty(\Omega)$.

Note that (4) is indeed a consequence of the (hypo)ellipticity of $P(\frac{d}{dx})$, but we prefer to prove it here without making use of the algebraic characterization of hypoelliptic operators by L. Hörmander.

Proof

(1) If $T' = 0$, then T vanishes on the hyperplane

$$H = \{\phi'; \phi \in \mathcal{D}(\Omega)\} = \{\psi \in \mathcal{D}(\Omega); \int_a^b \psi(x)\, dx = \langle \psi, 1 \rangle = 0\} = \ker 1$$

of $\mathcal{D}(\Omega)$. If we fix $\chi \in \mathcal{D}(\Omega)$ with $\langle \chi, 1 \rangle = 1$ and set $\mathrm{pr}\, \phi = \phi - \langle \phi, 1 \rangle \cdot \chi \in H$, then

$$\langle \phi, T \rangle = \langle \mathrm{pr}\, \phi + \langle \phi, 1 \rangle \cdot \chi, T \rangle = \langle \phi, 1 \rangle \cdot \langle \chi, T \rangle = \langle \phi, \langle \chi, T \rangle \rangle,$$

i.e., $T = \langle \chi, T \rangle \in \mathbf{C}$.
(2) This follows from (1) by induction.
(3) For χ and pr as in (1), we define

$$R : \mathcal{D}'(\Omega) \longrightarrow \mathcal{D}'(\Omega) : T \longmapsto \left(\phi \mapsto \langle - \int_a^x (\mathrm{pr}\, \phi)(t)\, dt, T \rangle \right)$$

and infer $(RT)' = T$ from

$$\int_a^x (\text{pr}\,\phi')(t)\,dt = \int_a^x \phi'(t)\,dt - \underbrace{\langle \phi', 1\rangle}_{0} \int_a^x \chi'(t)\,dt = \phi(x).$$

(4) This also follows inductively, once the first-order case $P(\frac{d}{dx}) = \frac{d}{dx} - \lambda$ is settled. In this case, $0 = (\frac{d}{dx} - \lambda)T = e^{\lambda x}\frac{d}{dx}(e^{-\lambda x}T)$ implies, by (1), that $e^{-\lambda x}T = c \in \mathbf{C}$ and $T = c\,e^{\lambda x} \in C^\infty(\Omega)$. $\square$

Let us yet give an elementary, but useful formula for the dependence of fundamental solutions on linear transformations of the operators, see Garnir [99], p. 284; Wagner [285], Satz 4, p. 10.

Proposition 1.3.19 *Let E be a fundamental solution of the operator $P(\partial) = \sum_{|\alpha| \le m} c_\alpha \partial^\alpha$ and $A = (a_{jk})_{1 \le j,k \le n} \in \mathrm{Gl}_n(\mathbf{R})$. Then $|\det A|^{-1} \cdot E \circ A^{-1T}$ is a fundamental solution of the composed operator $(P \circ A)(\partial) = \sum_{|\alpha| \le m} c_\alpha (A\partial)^\alpha$.*

Proof For $T \in \mathcal{D}'(\mathbf{R}^n)$, the chain rule yields

$$(A\partial)_j\big(T \circ A^{-1T}\big) = \left(\sum_{k=1}^n a_{jk}\partial_k\right)\big(T \circ A^{-1T}\big) = (\partial_j T) \circ A^{-1T}.$$

Therefore,

$$(P \circ A)(\partial)\big(E \circ A^{-1T}\big) = \big(P(\partial)E\big) \circ A^{-1T} = \delta \circ A^{-1T} = |\det A| \cdot \delta$$

by Example 1.2.8. $\square$

1.4 Distribution-Valued Functions

Definition 1.4.1 Let $\emptyset \ne \Omega \subset \mathbf{R}^n$ be open.

(1) For a metric space X, a mapping $f : X \longrightarrow \mathcal{D}'(\Omega)$ is called *continuous* if and only if the mappings

$$\langle \phi, f\rangle : X \longrightarrow \mathbf{C} : \lambda \longmapsto \langle \phi, f(\lambda)\rangle$$

are continuous for all $\phi \in \mathcal{D}(\Omega)$. The vector space of all such continuous functions $f : X \longrightarrow \mathcal{D}'(\Omega)$ is denoted by $C\big(X, \mathcal{D}'(\Omega)\big)$.

(Hence the "continuity" of f refers to the "weak topology" on $\mathcal{D}'(\Omega)$; due to the fact that $\mathcal{D}'(\Omega)$ is a Montel space and X is a metric space, this is equivalent to the continuity of f with respect to the "strong topology" on $\mathcal{D}'(\Omega)$, see Treves [273], 36.1.)

(2) Similarly, if $\emptyset \neq U \subset \mathbf{R}^l$ is open and $m \in \mathbf{N}_0 \cup \{\infty\}$, then $f : U \longrightarrow \mathcal{D}'(\Omega)$ is called m times *continuously differentiable* if this holds in the weak sense, i.e.,

$$\langle \phi, f \rangle : U \longrightarrow \mathbf{C} : \lambda \longmapsto \langle \phi, f(\lambda) \rangle \text{ is } C^m$$

for all $\phi \in \mathcal{D}(\Omega)$, and we set

$$C^m\big(U, \mathcal{D}'(\Omega)\big) := \{f : U \longrightarrow \mathcal{D}'(\Omega); f \, C^m\}.$$

(Note that $\partial f / \partial \lambda_j(\lambda) \in \mathcal{D}'(\Omega)$ for $f \in C^1\big(U, \mathcal{D}'(\Omega)\big)$ and $\lambda \in U$ by Prop. 1.1.5. Therefore, $\partial f / \partial \lambda_j \in C^{m-1}\big(U, \mathcal{D}'(\Omega)\big)$ for $f \in C^m\big(U, \mathcal{D}'(\Omega)\big)$ and $m \in \mathbf{N} \cup \{\infty\}, j = 1, \dots, l$).

(3) The space $C^m\big([0, \infty), \mathcal{D}'(\Omega)\big)$ is defined analogously, i.e., it consists of functions $f \in C^m\big((0, \infty), \mathcal{D}'(\Omega)\big)$ for which $f^{(j)}(t)$ converges if $t \searrow 0$ for all $j = 0, \dots, m$.

(As before, note that the notion of "weak differentiability" introduced above coincides with that of "strong differentiability" due to the Montel property of the space $\mathcal{D}'(\Omega)$, see Treves [273], Prop. 36.11, p. 377.)

Let us employ the notion of differentiable distribution-valued functions to give a useful formula for fundamental solutions of *powers* of differential operators with non-vanishing constant term.

Proposition 1.4.2 *Let $P_1(\partial), \dots, P_l(\partial)$ be linear differential operators in $\mathbf{R}^n$ with constant coefficients and assume that $U \subset \mathbf{R}^l$ open, $m \in \mathbf{N}$, and $E \in C^m\big(U, \mathcal{D}'(\mathbf{R}^n)\big)$ such that $E(\lambda)$ is a fundamental solution of $\prod_{j=1}^l \big(P_j(\partial) - \lambda_j\big)$. Then $\frac{1}{\alpha!} \partial_\lambda^\alpha E(\lambda)$ is a fundamental solution of $\prod_{j=1}^l \big(P_j(\partial) - \lambda_j\big)^{\alpha_j + 1}$ for $\lambda \in U$ and $\alpha \in \mathbf{N}_0^l$ with $|\alpha| \leq m$.*

Proof The assertion is shown in exactly the same way as part (2) of the proof of Proposition 1.3.7, which refers to the case $P_j(\partial) = \frac{d}{dx}$.   □

Example 1.4.3 If we apply Proposition 1.4.2 to the Helmholtz operator $\Delta_n + \lambda = \Delta_n - (-\lambda)$ considered in Example 1.3.14 (c), we obtain that

$$E_k(\lambda) = \frac{(-1)^{k-1}}{(k-1)!} \partial_\lambda^{k-1} \Big(\frac{\lambda^{n/4 - 1/2}}{2^{1 + n/2} \pi^{n/2 - 1}} |x|^{1 - n/2} N_{n/2 - 1}(\sqrt{\lambda}\, |x|) \Big)$$

$$= \frac{(-1)^{k-1} \lambda^{n/4 - k/2}}{2^{k + n/2} \pi^{n/2 - 1} (k-1)!} |x|^{k - n/2} N_{n/2 - k}(\sqrt{\lambda}\, |x|)$$

is a fundamental solution of $(\Delta_n + \lambda)^k$ for $\lambda > 0$, $k \in \mathbf{N}$, cf. Schwartz [246], (VII, 10; 17), p. 287. (For the differentiations with respect to λ, we used the recurrence formula 8.472.3 in Gradshteyn and Ryzhik [113], p. 968.)   □

Often in applications, one encounters operators of the form $\prod_{j=1}^l \big(P(\partial) - \lambda_j\big)^{\alpha_j + 1}$, where all the operators P_j in Proposition 1.4.2 coincide. Its fundamental solution can then be expressed through one of $P(\partial) - \lambda$, see Proposition 1.4.4 below, which goes back to Ortner [202], Prop. 1, p. 82. The special case of $P(\partial) = \frac{d}{dx}$ appeared already in Proposition 1.3.7.

Proposition 1.4.4 *Let $P(\partial)$ be a linear differential operator in $\mathbf{R}^n$ with constant coefficients and assume that $U \subset \mathbf{R}$ open, $m \in \mathbf{N}$, and $E \in \mathcal{C}^m\big(U, \mathcal{D}'(\mathbf{R}^n)\big)$ such that $E(\lambda)$ is a fundamental solution of $P(\partial) - \lambda$. Then*

$$\sum_{j=1}^{l} \frac{1}{\alpha_j!} \partial_{\lambda_j}^{\alpha_j}\big(c_j E(\lambda_j)\big), \qquad c_j := \prod_{\substack{k=1 \\ k \neq j}}^{l} (\lambda_j - \lambda_k)^{-\alpha_k - 1},$$

is a fundamental solution of $\prod_{j=1}^{l}\big(P(\partial) - \lambda_j\big)^{\alpha_j + 1}$ for pairwise different $\lambda_j \in U$, $j = 1, \dots, l$, and $\alpha \in \mathbf{N}_0^l$ with $\alpha_j \leq m$, $j = 1, \dots, l$.

Proof

(1) Let us first consider the case $\alpha = 0$. Then we have to check that

$$\Big(\prod_{j=1}^{l}(P(\partial) - \lambda_j)\Big) \sum_{j=1}^{l} \Big[E(\lambda_j) \cdot \prod_{\substack{k=1 \\ k \neq j}}^{l} (\lambda_j - \lambda_k)^{-1} \Big] = \delta.$$

Since $\big(P(\partial) - \lambda_j\big) E(\lambda_j) = \delta$, this is equivalent to

$$\sum_{j=1}^{l} \Big(\prod_{\substack{k=1 \\ k \neq j}}^{l} (\lambda_j - \lambda_k)^{-1} \big(P(\partial) - \lambda_k\big)\Big)\delta = \delta,$$

which in turn follows from the following resolution in partial fractions:

$$\sum_{j=1}^{l} \frac{1}{z - \lambda_j} \prod_{\substack{k=1 \\ k \neq j}}^{l} (\lambda_j - \lambda_k)^{-1} = \prod_{j=1}^{l} (z - \lambda_j)^{-1}.$$

(Note that $\mathrm{Res}_{z=\lambda_j} \prod_{k=1}^{l} (z - \lambda_k)^{-1} = \prod_{\substack{k=1 \\ k \neq j}}^{l} (\lambda_j - \lambda_k)^{-1}$.)

(2) The general formula in the proposition is now a consequence of Proposition 1.4.2 applied to the fundamental solution

$$F(\lambda) := \sum_{j=1}^{l} \Big[E(\lambda_j) \cdot \prod_{\substack{k=1 \\ k \neq j}}^{l} (\lambda_j - \lambda_k)^{-1} \Big]$$

of the operator $\prod_{j=1}^{l}(P(\partial) - \lambda_j)$. In fact,

$$\frac{1}{\alpha!}\,\partial_\lambda^\alpha F(\lambda) = \sum_{j=1}^{l}\frac{1}{\alpha_j!}\,\partial_{\lambda_j}^{\alpha_j}\Big[E(\lambda_j)\cdot\underbrace{\Big(\prod_{\substack{k=1\\k\neq j}}^{l}\frac{1}{\alpha_k!}\,\partial_{\lambda_k}^{\alpha_k}\Big)\frac{1}{\lambda_j - \lambda_k}}_{c_j}\Big].$$

Let us mention that—instead of the assumption $|\alpha| = |\alpha|_1 \leq m$ in Proposition 1.4.2—we can make do here with $|\alpha|_\infty \leq m$, since this is all one needs to perform the inductive steps in part (2) of the proof of Proposition 1.3.7.   □

Example 1.4.5 If we apply Proposition 1.4.4 to the Laplacean $P(\partial) = \Delta_n$, we obtain

$$E = \frac{(-1)^{|\alpha|+l-1}|x|^{1-n/2}}{4(2\pi)^{n/2-1}}\sum_{j=1}^{l}\frac{1}{\alpha_j!}\,\partial_{\lambda_j}^{\alpha_j}\big[c_j\lambda_j^{n/4-1/2}N_{n/2-1}(\lambda_j|x|)\big],$$

$$c_j = \prod_{\substack{k=1\\k\neq j}}^{l}(\lambda_j - \lambda_k)^{-\alpha_k-1},$$

as a fundamental solution of the rotationally invariant operator $\prod_{j=1}^{l}(\Delta_n + \lambda_j)^{\alpha_j+1}$, λ_j, $j = 1,\dots,l$, being pairwise different positive numbers, cf. Cheng et al. [49], (21), p. 189; Muhlisov [184], (3.17), p. 141; Paneyah [225], Teorema 2, p. 128.   □

Similarly as for differentiability, the *holomorphy* of distribution-valued functions is defined in the weak sense, see the next definition.

Definition 1.4.6 Let $\emptyset \neq \Omega \subset \mathbf{R}^n$ and $\emptyset \neq U \subset \mathbf{C}^l$ be open subsets.

(1) A mapping $f : U \longrightarrow \mathcal{D}'(\Omega)$ is called *holomorphic* (or *analytic*) if and only if

$$\langle\phi,f\rangle : U \longrightarrow \mathbf{C} : \lambda \longmapsto \langle\phi,f(\lambda)\rangle$$

is holomorphic for each $\phi \in \mathcal{D}(\Omega)$.
(2) If $l = 1$, i.e., $U \subset \mathbf{C}$, then f is called *meromorphic* if f is defined and holomorphic in $U \setminus D$ for a discrete set $D \subset U$, and if

$$\forall \lambda_0 \in D : \exists k \in \mathbf{N}_0 : (\lambda - \lambda_0)^k f(\lambda) \text{ can be continued holomorphically to } \lambda_0.$$

The *residue* and the *finite part* of f at $\lambda_0 \in D$ are then defined in the weak sense:

$$\langle\phi, \operatorname*{Res}_{\lambda=\lambda_0} f(\lambda)\rangle := \operatorname*{Res}_{\lambda=\lambda_0}\langle\phi,f(\lambda)\rangle, \quad \langle\phi, \operatorname*{Pf}_{\lambda=\lambda_0} f(\lambda)\rangle := \operatorname*{Pf}_{\lambda=\lambda_0}\langle\phi,f(\lambda)\rangle, \quad \phi \in \mathcal{D}(\Omega).$$

Proposition 1.4.7 *If $f : U \longrightarrow \mathcal{D}'(\Omega)$ is holomorphic, $\lambda_0 \in U$ and $r > 0$ such that $B_r = \{\lambda \in \mathbf{C}^l; |\lambda - \lambda_0| \le r\} \subset U$, then f can be developed into a convergent Taylor series in B_r, i.e., the series $\sum_{|\alpha| \le m} \frac{(\lambda-\lambda_0)^\alpha}{\alpha!} \partial^\alpha f(\lambda_0)$ converges uniformly to $f(\lambda)$ for $\lambda \in B_r$ if $m \to \infty$. Similarly, if $U \subset \mathbf{C}$ and f is meromorphic in U, then f can be developed into a Laurent series around any $\lambda_0 \in U$, and $\mathrm{Res}_{\lambda=\lambda_0} f(\lambda)$ and $\mathrm{Pf}_{\lambda=\lambda_0} f(\lambda)$ are the coefficients of $(\lambda - \lambda_0)^{-1}$ and $(\lambda - \lambda_0)^0$, respectively.*

For the proof in the case of $l = 1$, we refer to Horváth [143], (1.2.8), p. 65, and p. 75; Grothendieck [117], Thm. I, 5, pp. 37, 38; Ortner and Wagner [219], Prop. 1.5.5, p. 21.

Example 1.4.8 In Example 1.3.9, the distribution x_+^λ, $\lambda \in \mathbf{C} \setminus -\mathbf{N}$, was defined by differentiation. We now see that the mapping

$$\mathbf{C} \setminus -\mathbf{N} \longrightarrow \mathcal{D}'(\mathbf{R}^1) : \lambda \longmapsto x_+^\lambda$$

is holomorphic, since the integral

$$\int_0^\infty x^\lambda \phi(x)\,dx, \qquad \phi \in \mathcal{D}(\mathbf{R}), \qquad \mathrm{Re}\,\lambda > -1,$$

depends analytically on λ and hence the same holds for

$$\langle \phi, x_+^\lambda \rangle = \frac{(-1)^m}{(\lambda+1)\cdots(\lambda+m)} \int_0^\infty x^{\lambda+m} \phi^{(m)}(x)\,dx$$

for $m \in \mathbf{N}$, $\mathrm{Re}\,\lambda > -m-1$, and $\lambda \notin -\mathbf{N}$.

The distribution-valued function x_+^λ has simple poles at $\lambda = -k$, $k \in \mathbf{N}$, with the residues

$$\mathrm{Res}_{\lambda=-k} x_+^\lambda = \frac{(-1)^{k-1}}{(k-1)!} \delta^{(k-1)}, \qquad (1.4.1)$$

since

$$\lim_{\lambda\to-k} (\lambda+k)x_+^\lambda = \lim_{\lambda\to-k} \frac{(x_+^{\lambda+k})^{(k)}}{(\lambda+1)\cdots(\lambda+k-1)} = \frac{Y^{(k)}}{(1-k)\cdots(-1)}.$$

The Taylor series of $f(\lambda) = x_+^\lambda$ around $\lambda = 0$ is given by

$$x_+^\lambda = \sum_{k=0}^\infty \frac{f^{(k)}(0)}{k!} \lambda^k = \sum_{k=0}^\infty \frac{Y(x)\cdot\log^k x}{k!} \lambda^k, \qquad |\lambda| < 1.$$

This implies, in particular, that x_+^{-1}, as defined in Example 1.2.6, is the finite part of x_+^λ at $\lambda = -1$:

$$\Pf_{\lambda=-1} x_+^\lambda = \Pf_{\lambda=-1} \frac{d}{dx}\left(\frac{x_+^{\lambda+1}}{\lambda+1}\right) = \frac{d}{dx}\Pf_{\mu=0}\left(\frac{x_+^\mu}{\mu}\right) = \frac{d}{dx}\big(Y(x)\log x\big) = x_+^{-1},$$

see Example 1.3.9 for the last equation. More generally, let us define

$$x_+^{-k} := \Pf_{\lambda=-k} x_+^\lambda, \qquad k \in \mathbf{N}. \tag{1.4.2}$$

From the first two terms in the above Taylor series of x_+^λ, we then conclude that

$$x_+^{-k} = \Pf_{\lambda=-k} \frac{d^k}{dx^k} \frac{x_+^{\lambda+k}}{(\lambda+1)\cdots(\lambda+k)} = \frac{d^k}{dx^k} \Pf_{\mu=0} \frac{x_+^\mu}{\mu(\mu-1)\cdots(\mu-k+1)}$$

$$= \frac{d^k}{dx^k} \Pf_{\mu=0}\left[\frac{Y(x)}{\mu(\mu-1)\cdots(\mu-k+1)} + \frac{Y(x)\log x}{(\mu-1)\cdots(\mu-k+1)}\right]$$

$$= Y^{(k)} \cdot \frac{d}{d\mu}\left(\frac{1}{(\mu-1)\cdots(\mu-k+1)}\right)\bigg|_{\mu=0} + \frac{(-1)^{k-1}}{(k-1)!}\big(Y(x)\log x\big)^{(k)}$$

$$= \frac{(-1)^{k-1}}{(k-1)!}\big[(\psi(k)-\psi(1))\delta^{(k-1)} + (x_+^{-1})^{(k-1)}\big], \qquad \psi(z) = \frac{\Gamma'(z)}{\Gamma(z)}.$$

Let us yet derive a limit representation for x_+^{-k} corresponding to the one for x_+^λ, $\lambda \in \mathbf{C}\setminus -\mathbf{N}$, given in Example 1.3.9. From formula (1.3.7), we infer for $k \in \mathbf{N}$

$$x_+^{-k} = \Pf_{\lambda=-k} x_+^\lambda = \Pf_{\lambda=-k} \lim_{\epsilon\searrow 0}\left[Y(x-\epsilon)x^\lambda + \sum_{i=0}^{k-1} \frac{(-1)^i\epsilon^{\lambda+i+1}}{i!(\lambda+i+1)}\delta^{(i)}\right]$$

$$= \lim_{\epsilon\searrow 0}\Pf_{\lambda=-k}\left[Y(x-\epsilon)x^\lambda + \sum_{i=0}^{k-1} \frac{(-1)^i\epsilon^{\lambda+i+1}}{i!(\lambda+i+1)}\delta^{(i)}\right]$$

$$= \lim_{\epsilon\searrow 0}\left[Y(x-\epsilon)x^{-k} - \sum_{i=0}^{k-2} \frac{(-1)^i\epsilon^{i-k+1}}{i!(k-i-1)}\delta^{(i)} + \frac{(-1)^{k-1}\delta^{(k-1)}}{(k-1)!}\Pf_{\lambda=-k}\left(\frac{\epsilon^{\lambda+k}}{\lambda+k}\right)\right]$$

$$= \lim_{\epsilon\searrow 0}\left[Y(x-\epsilon)x^{-k} - \sum_{i=0}^{k-2} \frac{(-1)^i\epsilon^{i-k+1}}{i!(k-i-1)}\delta^{(i)} + \frac{(-1)^{k-1}\log\epsilon}{(k-1)!}\delta^{(k-1)}\right],$$

cf. Horváth [143], (2.2.5.6), p. 88.   □

Example 1.4.9 Next, let us subsume the extensions of homogeneous distributions of degree $-n$ from $\mathbf{R}^n\setminus\{0\}$ to $\mathbf{R}^n$ in Example 1.2.6 under the framework of analytic continuation of distribution-valued functions.

We consider $\mathbf{S}^{n-1}$ as a C^∞ submanifold of $\mathbf{R}^n$, we denote by $\mathcal{D}'(\mathbf{S}^{n-1})$ the dual of $\mathcal{D}(\mathbf{S}^{n-1}) = C^\infty(\mathbf{S}^{n-1})$ as in Example 1.3.17, and we embed $L^1(\mathbf{S}^{n-1}) = L^1_{\text{loc}}(\mathbf{S}^{n-1})$ by means of the surface measure, see (1.3.15). For $F \in \mathcal{D}'(\mathbf{S}^{n-1})$ and $\lambda \in \mathbf{C}$, we define $F \cdot |x|^\lambda \in \mathcal{D}'(\mathbf{R}^n)$ by the equation

$$\langle \phi, F \cdot |x|^\lambda \rangle = \langle \langle \phi(t\omega), F(\omega) \rangle, t_+^{\lambda+n-1} \rangle, \qquad \phi \in \mathcal{D}(\mathbf{R}^n). \tag{1.4.3}$$

Then the mapping

$$\{\lambda \in \mathbf{C}; \; -\lambda - n \notin \mathbf{N}_0\} \longrightarrow \mathcal{D}'(\mathbf{R}^n) : \lambda \longmapsto F \cdot |x|^\lambda$$

is holomorphic since the same holds for the function

$$\{\lambda \in \mathbf{C}; \; -\lambda - n \notin \mathbf{N}_0\} \longrightarrow \mathcal{D}'(\mathbf{R}^1) : \lambda \longmapsto t_+^{\lambda+n-1},$$

see Example 1.4.8.

The function $\lambda \mapsto F|x|^\lambda$ has at most simple poles in $\lambda = -n - k$, $k \in \mathbf{N}_0$, where the residues are given by

$$\langle \phi, \operatorname*{Res}_{\lambda=-n-k} F|x|^\lambda \rangle = \langle \langle \phi(t\omega), F(\omega) \rangle, \operatorname*{Res}_{\lambda=-n-k} t_+^{\lambda+n-1} \rangle$$

$$= \langle \langle \phi(t\omega), F(\omega) \rangle, \frac{(-1)^k}{k!} \delta^{(k)}(t) \rangle$$

$$= \frac{1}{k!} \langle ((\omega^T \nabla)^k \phi)(0), F(\omega) \rangle = \sum_{|\alpha|=k} \frac{\partial^\alpha \phi(0)}{\alpha!} \langle \omega^\alpha, F(\omega) \rangle,$$

i.e.,

$$\operatorname*{Res}_{\lambda=-n-k} F|x|^\lambda = (-1)^k \sum_{|\alpha|=k} \frac{\langle \omega^\alpha, F(\omega) \rangle}{\alpha!} \partial^\alpha \delta = \frac{(-1)^k}{k!} \langle (\omega^T \nabla)^k, F(\omega) \rangle \delta,$$

cf. Ortner and Wagner [219], Prop. 2.2.1, p. 35. In particular, $\lambda \mapsto F|x|^\lambda$ is holomorphic at $\lambda = -n$ iff $\langle 1, F \rangle = 0$, or, in other words, iff F fulfills the (generalized) mean-value zero condition, cf. (1.1.1).

Similarly, by (1.4.2), (1.4.3),

$$\langle \phi, F|x|^{-n-k} \rangle = \langle \langle \phi(t\omega), F(\omega) \rangle, t_+^{-k-1} \rangle$$

$$= \langle \langle \phi(t\omega), F(\omega) \rangle, \operatorname*{Pf}_{\lambda=-n-k} t_+^{\lambda+n-1} \rangle$$

$$= \langle \phi, \operatorname*{Pf}_{\lambda=-n-k} F|x|^\lambda \rangle, \qquad \phi \in \mathcal{D}(\mathbf{R}^n).$$

In particular, for $k = 0$, we have

$$\langle \phi, \operatorname*{Pf}_{\lambda=-n} F|x|^\lambda \rangle = \langle \langle \phi(t\omega), F(\omega) \rangle, t_+^{-1} \rangle$$

$$= \lim_{\epsilon \searrow 0}\left[\int_\epsilon^\infty \langle \phi(t\omega), F(\omega) \rangle \, \frac{\mathrm{d}t}{t} + \phi(0)\langle 1, F \rangle \log \epsilon \right],$$

i.e., the distribution

$$F \cdot |x|^{-n} = \operatorname*{Pf}_{\lambda=-n} F|x|^\lambda = \lim_{\epsilon \searrow 0}\left[F \cdot |x|^{-n} Y(|x| - \epsilon) + \langle 1, F \rangle (\log \epsilon) \delta \right] \qquad (1.4.4)$$

coincides with the extension $T \in \mathcal{D}'(\mathbf{R}^n)$ of $S = F \cdot |x|^{-n} \in \mathcal{D}'(\mathbf{R}^n \setminus \{0\})$ constructed directly in Example 1.2.6 (a) in case of $F = f \in L^1(\mathbf{S}^{n-1})$. If the mean-value zero condition $\langle 1, F \rangle = 0$ holds, then

$$F \cdot |x|^{-n} = \lim_{\epsilon \searrow 0}\left[F \cdot |x|^{-n} Y(|x| - \epsilon) \right] = \mathrm{vp}(F \cdot |x|^{-n}),$$

cf. Example 1.1.12.

Furthermore, we can transfer the limit representation for $x_+^\lambda \in \mathcal{D}'(\mathbf{R}^1)$ in (1.3.7) to this more general case. If $\lambda \in \mathbf{C}$ with $-\lambda - n \notin \mathbf{N}_0$ and $m \in \mathbf{N}_0$, $\operatorname{Re} \lambda > -m - n$, $\phi \in \mathcal{D}$, then

$$\langle \phi, F \cdot |x|^\lambda \rangle = \langle \langle \phi(t\omega), F(\omega) \rangle, t_+^{\lambda+n-1} \rangle$$

$$= \langle \langle \phi(t\omega), F(\omega) \rangle, \lim_{\epsilon \searrow 0}\left[Y(t - \epsilon)t^{\lambda+n-1} + \sum_{i=0}^{m-1} \frac{(-1)^i \epsilon^{\lambda+n+i}}{i!(\lambda + n + i)} \delta^{(i)} \right] \rangle$$

$$= \lim_{\epsilon \searrow 0}\left[\int_\epsilon^\infty \langle \phi(t\omega), F(\omega) \rangle \, t^{\lambda+n-1} \, \mathrm{d}t \right.$$

$$\left. + \sum_{i=0}^{m-1} \frac{\epsilon^{\lambda+n+i}}{\lambda + n + i} \sum_{|\alpha|=i} \langle \omega^\alpha, F(\omega) \rangle \frac{(\partial^\alpha \phi)(0)}{\alpha!} \right]$$

i.e.,

$$F \cdot |x|^\lambda = \lim_{\epsilon \searrow 0}\left[F \cdot |x|^\lambda Y(|x| - \epsilon) + \sum_{i=0}^{m-1} \frac{(-1)^i \epsilon^{\lambda+n+i}}{i!(\lambda + n + i)} \langle (\omega^T \nabla)^i, F(\omega) \rangle \delta \right]. \qquad (1.4.5)$$

(Of course, the distribution $F \cdot |x|^\lambda Y(|x| - \epsilon)$ is defined completely analogously as $F \cdot |x|^\lambda$ in (1.4.3). Note also that, conversely, formula (1.3.7) is contained in (1.4.5) upon taking $n = 1$ and $F : \mathbf{S}^0 = \{1, -1\} \longrightarrow \mathbf{C}$, $F(1) = 1$, $F(-1) = 0$.)

Similarly as in Example 1.4.8, formula (1.4.5) also yields limit representations for the finite parts in the poles $\lambda = -n - k$, $k \in \mathbf{N}_0$:

$$F \cdot |x|^{-n-k} = \Pf_{\lambda=-n-k} F \cdot |x|^\lambda = \lim_{\epsilon \searrow 0}\Big[F \cdot |x|^{-n-k} Y(|x| - \epsilon) -$$

$$- \sum_{i=0}^{k-1} \frac{(-1)^i \epsilon^{i-k}}{i!(k-i)} \langle (\omega^T \nabla)^i, F(\omega) \rangle \, \delta + \frac{(-1)^k \log \epsilon}{k!} \langle (\omega^T \nabla)^k, F(\omega) \rangle \, \delta \Big],$$

cf. Petersen [228], p. 40. (The special case of $k = 0$ is again formula (1.4.4).)

These limit representations can be transformed into integral representations for $\langle \phi, F \cdot |x|^\lambda \rangle$. If $\lambda \in \mathbf{C}$ with $-\lambda - n \notin \mathbf{N}_0$ and $m \in \mathbf{N}_0$, $\Re \lambda > -m - n, \phi \in \mathcal{D}$, then

$$\int_\epsilon^\infty \langle \phi(t\omega), F(\omega) \rangle \, t^{\lambda+n-1} \, dt = \int_1^\infty \langle \phi(t\omega), F(\omega) \rangle \, t^{\lambda+n-1} \, dt +$$

$$+ \int_\epsilon^1 \Big[\langle \phi(t\omega), F(\omega) \rangle - \sum_{i=0}^{m-1} \frac{t^i}{i!} \Big(\frac{d^i}{dt^i} \langle \phi(t\omega), F(\omega) \rangle \Big) \Big|_{t=0} \Big] t^{\lambda+n-1} \, dt$$

(1.4.6)

$$+ \sum_{i=0}^{m-1} \frac{1 - \epsilon^{\lambda+n+i}}{\lambda + n + i} \sum_{|\alpha|=i} \frac{1}{\alpha!} \langle \omega^\alpha, F(\omega) \rangle \, (\partial^\alpha \phi)(0).$$

Due to the substraction of regularization terms, the integral in (1.4.6) converges for $\epsilon \searrow 0$ and hence (1.4.5) furnishes

$$\langle \phi, F \cdot |x|^\lambda \rangle = \int_1^\infty \langle \phi(t\omega), F(\omega) \rangle \, t^{\lambda+n-1} \, dt +$$

$$+ \int_0^1 \Big[\langle \phi(t\omega), F(\omega) \rangle - \sum_{|\alpha|<m} \frac{t^\alpha}{\alpha!} (\partial^\alpha \phi)(0) \langle \omega^\alpha, F(\omega) \rangle \Big] t^{\lambda+n-1} \, dt +$$

(1.4.7)

$$+ \sum_{i=0}^{m-1} \frac{1}{\lambda + n + i} \sum_{|\alpha|=i} \frac{\langle \omega^\alpha, F(\omega) \rangle}{\alpha!} (\partial^\alpha \phi)(0)$$

if $\Re \lambda > -n - m, \lambda \notin -n - \mathbf{N}_0$.

In the case of $F \in L^1(\mathbf{S}^{n-1})$, this representation was given in Horváth [146], (2), p. 174. If we take the finite part of formula (1.4.7) in the poles $\lambda = -n - k$, $k \in \mathbf{N}_0$

(setting $m = k + 1$), we obtain

$$\langle \phi, F \cdot |x|^{-n-k} \rangle = \int_1^\infty \langle \phi(t\omega), F(\omega) \rangle \, t^{-k-1} \, dt +$$

$$+ \int_0^1 \Big[\langle \phi(t\omega), F(\omega) \rangle - \sum_{|\alpha| \le k} \frac{t^\alpha}{\alpha!} \, (\partial^\alpha \phi)(0) \langle \omega^\alpha, F(\omega) \rangle \Big] t^{-k-1} \, dt -$$

$$- \sum_{i=0}^{k-1} \frac{1}{k-i} \sum_{|\alpha|=i} \frac{\langle \omega^\alpha, F(\omega) \rangle}{\alpha!} \, (\partial^\alpha \phi)(0),$$

cf. Petersen [228], p. 40; Horváth [146], p. 175. Generalizations of these formulas to the quasihomogeneous case can be found in Ortner and Wagner [219], Prop. 2.2.1, p. 35.

Let us finally mention that each homogeneous distribution in $\mathcal{D}'(\mathbf{R}^n \setminus \{0\})$ can be cast in the form $F \cdot |x|^\lambda$ with a unique $F \in \mathcal{D}'(\mathbf{S}^{n-1})$, see Gårding [89], Lemmes 1.5, 4.1, pp. 393, 400; Ortner and Wagner [219], Thm. 2.5.1, p. 58, and Section 5.1 below.   □

Example 1.4.10 Let us investigate now the special case of the distributional gradient of a distribution which is homogeneous of the degree $1 - n$.

If $U = G \cdot |x|^{-n}$, $G \in \mathcal{D}'(\mathbf{S}^{n-1})$, see (1.4.3), then U is homogeneous of degree $-n$ in $\mathbf{R}^n \setminus \{0\}$, and

$$\forall c > 0 : U(cx) = c^{-n} U + c^{-n} (\log c) \langle 1, G \rangle \, \delta,$$

see Example 1.2.10 for the case of $G \in L^1(\mathbf{S}^{n-1})$. Hence U is homogeneous in $\mathbf{R}^n$ if and only if the mean-value zero condition $\langle 1, G \rangle = 0$ holds, i.e., iff $\lambda \mapsto G \cdot |x|^\lambda$ is holomorphic in $\lambda = -n$. In this case,

$$G \cdot |x|^{-n} = \Pf_{\lambda=-n} G \cdot |x|^\lambda = \mathrm{vp}(G \cdot |x|^{-n}),$$

cf. Examples 1.2.6 and 1.4.9.

Let us now consider the distribution $T = F \cdot |x|^{1-n}$, $F \in \mathcal{D}'(\mathbf{S}^{n-1})$, which is homogeneous of degree $1 - n$ in $\mathbf{R}^n$. Its gradient ∇T is homogeneous of degree $-n$ on $\mathbf{R}^n \setminus \{0\}$, and hence $\nabla T = G \cdot |x|^{-n}$ holds in $\mathcal{D}'(\mathbf{R}^n \setminus \{0\})$ for some $G \in \mathcal{D}'(\mathbf{S}^{n-1})^n$, cf. the remark at the end of Example 1.4.9. From Proposition 1.3.15 and Example 1.3.16, we conclude that

$$\exists c \in \mathbf{C}^n : \nabla T = G \cdot |x|^{-n} + c\delta \text{ in } \mathcal{D}'(\mathbf{R}^n).$$

Since T and ∇T are even homogeneous on $\mathbf{R}^n$, G must satisfy the mean-value zero condition $\langle 1, G \rangle = 0$. Therefore, $G \cdot |x|^{-n} = \mathrm{vp}(G \cdot |x|^{-n})$. Furthermore, in case

$F \in \mathcal{C}(\mathbf{S}^{n-1})$, the jump formula (1.3.9) implies

$$\nabla T = \nabla(F \cdot |x|^{1-n}) = \nabla \lim_{\epsilon \searrow 0}\big(F \cdot |x|^{1-n} Y(|x| - \epsilon)\big)$$

$$= \lim_{\epsilon \searrow 0}\big[G \cdot |x|^{-n} Y(|x| - \epsilon) + S_{\epsilon \mathbf{S}^{n-1}}\big(\epsilon^{-n} x F(\tfrac{x}{|x|})\big)\big] = \mathrm{vp}(G \cdot |x|^{-n}) + \langle \omega, F\rangle \delta.$$

In fact, for $\phi \in \mathcal{D}(\mathbf{R}^n)$,

$$\lim_{\epsilon \searrow 0}\langle \phi, S_{\epsilon \mathbf{S}^{n-1}}\big(\epsilon^{-n} x F(\tfrac{x}{|x|})\big)\rangle = \lim_{\epsilon \searrow 0} \epsilon^{-n} \int_{\epsilon \mathbf{S}^{n-1}} \phi(x) x F(\tfrac{x}{\epsilon})\, d\sigma(x)$$

$$= \lim_{\epsilon \searrow 0} \epsilon^{-n} \int_{\mathbf{S}^{n-1}} \phi(\epsilon \omega)\epsilon \omega F(\omega)\, \epsilon^{n-1} d\sigma(\omega) = \phi(0)\langle \omega, F(\omega)\rangle.$$

Of course, the formula

$$\nabla(F \cdot |x|^{1-n}) = \mathrm{vp}(G \cdot |x|^{-n}) + \langle \omega, F\rangle \delta \quad \text{in } \mathcal{D}'(\mathbf{R}^n), \tag{1.4.8}$$

where $\nabla(F \cdot |x|^{1-n}) = G \cdot |x|^{-n}$ in $\mathcal{D}'(\mathbf{R}^n \setminus \{0\})$,

persists for all $F \in \mathcal{D}'(\mathbf{S}^{n-1})$ by density, cf. Schwartz [246], p. 166, and Proposition 1.5.14 below. This generalizes Petersen [228], Thm. 15.8, p. 42.

The most notable special cases of (1.4.8) are $\nabla|x|^{1-n} = (1-n)\,\mathrm{vp}(x|x|^{-1-n})$ and

$$\nabla\nabla^T |x|^{2-n} = (2-n)\nabla(x^T |x|^{-n}) = (2-n)\,\mathrm{vp}\Big(\frac{|x|^2 I_n - n x x^T}{|x|^{n+2}}\Big) + \frac{2-n}{n}|\mathbf{S}^{n-1}| I_n \delta,$$

cf. Example 1.3.14 (a). The last equation implies, in particular,

$$\Delta_n(|x|^{2-n}) = \mathrm{tr}(\nabla\nabla^T |x|^{2-n}) = \mathrm{tr}\big(\tfrac{2-n}{n}|\mathbf{S}^{n-1}| I_n \delta\big) = (2-n)|\mathbf{S}^{n-1}|\delta,$$

in accordance with Example 1.3.14 (a). □

Example 1.4.11 As a concrete example for the analytic continuation of fundamental solutions with respect to parameters, let us deduce a fundamental solution of $\Delta_n + \lambda$, $\lambda \in \mathbf{C} \setminus \{0\}$, from that for $\lambda > 0$ which was constructed in Example 1.3.14 (c).

Let us start from the fundamental solution

$$F(\lambda) = -\mathrm{i}\, d_n(\lambda)|x|^{1-n/2} H^{(1)}_{n/2-1}(\sqrt{\lambda}\,|x|), \quad d_n(\lambda) = \frac{\lambda^{n/4-1/2}}{2^{1+n/2}\pi^{n/2-1}}, \quad \lambda > 0,$$

which differs from the fundamental solution $E(\lambda) = d_n(\lambda)|x|^{1-n/2} N_{n/2-1}(\sqrt{\lambda}\,|x|)$ verified in Example 1.3.14 (c) by the solution $-\mathrm{i}\, d_n(\lambda)|x|^{1-n/2} J_{n/2-1}(\sqrt{\lambda}\,|x|)$ of the homogeneous equation.

Both distributions $E(\lambda)$ and $F(\lambda)$ depend holomorphically on the parameter $\lambda \in \mathbf{C} \setminus (-\infty, 0]$ if $\sqrt{\lambda}$ and $\lambda^{n/4-1/2}$ are defined as usually on $U = \mathbf{C} \setminus (-\infty, 0]$. By analytic continuation, the equations $(\Delta_n + \lambda)E(\lambda) = (\Delta_n + \lambda)F(\lambda) = \delta$ carry over from positive λ to the whole slit plane U, and hence $E(\lambda), F(\lambda)$ are fundamental solutions of $\Delta_n + \lambda$ for $\lambda \in U$.

For $F(\lambda)$, we can more easily perform the limit on the branch cut $(-\infty, 0)$. If $\lambda_0 < 0$, then Gradshteyn and Ryzhik [113], Eq. 8.407.1, p. 952, yields

$$\lim_{\epsilon \searrow 0}(\lambda_0 + i\epsilon)^{n/4-1/2}H^{(1)}_{n/2-1}(\sqrt{\lambda_0 + i\epsilon}\,|x|) = |\lambda_0|^{n/4-1/2}\,e^{i\pi(n-2)/4}H^{(1)}_{n/2-1}(i\,\sqrt{|\lambda_0|}\,|x|)$$

$$= \frac{2|\lambda_0|^{n/4-1/2}}{i\pi}K_{n/2-1}(\sqrt{|\lambda_0|}\,|x|),$$

and hence

$$G(\mu) := -\frac{|x|^{1-n/2}\mu^{n/4-1/2}}{(2\pi)^{n/2}}\,K_{n/2-1}(\sqrt{\mu}\,|x|)$$

is a fundamental solution of $\Delta_n - \mu$, $\mu > 0$.

Either by applying the same process of analytic continuation to the fundamental solution of $(\Delta_n + \lambda)^k$, $\lambda > 0$, calculated in Example 1.4.3, or by differentiation with respect to μ according to Proposition 1.4.2, we deduce that

$$\frac{1}{(k-1)!}\frac{\partial^{k-1}}{\partial\mu^{k-1}}G(\mu) = \frac{(-1)^k|x|^{k-n/2}\mu^{n/4-k/2}}{2^{n/2+k-1}\pi^{n/2}(k-1)!}\,K_{n/2-k}(\sqrt{\mu}\,|x|) \qquad (1.4.9)$$

is a fundamental solution of $(\Delta_n - \mu)^k$, $\mu > 0$, $k \in \mathbf{N}$. The expression in (1.4.9) coincides with formula (VII, 10; 15) in Schwartz [246], where the same method of analytic continuation combined with a limit on the boundary is used in reverse order, i.e., in passing from $(\Delta_n - \mu)^k$ to $(\Delta_n + \lambda)^k$ for $\lambda, \mu > 0$.   □

Example 1.4.12

(a) Let $A = (a_{ij}) = A^T \in \mathbf{R}^{n \times n}$ be a symmetric, positive definite matrix. By linear transformation, we can derive a fundamental solution of the elliptic operator $\nabla^T A \nabla = \sum_{i,j=1}^n a_{ij}\partial_i\partial_j$ from the one of Δ_n we considered in Example 1.3.14 (a), i.e.,

$$E = \begin{cases} \dfrac{1}{2\pi}\log|x|, \text{ if } n = 2, \\ c_n|x|^{2-n}, \quad \text{ if } n \neq 2 \end{cases}, \qquad c_n = \frac{\Gamma(\frac{n}{2})}{(2-n)2\pi^{n/2}}.$$

By Proposition 1.3.19, it follows from $x^T A x = |x|^2 \circ \sqrt{A}$ that $(\det A)^{-1/2}E \circ A^{-1/2}$ is a fundamental solution of $\nabla^T A \nabla$. In particular, if $n \in \mathbf{N} \setminus \{2\}$ and

$\lambda_1 > 0, \ldots, \lambda_n > 0$, we obtain that

$$E(\lambda) = c_n(\lambda_1 \cdots \lambda_n)^{-1/2} \Big(\sum_{j=1}^{n} \frac{x_j^2}{\lambda_j} \Big)^{(2-n)/2}$$

is a fundamental solution of $\sum_{j=1}^{n} \lambda_j \partial_j^2$.

Let us suppose now that $n \geq 3$ in order to avoid the appearance of logarithms. The set

$$\Gamma_n = \{\lambda \in \mathbf{C}^n; \forall \xi \in \mathbf{R}^n \setminus \{0\} \sum_{j=1}^{n} \lambda_j \xi_j^2 \neq 0\}$$

$$= \{\lambda \in \mathbf{C}^n; 0 \notin \text{convex hull of } \{\lambda_1, \ldots, \lambda_n\} \text{ in } \mathbf{C}\}$$

is an open cone in $\mathbf{C}^n$, and it is arcwise connected. The distribution-valued function

$$\lambda \longmapsto E(\lambda) = \frac{c_n}{\sqrt{\prod_{j=1}^{n} \lambda_j}} \Big(\sum_{j=1}^{n} \frac{x_j^2}{\lambda_j} \Big)^{(2-n)/2} \in \mathcal{D}'(\mathbf{R}^n)$$

can be continued analytically along any path in Γ_n starting at some $\lambda \in (0, \infty)^n$ and yields a homogeneous fundamental solution of $\sum_{j=1}^{n} \lambda_j \partial_j^2$. Since such a fundamental solution is unique, as one sees by employing the Fourier transformation (see Example 1.6.11 (b) or Proposition 2.4.8 below), $E(\lambda)$ does not depend on the path chosen, i.e., the function

$$\Gamma_n \longrightarrow \mathcal{D}'(\mathbf{R}^n) : \lambda \longmapsto E(\lambda)$$

is well defined and analytic. For example, for the operator $\partial_1^2 + i(\partial_2^2 + \partial_3^2)$, we have $\lambda_0 = (1, i, i)$, and by extending $E(\lambda)$ analytically from $(1, 1, 1)$ to λ_0 along the quarter circle $\lambda = (1, e^{i\varphi}, e^{i\varphi})$, $0 \leq \varphi \leq \frac{\pi}{2}$, we obtain the fundamental solution

$$E(\lambda) = \frac{i}{4\pi \sqrt{x_1^2 - i(x_2^2 + x_3^2)}},$$

where the square root has its usual meaning in the complex right half-plane.

Let us observe, incidentally, that, for *even* n, the function $\lambda \mapsto \sqrt{\lambda_1 \cdots \lambda_n}$ can be extended analytically from $(0, \infty)^n$ to Γ_n since $(\lambda_1, \lambda_2) \mapsto \sqrt{\lambda_1 \lambda_2}$ can be defined analytically on

$$\Gamma_2 = \{\lambda \in \mathbf{C}^2; \forall t \in [0, 1] : t\lambda_1 + (1 - t)\lambda_2 \neq 0\},$$

and since, obviously, $\Gamma_{2k} \subset (\Gamma_2)^k$, $k \in \mathbf{N}$. (In fact, we can set $\sqrt{\lambda_1\lambda_2} = \lambda_1\sqrt{\lambda_2/\lambda_1}$ due to $\lambda_2/\lambda_1 \in \mathbf{C} \setminus (-\infty, 0]$.) In contrast, for n *odd*, such an analytic extension to Γ_n is impossible. However, the functions $\lambda \mapsto (\lambda_1 \cdots \lambda_n \sum_{j=1}^n \frac{x_j^2}{\lambda_j})^{1/2}$ and $\lambda \mapsto E(\lambda)$ can be extended from $(0, \infty)^n$ to Γ_n for *each* $n \in \mathbf{N}$.

(b) Let us next deduce fundamental solutions of the wave operators $\partial_t^2 - \Delta_2$ and $\partial_t^2 - \Delta_3$ from those of the Laplaceans Δ_3 and Δ_4, respectively.

For $z \in \mathbf{C} \setminus (-\infty, 0]$, we set $\lambda = (1, \ldots, 1, z) \in \Gamma_n$, and we consider the limit for $z = -1 + i\epsilon, \epsilon \searrow 0$. For $n = 3$, we have

$$F(z) := E(\lambda) = \frac{-1}{4\pi\sqrt{z|x'|^2 + x_3^2}}, \qquad x' = \begin{pmatrix} x_1 \\ x_2 \end{pmatrix},$$

and

$$\lim_{\epsilon \searrow 0} F(-1 + i\epsilon) = -\frac{Y(x_3^2 - |x'|^2)}{4\pi\sqrt{x_3^2 - |x'|^2}} + i\,\frac{Y(|x'|^2 - x_3^2)}{4\pi\sqrt{|x'|^2 - x_3^2}}.$$

This is a fundamental solution of $\Delta_2 - \partial_3^2$, and, obviously, the same holds for the real part

$$-\frac{Y(x_3^2 - |x'|^2)}{4\pi\sqrt{x_3^2 - |x'|^2}} \in L_{\mathrm{loc}}^1(\mathbf{R}^3).$$

Its support consists of both the forward and the backward light cone, i.e., of $\{x \in \mathbf{R}^3; |x_3| \geq |x'|\}$.

In order to deduce therefrom the "forward" fundamental solution

$$G := \frac{1}{2\pi} \frac{Y(t - |x|)}{\sqrt{t^2 - |x|^2}} \in L_{\mathrm{loc}}^1(\mathbf{R}_{t,x_1,x_2}^3),$$

of the (hyperbolic) wave operator $\partial_t^2 - \Delta_2$, (G being uniquely determined by the condition $G = 0$ for $t < 0$, see Hörmander [138], Thm. 12.5.1, p. 120), we note that we have

$$(\partial_t^2 - \Delta_2)F = \delta, \qquad F := \frac{Y(t^2 - |x|^2)}{4\pi\sqrt{t^2 - |x|^2}},$$

and $F(t, x) = \frac{1}{2}[G(t, x) + G(-t, x)]$. But then the inclusion relations $M := \mathrm{supp}\,(\partial_t^2 - \Delta_2)G \subset \{(t, x) \in \mathbf{R}^3; t \geq |x|\}$ and

$$M = \mathrm{supp}\left[-(\partial_t^2 - \Delta_2)G(-t, x) + 2(\partial_t^2 - \Delta_2)F\right] \subset \{(-t, x) \in \mathbf{R}^3; t \geq |x|\}$$

imply that $(\partial_t^2 - \Delta_2)G$ vanishes outside the origin. From the homogeneity of $(\partial_t^2 - \Delta_2)G$, we conclude that $(\partial_t^2 - \Delta_2)G = c\delta$ for some $c \in \mathbf{R}$. Finally, $c = 1$ follows from $(\partial_t^2 - \Delta_2)G(-t, x) = c\delta(-t, x) = c\delta$ and $(\partial_t^2 - \Delta_2)F = \delta$.

Similarly, for $n = 4$ and $\lambda = (1, 1, 1, z), z \in \mathbf{C} \setminus (-\infty, 0]$, we have

$$F(z) := E(\lambda) = \frac{-1}{4\pi^2(\sqrt{z}\,|x'|^2 + x_4^2/\sqrt{z})}, \qquad x' = (x_1, x_2, x_3).$$

This yields

$$\lim_{\epsilon \searrow 0} F(-1 + i\epsilon) = \frac{i}{4\pi^2} \lim_{\rho \searrow 0} \frac{1}{|x'|^2 - x_4^2 - i\rho} = \frac{i}{4\pi^2}\left(\mathrm{vp}\,\frac{1}{|x'|^2 - x_4^2} + i\pi\delta(|x'|^2 - x_4^2)\right)$$

$$(1.4.10)$$

as a fundamental solution of $\Delta_3 - \partial_4^2$. Outside the origin, Eq. (1.4.10) follows from Sokhotski's formula (1.1.2) $\lim_{\rho \searrow 0}(t - i\rho)^{-1} = \mathrm{vp}\,\frac{1}{t} + i\pi\delta$ in $\mathcal{D}'(\mathbf{R}_t^1)$ and by employing the pull-back

$$h : \mathbf{R}^4 \setminus \{0\} \longrightarrow \mathbf{R} : x \longmapsto t = |x'|^2 - x_4^2,$$

cf. Definition 1.2.12, Proposition 1.2.13. Since $\mathrm{vp}\,\frac{1}{|x'|^2 - x_4^2}, \delta(|x'|^2 - x_4^2) \in \mathcal{D}'(\mathbf{R}^4 \setminus \{0\})$ are homogeneous of degree -2, they can be written in the form $H(\omega) \cdot |x|^{-2}$ (see the remark at the end of Example 1.4.9), and this shows that they can uniquely be extended in $\mathcal{D}'(\mathbf{R}^4)$ as homogeneous distributions of degree-2. For example, for $\delta(|x'|^2 - x_4^2)$, this extension is given by

$$\langle \phi, \delta(|x'|^2 - x_4^2) \rangle = \int_{\mathbf{R}^3} \frac{\phi(x', |x'|) + \phi(x', -|x'|)}{2|x'|}\, dx', \qquad \phi \in \mathcal{D}(\mathbf{R}^4).$$

The limit relation in (1.4.10), which originally holds in $\mathcal{D}'(\mathbf{R}^4 \setminus \{0\})$ only, must persist in $\mathcal{D}'(\mathbf{R}^4)$, since the bijection

$$\mathcal{D}'(\mathbf{S}^{n-1}) \longrightarrow \{T \in \mathcal{D}'(\mathbf{R}^n \setminus \{0\}); T \text{ homogeneous of degree } \lambda\} : H \longmapsto H \cdot |x|^{\lambda}$$

is of course an isomorphism of topological vector spaces, cf. Ortner and Wagner [219], Thm. 2.5.1, p. 58. Hence, if $F(-1 + i\epsilon) = H_\epsilon \cdot |x|^{-2}$ converges in $\mathcal{D}'(\mathbf{R}^4 \setminus \{0\})$, then H_ϵ converges in $\mathcal{D}'(\mathbf{S}^3)$, and $F(-1 + i\epsilon)$ converges also in $\mathcal{D}'(\mathbf{R}^4)$ for $\epsilon \searrow 0$.

Finally, by the same procedure as for $n = 3$, we obtain the forward fundamental solution

$$G = \frac{1}{4\pi t}\delta(t - |x|) \in \mathcal{D}'(\mathbf{R}_{t,x}^4)$$

of the wave operator $\partial_t^2 - \Delta_3$. Here

$$\langle \phi, \frac{1}{4\pi t} \delta(t - |x|) \rangle = \frac{1}{4\pi} \int_{\mathbf{R}^3} \frac{\phi(|x|, x)}{|x|} \, \mathrm{d}x, \qquad \phi \in \mathcal{D}(\mathbf{R}_{t,x}^4). \tag{1.4.11}$$

(This is the unique extension by zero of $\frac{1}{4\pi t} \delta(t - |x|)$, which is defined by pullback in $\mathcal{D}'(\mathbf{R}^4 \setminus \{(t, 0); \, t \in \mathbf{R}\})$.)

We shall take up the calculation of fundamental solutions for the wave operator $\partial_t^2 - \Delta_n$, $n \geq 4$, in Example 1.6.17 below. $\qquad\qquad\qquad\qquad\qquad\qquad \square$

Example 1.4.13 Let us finally deduce a fundamental solution of the *Schrödinger operator* $\partial_t - i\Delta_n$ from that of the heat operator $\partial_t - \Delta_n$, cf. Examples 1.3.14 (d), and 1.6.16 below.

As in Hörmander [139], Section 3.3, let us consider more generally the operator $\partial_t - \nabla^T A \nabla$ for a symmetric matrix $A \in \mathbf{C}^{n \times n}$. First, if $A = A^T \in \mathbf{R}^{n \times n}$ is positive definite, then $|x|^2 \circ A^{1/2} = x^T A x$ and hence we obtain from (1.3.14) the fundamental solution

$$(\det A)^{-1/2} E(t, -) \circ A^{-1/2} = \frac{Y(t)}{(4\pi t)^{n/2} \sqrt{\det A}} \, \mathrm{e}^{-x^T A^{-1} x/(4t)}$$

of the operator $\partial_t - \nabla^T A \nabla$, see Proposition 1.3.19.

Let us next consider the set

$$U = \{A = A^T \in \mathbf{C}^{n \times n}; \, \mathrm{Re}\, A \text{ is positive definite}\},$$

which is an open subset of the linear subspace of all symmetric matrices in $\mathbf{C}^{n \times n}$, and let us show that the mapping

$$U \longrightarrow \mathcal{D}'(\mathbf{R}^{n+1}) : A \longmapsto E_A := \frac{Y(t)}{(4\pi t)^{n/2} \sqrt{\det A}} \, \mathrm{e}^{-x^T A^{-1} x/(4t)}$$

is well defined and holomorphic. In fact, U is convex and hence simply connected, and $\det A \neq 0$ for $A \in U$, since

$$\left(\int_{\mathbf{R}^n} \mathrm{e}^{-x^T A x} \mathrm{d}x \right)^2 \cdot \det A = \pi^n \neq 0$$

holds for positive definite $A \in \mathbf{R}^{n \times n}$ (by linear transformation from the Eulerian integral) and thus on U by analytic continuation. Therefore $\sqrt{\det A}$ and E_A are uniquely defined on U by analytic continuation from their classical values at $A = I_n$. More specifically, we have

$$\sqrt{\det A} = \pi^{n/2} \cdot \left(\int_{\mathbf{R}^n} \mathrm{e}^{-x^T A x} \mathrm{d}x \right)^{-1}$$

for $A \in U$. (Cf. also the method of analytic continuation of $\sqrt{\det A}$ in Hörmander [139], 3.4, p. 85.) For example, for $A = zI_n$, $z \in \mathbf{C}$, $|z| = 1$, $\mathrm{Re}\, z > 0$, we have $\sqrt{\det A} = e^{(n/2)\log z}$ with the usual determination of $\log z$ in the complex right half-plane.

Also, by analytic continuation, the equation $(\partial_t - \nabla^T A \nabla) E_A = \delta$ holds for each $A \in U$. Let us finally extend this equation by continuity to the set

$$U_1 = \{A \in \mathbf{C}^{n \times n};\ A = A^T,\ \det A \neq 0,\ \mathrm{Re}\, A \geq 0 \text{ i.e., } \forall x \in \mathbf{R}^n : x^T (\mathrm{Re}\, A) x \geq 0\}.$$
(1.4.12)

On the one hand, $\sqrt{\det A}$ is still uniquely defined by continuity on U_1, cf. the discussion in Example 1.6.14 below. (In particular, for the Schrödinger operator $\partial_t - i\Delta_n$, we have $A = iI_n = \lim_{\epsilon \searrow 0} (\epsilon + i) I_n$ and hence $\sqrt{\det A} = e^{in\pi/4}$.)

On the other hand, $Y(t) t^{-n/2} e^{-x^T A^{-1} x/(4t)}$ ceases to be a locally integrable function if $\mathrm{Re}\, A$ is no longer positive definite. When approximating A by $A + \epsilon I_n$, $\epsilon \searrow 0$, we obtain for the distributional limit

$$Y(t) t^{-n/2} e^{-x^T A^{-1} x/(4t)} \in \mathcal{C}\big([0,\infty), \mathcal{D}'(\mathbf{R}_x^n)\big) \subset \mathcal{D}'(\mathbf{R}_{t,x}^{n+1}),$$

i.e.,

$$\langle \phi, Y(t) t^{-n/2} e^{-x^T A^{-1} x/(4t)} \rangle = \int_0^\infty \bigg(\int_{\mathbf{R}^n} \frac{\phi(t,x)}{t^{n/2}} e^{-x^T A^{-1} x/(4t)}\, dx \bigg) dt, \quad \phi \in \mathcal{D}(\mathbf{R}^{n+1}).$$

This can be justified by regularization with respect to t, cf. Hörmander [139], Thms. 3.3.4, 3.3.5, p. 82. More easily, this follows by Fourier transformation, see Example 1.6.14 below. In fact, for $t > 0$, we have

$$\int_{\mathbf{R}^n} \phi(t,x) t^{-n/2} e^{-x^T A^{-1} x/(4t)}\, dx = \pi^{-n/2} \sqrt{\det A} \int_{\mathbf{R}^n} \psi(t,y)\, e^{-t y^T A y}\, dy, \quad (1.4.13)$$

where $\psi(t,y) = \mathcal{F}_x(\phi(t,x)) = \int_{\mathbf{R}^n} \phi(t,x) e^{-ixy}\, dx \in \mathcal{S}(\mathbf{R}_{t,y}^{n+1})$. Note that, by Lebesgue's theorem on dominated convergence, the integral on the right-hand side of (1.4.13) continuously depends on $A \in U_1$ and $t \geq 0$.

Summarizing we conclude that

$$E_A = \frac{Y(t)}{(4\pi t)^{n/2} \sqrt{\det A}}\, e^{-x^T A^{-1} x/(4t)} \in \mathcal{C}\big([0,\infty), \mathcal{D}'(\mathbf{R}_x^n)\big) \qquad (1.4.14)$$

yields a fundamental solution of $\partial_t - \nabla^T A \nabla$ for $A \in U_1$. In particular, if $A = iI_n$, we obtain the fundamental solution

$$F = Y(t) e^{-in\pi/4} (4\pi t)^{-n/2} e^{i|x|^2/(4t)} \in \mathcal{C}\big([0,\infty), \mathcal{D}'(\mathbf{R}_x^n)\big)$$

of the Schrödinger operator $\partial_t - i\Delta_n$. Applied to a test function $\phi \in \mathcal{D}(\mathbf{R}^{n+1})$, this means

$$\langle \phi, F \rangle = \frac{e^{-in\pi/4}}{2^n \pi^{n/2}} \int_0^\infty \left(\int_{\mathbf{R}^n} \phi(t,x) e^{i|x|^2/(4t)} \, dx \right) \frac{dt}{t^{n/2}},$$

cf. Schwartz [246], (VII, 10; 31), p. 290, where some signs have to be corrected; Treves and Zerner [275], p. 184; Vladimirov [279], Sect. 10, 12 Ex. (e), p. 156; Rauch [232], Sect. 4.2 (8), p. 138.   □

1.5 Tensor Product and Convolution

Definition 1.5.1 For $\emptyset \neq \Omega_1 \subset \mathbf{R}^m$, $\emptyset \neq \Omega_2 \subset \mathbf{R}^n$ open and $S \in \mathcal{D}'(\Omega_1)$, $T \in \mathcal{D}'(\Omega_2)$, we define the *tensor product* $S \otimes T \in \mathcal{D}'(\Omega_1 \times \Omega_2)$ by

$$\langle \phi, S \otimes T \rangle = \langle \langle \phi(x,y), S_x \rangle, T_y \rangle, \qquad \phi \in \mathcal{D}(\Omega_1 \times \Omega_2).$$

The next proposition will show that $S \otimes T$ indeed belongs to $\mathcal{D}'(\Omega_1 \times \Omega_2)$, and that

$$\langle \phi, S \otimes T \rangle = \langle \langle \phi(x,y), T_y \rangle, S_x \rangle$$

holds as well.

Proposition 1.5.2 *Under the assumptions of* Definition 1.5.1, *the following holds:*

(1) $\forall \phi \in \mathcal{D}(\Omega_1 \times \Omega_2) : (y \mapsto \langle \phi(x,y), S_x \rangle) \in \mathcal{D}(\Omega_2)$;
(2) $S \otimes T \in \mathcal{D}'(\Omega_1 \times \Omega_2)$;
(3) $\forall \phi \in \mathcal{D}(\Omega_1 \times \Omega_2) : \langle \phi, S \otimes T \rangle = \langle \langle \phi(x,y), T_y \rangle, S_x \rangle$.

Proof

(1) If $y_k \to y$ in Ω_2 for $k \to \infty$, then $\phi(x, y_k) \to \phi(x,y)$ in $\mathcal{D}(\Omega_{1x})$ and hence

$$f : \Omega_2 \longrightarrow \mathbf{C} : y \longmapsto \langle \phi(x,y), S_x \rangle$$

is a continuous function. Furthermore, if $y \in \Omega_2$ and $y' = (y_2, \ldots, y_n)$, then

$$\frac{1}{h}[\phi(x, y_1 + h, y') - \phi(x,y)] \to \frac{\partial \phi}{\partial y_1} \quad \text{in} \quad \mathcal{D}(\Omega_{1x})$$

for $h \to 0$ in $\mathbf{R}$, and hence f is differentiable with $\frac{\partial f}{\partial y_j} = \langle \frac{\partial \phi}{\partial y_j}, S_x \rangle$ as partial derivatives. Inductively, we obtain in this way $f \in C^\infty(\Omega_2)$ and $\partial_y^\beta f = \langle \partial_y^\beta \phi, S_x \rangle$ for $\beta \in \mathbf{N}_0^n$.

Finally, if we denote by

$$\mathrm{pr}_2 : \Omega_1 \times \Omega_2 \longrightarrow \Omega_2 : (x, y) \longmapsto y$$

the projection on the second group of variables, then $\mathrm{supp} f \subset \mathrm{pr}_2(\mathrm{supp}\,\phi) \subset \Omega_2$. Therefore f has compact support and belongs to $\mathcal{D}(\Omega_2)$.

(2) The map

$$S \otimes T : \mathcal{D}(\Omega_1 \times \Omega_2) \longrightarrow \mathbf{C} : \phi \longmapsto \langle \langle \phi(x, y), S_x \rangle, T_y \rangle$$

is well-defined by (1) and it is obviously linear.

If $\phi_k \to 0$ in $\mathcal{D}(\Omega_1 \times \Omega_2)$ for $k \to \infty$, then there exists a fixed compact set $K \subset \Omega_1 \times \Omega_2$ such that $\mathrm{supp}\,\phi_k \subset K$ for all $k \in \mathbf{N}$. Denoting, as above, $f_k(y) := \langle \phi_k(x, y), S_x \rangle$ we obtain, for all $k \in \mathbf{N}$, that $\mathrm{supp} f_k \subset \mathrm{pr}_2(K)$, which is a compact subset of Ω_2. In order to show that $f_k \to 0$ in $\mathcal{D}(\Omega_2)$ for $k \to \infty$, we still have to verify that the functions $\partial_y^\beta f_k = \langle \partial_y^\beta \phi_k, S_x \rangle$, $\beta \in \mathbf{N}_0^n$, converge uniformly to 0. Obviously, it is sufficient to consider the case $\beta = 0$, and this follows from the inequality

$$|\langle \phi_k(x, y), S_x \rangle| \leq C \max_{\substack{\alpha \in \mathbf{N}_0^m \\ |\alpha| \leq l}} \max_{x \in \Omega_1} |\partial_x^\alpha \phi_k(x, y)|,$$

where C and l depend on S and $\mathrm{pr}_1(K)$ only, but not on k and y, cf. Proposition 1.5.3 below. Hence

$$\langle \phi_k, S \otimes T \rangle = \langle f_k, T \rangle \to 0 \quad \text{for} \quad k \to \infty$$

and, consequently, $S \otimes T$ defines a distribution in $\mathcal{D}'(\Omega_1 \times \Omega_2)$.

(3) By exchanging the rôles of x and y, we conclude from (2) that the mapping

$$S \tilde{\otimes} T : \mathcal{D}(\Omega_1 \times \Omega_2) \longrightarrow \mathbf{C} : \phi \longmapsto \langle \langle \phi(x, y), T_y \rangle, S_x \rangle$$

also defines a distribution in $\mathcal{D}'(\Omega_1 \times \Omega_2)$. Obviously, $S \otimes T$ and $S \tilde{\otimes} T$ coincide on all test functions of the form $\psi(x)\rho(y)$ with $\psi \in \mathcal{D}(\Omega_1)$ and $\rho \in \mathcal{D}(\Omega_2)$. Let $U = S \otimes T - S \tilde{\otimes} T \in \mathcal{D}'(\Omega_1 \times \Omega_2)$. In order to conclude that U vanishes, it is, by Proposition 1.2.3, enough to show that $\phi_1(x)\phi_2(y)U = 0$ for each $(\phi_1, \phi_2) \in \mathcal{D}(\Omega_1) \times \mathcal{D}(\Omega_2)$. Since $\phi_1(x)\phi_2(y)U \in \mathcal{E}'(\mathbf{R}^{m+n})$, the fact that this distribution vanishes is a consequence of the injectivity of the Fourier transform on $\mathcal{S}' \supset \mathcal{E}'$, see Proposition 1.6.5 below:

$$\mathcal{F}(\phi_1(x)\phi_2(y)U)(\xi, \eta) = \langle \phi_1(x)e^{-ix\xi}\phi_2(y)e^{-iy\eta}, U \rangle = 0. \qquad \square$$

The next proposition characterizes the continuity condition distributions must satisfy (see Definition 1.1.4) by inequalities, cf. Hörmander [139], Thm. 2.1.4, p. 35; Vo-Khac Khoan [282], BC I, Prop., p. 163.

Proposition 1.5.3 *Let* $\emptyset \neq \Omega \subset \mathbf{R}^n$ *open. A linear map* $T : \mathcal{D}(\Omega) \longrightarrow \mathbf{C}$ *belongs to* $\mathcal{D}'(\Omega)$ *if and only if*

$$\forall K \subset \Omega \ compact : \exists C > 0 : \exists m \in \mathbf{N} : \forall \phi \in \mathcal{D}(\Omega) \ with \ \mathrm{supp} \, \phi \subset K :$$

$$|\langle \phi, T \rangle| \leq C \max_{\substack{\alpha \in \mathbf{N}_0^n \\ |\alpha| \leq m}} \max_{x \in \Omega} |\partial^\alpha \phi(x)| = C \max_{|\alpha| \leq m} \|\partial^\alpha \phi\|_\infty. \qquad (1.5.1)$$

Proof Let us suppose first that T fulfills condition (1.5.1). If $\phi_k \to 0$ in $\mathcal{D}(\Omega)$ for $k \to \infty$, then the supports of ϕ_k, $k \in \mathbf{N}$, are contained in a compact set $K \subset \Omega$ and hence $\lim_{k\to\infty} \langle \phi_k, T \rangle = 0$ by (1.5.1), since the derivatives $\partial^\alpha \phi_k$ converge uniformly to 0 for $k \to \infty$. Hence $T \in \mathcal{D}'(\Omega)$ holds.

Conversely, let us take $T \in \mathcal{D}'(\Omega)$ and assume that (1.5.1) does not hold for some compact $K \subset \Omega$. Then there exist $\psi_k \in \mathcal{D}(\Omega)$ with $\mathrm{supp} \, \psi_k \subset K$ and

$$\forall k \in \mathbf{N} : |\langle \psi_k, T \rangle| > k \underbrace{\max_{|\alpha| \leq k} \|\partial^\alpha \psi_k\|_\infty}_{=: a_k}.$$

Since ψ_k does not vanish identically, we have $a_k > 0$, we can define $\phi_k = \psi_k/a_k$, and we obtain a contradiction from $|\langle \phi_k, T \rangle| > 1$ and $\max_{|\alpha| \leq k} \|\partial^\alpha \phi_k\|_\infty = \frac{1}{k}$, which implies $\phi_k \to 0$ in $\mathcal{D}(\Omega)$. $\qquad \square$

In the next example, we will see that the tensor product of distributions is consistent with the usual tensor product in the case of locally integrable functions.

Example 1.5.4 For $f \in L^1_{\mathrm{loc}}(\Omega_1), g \in L^1_{\mathrm{loc}}(\Omega_2)$, we have $T_f \otimes T_g = T_{f(x)g(y)}$ since

$$\langle \phi, T_f \otimes T_g \rangle = \int_{\Omega_2} \left(\int_{\Omega_1} \phi(x, y) f(x) \, \mathrm{d}x \right) g(y) \, \mathrm{d}y = \int_{\Omega_1 \times \Omega_2} \phi(x, y) f(x) g(y) \, \mathrm{d}x \mathrm{d}y$$

for $\phi \in \mathcal{D}(\Omega_1 \times \Omega_2)$ by Fubini's theorem.

Another obvious example of a tensor product of distributions is the following:

$$\delta_{x_0} \otimes \delta_{y_0} = \delta_{(x_0, y_0)} \in \mathcal{D}'(\mathbf{R}^{m+n}) \quad \text{if} \quad x_0 \in \mathbf{R}^m, y_0 \in \mathbf{R}^n. \qquad \square$$

Example 1.5.5

(a) If, as in Definition 1.5.1, $\emptyset \neq \Omega_1 \subset \mathbf{R}^m$, $\emptyset \neq \Omega_2 \subset \mathbf{R}^n$ open, $S \in \mathcal{D}'(\Omega_1)$, $T \in \mathcal{D}'(\Omega_2)$, and $\alpha \in \mathbf{N}_0^m$, $\beta \in \mathbf{N}_0^n$, then, obviously,

$$\partial^{(\alpha, \beta)}(S \otimes T) = \partial^\alpha S \otimes \partial^\beta T.$$

Therefore, for $\alpha \in \mathbf{N}^n$, a fundamental solution of ∂^α in $\mathbf{R}^n$ is

$$\frac{t^{\alpha_1-1}}{(\alpha_1-1)!} Y(t) \otimes \cdots \otimes \frac{t^{\alpha_n-1}}{(\alpha_n-1)!} Y(t) = \frac{x^{\alpha-e}}{(\alpha-e)!} Y(x_1) \ldots Y(x_n), \quad e = (1, \ldots, 1).$$

(b) In particular, the operator $(\partial_1 \partial_2)^m$ in $\mathbf{R}^2$ (with $m \in \mathbf{N}$) has the fundamental solution

$$E = \frac{(x_1 x_2)^{m-1}}{(m-1)!^2} Y(x_1) Y(x_2).$$

If the linear transformation $A : \mathbf{R}^2 \longrightarrow \mathbf{R}^2$ is determined by the matrix $A = \begin{pmatrix} 1 & c \\ 1 & -c \end{pmatrix}$, $c > 0$, then $x_1 x_2 \circ A = x_1^2 - c^2 x_2^2$, and hence, by Proposition 1.3.19, the *iterated one-dimensional wave operator* $(\partial_1^2 - c^2 \partial_2^2)^m$ has the fundamental solution

$$\frac{1}{|\det A|} E \circ A^{-1T} = \frac{1}{c} \frac{(x_1^2 - \frac{x_2^2}{c^2})^{m-1}}{2^{2m-1}(m-1)!^2} Y\left(x_1 - \frac{|x_2|}{c}\right).$$

In the traditional physical variables (t, x) instead of (x_1, x_2), we obtain that

$$F = \frac{(t^2 - \frac{x^2}{c^2})^{m-1}}{c \cdot 2^{2m-1}(m-1)!^2} Y\left(t - \frac{|x|}{c}\right)$$

is a fundamental solution of $(\partial_t^2 - c^2 \partial_x^2)^m$, $m \in \mathbf{N}$. It is uniquely determined by the condition

$$\operatorname{supp} F \subset \{(t, x) \in \mathbf{R}^2; \, t \geq 0\},$$

since $(\partial_t^2 - c^2 \partial_x^2)^m$ is hyperbolic with respect to $(1, 0)$, cf. Hörmander [138], Thm. 12.5.1, p. 120, or Prop. 2.4.11 below. $\square$

The *classical convolution* of absolutely integrable functions on $\mathbf{R}^n$ is defined by

$$(f * g)(x) = \int_{\mathbf{R}^n} f(x - y) g(y) \, dy$$

and renders $L^1(\mathbf{R}^n)$ a Banach algebra. In order to generalize this convolution to distributions, let us apply the distribution T_{f*g} to a test function $\phi \in \mathcal{D}(\mathbf{R}^n)$:

$$\langle \phi, T_{f*g} \rangle = \int_{\mathbf{R}^n} \phi(x)(f * g)(x) \, dx = \int_{\mathbf{R}^{2n}} \phi(x) f(x - y) g(y) \, dxdy$$

$$= \int_{\mathbf{R}^{2n}} \phi(x + y) f(x) g(y) \, dxdy = \langle 1, \phi^\Delta (T_f \otimes T_g) \rangle,$$

where we used Fubini's theorem, a linear transformation, and the abbreviation $\phi^\Delta(x, y) = \phi(x + y)$, $x, y \in \mathbf{R}^n$. For arbitrary distributions $S, T \in \mathcal{D}'(\mathbf{R}^n)$, the product $\phi^\Delta(S \otimes T) \in \mathcal{D}'(\mathbf{R}^{2n})$ is always well defined, but it can be applied to the constant function 1 on $\mathbf{R}^{2n}$ only if it belongs to the space $\mathcal{D}'_{L^1}(\mathbf{R}^{2n})$ of integrable distributions.

Definition 1.5.6 The space of *integrable distributions* is defined by

$$\mathcal{D}'_{L^1} = \mathcal{D}'_{L^1}(\mathbf{R}^n) = \left\{ \sum_{|\alpha| \le m} \partial^\alpha f_\alpha; f_\alpha \in L^1(\mathbf{R}^n), \; m \in \mathbf{N}_0 \right\}.$$

Example 1.5.7 Apparently, $L^1(\mathbf{R}^n) \subset \mathcal{D}'_{L^1}(\mathbf{R}^n) \subset \mathcal{D}'(\mathbf{R}^n)$. However, note that $\mathcal{D}'_{L^1}$ also contains non-integrable functions as, e.g., $e^{ix^2} \in \mathcal{C}^\infty(\mathbf{R}^n)$. In fact, the equation

$$\left(\frac{e^{ix^2} - 1}{x^2} \right)'' = -4 \, e^{ix^2} + 6 \, \frac{e^{ix^2} - 1 - ix^2 e^{ix^2}}{x^4}$$

shows that $e^{ix^2} = f_1 + f_2''$ with $f_j \in L^1(\mathbf{R})$, $j = 1, 2$.

We shall see in Proposition 1.5.8 below that distributions in $\mathcal{D}'_{L^1}$ can always be "integrated," i.e., applied to the constant function 1. For example, for e^{ix^2}, the above representation yields

$$\langle 1, e^{ix^2} \rangle = \frac{3}{2} \int_{-\infty}^{\infty} \frac{e^{ix^2} - 1 - ix^2 e^{ix^2}}{x^4} \, dx.$$

By contour integration and partial integration, we obtain

$$\frac{3}{2} \int_{-\infty}^{\infty} \frac{e^{ix^2} - 1 - ix^2 e^{ix^2}}{x^4} \, dx = -3\sqrt{i} \int_0^\infty \frac{e^{-t^2}(1 + t^2) - 1}{t^4} \, dt$$

$$= -\frac{3\sqrt{i}}{2} \int_0^\infty [e^{-s}(1 + s) - 1] s^{-5/2} \, ds = \sqrt{i} \int_0^\infty e^{-s} \frac{ds}{\sqrt{s}} = \frac{1 + i}{\sqrt{2}} \sqrt{\pi},$$

which value coincides of course with the well-known Fresnel integral $\int_{-\infty}^{\infty} e^{ix^2} \, dx$. $\square$

Definition and Proposition 1.5.8 *For $1 \le p \le \infty$ let*

$$D_{L^p} = D_{L^p}(\mathbf{R}^n) = \{ \phi \in \mathcal{E}(\mathbf{R}^n); \; \forall \alpha \in \mathbf{N}_0^n : \partial^\alpha \phi \in L^p(\mathbf{R}^n) \}.$$

If $T \in \mathcal{D}'_{L^1}$ with $T = \sum_{|\alpha| \le m} \partial^\alpha f_\alpha$, $f_\alpha \in L^1(\mathbf{R}^n)$, and $\phi \in D_{L^\infty}(\mathbf{R}^n)$, then the expression

$$_{D_{L^\infty}}\langle \phi, T \rangle_{\mathcal{D}'_{L^1}} = \sum_{|\alpha| \le m} (-1)^{|\alpha|} \int_{\mathbf{R}^n} f_\alpha(x) \cdot \partial^\alpha \phi(x) \, dx$$

does not depend on the choice of the representation of T.

Proof We have to show that $\sum_{|\alpha|\leq m} \partial^\alpha f_\alpha = 0$ in $\mathcal{D}'(\mathbf{R}^n)$ for $f_\alpha \in L^1(\mathbf{R}^n)$, i.e.,

$$\forall \phi \in \mathcal{D}(\mathbf{R}^n): \sum_{|\alpha|\leq m} (-1)^{|\alpha|} \int_{\mathbf{R}^n} f_\alpha(x) \cdot \partial^\alpha \phi(x)\, dx = 0,$$

implies the same equation for all $\phi \in \mathcal{D}_{L^\infty}(\mathbf{R}^n)$. But this follows from Lebesgue's theorem on dominated convergence if we approximate $\phi \in \mathcal{D}_{L^\infty}(\mathbf{R}^n)$ by the sequence of test functions $\phi_k(x) := \phi(x)\psi(\frac{x}{k}) \in \mathcal{D}(\mathbf{R}^n)$ with fixed $\psi \in \mathcal{D}(\mathbf{R}^n)$ satisfying $\psi = 1$ in a neighborhood of 0:

$$0 = \lim_{k\to\infty} \sum_{|\alpha|\leq m} (-1)^{|\alpha|} \int_{\mathbf{R}^n} f_\alpha \cdot \partial^\alpha \phi_k\, dx = \sum_{|\alpha|\leq m} (-1)^{|\alpha|} \int_{\mathbf{R}^n} f_\alpha \cdot \partial^\alpha \phi\, dx. \qquad \square$$

The Mackey topology τ on $\mathcal{D}_{L^\infty}$ with respect to $\mathcal{D}'_{L^1}$ is the finest locally convex topology which coincides with the Fréchet space topology of $\mathcal{E}$ (see Definition 1.1.1) on bounded subsets of $\mathcal{D}_{L^\infty}$, cf. Schwartz [246], Ch. VI, Sect. 8, p. 203, and $\mathcal{D}'_{L^1}$ is the dual of $\mathcal{D}_{L^\infty}$ equipped with the Mackey topology, i.e., the integrable distributions are just the continuous linear functionals on $(\mathcal{D}_{L^\infty}, \tau)$. (In fact, $\phi_k \to \phi$ in $(\mathcal{D}_{L^\infty}, \tau)$ for $k \to \infty$ is equivalent to the uniform boundedness with respect to k of each derivative $\partial^\alpha \phi_k$ and the convergence of ϕ_k to ϕ in $\mathcal{E}$, and this implies

$$\int_{\mathbf{R}^n} f_\alpha \cdot \partial^\alpha \phi_k\, dx \to \int_{\mathbf{R}^n} f_\alpha \cdot \partial^\alpha \phi\, dx$$

for $f_\alpha \in L^1(\mathbf{R}^n)$. The converse is more involved, cf. Ortner and Wagner [219], Section 1.3, p. 11.)

Furthermore, the sequential convergence in $\mathcal{D}'_{L^1}$ is given by evaluation on $\mathcal{D}_{L^\infty}$, i.e., $T_k \to T$ in $\mathcal{D}'_{L^1}$ for $k \to \infty$ if and only if $\langle \phi, T_k \rangle \to \langle \phi, T \rangle$ for all $\phi \in \mathcal{D}_{L^\infty}$.

In accordance with the formula (see the motivation before Definition 1.5.6)

$$\langle \phi, T_{f*g} \rangle = \langle 1, \phi^\Delta (T_f \otimes T_g) \rangle, \qquad f, g \in L^1, \ \phi \in \mathcal{D},$$

we now define the convolution of distributions.

Definition 1.5.9

(1) Two distributions $S, T \in \mathcal{D}'(\mathbf{R}^n)$ are called *convolvable* if and only if $\phi^\Delta \cdot (S \otimes T) \in \mathcal{D}'_{L^1}(\mathbf{R}^{2n})$ for each $\phi \in \mathcal{D}(\mathbf{R}^n)$. (Herein, $\phi^\Delta \in \mathcal{E}(\mathbf{R}^{2n})$ is defined by $\phi^\Delta(x, y) = \phi(x + y)$.)

(2) The *convolution* $S * T \in \mathcal{D}'(\mathbf{R}^n)$ of two convolvable distributions $S, T \in \mathcal{D}'(\mathbf{R}^n)$ is defined by

$$\langle \phi, S * T \rangle = {}_{\mathcal{D}_{L^\infty}}\langle 1, \phi^\Delta \cdot (S \otimes T) \rangle_{\mathcal{D}'_{L^1}}, \qquad \phi \in \mathcal{D}(\mathbf{R}^n).$$

Let us remark that $\phi_k \to \phi$ in $\mathcal{D}$ implies $\phi_k^\Delta \to \phi^\Delta$ in $\mathcal{D}_{L^\infty}(\mathbf{R}^{2n})$ and hence, by the closed graph theorem, $\phi_k^\Delta \cdot (S \otimes T) \to \phi^\Delta \cdot (S \otimes T)$ in $\mathcal{D}'_{L^1}(\mathbf{R}^{2n})$. This furnishes that $S * T$ is a continuous linear form on $\mathcal{D}(\mathbf{R}^n)$, i.e., $S * T \in \mathcal{D}'(\mathbf{R}^n)$. Furthermore, $S * T = T * S$ is immediate from the definition.

It can also be shown that S, T are convolvable iff $(\phi * \check{S}) \cdot T \in \mathcal{D}'_{L^1}(\mathbf{R}^n)$ for each $\phi \in \mathcal{D}(\mathbf{R}^n)$. (Herein ϕ and $\check{S}$ are convolvable due to Example 1.5.11 below.) For Definition 1.5.9 and the last mentioned property, cf. Schwartz [242], Exp. 22; Horváth [141], p. 381; Horváth [144], Déf., p. 185; Roider [237]; Shiraishi [252], Def. 1, p. 22; Ortner and Wagner [219]; Ortner [204].

Furthermore, from the very definition of convergence in $(\mathcal{D}_{L^\infty}, \tau)$, it follows that S, T are convolvable iff $\lim_{k\to\infty}\langle\phi^\Delta\chi_k, S \otimes T\rangle$ exists for each $\phi \in \mathcal{D}(\mathbf{R}^n)$ and for each *special approximate unit* $(\chi_k)_{k\in\mathbf{N}}$, i.e., a sequence of test functions $\chi_k \in \mathcal{D}(\mathbf{R}^{2n})$, $k \in \mathbf{N}$, such that the set $\{\chi_k; k \in \mathbf{N}\}$ is bounded in $\mathcal{D}_{L^\infty}(\mathbf{R}^{2n})$ and $\forall N > 0 : \exists m \in \mathbf{N} : \forall k \geq m : \chi_k(x) = 1$ for $x \in \mathbf{R}^{2n}$ with $|x| \leq N$. In this case, the above limit does not depend on the choice of χ_k and yields $\langle\phi, S * T\rangle$, cf. Vladimirov [279], Ch. 2, 7.4, pp. 102–105; [280], 4.1, pp. 59–63; Dierolf and Voigt [58], Thm. 1.3, p. 190.

Proposition 1.5.10 *Let $S, T \in \mathcal{D}'(\mathbf{R}^n)$ be convolvable and $P(\partial) = \sum_{|\alpha|\leq m} a_\alpha \partial^\alpha$ be a differential operator with constant coefficients. Then the distributions $P(\partial)S, T$ and $S, P(\partial)T$ are convolvable and*

$$P(\partial)(S * T) = \big(P(\partial)S\big) * T = S * \big(P(\partial)T\big).$$

In particular, if E is a fundamental solution of $P(\partial)$, i.e., $P(\partial)E = \delta$, then $U = E * T$ solves the inhomogeneous equation $P(\partial)U = T$ if E and T are convolvable.

Proof The convolvability of $\partial_j S$ and T follows from the equation

$$\phi^\Delta(\partial_j S \otimes T) = \partial_j\big(\phi^\Delta(S \otimes T)\big) - (\partial_j\phi)^\Delta(S \otimes T),$$

$j = 1, \ldots, n$. This also yields

$$\langle\phi, (\partial_j S) * T\rangle = \langle 1, \phi^\Delta(\partial_j S \otimes T)\rangle$$

$$= \langle 1, \partial_j\big(\phi^\Delta(S \otimes T)\big)\rangle - \langle\partial_j\phi, S * T\rangle = \langle\phi, \partial_j(S * T)\rangle,$$

since $\langle 1, \partial_j V\rangle = 0$ for $V \in \mathcal{D}'_{L^1}$ by 1.5.8. $\square$

Example 1.5.11 The condition $\phi^\Delta \cdot (S \otimes T) \in \mathcal{D}'_{L^1}(\mathbf{R}^{2n})$ is satisfied if $\phi^\Delta \cdot (S \otimes T) \in \mathcal{E}'(\mathbf{R}^{2n})$, since $\mathcal{E}' \subset \mathcal{D}'_{L^1}$. (This inclusion is a consequence of the continuity of the imbedding $(\mathcal{D}_{L^\infty}, \tau) \hookrightarrow \mathcal{E}$, cf. Definition 1.5.8, or, alternatively, of the structure theorem for distributions with compact support, see Schwartz [246], Ch. III, Thm. XXVI, p. 91. In a similar vein, $\mathcal{M}^1 \subset \mathcal{D}'_{L^1}$ since integrable Radon measures also yield continuous linear functionals on $(\mathcal{D}_{L^\infty}, \tau)$.)

If $\phi^\Delta \cdot (S \otimes T) \in \mathcal{E}'(\mathbf{R}^{2n})$ for each $\phi \in \mathcal{D}(\mathbf{R}^n)$, we shall sometimes call S, T *convolvable by support.* Since

$$\mathrm{supp}(\phi^\Delta \cdot (S \otimes T)) = (\mathrm{supp}\, S \times \mathrm{supp}\, T) \cap \{(x,y) \in \mathbf{R}^{2n}; x + y \in \mathrm{supp}\,\phi\},$$

we conclude that S, T are convolvable by support if and only if the mapping

$$F : \mathrm{supp}\, S \times \mathrm{supp}\, T \longrightarrow \mathbf{R}^n : (x,y) \longmapsto x + y$$

is proper. For example, this is the case if one of the distributions S, T belongs to $\mathcal{E}'$, or if one of them has its support contained in the half-space $\{x \in \mathbf{R}^n; x_n \geq 0\}$ and the other one in the cone $\{x \in \mathbf{R}^n; x_n \geq c|x'|\}$ for some $c > 0$ and with the abbreviation $x' = (x_1, \dots, x_{n-1})$.

Note that, generally,

$$\mathrm{supp}\,(S * T) \subset \overline{\mathrm{supp}\, S + \mathrm{supp}\, T} \tag{1.5.2}$$

if S, T are convolvable. In contrast, if S, T are convolvable by support, then the mapping F above must be closed (as a proper mapping between locally compact spaces), and hence its image is closed and $\mathrm{supp}\,(S * T) \subset \mathrm{supp}\, S + \mathrm{supp}\, T$, cf. Horváth [141], Ch. V, Sect. 9, Lemma 1, p. 385. The two distributions $S = 1 \otimes \delta, T = x_2 \delta(x_1 x_2 - 1) \in \mathcal{D}'(\mathbf{R}^2)$ provide an example of convolvable distributions, which are not convolvable by support, and where

$$\mathrm{supp}\,(S * T) = \mathrm{supp}\,(\mathrm{sign}\, x_2) = \mathbf{R}^2 \supsetneq \mathrm{supp}\, S + \mathrm{supp}\, T = \mathbf{R} \times (\mathbf{R} \setminus \{0\}).$$

Indeed, for $\phi \in \mathcal{D}(\mathbf{R}^2)$, we have $\mu = \phi^\Delta \cdot (S \otimes T) \in \mathcal{M}^1(\mathbf{R}^4) \subset \mathcal{D}'_{L^1}(\mathbf{R}^4)$ since

$$\int_{\mathbf{R}^4} |\mu| = \langle 1_{y_1} \otimes \int_{-\infty}^\infty |\phi(t, y_2)|\, dt, |y_2|\delta(y_1 y_2 - 1)\rangle = \int_{\mathbf{R}^2} |\phi(y)|\, dy < \infty.$$

Similarly, we obtain $S * T = 1 \otimes \mathrm{sign} \in \mathcal{D}'(\mathbf{R}^2)$.

The inclusion (1.5.2) implies in particular that the three spaces $L^1_c \subset \mathcal{M}_c \subset \mathcal{E}'$ are convolution algebras, and the same holds as well for $\mathcal{D}'_\Gamma := \{S \in \mathcal{D}'; \mathrm{supp}\, S \subset \Gamma\}$, where Γ is an acute closed convex cone with vertex in 0.

As we shall see in Example 1.5.13, also the spaces $L^1 \subset \mathcal{M}^1 \subset \mathcal{D}'_{L^1}$ are convolution algebras, and hence all the spaces in the diagram

$$\begin{array}{ccccc} L^1_c & \subset & \mathcal{M}_c & \subset & \mathcal{E}' \\ \cap & & \cap & & \cap \\ L^1 & \subset & \mathcal{M}^1 & \subset & \mathcal{D}'_{L^1} \end{array}$$

are convolution algebras.

A simple concrete example of convolution in the convolution algebra $\mathcal{D}'_{[0,\infty)}(\mathbf{R}) = \{T \in \mathcal{D}'(\mathbf{R}^1); \operatorname{supp} T \subset [0,\infty)\}$ is provided by $x_+^{\lambda-1} * x_+^{\mu-1}$ for $\lambda, \mu \in \mathbf{C}$, cf. Example 1.4.8. These distributions are convolvable by support. If $\operatorname{Re}\lambda > 0$ and $\operatorname{Re}\mu > 0$, then $x_+^{\lambda-1}, x_+^{\mu-1} \in L^1_{\text{loc}}(\mathbf{R})$, and classical integration yields

$$\frac{x_+^{\lambda-1}}{\Gamma(\lambda)} * \frac{x_+^{\mu-1}}{\Gamma(\mu)} = \frac{x_+^{\lambda+\mu-1}}{\Gamma(\lambda+\mu)}. \tag{1.5.3}$$

The distribution-valued function

$$F : \mathbf{C} \longrightarrow \mathcal{D}'_{[0,\infty)}(\mathbf{R}) : \lambda \longmapsto \begin{cases} \dfrac{x_+^{\lambda-1}}{\Gamma(\lambda)}, & \text{if } \lambda \in \mathbf{C} \setminus -\mathbf{N}_0, \\[2mm] \delta^{(k)}, & \text{if } \lambda = -k, \, k \in \mathbf{N}_0, \end{cases}$$

is holomorphic since, by (1.4.1),

$$\lim_{\lambda \to -k} F(\lambda) = \frac{\operatorname{Res}_{\lambda=-k} x_+^{\lambda-1}}{\operatorname{Res}_{\lambda=-k} \Gamma(\lambda)} = \frac{\frac{(-1)^k}{k!}\delta^{(k)}}{\frac{(-1)^k}{k!}} = \delta^{(k)}, \qquad k \in \mathbf{N}_0.$$

By analytic continuation, the convolution equation (1.5.3) persists therefore in the form $F(\lambda) * F(\mu) = F(\lambda + \mu)$ for all $\lambda, \mu \in \mathbf{C}$. Sometimes, we shall express the fact that $F : \mathbf{C} \to \mathcal{D}'_{[0,\infty)}(\mathbf{R})$ is a group homomorphism by saying that $\lambda \mapsto F(\lambda)$ is a *convolution group*. For example, the equation $F(-1) * F(\lambda) = F(\lambda - 1)$, which is equivalent to $\frac{\mathrm{d}}{\mathrm{d}x}F(\lambda) = F(\lambda - 1)$, comprises the original recurrent definition of x_+^λ in Example 1.3.9.

When going back to $x_+^{\lambda-1}, x_+^{\mu-1}$ we conclude that

$$x_+^{\lambda-1} * x_+^{\mu-1} = B(\lambda, \mu)x_+^{\lambda+\mu-1}, \qquad \lambda, \mu \in \mathbf{C} \setminus -\mathbf{N}_0,$$

if B denotes the *beta function*, i.e., Euler's integral of the second kind. In particular, for $0 < \operatorname{Re}\alpha < 1$, *Abel's integral equation*

$$g(x) = \int_0^x \frac{f(y)}{(x-y)^\alpha}\, \mathrm{d}y$$

can be cast in the form $g = \Gamma(1 - \alpha) \cdot f * F(1 - \alpha)$, and hence it has the solution

$$f = \frac{1}{\Gamma(1-\alpha)} F(\alpha - 1) * g = \frac{1}{\Gamma(1-\alpha)} F(\alpha) * g'$$

$$= \frac{x_+^{\alpha-1}}{\Gamma(1-\alpha)\Gamma(\alpha)} * g' = \frac{\sin(\alpha\pi)}{\pi} \int_0^x \frac{g'(y)}{(x-y)^{1-\alpha}}\, \mathrm{d}y$$

where we assume that $g \in \mathcal{C}^1([0,\infty))$ and $g(0) = 0$.

In contrast, if e.g., $\lambda = 0$ and $\mu \in \mathbf{C} \setminus -\mathbf{N}_0$, then

$$x_+^{-1} * x_+^{\mu-1} = \left(\Pr_{\lambda=0} x_+^{\lambda-1}\right) * x_+^{\mu-1} = \Pr_{\lambda=0}\left(B(\lambda,\mu)x_+^{\lambda+\mu-1}\right)$$

$$= \left(\Pr_{\lambda=0} B(\lambda,\mu)\right) \cdot x_+^{\mu-1} + \left(\operatorname*{Res}_{\lambda=0} B(\lambda,\mu)\right) \cdot \frac{\partial}{\partial\lambda} x_+^{\lambda+\mu-1}\Big|_{\lambda=0}$$

$$= \frac{\partial}{\partial\lambda}(\lambda B(\lambda,\mu))\Big|_{\lambda=0} \cdot x_+^{\mu-1} + \log x \cdot x_+^{\mu-1}$$

$$= (\psi(1) - \psi(\mu))x_+^{\mu-1} + \log x \cdot x_+^{\mu-1}. \tag{1.5.4}$$

Here we used the definition

$$\log x \cdot x_+^{\lambda} := \frac{d}{d\lambda} x_+^{\lambda}, \qquad \lambda \in \mathbf{C} \setminus -\mathbf{N},$$

which furnishes a meromorphic function of λ with double poles for $\lambda \in -\mathbf{N}$, cf. Example 1.4.8. As in formula (1.4.2), we also define

$$\log x \cdot x_+^{-k} := \Pr_{\lambda=-k}\left(\log x \cdot x_+^{\lambda}\right), \qquad k \in \mathbf{N}. \tag{1.5.5}$$

An analogous calculation as in Example 1.4.8 then yields, e.g., that $\log x \cdot x_+^{-1} = \frac{d}{dx}\left(\frac{1}{2} Y(x) \cdot \log^2 x\right)$, and

$$\langle\phi, \log x \cdot x_+^{-1}\rangle = \int_0^1 \frac{\phi(x) - \phi(0)}{x} \cdot \log x\, dx + \int_1^\infty \frac{\phi(x)\log x}{x}\, dx, \qquad \phi \in \mathcal{D}(\mathbf{R}).$$

Finally, in order to calculate, e.g., $x_+^{-1} * x_+^{-1}$, one employs the formula

$$\Pr_{z=z_0}\left(f(z)g(z)\right) = \operatorname*{Res}_{z=z_0} f(z) \cdot \Pr_{z=z_0} g'(z) + \Pr_{z=z_0} f(z) \cdot \Pr_{z=z_0} g(z) + \Pr_{z=z_0} f'(z) \cdot \operatorname*{Res}_{z=z_0} g(z)$$

for meromorphic functions f, g having both a simple pole in z_0. Then (1.5.4), (1.5.5), and Gradshteyn and Ryzhik [113], Eq. 8.366.8 imply

$$x_+^{-1} * x_+^{-1} = \Pr_{\mu=0}\left[(\psi(1) - \psi(\mu))x_+^{\mu-1}\right] + \Pr_{\mu=0}\left[\log x \cdot x_+^{\mu-1}\right]$$

$$= \Pr_{\mu=0}\left[\left(\frac{1}{\mu} - \mu \cdot \psi'(1)\right) \cdot \left(\frac{\delta}{\mu} + x_+^{-1} + \mu \cdot \log x \cdot x_+^{-1}\right)\right] + \log x \cdot x_+^{-1}$$

$$= 2\log x \cdot x_+^{-1} - \frac{\pi^2}{6}\delta. \tag{1.5.6}$$

$\square$

Let us define now $\mathcal{D}'_{L^p}$ in analogy with Definition 1.5.6.

Definition 1.5.12 For $1 \le p \le \infty$, we set

$$\mathcal{D}'_{L^p} = \mathcal{D}'_{L^p}(\mathbf{R}^n) = \Big\{ \sum_{|\alpha| \le m} \partial^\alpha f_\alpha; \ f_\alpha \in L^p(\mathbf{R}^n), \ m \in \mathbf{N}_0 \Big\}.$$

Example 1.5.13 For $\frac{1}{p} + \frac{1}{q} \ge 1$, the convolution

$$\mathcal{D}'_{L^p}(\mathbf{R}^n) \times \mathcal{D}'_{L^q}(\mathbf{R}^n) \longrightarrow \mathcal{D}'_{L^r}(\mathbf{R}^n) : (S, T) \longmapsto S * T$$

is well defined if $\frac{1}{p} + \frac{1}{q} = 1 + \frac{1}{r}$, $r \in [1, \infty]$.

In fact, if $\phi \in \mathcal{D}$ and $S \in \mathcal{D}'_{L^p}$, $T \in \mathcal{D}'_{L^q}$, then $S = \sum_{|\alpha| \le m} \partial^\alpha f_\alpha$, $f_\alpha \in L^p$, and hence $(\phi * \check{S}) \cdot T \in \mathcal{D}_{L^p} \cdot \mathcal{D}'_{L^q} \subset \mathcal{D}'_{L^1}$ by Hölder's inequality. Moreover, from Proposition 1.5.10 and $L^p * L^q \subset L^r$ due to Young's inequality, we obtain $S * T \in \mathcal{D}'_{L^r}$.

In particular, $\mathcal{D}'_{L^1}$ is also a convolution algebra, as well as L^1 and $\mathcal{M}^1$, cf. Example 1.5.11.

As a non-trivial example, let us consider the convolution of $x_+^{\lambda-1}$ and $x_-^{\mu-1}$, $\lambda, \mu \in \mathbf{C}$. Just as x_+^λ, also $\lambda \mapsto x_-^\lambda := (x_+^\lambda)^{\vee}$ is meromorphic, has simple poles for $\lambda \in \mathbf{N}$, and

$$x_-^{-k} := \Pf_{\lambda = -k} x_-^\lambda = (x_+^{-k})^{\vee}, \qquad k \in \mathbf{N},$$

cf. (1.4.2). For $\operatorname{Re} \lambda < \frac{1}{2}$ and $\operatorname{Re} \mu < \frac{1}{2}$, the two distributions $x_+^{\lambda-1}$, $x_-^{\mu-1}$ belong to $\mathcal{D}'_{L^2}(\mathbf{R}^1)$ and are therefore convolvable. More generally, $x_+^{\lambda-1}$, $x_-^{\mu-1}$ are convolvable if and only if $\operatorname{Re}(\lambda + \mu) < -1$, see Ortner [199]. For $\operatorname{Re} \lambda, \operatorname{Re} \mu \in (0, \frac{1}{2})$, we have $x_+^{\lambda-1}$, $x_-^{\mu-1} \in L_c^1 + L^2$, and hence the convolution can be calculated classically. This yields

$$\frac{x_+^{\lambda-1}}{\Gamma(\lambda)} * \frac{x_-^{\mu-1}}{\Gamma(\mu)} = \frac{\sin(\lambda\pi)x_+^{\lambda+\mu-1} + \sin(\mu\pi)x_-^{\lambda+\mu-1}}{\Gamma(\lambda+\mu)\sin((\lambda+\mu)\pi)}, \qquad \operatorname{Re} \lambda, \operatorname{Re} \mu \in (0, \tfrac{1}{2}).$$

$$(1.5.7)$$

By analytic continuation, this formula remains true as long as $\operatorname{Re}(\lambda + \mu) < -1$, cf. also Brédimas [22], where (1.5.7) appears as a definition for $\lambda, \mu \notin \mathbf{N}_0$ and $\lambda + \mu \notin \mathbf{Z}$. (Note that $\lambda \mapsto \frac{x_+^{\lambda-1}}{\Gamma(\lambda)}$ and $\mu \mapsto \frac{x_-^{\mu-1}}{\Gamma(\mu)}$ are *entire* functions, and therefore the right-hand side of (1.5.7) depends analytically on λ, μ satisfying $\operatorname{Re}(\lambda + \mu) < -1$.)

Again, as in Example 1.5.11, we can deduce from (1.5.7) a formula for $x_+^{-1} * x_-^{-1}$ in two steps:

$$x_+^{-1} * x_-^{\mu-1} = \Pf_{\lambda=0}\left(x_+^{\lambda-1} * x_-^{\mu-1}\right)$$

$$= \frac{\pi}{\sin(\mu\pi)} x_+^{\mu-1} + \left(\psi(1) - \psi(\mu) - \pi\cot(\mu\pi)\right)x_-^{\mu-1} + \log|x| \cdot x_-^{\mu-1},$$

for Re $\mu < 1$, $\mu \notin -\mathbf{N}_0$, and this implies

$$x_+^{-1} * x_-^{-1} = \Pf_{\mu=0}\left(x_+^{-1} * x_-^{\mu-1}\right) = \log|x| \cdot |x|^{-1} + \frac{\pi^2}{3}\delta, \qquad (1.5.8)$$

where $\log|x| \cdot |x|^{-1} := \Pf_{\lambda=-1}(\log|x| \cdot |x|^\lambda) = \Pf_{\lambda=-1}(\frac{d}{d\lambda}|x|^\lambda)$.

If we combine the formulas (1.5.6) and (1.5.8) for $x_\pm^{-1} * x_\pm^{-1}$, we obtain

$$|x|^{-1} * |x|^{-1} = (x_+^{-1} + x_-^{-1}) * (x_+^{-1} + x_-^{-1}) = 4\log|x| \cdot |x|^{-1} + \frac{\pi^2}{3}\delta$$

in accordance with Wagner [286], Satz 7, p. 478.

Note that the inversion formula for the one-dimensional Hilbert transformation, i.e.,

$$\vp\frac{1}{x} * \vp\frac{1}{x} = -\pi^2\delta,$$

is also an immediate consequence of the above formulas:

$$\vp\frac{1}{x} * \vp\frac{1}{x} = (x_+^{-1} - x_-^{-1}) * (x_+^{-1} - x_-^{-1}) = x_+^{-1} * x_+^{-1} + x_-^{-1} * x_-^{-1} - 2x_+^{-1} * x_-^{-1}$$

$$= 2\log|x| \cdot |x|^{-1} - \frac{\pi^2}{3}\delta - 2\left[\log|x| \cdot |x|^{-1} + \frac{\pi^2}{3}\delta\right] = -\pi^2\delta.$$

$\square$

Similarly to the possibility of extending equations by analytic continuation with respect to parameters (see above), it is also possible to extend equations and operations by *density* from the level of functions to that of distributions. The most direct way to show that distributions can be approximated by $\mathcal{C}^\infty$ functions is by *regularization*, i.e. by convolution with δ-sequences.

Proposition 1.5.14 Let $\psi \in \mathcal{D}(\mathbf{R}^n)$ with $\int \psi(x)\,dx = 1$ and set $\psi_k(x) = k^n\psi(kx)$ for $k \in \mathbf{N}$. Then $\psi_k * T \in \mathcal{E}(\mathbf{R}^n)$ and $\psi_k * T \to T$ in $\mathcal{D}'(\mathbf{R}^n)$ for $T \in \mathcal{D}'(\mathbf{R}^n)$.

Proof

(1) ψ_k and T are convolvable by support, see Example 1.5.11. Furthermore, for $\phi \in \mathcal{D}$, the functions

$$(\phi * \check\psi_k)(x) = \int \phi(x+y)\psi_k(y)\,dy$$

belong to $\mathcal{D}$, and converge in $\mathcal{D}$ to ϕ for $k \to \infty$, since

$$\left|\int \phi(x+y)\psi_k(y)\,dy - \phi(x)\right| \le \int |\phi(x+\tfrac{u}{k}) - \phi(x)| \cdot |\psi(u)|\,du$$

$$\le \frac{1}{k}\|\nabla\phi\|_\infty \cdot \||x|\psi(x)\|_1$$

by the mean value theorem. From this limit in $\mathcal{D}$ follows $\psi_k * T \to T$ in $\mathcal{D}'$ due to

$$\langle \phi, \psi_k * T \rangle = {}_{\mathcal{E}}\langle \phi^{\Delta}, T \otimes \psi_k \rangle_{\mathcal{E}'} = \langle \phi * \check{\psi}_k, T \rangle \to \langle \phi, T \rangle, \qquad k \to \infty.$$

(2) Let us finally show that $(h * T)(x) = \langle h(x - y), T_y \rangle$ for $h \in \mathcal{D}$, which implies $h * T \in \mathcal{E}$ by the proof of Proposition 1.5.2.

Let $\phi \in \mathcal{D}$ and take $\chi \in \mathcal{D}$ such that $\chi = 1$ in a neighborhood of $\operatorname{supp} \phi - \operatorname{supp} h$. Then Definition 1.5.9 and Proposition 1.5.2 imply

$$\langle \phi, h * T \rangle = {}_{\mathcal{E}}\langle 1, \phi^{\Delta}(h \otimes T) \rangle_{\mathcal{E}'} = {}_{\mathcal{D}}\langle \phi^{\Delta}\chi(y), h(x) \otimes T_y \rangle_{\mathcal{D}'}$$

$$= \langle \int \phi(x + y) h(x) \chi(y) \, dx, T_y \rangle = \langle \chi(y) \int \phi(u) h(u - y) \, du, T_y \rangle$$

$$= \langle h(x - y)\chi(y), \phi(x) \otimes T_y \rangle = \langle \langle h(x - y)\chi(y), T_y \rangle, \phi(x) \rangle$$

$$= \langle \langle h(x - y), T_y \rangle, \phi(x) \rangle,$$

and thus $(h * T)(x) = \langle h(x - y), T_y \rangle$. $\square$

Let us turn now to examples of convolution products in several variables.

Example 1.5.15 Since the C^{∞} mapping

$$h : \mathbf{R}^n \setminus \{0\} \longrightarrow (0, \infty) : x \longmapsto |x|$$

is submersive, $T(|x|) := h^* T$ is well defined for $T \in \mathcal{D}'((0, \infty))$, see Definition 1.2.12, and it is obvious that $T_1 = T(|x|)$ is *radially symmetric*, i.e., $T_1 \circ A = T_1$ for all orthogonal linear mappings $A : \mathbf{R}^n \to \mathbf{R}^n$. (Conversely, it can be shown that each radially symmetric distribution in $\mathcal{D}'(\mathbf{R}^n \setminus \{0\})$ is a pull-back of some $T \in \mathcal{D}'((0, \infty))$.)

Let us consider now the convolution of $f(|x|), g(|x|)$ assuming first that $f, g \in L_c^1((0, \infty))$. For $n \geq 2$, we obtain the following:

$$f(|x|) * g(|x|) = \int_{\mathbf{R}^n} f(|x - \xi|) g(|\xi|) \, d\xi$$

$$= |\mathbf{S}^{n-2}| \int_0^{\infty} \int_0^{\pi} f(\sqrt{|x|^2 + \sigma^2 - 2|x|\sigma \cos \theta}) g(\sigma) \sigma^{n-1} \sin^{n-2} \theta \, d\theta d\sigma$$

$$= \frac{2^{n-3} |\mathbf{S}^{n-2}|}{|x|^{n-2}} \int_0^{\infty} \int_0^{\infty} \rho\sigma f(\rho) g(\sigma) \Delta(|x|, \rho, \sigma)^{n-3} \, d\rho d\sigma,$$

$$(1.5.9)$$

where $\Delta(|x|, \rho, \sigma)$ denotes the area of a triangle with side lengths $|x|, \rho, \sigma$ if such a triangle exists, and $\Delta(|x|, \rho, \sigma) = 0$ otherwise, i.e.,

$$\Delta(|x|, \rho, \sigma) = \frac{1}{4} Y(\rho + \sigma - |x|) Y(|x| - |\rho - \sigma|) \times$$

$$\times \sqrt{(\rho + \sigma - |x|)(|x| + \rho - \sigma)(|x| + \sigma - \rho)(|x| + \rho + \sigma)},$$

cf. John [151], Ch. IV, p. 80; Trimèche [277], Ch. 2, p. 90.

Since the pull-back is continuous, see Proposition 1.2.13, an approximation of δ_ρ, $\rho > 0$ by functions in $\mathcal{D}((0, \infty)) \subset L^1_c((0, \infty))$ implies the following formula for the convolution of the spherical layers $\delta_\rho \circ |x| = \delta(|x| - \rho) = S_{\rho S^{n-1}}(1)$ (cf. Example 1.2.14):

$$\delta(|x| - \rho) * \delta(|x| - \sigma) = \frac{2^{n-3} \rho \sigma |S^{n-2}|}{|x|^{n-2}} \Delta(|x|, \rho, \sigma)^{n-3} \in L^1_{loc}(\mathbf{R}^n), \qquad \rho, \sigma > 0,$$

cf. Hörmander [137], Ex. 4.2.8, pp. 381, 404; Ortner and Wagner [206], p. 585.

Let us deduce now from (1.5.9) a solution of $\Delta_n U = \delta(|x| - \rho)$, $\rho > 0$. By Proposition 1.5.10, we obtain such a solution U by setting $U = E * \delta(|x| - \rho)$. For $n = 3$, U can be physically interpreted as the electrostatic potential of a uniformly charged sphere if we take for E the only radially symmetric fundamental solution which decreases at infinity, i.e., $E = -\frac{1}{4\pi|x|}$, cf. Example 1.3.14. More generally, if $n \geq 3$, and $E = c_n |x|^{2-n}$, $c_n = \frac{1}{(2-n)|S^{n-1}|}$, as in Example 1.3.14, then (1.5.9) furnishes

$$U = \frac{2^{n-3} |S^{n-2}| \rho}{(2-n) |S^{n-1}| \cdot |x|^{n-2}} \int_0^\infty \sigma^{3-n} \Delta(|x|, \rho, \sigma)^{n-3} \mathrm{d}\sigma.$$

With the substitution $\tau = \frac{(\rho + |x|)^2 - \sigma^2}{4\rho|x|}$ and from the definition of Gauß' hypergeometric function (Gradshteyn and Ryzhik [113], Eq. 9.111), we infer

$$U = \frac{2^{3-n} \Gamma(\frac{n}{2}) \rho}{(2-n) \sqrt{\pi} \, \Gamma(\frac{n-1}{2}) \cdot |x|^{n-2}} \times$$

$$\times \int_{|\rho - |x||}^{\rho + |x|} \sigma^{3-n} \left([(\rho + |x|)^2 - \sigma^2][\sigma^2 - (\rho - |x|)^2]\right)^{(n-3)/2} \mathrm{d}\sigma$$

$$= \frac{2^{n-2} \Gamma(\frac{n}{2}) \rho^{n-1}}{(2-n) \sqrt{\pi} \, \Gamma(\frac{n-1}{2})} \int_0^1 [(\rho + |x|)^2 - 4\rho|x|\tau]^{-n/2+1} (\tau(1-\tau))^{(n-3)/2} \mathrm{d}\tau$$

$$\tag{1.5.10}$$

$$= \frac{\rho^{n-1}}{(2-n)(\rho + |x|)^{n-2}} \, {}_2F_1\left(\frac{n-1}{2}, \frac{n}{2} - 1; n - 1; \frac{4\rho|x|}{(\rho + |x|)^2}\right).$$

If we use one of Kummer's quadratic transformations (see Gradshteyn and Ryzhik [113], Eq. 9.135), namely

$$_2F_1(\alpha, \beta; \alpha + \beta + \tfrac{1}{2}; \sin^2 \varphi) = {_2F_1}(2\alpha, 2\beta; \alpha + \beta + \tfrac{1}{2}; \sin^2 \tfrac{\varphi}{2})$$

with

$$\sin^2 \varphi = \frac{4\rho|x|}{(\rho + |x|)^2} \quad \text{and} \quad \sin^2 \tfrac{\varphi}{2} = \frac{1}{2}\left(1 - \frac{|\rho - |x||}{\rho + |x|}\right),$$

we finally obtain

$$
\begin{aligned}
U &= \frac{\rho^{n-1}}{(2-n)(\rho + |x|)^{n-2}} \, {_2F_1}\left(n-1, n-2; n-1; \frac{1}{2}\left(1 - \frac{|\rho - |x||}{\rho + |x|}\right)\right) \\
&= \frac{\rho^{n-1}}{2-n}\left[\frac{\rho + |x| + |\rho - |x||}{2}\right]^{2-n} \\
&= \frac{\rho}{2-n}\left[Y(\rho - |x|) + Y(|x| - \rho) \cdot \left(\frac{\rho}{|x|}\right)^{n-2}\right].
\end{aligned}
\tag{1.5.11}
$$

Of course, the final formula can be deduced much more rapidly observing that U is rotationally symmetric and harmonic for $|x| \neq \rho$ and hence

$$
U = \left\{
\begin{array}{ll}
C_1, & \text{if } |x| < \rho, \\
C_2|x|^{2-n} + C_3, & \text{if } |x| > \rho
\end{array}
\right\}.
$$

Moreover, by Lebesgue's theorem on dominated convergence, we deduce from (1.5.10) that U is continuous along the sphere $|x| = \rho$ and vanishes at infinity, facts which are also evident from physical reasons for $n = 3$. Hence, for $n \geq 3$,

$$U(x) = C_1\left[Y(\rho - |x|) + Y(|x| - \rho) \cdot \left(\frac{\rho}{|x|}\right)^{n-2}\right]$$

and

$$C_1 = U(0) = \int_{|x|=\rho} E(x)\, d\sigma(x) = \frac{1}{(2-n)|S^{n-1}| \cdot \rho^{n-2}} \cdot |S^{n-1}|\rho^{n-1} = \frac{\rho}{2-n},$$

cf. also Donoghue [61], Sect. 8, p. 39, and the verification of (1.5.11) in Hirsch and Lacombe [131], Ex. 1, p. 344.

Note that, for $c = |S^{n-1}| \cdot \rho^{n-1}$, the distribution

$$\mu := U - cE = \frac{\rho}{2-n} \, Y(\rho - |x|)\left[1 - \left(\frac{\rho}{|x|}\right)^{n-2}\right] \in L_c^1 \subset \mathcal{E}'$$

satisfies the equation

$$\Delta_n \mu = \delta(|x| - \rho) - |\mathbf{S}^{n-1}| \rho^{n-1} \cdot \delta.$$

Hence $\Delta_n V = T$ for arbitrary $T, V \in \mathcal{D}'$ implies

$$T * \mu = \Delta_n V * \mu = V * \Delta_n \mu = V * \delta(|x| - \rho) - cV,$$

which extends the mean value property $V * \delta(|x| - \rho) = cV$ fulfilled for harmonic distributions (where $T = 0$) to the inhomogeneous case, cf. Schwartz [246], Ch. VI, Sect. 10, p. 217; Hirsch and Lacombe [131], Ex. 3, p. 332; Ortner and Wagner [216], p. 836. □

Example 1.5.16 As our final example concerning convolution, let us calculate the *Liénard–Wiechert potentials and fields*, i.e. the electromagnetic potentials and fields induced by a moving point charge in $\mathbf{R}^3$.

(a) In general, the potentials $\Phi \in \mathcal{D}'(\mathbf{R}^4)$, $A \in \mathcal{D}'(\mathbf{R}^4)^3$ fulfill the wave equations

$$(\tfrac{1}{c^2}\partial_t^2 - \Delta_3)\Phi = 4\pi\rho, \qquad (\tfrac{1}{c^2}\partial_t^2 - \Delta_3)A = \tfrac{4\pi}{c}J,$$

where ρ, J denote the charge and current densities, respectively, and $c > 0$ the speed of light, cf. Heald and Marion [124], (4.55), (4.56), p. 142; Becker and Sauter [10], (66.3), p. 194.

By Example 1.4.12 (b) and Proposition 1.3.19, the wave operator $\frac{1}{c^2}\partial_t^2 - \Delta_3$ has the fundamental solution $G = \frac{1}{4\pi t}\delta(ct - |x|)$. According to (1.4.11), $G \in \mathcal{M}(\mathbf{R}^n)$ is the Radon measure given by

$$\langle \phi, G \rangle = \langle \phi, \frac{1}{4\pi t}\delta(ct - |x|) \rangle = \frac{1}{4\pi}\int_{\mathbf{R}^3} \frac{\phi(|x|/c, x)}{|x|}\,dx, \qquad \phi \in \mathcal{D}(\mathbf{R}^4_{t,x}),$$

and, by the hyperbolicity of the wave operator, G is the only fundamental solution with support in the half-space $t \geq 0$, see Hörmander [138], Thm. 12.5.1, p. 120.

By convolution with G, we obtain the *retarded potentials*

$$\Phi = \rho * \frac{1}{t}\delta(ct - |x|), \qquad A = \frac{1}{c}J * \frac{1}{t}\delta(ct - |x|).$$

If ρ, J result from the point charge ρ_0 moving on the trajectory $x = u(t)$ (with $u : \mathbf{R} \longrightarrow \mathbf{R}^3$ in $\mathcal{C}^1$), i.e., $\rho = \rho_0\delta(x - u(t))$, $J = \rho_0\dot{u}\delta(x - u(t))$, (cf. Heald and Marion [124], p. 266), then Φ, A are called the *Liénard–Wiechert potentials*.

Let us assume that the speed of the point charge remains below a bound strictly smaller than c, i.e.,

$$\exists c_0 < c : \forall t \in \mathbf{R} : |\dot{u}(t)| \leq c_0. \tag{1.5.12}$$

Then G, ρ as well as G, J are convolvable by support (see Example 1.5.11). In fact, in order to show that $\phi^\Delta \cdot (G \otimes \rho)$ has compact support for $\phi \in \mathcal{D}(\mathbf{R}^4)$, it suffices to show that the sets

$$M_N = \{(t, y, s, u(s)) \in \mathbf{R}^8; \ s, t \in \mathbf{R}, \ y \in \mathbf{R}^3$$
$$\text{with } |y| = ct, \ |s + t| \le N, \ |y + u(s)| \le N\}$$

are bounded if $N \in \mathbf{N}$. But this follows from the boundedness of t on M_N, which boundedness is in turn implied by

$$0 \le (c - c_0)t = |y| - c_0 t \le N + |u(s)| - c_0(|s| - N)$$
$$\le N + c_0|s| + |u(0)| - c_0|s| + c_0 N = N(1 + c_0) + |u(0)|.$$

Hence the inequality (1.5.12) is a sufficient condition for the convolvability of G and $\delta(x - u(t))$, and thus for the existence of the Liénard–Wiechert potentials Φ, A.

(b) Let us now calculate $\Phi = G * 4\pi\rho$ and $A = G * \frac{4\pi}{c}J$ under the condition $\|\dot{u}\|_\infty \le c_0$ stipulated in (1.5.12). For $\phi \in \mathcal{D}(\mathbf{R}^4)$, we have

$$\langle \phi, \Phi \rangle = \rho_0 \langle \phi, \delta(x - u(t)) * \tfrac{1}{t}\delta(ct - |x|) \rangle$$
$$= \rho_0 \langle \langle \phi(s + t, x + y), \delta(x - u(s)) \rangle, \tfrac{1}{t}\delta(ct - |y|) \rangle$$
$$= \rho_0 \langle \int_{-\infty}^{\infty} \phi(s + t, u(s) + y)\, ds, \tfrac{1}{t}\delta(ct - |y|) \rangle$$
$$= \rho_0 \int_{\mathbf{R}^4} \frac{1}{|y|} \phi(s + \tfrac{|y|}{c}, u(s) + y)\, ds dy.$$

The substitution

$$h : \mathbf{R}^4 \longrightarrow \mathbf{R}^4 : (s, y) \longmapsto (t, x) = (s + \tfrac{|y|}{c}, u(s) + y)$$

is bijective, since the equation $c(t - s) = |y| = |x - u(s)|$ has a unique solution $s(t, x)$. This is due to the strict monotonicity of $f : s \mapsto |x - u(s)| - c(t - s)$ because of $\|\dot{f}\|_\infty \ge c - c_0 > 0$. Furthermore, h is C^1 except for $y = 0$ and h^{-1} is continuous since h is proper. For $y \ne 0$, the determinant of the Jacobian of h is

$$\det h' = \frac{\partial(t, x)}{\partial(s, y)} = \det \begin{pmatrix} 1 & \frac{y^T}{c|y|} \\ \dot{u}(s) & I_3 \end{pmatrix} = 1 - \frac{y^T \cdot \dot{u}(s)}{c|y|} > 0,$$

and hence

$$\langle \phi, \Phi \rangle = c\rho_0 \int_{\mathbf{R}^4} \phi(t,x) \frac{dt\,dx}{c|y| - y^T \cdot \dot{u}(s)}$$

$$= c\rho_0 \int_{\mathbf{R}^4} \phi(t,x) \frac{dt\,dx}{c|x - u(s)| - \dot{u}(s)^T \cdot (x - u(s))},$$

i.e.,

$$\Phi(t,x) = \frac{\rho_0}{|x - u(s)| - \frac{1}{c}\dot{u}(s)^T \cdot (x - u(s))},$$

wherein $s(t,x)$ is the *retarded time* determined by the equation $c(t - s) = |x - u(s)|$, cf. Heald and Marion [124], (8.50), p. 267; Feynman et al. [74], (21.33). By the same token, we obtain $A(t,x) = \frac{1}{c}\Phi(t,x) \cdot \dot{u}(s(t,x))$. An extension of the formula for Φ in a curved space-time can be found in Friedlander [83], Eq. (5.6.7), Thm. 5.6.1, p. 214.

Note that $\Phi(t,x)$ is finite and positive except on the curve $x = u(t)$ where Φ becomes infinite. Moreover, Φ is locally integrable, since, for $K \subset \mathbf{R}^4$ compact,

$$\int_K \Phi(t,x)\,dt dx = \rho_0 \int_{h^{-1}(K)} |y|^{-1}\,ds dy < \infty.$$

(c) Let us finally calculate the so-called *Liénard–Wiechert fields* $\mathcal{E}, \mathcal{B}$ from its potentials Φ, A. We suppose here that u is C^2. Generally,

$$\mathcal{E} = -\operatorname{grad}\Phi - \frac{1}{c}\partial_t A \qquad \text{and} \qquad \mathcal{B} = \operatorname{curl} A,$$

see Heald and Marion [124], (4.40), (4.42), pp. 139, 140.

If we consider $r = |x - u(s)|$ and $w = \frac{1}{r}(x - u(s))$ as depending on t, x (with $s(t,x)$ as in (b)), then $c(t - s) = |x - u(s)| = r$ implies

$$c\left(1 - \frac{\partial s}{\partial t}\right) = \frac{\partial r}{\partial t} = -w^T \cdot \dot{u}(s)\frac{\partial s}{\partial t},$$

and hence

$$\left(c - w^T \cdot \dot{u}(s)\right)\frac{\partial s}{\partial t} = c, \qquad \Phi = \frac{\rho_0}{r}\frac{\partial s}{\partial t}, \qquad A = -\frac{\rho_0}{cr}\frac{\partial(rw)}{\partial t}. \qquad (1.5.13)$$

Furthermore,

$$\nabla r = -c\nabla s = \nabla|x - u(s)| = w - (w^T \cdot \dot{u}(s))\nabla s = w + \frac{1}{c}(w^T \cdot \dot{u}(s))\nabla r,$$

and hence

$$\nabla r = \frac{cw}{c - w^T \cdot \dot{u}(s)} = w\frac{\partial s}{\partial t}, \qquad \nabla s = -\frac{w}{c}\frac{\partial s}{\partial t}.$$

This implies the following:

$$\begin{aligned}
\nabla \Phi = \rho_0 \nabla\left(\frac{1}{r}\frac{\partial s}{\partial t}\right) &= -\frac{\rho_0 \nabla r}{r^2} \cdot \frac{\partial s}{\partial t} + \frac{\rho_0}{r}\frac{\partial}{\partial t}(\nabla s) \\
&= -\frac{\rho_0 w}{r^2}\left(\frac{\partial s}{\partial t}\right)^2 - \frac{\rho_0}{cr}\frac{\partial}{\partial t}\left(w\frac{\partial s}{\partial t}\right) \\
&= -\frac{\rho_0 w}{r^2}\left(\frac{\partial s}{\partial t}\right)^2 - \frac{\rho_0 w}{cr}\frac{\partial^2 s}{\partial t^2} - \frac{\rho_0}{cr}\frac{\partial w}{\partial t}\frac{\partial s}{\partial t}
\end{aligned}$$

and

$$\begin{aligned}
\frac{1}{c}\partial_t A &= -\frac{\rho_0}{c^2}\frac{\partial}{\partial t}\left(\frac{1}{r}\frac{\partial(rw)}{\partial t}\right) \\
&= \frac{\rho_0}{c^2 r^2}\frac{\partial r}{\partial t}\frac{\partial(rw)}{\partial t} - \frac{\rho_0}{c^2 r}\frac{\partial^2(rw)}{\partial t^2} \\
&= \frac{\rho_0 w}{c^2 r^2}\left(\frac{\partial r}{\partial t}\right)^2 - \frac{\rho_0}{c^2 r}\frac{\partial r}{\partial t}\frac{\partial w}{\partial t} - \frac{\rho_0 w}{c^2 r}\frac{\partial^2 r}{\partial t^2} - \frac{\rho_0}{c^2}\frac{\partial^2 w}{\partial t^2}.
\end{aligned}$$

Because of $\frac{\partial s}{\partial t} = 1 - \frac{1}{c}\frac{\partial r}{\partial t}$, we finally obtain

$$\mathcal{E} = -\operatorname{grad}\Phi - \frac{1}{c}\partial_t A = \rho_0\left[\frac{w}{r^2} + \frac{r}{c}\frac{\partial(w/r^2)}{\partial t} + \frac{1}{c^2}\frac{\partial w^2}{\partial t^2}\right], \qquad (1.5.14)$$

which is commonly called Feynman's formula, but in fact goes back to O. Heaviside, cf. Feynman et al. [74], (21.1), and footnote on p. 21-11; Heald and Marion [124], (8.54a), p. 268.

For completeness, let us also calculate the magnetic induction. From formula (1.5.13), we infer

$$\begin{aligned}
B = \operatorname{curl}A = \nabla \times A &= -\frac{\rho_0}{c}\nabla \times \left(\frac{1}{r}\frac{\partial(rw)}{\partial t}\right) \\
&= -\frac{\rho_0}{c}\left[\left(\nabla\frac{1}{r}\right) \times \frac{\partial(rw)}{\partial t} + \frac{1}{r}\frac{\partial}{\partial t}(\nabla \times (rw))\right].
\end{aligned}$$

Since

$$\begin{aligned}
\nabla \times (rw) = \nabla \times (x - u(s)) = -\nabla \times u(s) &= -\nabla s \times \dot{u}(s) = \frac{1}{c}\frac{\partial s}{\partial t} \cdot w \times \dot{u}(s) \\
&= \frac{1}{c}w \times \frac{\partial u}{\partial t} = \frac{1}{c}w \times \frac{\partial}{\partial t}(x - rw) = -\frac{rw}{c} \times \frac{\partial w}{\partial t},
\end{aligned}$$

we obtain

$$\mathcal{B} = \frac{\rho_0}{cr^2}\nabla r \times \frac{\partial (rw)}{\partial t} + \frac{\rho_0}{c^2 r}\frac{\partial}{\partial t}\left(rw \times \frac{\partial w}{\partial t}\right)$$

$$= \frac{\rho_0}{cr^2}\frac{\partial s}{\partial t}\cdot w \times \frac{\partial (rw)}{\partial t} + \frac{\rho_0}{c^2 r}\frac{\partial r}{\partial t}\cdot w \times \frac{\partial w}{\partial t} + \frac{\rho_0}{c^2}w \times \frac{\partial^2 w}{\partial t^2}$$

$$= \frac{\rho_0}{cr}\left(1 - \frac{1}{c}\frac{\partial r}{\partial t}\right)\cdot w \times \frac{\partial w}{\partial t} + \frac{\rho_0}{c^2 r}\frac{\partial r}{\partial t}\cdot w \times \frac{\partial w}{\partial t} + \frac{\rho_0}{c^2}w \times \frac{\partial^2 w}{\partial t^2}$$

$$= \rho_0 w \times \left(\frac{1}{cr}\frac{\partial w}{\partial t} + \frac{1}{c^2}\frac{\partial^2 w}{\partial t^2}\right) = w \times \mathcal{E},$$

cf. Feynman et al. [74], (21.1'); Heald and Marion [124], (8.64), p. 269; Becker and Sauter [10], 69, 3), p. 204. □

1.6 The Fourier Transformation

First let us motivate—following Schwartz [246], Ch. VII, Sect. 2, p. 233—why the classical Fourier transformation

$$\mathcal{F} : L^1(\mathbf{R}^n) \longrightarrow \mathcal{BC}(\mathbf{R}^n) : f \longmapsto \left(\xi \mapsto \int_{\mathbf{R}^n} f(x)\mathrm{e}^{-\mathrm{i}\xi x}\,\mathrm{d}x\right), \qquad (1.6.1)$$

cf. Proposition 1.1.8, cannot be extended continuously to yield a mapping from $\mathcal{D}'(\mathbf{R}^n)$ to $\mathcal{D}'(\mathbf{R}^n)$.

In fact, the classical Fourier transform (1.6.1) yields $\mathcal{F}(\mathrm{e}^{-\epsilon|x|^2}) = (\frac{\pi}{\epsilon})^{n/2}\mathrm{e}^{-|x|^2/(4\epsilon)}$ for $\epsilon > 0$, see, e.g., Example 1.6.14 below. Therefore, due to $\lim_{\epsilon \searrow 0}(\pi\epsilon)^{-n/2}\mathrm{e}^{-|x|^2/\epsilon} = \delta$, see Example 1.1.11, (iii), we must have

$$\mathcal{F}1 = \lim_{\epsilon \searrow 0}\mathcal{F}(\mathrm{e}^{-\epsilon|x|^2}) = \lim_{\epsilon \searrow 0}\left(\frac{\pi}{\epsilon}\right)^{n/2}\mathrm{e}^{-|x|^2/(4\epsilon)} = (2\pi)^n\delta,$$

and, consequently, $\mathcal{F}x^\alpha = (\mathrm{i}\partial)^\alpha\mathcal{F}1 = (2\pi)^n(\mathrm{i}\partial)^\alpha\delta$. Hence, if $\mathcal{F}(\mathrm{e}^x)$ could be defined by continuous extension on $\mathcal{D}'(\mathbf{R})$, then we had

$$\mathcal{F}(\mathrm{e}^x) = \mathcal{F}\left(\lim_{N\to\infty}\sum_{k=0}^N \frac{x^k}{k!}\right) = 2\pi\lim_{N\to\infty}\sum_{k=0}^N \frac{\mathrm{i}^k}{k!}\delta^{(k)}.$$

However, the last series diverges in $\mathcal{D}'(\mathbf{R})$ since, by E. Borel's theorem (see Treves [273], Ch. 38, Thm. 38.1, p. 390; Zuily [309], Ch. 1, Exercise 1, pp. 16, 18, 19),

$$\exists\phi \in \mathcal{D}(\mathbf{R}) : \forall k \in \mathbf{N}_0 : \phi^{(k)}(0) = k!.$$

Following L. Schwartz, we therefore define the Fourier transform on the space of *temperate distributions* $S'(\mathbf{R}^n)$, which arises as the "closure" of $L^1(\mathbf{R}^n)$ with respect to differentiation and multiplication by polynomials.

Definition 1.6.1

(1) The vector space

$$S = S(\mathbf{R}^n) = \{\phi \in \mathcal{E}(\mathbf{R}^n); \ \forall \alpha, \beta \in \mathbf{N}_0^n : x^\alpha \partial^\beta \phi \in L^1(\mathbf{R}^n)\}$$

is called the space of *rapidly decreasing C^∞ functions*.

The sequence $(\phi_k)_{k \in \mathbf{N}} \in S(\mathbf{R}^n)^{\mathbf{N}}$ *converges to ϕ in $S(\mathbf{R}^n)$* iff, for all $\alpha, \beta \in \mathbf{N}_0^n$, the sequences $x^\alpha \partial^\beta (\phi_k - \phi)$ converge to 0 in $L^1(\mathbf{R}^n)$ for $k \to \infty$. (In other words, the topology on S is generated by the seminorms $\phi \mapsto \|x^\alpha \partial^\beta \phi\|_1$; they render S a Fréchet space, i.e., a complete metrizable locally convex topological vector space.)

(2) The space of *temperate distributions* S' (originally called "distributions sphériques" by L. Schwartz) is the dual of S, i.e.,

$$S' = S'(\mathbf{R}^n) = \{T : S(\mathbf{R}^n) \longrightarrow \mathbf{C} \text{ linear}; \ T(\phi_k) \to 0 \text{ if } \phi_k \to 0 \text{ in } S \text{ for } k \to \infty\}.$$

As for distributions, we write $T(\phi) = \langle \phi, T \rangle$ for $\phi \in S$, $T \in S'$, and $T_k \to T$ in S' iff $\lim_{k \to \infty} \langle \phi, T_k \rangle = \langle \phi, T \rangle$ for each $\phi \in S$.

Proposition 1.6.2 $\mathcal{D}$ *is dense in S. Hence $S' \to \mathcal{D}' : T \mapsto T|_{\mathcal{D}}$ is injective and S' can be identified with a subspace of $\mathcal{D}'$. The spaces $\mathcal{D}'_{L^p}(\mathbf{R}^n)$, $1 \le p \le \infty$, (see Definition 1.5.12) are subspaces of S'; in particular, $\mathcal{M}^1$ and $\mathcal{E}'$ are subspaces of S'. Furthermore, S' is stable under differentiation and under multiplication with polynomials:*

$$\forall T \in S' : \forall \alpha \in \mathbf{N}_0^n : \partial^\alpha T \in S', \ x^\alpha T \in S'.$$

Proof

(1) Let $\chi \in \mathcal{D}$ with $\chi(x) = 1$ for x in a neighborhood of 0. If $\phi \in S$, then $\phi_k(x) := \phi(x)\chi(\frac{x}{k}) \in \mathcal{D}$ and $\phi_k \to \phi$ in S for $k \to \infty$. Hence $\mathcal{D} \subset S$ is dense and $S' \subset \mathcal{D}'$.

(2) The stability of S' under differentiation and under multiplication by polynomials follows from that of S. Since

$$L^p(\mathbf{R}^n) \hookrightarrow S' : f \longmapsto \left(\phi \mapsto \int_{\mathbf{R}^n} f(x)\phi(x)\,dx\right), \qquad 1 \le p \le \infty,$$

is a continuous embedding, we also obtain that $\mathcal{D}'_{L^p} = \{\sum_{|\alpha| \le m} \partial^\alpha f_a; \ m \in \mathbf{N}_0, \ f_\alpha \in L^p(\mathbf{R}^n)\}$ is contained in S'. $\square$

Let us summarize the various inclusions in a table, cf. Table 1.1.10, and Schwartz [246], p. 420. (Note that the inclusion $\mathcal{D}_{L^p} \subset \mathcal{D}_{L^q}$ for $p < q$ follows from the boundedness of the functions in $\mathcal{D}_{L^p}$. The inclusion $\mathcal{D}'_{L^p} \subset \mathcal{D}'_{L^q}$ for $p < q$ is non-trivial if our definition of the spaces $\mathcal{D}'_{L^p}$ in Definition 1.5.12 is taken as basis; however, it is immediate if $\mathcal{D}'_{L^p}$ is defined by duality as in Schwartz [246], Ch. VI, Sect. 8, p. 199. The two empty places in the table correspond to the topological vector spaces $\mathcal{O}_M$ and $\mathcal{O}'_C$, which are investigated in Schwartz [246], Ch. VII, Sect. 5, p. 243.)

Table 1.6.3 If $1 < p < q < \infty$, then the following inclusions hold:

$$\mathcal{D} \subset \mathcal{S} \subset \mathcal{D}_{L^1} \subset \mathcal{D}_{L^p} \subset \mathcal{D}_{L^q} \subset \mathcal{D}_{L^\infty} \subset \qquad \subset \mathcal{E}$$
$$\cap \qquad \cap \qquad \cap \qquad \cap \qquad \cap \qquad \qquad \cap$$
$$\mathcal{E}' \subset \qquad \subset \mathcal{D}'_{L^1} \subset \mathcal{D}'_{L^p} \subset \mathcal{D}'_{L^q} \subset \mathcal{D}'_{L^\infty} \subset \mathcal{S}' \subset \mathcal{D}'$$

Definition and Proposition 1.6.4 *For $\phi \in \mathcal{S}(\mathbf{R}^n) \subset L^1(\mathbf{R}^n)$, the Fourier transform $\mathcal{F}\phi$ defined in (1.6.1) also belongs to $\mathcal{S}(\mathbf{R}^n)$ and the mapping $\mathcal{F} : \mathcal{S} \longrightarrow \mathcal{S}$ is continuous. Therefore the adjoint*

$$\mathcal{F} : \mathcal{S}'(\mathbf{R}^n) \longrightarrow \mathcal{S}'(\mathbf{R}^n) : T \longmapsto (\phi \mapsto \langle \mathcal{F}\phi, T \rangle)$$

is well defined and continuous. It is called the Fourier transformation on the space of temperate distributions. For integrable functions, the two definitions coincide, i.e., $\mathcal{F}T_f = T_{\mathcal{F}f}$ for $f \in L^1(\mathbf{R}^n)$.

Proof

(1) Since $x^\alpha \partial^\beta \phi \in L^1(\mathbf{R}^n)$ for $\phi \in \mathcal{S}(\mathbf{R}^n)$ and $\alpha, \beta \in \mathbf{N}_0^n$, we conclude that $\mathcal{F}(x^\alpha \partial^\beta \phi) = (\mathrm{i}\partial)^\alpha (\mathrm{i}\xi)^\beta \mathcal{F}\phi \in BC(\mathbf{R}^n)$. This implies $\xi^\alpha \partial^\beta \mathcal{F}\phi \in L^1(\mathbf{R}^n)$ for all $\alpha, \beta \in \mathbf{N}_0^n$ and hence $\mathcal{F}\phi \in \mathcal{S}$.

(2) Finally, if $f \in L^1(\mathbf{R}^n)$, then Fubini's theorem yields

$$\langle \phi, \mathcal{F}T_f \rangle = \int_{\mathbf{R}^n} (\mathcal{F}\phi)(\xi) f(\xi) \, \mathrm{d}\xi = \int_{\mathbf{R}^{2n}} \phi(x) f(\xi) \mathrm{e}^{-\mathrm{i}x\xi} \, \mathrm{d}x \mathrm{d}\xi = \langle \phi, T_{\mathcal{F}f} \rangle. \qquad \square$$

As used already above, let us denote the reflection about the origin by ˇ, i.e.,

$$\check{} : \mathcal{D}(\mathbf{R}^n) \longrightarrow \mathcal{D}(\mathbf{R}^n) : \phi \longmapsto \left(x \mapsto \check{\phi}(x) = \phi(-x)\right),$$
$$\check{} : \mathcal{D}'(\mathbf{R}^n) \longrightarrow \mathcal{D}'(\mathbf{R}^n) : T \longmapsto \left(\phi \mapsto \check{T}(\phi) = \langle \check{\phi}, T \rangle\right),$$

and similarly ˇ: $\mathcal{S} \to \mathcal{S}$ and ˇ: $\mathcal{S}' \to \mathcal{S}'$. The next proposition, which is often called the *Fourier inversion theorem*, will show that the Fourier transformation is, up to a constant, the complex conjugate of its inverse.

Proposition 1.6.5 *For $T \in \mathcal{S}'(\mathbf{R}^n)$, we have $\mathcal{F}(\mathcal{F}T) = (2\pi)^n \check{T}$, and hence $\mathcal{F}$: $\mathcal{S}' \longrightarrow \mathcal{S}'$ is an isomorphism (of topological vector spaces) with the inverse $\mathcal{F}^{-1} = (2\pi)^{-n} \check{} \circ \mathcal{F}$.*

Proof It suffices to show that $\mathcal{F}(\mathcal{F}\phi) = (2\pi)^n \check{\phi}$ for $\phi \in \mathcal{S}$ and to apply transposition for $T \in \mathcal{S}'$. We shall employ $\mathcal{F}1 = (2\pi)^n \delta$, as explained in the introduction, and the Fourier exchange formula $\mathcal{F}(\phi * \psi) = (\mathcal{F}\phi) \cdot (\mathcal{F}\psi)$ in the case of $\phi, \psi \in \mathcal{S}$, where it is an immediate consequence of Fubini's theorem. We then obtain

$$\langle \psi, \mathcal{F}^2\phi \rangle = \langle \mathcal{F}\psi, \mathcal{F}\phi \rangle = {}_{\mathcal{S}}\langle (\mathcal{F}\phi) \cdot (\mathcal{F}\phi), 1 \rangle_{\mathcal{S}'}$$

$$= \langle \mathcal{F}(\phi * \psi), 1 \rangle = \langle \phi * \psi, \mathcal{F}1 \rangle = (2\pi)^n \langle \phi * \psi, \delta \rangle = (2\pi)^n \langle \psi, \check{\phi} \rangle.$$

$\square$

Let us observe that Proposition 1.6.5 also furnishes the injectivity of the map $\mathcal{F}$: $L^1(\mathbf{R}^n) \to \mathcal{BC}(\mathbf{R}^n)$, which has been proved directly in Proposition 1.1.8.

In the next proposition, we collect some further important properties of the Fourier transformation.

Proposition 1.6.6

(1) *For $T \in \mathcal{S}'(\mathbf{R}^n)$, $\alpha \in \mathbf{N}_0^n$, $A \in \mathrm{Gl}_n(\mathbf{R})$, we have*

$$\mathcal{F}(\partial^\alpha T) = (i\xi)^\alpha \mathcal{F}T, \quad \mathcal{F}(x^\alpha T) = (i\partial)^\alpha \mathcal{F}T, \quad \mathcal{F}(T \circ A) = |\det A|^{-1}(\mathcal{F}T) \circ A^{-1T}.$$

(2) *If $T \in \mathcal{D}'_{L^1}(\mathbf{R}^n)$, then $\mathcal{F}T$ is a continuous function with at most polynomial growth, and $\mathcal{F}T$ is given by the formula $\mathcal{F}T(\xi) = {}_{\mathcal{D}_{L^\infty}}\langle e^{-ix\xi}, T_x \rangle_{\mathcal{D}'_{L^1}}$.*

(3) *If $T \in \mathcal{E}'(\mathbf{R}^n)$, then $\mathcal{F}T$ is the restriction to $\mathbf{R}^n$ of the entire function*

$$\mathbf{C}^n \longrightarrow \mathbf{C} : \zeta \longmapsto {}_{\mathcal{E}}\langle e^{-ix\zeta}, T_x \rangle_{\mathcal{E}'}.$$

(4) *If $f \in L^2(\mathbf{R}^n)$, then $\mathcal{F}f \in L^2(\mathbf{R}^n)$ and $\|\mathcal{F}f\|_2 = (2\pi)^{n/2}\|f\|_2$, i.e., $(2\pi)^{-n/2}\mathcal{F}$: $L^2(\mathbf{R}^n) \longrightarrow L^2(\mathbf{R}^n)$ is an isomorphism of Hilbert spaces. Moreover,*

$$\mathcal{F}\mathcal{D}'_{L^2}(\mathbf{R}^n) = \{g \in L^1_{\mathrm{loc}}(\mathbf{R}^n); \exists m \in \mathbf{N}_0 : g(\xi)(1 + |\xi|)^{-m} \in L^2(\mathbf{R}^n)\}.$$

In particular, this implies $\mathcal{D}'_{L^1} \subset \mathcal{D}'_{L^2}$.

(5) *The Fourier exchange theorem $\mathcal{F}(S * T) = \mathcal{F}S \cdot \mathcal{F}T$ holds in the following two special cases: (a) $S, T \in \mathcal{D}'_{L^2}$; (b) $S \in \mathcal{E}'$, $T \in \mathcal{S}'$.*

Proof

(1) follows by application to test functions.

(2) If $T = \sum_{|\alpha| \le m} \partial^\alpha f_\alpha$ with $f_\alpha \in L^1$, then, by (1) and Proposition 1.6.4, $\mathcal{F}T = \sum_{|\alpha| \le m} (i\xi)^\alpha \mathcal{F}f_\alpha$ is continuous and of polynomial growth. Furthermore,

Definition 1.5.8 implies

$$\mathcal{D}_{L^\infty} \langle e^{-ix\xi}, T_x \rangle_{\mathcal{D}'_{L^1}} = \sum_{|\alpha| \le m} (-1)^{|\alpha|} \int_{\mathbf{R}^n} \partial_x^\alpha (e^{-ix\xi}) f_\alpha(x) \, dx = \sum_{|\alpha| \le m} (i\xi)^\alpha \mathcal{F} f_\alpha = \mathcal{F} T.$$

(3) For $T \in \mathcal{E}'$, the function $g(\zeta) = \varepsilon \langle e^{-ix\xi}, T_x \rangle_{\mathcal{E}'}$, $\zeta \in \mathbf{C}^n$, is analytic since the Cauchy–Riemann equations

$$\frac{\partial g}{\partial \bar{\zeta}_j} = \varepsilon \langle \frac{\partial}{\partial \bar{\zeta}_j} e^{-ix\xi}, T_x \rangle_{\mathcal{E}'} = 0, \qquad j = 1, \dots, n,$$

hold. Furthermore, for $\xi \in \mathbf{R}^n$, $g(\xi) = (\mathcal{F} T)(\xi)$ by (2) since $\mathcal{E}' \subset \mathcal{D}'_{L^1}$.

(4) For $\phi \in \mathcal{S}$,

$$\|\mathcal{F}\phi\|_2^2 = s\langle \mathcal{F}\phi, \overline{\mathcal{F}\phi} \rangle_{\mathcal{S}'} = s\langle \mathcal{F}\phi, \mathcal{F}\check{\bar{\phi}} \rangle_{\mathcal{S}'} = \langle \mathcal{F}^2\phi, \check{\bar{\phi}} \rangle = (2\pi)^n s\langle \phi, \check{\bar{\phi}} \rangle_{\mathcal{S}'}$$
$$= (2\pi)^n \|\phi\|_2^2.$$

If $f \in L^2(\mathbf{R}^n)$, then the linear form

$$\mathcal{S}(\mathbf{R}^n) \longrightarrow \mathbf{C} : \phi \longmapsto \langle \phi, \mathcal{F}f \rangle$$

is continuous with respect to the L^2-norm on $\mathcal{S}(\mathbf{R}^n)$ since

$$|\langle \phi, \mathcal{F}f \rangle| = |\langle \mathcal{F}\phi, f \rangle| \le \|\mathcal{F}\phi\|_2 \cdot \|f\|_2 = (2\pi)^{n/2} \|\phi\|_2 \cdot \|f\|_2.$$

Therefore, by Riesz' representation theorem, $\mathcal{F}f \in L^2(\mathbf{R}^n)$. The equality $\|\mathcal{F}f\|_2 = (2\pi)^{n/2} \|f\|_2$, which by the above is valid for $f \in \mathcal{S}$, then follows generally for $f \in L^2$ from the density of $\mathcal{S}(\mathbf{R}^n)$ in $L^2(\mathbf{R}^n)$.

Obviously, if $T = \sum_{|\alpha| \le m} \partial^\alpha f_\alpha \in \mathcal{D}'_{L^2}(\mathbf{R}^n)$, $f_\alpha \in L^2(\mathbf{R}^n)$, then $\mathcal{F} f_\alpha \in L^2(\mathbf{R}^n)$, $\mathcal{F} T = \sum_{|\alpha| \le m} (i\xi)^\alpha \mathcal{F} f_\alpha$, and $(1 + |\xi|)^{-m} \mathcal{F} T \in L^2(\mathbf{R}^n)$. Conversely, if $g \in L^1_{\text{loc}}(\mathbf{R}^n)$ and $g(\xi)(1 + |\xi|)^{-m} \in L^2(\mathbf{R}^n)$, then also $h := (1 + |\xi|^2)^{-m} g \in L^2(\mathbf{R}^n)$ and $T = \mathcal{F}^{-1} g = (1 - \Delta_n)^m \mathcal{F}^{-1} h \in \mathcal{D}'_{L^2}$.

In particular,

$$\mathcal{D}'_{L^1} \subset \mathcal{F}^{-1}\big(\{g \in \mathcal{C}(\mathbf{R}^n); \exists m \in \mathbf{N}_0 : g(\xi)(1 + |\xi|)^{-m} \in \mathcal{BC}(\mathbf{R}^n)\}\big) \subset \mathcal{D}'_{L^2}.$$

(5) (a) If $f, g \in L^2(\mathbf{R}^n)$, then $(\mathcal{F}f) \cdot \mathcal{F}g \in L^2 \cdot L^2 \subset L^1$, and hence, by Proposition 1.6.4, $\mathcal{F}^{-1}\big((\mathcal{F}f) \cdot \mathcal{F}g\big) \in \mathcal{BC}$ and

$$\mathcal{F}^{-1}\big((\mathcal{F}f) \cdot \mathcal{F}g\big)(x) = (2\pi)^{-n} \int e^{ix\xi} (\mathcal{F}f)(\xi)(\mathcal{F}g)(\xi) \, d\xi$$

$$= (2\pi)^{-n} \int \mathcal{F}(\tau_{-x} f)(\xi)(\mathcal{F}g)(\xi) \, d\xi$$

$$= (2\pi)^{-n} \big(\mathcal{F}(\tau_{-x} f), \overline{\mathcal{F}g}\big)$$

where $(\tau_{-x}f)(y) = f(x+y)$ and $(\cdot, \cdot)$ denotes the inner product in $L^2(\mathbf{R}^n)$. Since $(2\pi)^{-n/2}\mathcal{F} : L^2(\mathbf{R}^n) \to L^2(\mathbf{R}^n)$ is a Hilbert space isomorphism, we obtain

$$\mathcal{F}^{-1}((\mathcal{F}f) \cdot \mathcal{F}g)(x) = (\tau_{-x}f, \check{\bar{g}}) = \int f(x+y)g(-y)\,\mathrm{d}y = (f * g)(x).$$

This implies $\mathcal{F}(f * g) = (\mathcal{F}f) \cdot \mathcal{F}g$ for $f, g \in L^2$, and the same holds also for $S, T \in \mathcal{D}'_{L^2}$ due to $\mathcal{F}(\partial^\alpha T) = (\mathrm{i}\xi)^\alpha \mathcal{F}T$.

(5) (b) Analogously to part (2) of the proof of Proposition 1.5.14, we obtain $\phi * S \in \mathcal{E}$ and $(\phi * S)(x) = \langle \phi(x-y), S_y \rangle$ for $\phi \in \mathcal{E}$, $S \in \mathcal{E}'$. Furthermore, if $\phi \in \mathcal{S}$ and $\alpha \in \mathbf{N}_0$, then $x^\alpha \phi(x-y) \to 0 \in \mathcal{E}(\mathbf{R}^n_y)$ for $|x| \to \infty$ and hence $\phi * S \in \mathcal{S}$ if $\phi \in \mathcal{S}$, $S \in \mathcal{E}'$. This shows that the mapping

$$\mathcal{S} \longrightarrow \mathbf{C} : \phi \longmapsto \langle \phi * \check{S}, T \rangle$$

is well defined and continuous for $S \in \mathcal{E}'$, $T \in \mathcal{S}'$. Since

$$\langle \phi * \check{S}, T \rangle = \langle \langle \phi(x+y), S_y \rangle, T_x \rangle = \langle \phi, S * T \rangle, \qquad \phi \in \mathcal{D},$$

we obtain $S * T \in \mathcal{S}'$ for $S \in \mathcal{E}'$, $T \in \mathcal{S}'$ and $\langle \phi * \check{S}, T \rangle = \langle \phi, S * T \rangle$ for $\phi \in \mathcal{S}$. Finally, for $\phi \in \mathcal{S}$,

$$\langle \phi, \mathcal{F}(S * T) \rangle = \langle \mathcal{F}\phi, S * T \rangle = \langle (\mathcal{F}\phi) * \check{S}, T \rangle;$$

on the other hand, since $\mathcal{E}' \subset \mathcal{D}'_{L^2}$, part (a) furnishes

$$(\mathcal{F}\phi) * \check{S} = \mathcal{F}^{-1}((2\pi)^n \check{\phi} \cdot \mathcal{F}\check{S}) = \mathcal{F}(\phi \cdot \mathcal{F}S)$$

and hence

$$\langle \phi, \mathcal{F}(S * T) \rangle = \langle \mathcal{F}(\phi \cdot \mathcal{F}S), T \rangle = \langle \phi \cdot \mathcal{F}S, \mathcal{F}T \rangle = \langle \phi, (\mathcal{F}S) \cdot \mathcal{F}T \rangle.$$

Therefore, $\mathcal{F}(S * T) = (\mathcal{F}S) \cdot \mathcal{F}T$ for $S \in \mathcal{E}'$, $T \in \mathcal{S}'$, and the proof is complete. $\square$

Example 1.6.7 Let us calculate the Fourier transforms of the distributions x_+^λ, $\lambda \in \mathbf{C}$, defined in Example 1.3.9 and reconsidered in Example 1.4.8.

If $\operatorname{Re}\lambda > -1$, then $\mathrm{e}^{-\epsilon x}x_+^\lambda \in L^1(\mathbf{R}^1)$ for $\epsilon > 0$, and $x_+^\lambda = \lim_{\epsilon \searrow 0}\mathrm{e}^{-\epsilon x}x_+^\lambda$ in $\mathcal{S}'(\mathbf{R}^1)$, and hence

$$\mathcal{F}x_+^\lambda = \lim_{\epsilon \searrow 0}\mathcal{F}(\mathrm{e}^{-\epsilon x}x_+^\lambda) = \lim_{\epsilon \searrow 0}\int_0^\infty x^\lambda \mathrm{e}^{-(\epsilon+\mathrm{i}\xi)x}\,\mathrm{d}x = \Gamma(1+\lambda)\lim_{\epsilon \searrow 0}(\epsilon+\mathrm{i}\xi)^{-\lambda-1},$$

by Euler's definition of the gamma function. (The limits in these equations must be performed in $\mathcal{S}'(\mathbf{R}^1_\xi)$, and $z^w = \exp(w \cdot \log z)$ is defined with the usual cut along the negative real axis, i.e., for $w \in \mathbf{C}$, $z \in \mathbf{C} \setminus -[0, \infty)$.) For $-1 < \operatorname{Re} \lambda < 0$, the last limit yields a locally integrable function, i.e.,

$$(\mathcal{F}x_+^\lambda)(\xi) = \Gamma(1 + \lambda)\left[e^{-(\lambda+1)i\pi/2}Y(\xi)\xi^{-\lambda-1} + e^{(\lambda+1)i\pi/2}Y(-\xi)|\xi|^{-\lambda-1}\right].$$

If we set $\xi_-^\mu = (\xi_+^\mu)^\vee$ for $\mu \in \mathbf{C}$ as in Example 1.5.13, then analytic continuation with respect to λ (see Sect. 1.4) yields

$$(\mathcal{F}x_+^\lambda)(\xi) = \Gamma(1 + \lambda)\left[e^{-(\lambda+1)i\pi/2}\,\xi_+^{-\lambda-1} + e^{(\lambda+1)i\pi/2}\,\xi_-^{-\lambda-1}\right], \qquad \lambda \in \mathbf{C} \setminus \mathbf{Z}, \tag{1.6.2}$$

cf. Gel'fand and Shilov [104], p. 360.

For $\lambda = m \in \mathbf{Z}$, we have

$$\mathcal{F}x_+^m = \mathcal{F}\left(\operatorname*{Pf}_{\lambda=m} x_+^\lambda\right) = \operatorname*{Pf}_{\lambda=m} \mathcal{F}x_+^\lambda \tag{1.6.3}$$

$$= \operatorname*{Pf}_{\lambda=m}\left(\Gamma(1 + \lambda)\left[e^{-(\lambda+1)i\pi/2}\,\xi_+^{-\lambda-1} + e^{(\lambda+1)i\pi/2}\,\xi_-^{-\lambda-1}\right]\right), \qquad m \in \mathbf{Z}.$$

If $\operatorname{Re} \lambda < 0$ and $\lambda \notin -\mathbf{N}$, then (1.6.2) implies that $\mathcal{F}x_+^\lambda$ is locally integrable and given by

$$(\mathcal{F}x_+^\lambda)(\xi) = \Gamma(1 + \lambda)(i\xi)^{-\lambda-1} \tag{1.6.4}$$

$$= \Gamma(1 + \lambda)|\xi|^{-\lambda-1}e^{-i\pi(\operatorname{sign}\xi)\cdot(\lambda+1)/2} \in L_{\text{loc}}^1(\mathbf{R}), \qquad \operatorname{Re} \lambda < 0, \ \lambda \notin -\mathbf{N}.$$

If $\lambda = -m$, $m \in \mathbf{N}$, then (1.6.3) yields

$$\mathcal{F}x_+^{-m} = \left(\operatorname*{Pf}_{\lambda=-m} \Gamma(1 + \lambda)\right)(i\xi)^{m-1} + \left(\operatorname*{Res}_{\lambda=-m} \Gamma(1 + \lambda)\right)\left(\frac{\mathrm{d}}{\mathrm{d}\lambda}(i\xi)^{-\lambda-1}\right)\bigg|_{\lambda=-m}$$

$$= \frac{(-i\xi)^{m-1}}{(m-1)!}\left[\psi(m) - \log(i\xi)\right] = \frac{(-i\xi)^{m-1}}{(m-1)!}\left[\psi(m) - \log|\xi| - \tfrac{i\pi}{2}\operatorname{sign}\xi\right], \tag{1.6.5}$$

cf. Lavoine [160], p. 85. In particular

$$\mathcal{F}(\operatorname{vp}\tfrac{1}{x}) = \mathcal{F}(x_+^{-1} - x_-^{-1}) = \log(-i\xi) - \log(i\xi) = -i\pi \operatorname{sign}\xi. \tag{1.6.6}$$

Formula (1.6.6) can also be deduced in the following way: $x \cdot \operatorname{vp}\tfrac{1}{x} = 1$ implies

$$2\pi\delta = \mathcal{F}1 = \mathcal{F}(x \cdot \operatorname{vp}\tfrac{1}{x}) = i\frac{\mathrm{d}}{\mathrm{d}\xi}\mathcal{F}(\operatorname{vp}\tfrac{1}{x}),$$

and hence $\mathcal{F}(\operatorname{vp}\tfrac{1}{x}) = -i\pi \operatorname{sign}\xi + C$. The constant C has to vanish since $\operatorname{sign}\xi$ and $\mathcal{F}(\operatorname{vp}\tfrac{1}{x})$ are odd. For (1.6.6), cf. Zuily [309], ex. 73, pp. 144, 159; Friedlander and Joshi [84], p. 101; Schwartz [246], (VII, 7; 19), p. 259.

We have showed at the beginning that

$$\frac{\mathcal{F}x_+^\lambda}{\Gamma(1+\lambda)} = \lim_{\epsilon \searrow 0}(i\xi + \epsilon)^{-\lambda-1} \qquad (1.6.7)$$

holds for $\operatorname{Re}\lambda > -1$. The fraction on the left-hand side is an entire function of λ, since the factor $1/\Gamma(1+\lambda)$ cancels the simple poles of $\lambda \mapsto \mathcal{F}x_+^\lambda$ at $\lambda = -1, -2, \ldots$, cf. Example 1.4.8. For $\operatorname{Re}\lambda < 0$, the limit with respect to ϵ on the right-hand side yields a locally integrable function, which obviously depends analytically on λ. Therefore, the map

$$\mathbf{C} \longrightarrow \mathcal{S}'(\mathbf{R}) : \mu \longmapsto (i\xi + 0)^\mu := \lim_{\epsilon \searrow 0}(i\xi + \epsilon)^\mu$$

is well-defined and entire. Analytic continuation of (1.6.7) then implies

$$\mathcal{F}x_+^\lambda = \Gamma(1+\lambda)(i\xi + 0)^{-\lambda-1}, \qquad \lambda \in \mathbf{C} \setminus -\mathbf{N}. \qquad (1.6.8)$$

In particular,

$$\mathcal{F}Y = (i\xi + 0)^{-1} = \lim_{\epsilon \searrow 0}(i\xi + \epsilon)^{-1} = -i\operatorname{vp}\frac{1}{\xi} + \pi\delta \qquad (1.6.9)$$

by Sokhotski's formula (1.1.2) in Example 1.1.12. Note that formula (1.6.9) follows also from (1.6.6):

$$-2\pi \operatorname{vp}\frac{1}{\xi} = \mathcal{F}\mathcal{F}\operatorname{vp}\frac{1}{\xi} = -i\pi\mathcal{F}(\operatorname{sign}x)$$

and hence

$$\mathcal{F}Y = \tfrac{1}{2}\mathcal{F}(1 + \operatorname{sign}x) = \tfrac{1}{2}\big(2\pi\delta - 2i\operatorname{vp}\frac{1}{\xi}\big) = -i\operatorname{vp}\frac{1}{\xi} + \pi\delta.$$

By multiplication with monomials, formula (1.6.9) yields

$$\mathcal{F}x_+^m = \mathcal{F}(x^m \cdot Y) = \Big(i\frac{d}{d\xi}\Big)^m \mathcal{F}Y = i^{m-1}\big(\operatorname{vp}\frac{1}{\xi}\big)^{(m)} + i^m\pi\delta^{(m)}, \qquad m \in \mathbf{N}.$$

For the sake of subsuming the example of $\mathcal{F}x_+^\lambda$ into a more general framework, let us mention that the Fourier transform of a temperate distribution with support in an acute closed convex cone Γ in $\mathbf{R}^n$ (which is $[0,\infty) \subset \mathbf{R}$ in the case above) is always representable as the distributional boundary value of a function holomorphic in a tube domain (which is $\{\xi - i\epsilon;\ \xi \in \mathbf{R},\ \epsilon > 0\}$ in the case above). In this situation, the Fourier–Laplace transform $(\mathcal{L}S)(\vartheta + i\xi) = \mathcal{F}(e^{-\vartheta x}S_x)$ provides an algebra isomorphism of the convolution algebra consisting of the temperate distributions S with support in Γ with an algebra of holomorphic functions on the tube domain $C + i\mathbf{R}^n$, $C = \text{int } \Gamma^*$, cf. Definition and Proposition 4.1.2 below and Vladimirov [280], Sect. 12, p. 185, where $\mathcal{F}, \mathcal{L}$ are defined slightly differently.   □

Example 1.6.8 Let us next investigate a one-dimensional Fourier transform which strikingly shows that the spaces $\mathcal{F}L^p$, $p \neq 2$, cannot be characterized by regularity or growth properties.

For this purpose, let us consider the entire function

$$T : \mathbf{C} \longrightarrow \mathcal{E}(\mathbf{R}^1) : \lambda \longmapsto T_\lambda = \exp(\lambda x + i e^x).$$

(a) For $\text{Re } \lambda \geq 0$, we have $T_\lambda \in \mathcal{S}'$ since, on the one hand, $Y(-x)T_\lambda \in L^\infty \subset \mathcal{S}'$ for $\text{Re } \lambda \geq 0$; on the other hand, $Y(x)T_\lambda \in L^1 \subset \mathcal{D}'_{L^1}$ for $\text{Re } \lambda < 0$, and the equation

$$\left(\frac{\mathrm{d}}{\mathrm{d}x} - \lambda\right)\left(Y(x)T_\lambda\right) = e^i \delta + iY(x)T_{\lambda+1} \qquad (1.6.10)$$

shows inductively that $Y(x)T_\lambda \in \mathcal{D}'_{L^1} \cap L^1_{\text{loc}}$ for each $\lambda \in \mathbf{C}$.

Let us observe that the evaluation of T_λ on a test function $\phi \in \mathcal{S}$ is, in general, *not* given by the integral $\int_\mathbf{R} \phi(x)T_\lambda(x) \, \mathrm{d}x$. Indeed, e.g. for $\lambda = 2$ and $\phi(x) = 1/\cosh(x) \in \mathcal{S}(\mathbf{R})$, obviously

$$\phi \cdot T_2 = \frac{\exp(2x + i e^x)}{\cosh x} \notin L^1(\mathbf{R}).$$

What is more conspicuous, even the improper Riemann integral

$$\int_{-\infty}^\infty \phi(x)T_2(x) \, \mathrm{d}x = \int_{-\infty}^\infty \frac{\exp(2x + i e^x)}{\cosh x} \, \mathrm{d}x = 2 \int_0^\infty \frac{t^2 e^{it}}{t^2 + 1} \, \mathrm{d}t$$

is apparently divergent, cf. the remark in Petersen [228], p. 84: "The extension …to $\mathcal{S}$, however, is not necessarily given by an integral." More precisely, for a temperate Radon measure $\mu \in \mathcal{S}'(\mathbf{R}^n) \cap \mathcal{M}(\mathbf{R}^n)$, the evaluation $\langle \phi, \mu \rangle$ can be written as $\int_{\mathbf{R}^n} \phi(x)\mu(x)$ for each $\phi \in \mathcal{S}$ if and only if μ is, *as a measure*, of polynomial growth, i.e., $\exists k \in \mathbf{N}_0 : \mu/(1 + |x|)^k \in \mathcal{M}^1(\mathbf{R}^n)$, see Schwartz [246], (VII, 4; 6), p. 241, or, equivalently, iff the function $f(r) = \int_{|x| \leq r} |\mu|(x)$ increases not faster than a polynomial, see Schwartz [246], (VII, 4; 7), p. 242; Strichartz [266], p. 44. This condition is not fulfilled for T_λ if $\text{Re } \lambda \neq 0$.

(b) For $0 < \operatorname{Re}\lambda < 1$, we have $Y(-x)T_\lambda \in L^1$ and $Y(x)T_{\lambda-1} \in L^1$, and hence, substituting $t = \mathrm{e}^x$ and using (1.6.10) and partial integration we obtain

$$\mathcal{F}T_\lambda = \mathcal{F}\Big[Y(-x)T_\lambda + \mathrm{i}\mathrm{e}^{\mathrm{i}}\delta - \mathrm{i}\Big(\frac{\mathrm{d}}{\mathrm{d}x} - \lambda + 1\Big)\big(Y(x)T_{\lambda-1}\big)\Big]$$

$$= \mathcal{F}\big(Y(-x)T_\lambda\big) + \mathrm{i}\mathrm{e}^{\mathrm{i}} + \big(\xi + \mathrm{i}(\lambda-1)\big)\mathcal{F}\big(Y(x)T_{\lambda-1}\big)$$

$$= \int_{-\infty}^{0} \exp(\lambda x + \mathrm{i}\mathrm{e}^x - \mathrm{i}x\xi)\,\mathrm{d}x + \mathrm{i}\mathrm{e}^{\mathrm{i}}$$

$$\quad + \big(\xi + \mathrm{i}(\lambda-1)\big)\int_{0}^{\infty} \exp\big((\lambda-1)x + \mathrm{i}\mathrm{e}^x - \mathrm{i}x\xi\big)\,\mathrm{d}x$$

$$= \int_{0}^{1} t^{\lambda-\mathrm{i}\xi-1}\mathrm{e}^{\mathrm{i}t}\,\mathrm{d}t + \mathrm{i}\mathrm{e}^{\mathrm{i}} + \big(\xi + \mathrm{i}(\lambda-1)\big)\int_{1}^{\infty} t^{\lambda-\mathrm{i}\xi-2}\mathrm{e}^{\mathrm{i}t}\,\mathrm{d}t$$

$$= \int_{0}^{\infty} t^{\lambda-\mathrm{i}\xi-1}\mathrm{e}^{\mathrm{i}t}\,\mathrm{d}t + \mathrm{i}\mathrm{e}^{\mathrm{i}} + \mathrm{i}t^{\lambda-\mathrm{i}\xi-1}\mathrm{e}^{\mathrm{i}t}\big|_{1}^{\infty}$$

$$= \int_{0}^{\infty} t^{\lambda-\mathrm{i}\xi-1}\mathrm{e}^{\mathrm{i}t}\,\mathrm{d}t = \Gamma(\lambda - \mathrm{i}\xi)\,\mathrm{i}^{\lambda-\mathrm{i}\xi} = \Gamma(\lambda - \mathrm{i}\xi)\,\mathrm{e}^{(\xi+\mathrm{i}\lambda)\pi/2}.$$

We point out that the integral $\int_{0}^{\infty} t^{\lambda-\mathrm{i}\xi-1}\mathrm{e}^{\mathrm{i}t}\,\mathrm{d}t$ is an improper Riemann integral, in contrast to the absolutely convergent integrals, which represent Fourier transforms of L^1-functions, in the beginning of the calculation.

Since the mappings

$$\{\lambda \in \mathbf{C};\ \operatorname{Re}\lambda > 0\} \longrightarrow \mathcal{D}'(\mathbf{R}) : \lambda \mapsto \mathcal{F}T_\lambda \ \text{and}\ \lambda \mapsto \Gamma(\lambda - \mathrm{i}\xi)\mathrm{e}^{(\xi+\mathrm{i}\lambda)\pi/2}$$

are both holomorphic and coincide for $0 < \operatorname{Re}\lambda < 1$, we conclude by analytic continuation that

$$\mathcal{F}T_\lambda = \Gamma(\lambda - \mathrm{i}\xi)\mathrm{e}^{(\xi+\mathrm{i}\lambda)\pi/2}, \qquad \operatorname{Re}\lambda > 0. \qquad (1.6.11)$$

By the continuity of the mapping

$$\{\lambda \in \mathbf{C};\ \operatorname{Re}\lambda \geq 0\} \longrightarrow \mathcal{S}'(\mathbf{R}) : \lambda \mapsto T_\lambda$$

we furthermore obtain from (1.1.2) that

$$\mathcal{F}T_0 = \mathcal{F}\exp(\mathrm{i}\mathrm{e}^x) = \lim_{\lambda\searrow 0} \Gamma(\lambda - \mathrm{i}\xi)\mathrm{e}^{(\xi+\mathrm{i}\lambda)\pi/2}$$

$$= \lim_{\lambda\searrow 0} \frac{\Gamma(\lambda + 1 - \mathrm{i}\xi)}{\lambda - \mathrm{i}\xi}\,\mathrm{e}^{(\xi+\mathrm{i}\lambda)\pi/2}$$

$$= \mathrm{i}\,\Gamma(1 - \mathrm{i}\xi)\mathrm{e}^{\pi\xi/2} \lim_{\lambda\searrow 0} \frac{1}{\xi + \mathrm{i}\lambda} = \mathrm{i}\,\Gamma(1 - \mathrm{i}\xi)\mathrm{e}^{\pi\xi/2}\,\mathrm{vp}\tfrac{1}{\xi} + \pi\delta.$$

$$(1.6.12)$$

(c) By means of Stirling's formula, see Gradshteyn and Ryzhik [113], Eq. 8.328.1, we can estimate the growth of $\mathcal{F}T_\lambda$ precisely:

$$\lim_{|\xi|\to\infty} |\Gamma(\lambda - i\xi)| e^{\pi |\operatorname{Im}\lambda - \xi|/2} |\operatorname{Im}\lambda - \xi|^{1/2 - \operatorname{Re}\lambda} = \sqrt{2\pi}.$$

In particular, (1.6.12) implies

$$\lim_{\xi\to\infty} \sqrt{\xi}\, |(\mathcal{F}T_0)(\xi)| = \sqrt{2\pi}, \qquad \text{and} \qquad \lim_{\xi\to-\infty} \sqrt{-\xi}\, |(\mathcal{F}T_0)(\xi)| = 0.$$

Therefore, the locally integrable function

$$f(\xi) := \mathcal{F}T_0 - i\,\mathrm{vp}\,\tfrac{1}{\xi} - \pi\delta = i\,\frac{\Gamma(1 - i\xi)e^{\pi\xi/2} - 1}{\xi}$$

belongs to $\cap_{q>2} L^q(\mathbf{R}) \cap C_0(\mathbf{R})$, whereas its inverse Fourier transform

$$(\mathcal{F}^{-1}f)(x) = T_0 - \check{Y} = \exp(ie^x) - Y(-x) \in L^\infty(\mathbf{R})$$

is not contained in any L^p-space, $1 \le p < \infty$. This shows that $\mathcal{F}L^1$ is a proper subspace of C_0 and that the Fourier transform according to the Hausdorff–Young theorem

$$\mathcal{F}: L^p(\mathbf{R}^n) \longrightarrow L^q(\mathbf{R}^n), \quad 1 \le p < 2, \ \tfrac{1}{p} + \tfrac{1}{q} = 1,$$

is not surjective.

   A much more involved concrete example with these properties is due to R. Salem, see Donoghue [61], Sect. 52, p. 265; Goldberg [111], pp. 8, 9; Stein and Weiss [264], remark 4.1, p. 31; Hörmander [139], Ex. 7.1.13, p. 385.

(d) The non-surjectivity of $\mathcal{F}: L^1(\mathbf{R}) \to C_0(\mathbf{R})$ can also be seen more easily by employing the function $g(\xi) = (\exp(i\xi^2) - 1)/\xi \in C_0(\mathbf{R})$ with the non-integrable inverse Fourier transform

$$\mathcal{F}^{-1}g = \left[\mathcal{F}^{-1}(e^{i\xi^2} - 1)\right] * \left[\mathcal{F}^{-1}(\mathrm{vp}\,\tfrac{1}{\xi})\right] = \left(\tfrac{1}{2}\sqrt{\tfrac{i}{\pi}}\, e^{-ix^2/4} - \delta\right) * \tfrac{i}{2}\,\mathrm{sign}\,x$$

$$= -\frac{i}{2}\,\mathrm{sign}\,x + \frac{1}{2}i\sqrt{\frac{i}{\pi}}\int_0^x e^{-it^2/4}\,dt = -\frac{i}{2}\,\mathrm{sign}\,x$$

$$+ \frac{i^{3/2}}{\sqrt{2}}\left[C\!\left(\tfrac{x}{2}\right) - iS\!\left(\tfrac{x}{2}\right)\right],$$

see Gradshteyn and Ryzhik [113], Section 8.25, for the definition and the asymptotic of Fresnel's integrals S, C. (For $\mathcal{F}^{-1}(e^{i\xi^2})$ see Example 1.6.9 (d) below; in order to justify the use of the Fourier exchange theorem, one could recur to Schwartz [246], Ch. VII, Thm. XV, p. 268, since $e^{-ix^2/4} \in \mathcal{O}'_C$, $\mathrm{sign}\,x \in \mathcal{S}'$.)

A different proof of the non-surjectivity of $\mathcal{F} : L^1(\mathbf{R}) \to \mathcal{C}_0(\mathbf{R})$ relies on Banach's homomorphism theorem. Since this mapping is continuous and injective, surjectivity would imply that the inverse mapping $\mathcal{F}^{-1} : \mathcal{C}_0(\mathbf{R}) \to L^1(\mathbf{R})$ was likewise continuous, i.e.,

$$\exists C > 0 : \forall f \in L^1(\mathbf{R}) : \|f\|_1 \leq C\|\mathcal{F}f\|_\infty.$$

A sequence of functions contradicting this inequality is

$$f_k(x) = \frac{1 - \cos(kx)}{x}\, e^{-|x|} \in L^1(\mathbf{R}), \qquad k \in \mathbf{N}.$$

By Proposition 1.6.6, (5), the Fourier transform of f_k can be calculated:

$$(\mathcal{F}f_k)(\xi) = -2\pi \mathcal{F}^{-1}\Big(\frac{1 - \cos(kx)}{x}\, e^{-|x|}\Big) = -2\pi \mathcal{F}^{-1}\Big(\frac{1 - \cos(kx)}{x}\Big) * \mathcal{F}^{-1}\big(e^{-|x|}\big)$$

$$= i\pi\big[Y(k - |\xi|)\,\text{sign}\,\xi\big] * \frac{1}{\pi(1 + \xi^2)}$$

$$= i\big(\arctan(\xi + k) + \arctan(\xi - k) - 2\arctan\xi\big).$$

Whereas $\|\mathcal{F}f_k\|_\infty$ is bounded by π for each $k \in \mathbf{N}$, we have

$$\|f_k\|_1 = 2\int_0^\infty \frac{1 - \cos(kx)}{x}\, e^{-x}\,dx = \log(1 + k^2) \to \infty, \qquad k \to \infty,$$

by Gröbner and Hofreiter [115], 336.8c, p. 139. Other slightly less explicit sequences of this kind can be found in Jörgens [153], ex. 13.5, p. 200; Rudin [240], Sect. 9, ex. 2, p. 193; Larsen [159], Thm. 7.8.1, p. 198.   □

Example 1.6.9 Let us next investigate the calculation of Fourier transforms of integrable distributions by means of approximation.

(a) Smooth approximation.
    If $T = \sum_{|\alpha| \leq m} \partial^\alpha f_\alpha \in \mathcal{D}'_{L^1}$, $f_\alpha \in L^1$, then $\mathcal{F}T$ is a continuous function of polynomial growth (see Proposition 1.6.6), and

$$(\mathcal{F}T)(\xi) = {}_{\mathcal{D}_{L^\infty}}\langle e^{-ix\xi}, T\rangle_{\mathcal{D}'_{L^1}} = \lim_{R \to \infty}\langle e^{-ix\xi}, \phi(\tfrac{x}{R})T\rangle$$

for arbitrary $\phi \in \mathcal{D}_{L^\infty}$ with $\phi(0) = 1$. In fact,

$$\langle e^{-ix\xi}, \phi(\tfrac{x}{R})T\rangle = \sum_{|\alpha| \leq m} (-1)^{|\alpha|}\int_{\mathbf{R}^n} f_\alpha(x)\partial_x^\alpha\big(e^{-ix\xi}\phi(\tfrac{x}{R})\big)\,dx$$

$$\to \sum_{|\alpha| \leq m} (i\xi)^\alpha \int_{\mathbf{R}^n} f_\alpha(x)e^{-ix\xi}\,dx = (\mathcal{F}T)(\xi)$$

by Lebesgue's theorem.

(b) Approximation by cut-off ("partial summation").

Let us suppose here that $f \in \mathcal{D}'_{L^1}(\mathbf{R}^n) \cap L^1_{\mathrm{loc}}(\mathbf{R}^n)$ and that $K \subset \mathbf{R}^n$ is a measurable neighborhood of 0 such that $\chi_{RK} f \to f$ in $\mathcal{D}'_{L^1}$ for $R \to \infty$. (For a set $A \subset \mathbf{R}^n$, we denote the characteristic function of A by χ_A.) Then

$$(\mathcal{F}f)(\xi) = {}_{\mathcal{D}_{L^\infty}}\langle e^{-ix\xi}, f \rangle_{\mathcal{D}'_{L^1}} = \lim_{R \to \infty} \langle e^{-ix\xi}, \chi_{RK} f \rangle = \lim_{R \to \infty} \int_{RK} e^{-ix\xi} f(x)\, dx,$$

i.e., the Fourier transform of f is given by the conditionally convergent integral

$$\int_{\mathbf{R}^n} e^{-ix\xi} f(x)\, dx := \lim_{R \to \infty} \int_{RK} e^{-ix\xi} f(x)\, dx.$$

(c) Let us illustrate the above procedure first in the case of $T_\lambda = \exp(\lambda x + ie^x)$ considered in Example 1.6.8. For $\lambda = 1, f = T_1 \in \mathcal{D}'_{L^1} \cap L^1_{\mathrm{loc}}$ by (1.6.10). But if we set $K = [-1, 1]$, and $\xi = 0$, then the corresponding limit

$$\lim_{R \to \infty} \int_{RK} e^{-ix\xi} f(x)\, dx = \lim_{R \to \infty} \int_{-R}^{R} \exp(x + ie^x)\, dx = \lim_{R \to \infty} \int_{\exp(-R)}^{\exp R} e^{it}\, dt$$

does not exist, and hence $\mathcal{F}f(0) = i$ (see formula (1.6.11)) cannot be calculated by cut-off, and $\chi_{[-R,R]} f = Y(R - |x|)\exp(x + ie^x)$ cannot converge to $f = \exp(x + ie^x)$ in $\mathcal{D}'_{L^1}$.

In contrast, for $0 < \operatorname{Re}\lambda < 1$, the cut-offs $Y(R - |x|)T_\lambda$ converge to T_λ in $\mathcal{D}'_{L^1}$. In fact, $Y(-x)T_\lambda \in L^1$ and hence $Y(-R - x)T_\lambda$ converges to 0 in L^1, and all the more so in $\mathcal{D}'_{L^1}$. On the other hand, as in (1.6.10),

$$Y(x - R)T_\lambda = Y(x - R)\exp(\lambda x + ie^x)$$

$$= -i\Big(\frac{d}{dx} - \lambda + 1\Big)\big[Y(x - R)\exp((\lambda - 1)x + ie^x)\big]$$

$$+ i\delta_R \cdot \exp((\lambda - 1)R + ie^R)$$

converges to 0 in $\mathcal{D}'_{L^1}$ for $R \to \infty$ since $Y(x)\exp((\lambda - 1)x + ie^x) \in L^1$ and $|\phi(R)|\exp((\operatorname{Re}\lambda - 1)R) \to 0$ for $\phi \in \mathcal{D}_{L^\infty}$.

Therefore, for $0 < \operatorname{Re}\lambda < 1$, the Fourier transform of T_λ is given by the improper Riemann integral

$$(\mathcal{F}T_\lambda)(\xi) = \int_{-\infty}^{\infty} \exp((\lambda - i\xi)x + ie^x)\, dx = \int_{0}^{\infty} e^{it} t^{\lambda - i\xi - 1}\, dt$$

as has been proved directly in Example 1.6.8 (b).

(d) Finally, let us consider $f(x) = \exp(i|x|^2) \in \mathcal{D}'_{L^1}(\mathbf{R}^n) \cap L^1_{\text{loc}}(\mathbf{R}^n)$, see Example 1.5.7. If we put $K = [-1, 1]^n$, then

$$\chi_{RK} f = \prod_{j=1}^{n} Y(R - |x_j|)\, e^{ix_j^2}$$

converges to f in $\mathcal{D}'_{L^1}(\mathbf{R}^n)$ for $R \to 0$. To verify this, it suffices to consider the case $n = 1$. By Example 1.5.7, we have

$$f = f_1 + f_2'', \ f_1(x) = \frac{3}{2}\frac{e^{ix^2} - 1 - ix^2 e^{ix^2}}{x^4} \in L^1(\mathbf{R}), \ f_2(x) = \frac{1 - e^{ix^2}}{4x^2} \in L^1(\mathbf{R}).$$

Therefore, by the jump formula (1.3.5),

$$Y(|x| - R)f = Y(|x| - R)f_1 + \big(Y(|x| - R)f_2\big)'' - f_2(R)\delta_R' + f_2(-R)\delta_{-R}'$$
$$- f_2'(R)\delta_R + f_2'(-R)\delta_{-R}$$

converges to 0 in $\mathcal{D}'_{L^1}(\mathbf{R})$ due to $f_2, f_2' \in \mathcal{C}_0(\mathbf{R})$.

Hence, according to what has been said in (b), the Fourier transform of f is given by the following improper integral:

$$(\mathcal{F}e^{i|x|^2})(\xi) = \prod_{j=1}^{n} \int_{-\infty}^{\infty} e^{i|x|^2 - ix_j\xi_j}\, dx_j = e^{-i|\xi|^2/4}\left(\int_{-\infty}^{\infty} e^{it^2}\, dt\right)^n = (i\pi)^{n/2} e^{-i|\xi|^2/4},$$

see Gröbner and Hofreiter [115], 334.1a/2a, p. 131.

Let us observe—following A. Cayley, cf. Bass [9], p. 624—that, in contrast, for $n \geq 2$ and $K = \{x \in \mathbf{R}^n; |x| \leq 1\}$, the sequence $\chi_{RK} f$ does *not* converge to f in $\mathcal{D}'_{L^1}(\mathbf{R}^n)$ for $R \to \infty$, and $\mathcal{F}f$ cannot be calculated as the *pointwise limit* $\lim_{R \to \infty} \int_{|x| \leq R} f(x) e^{-ix\xi}\, dx$. In fact, the improper integral

$$\lim_{R \to \infty} \int_{|x| \leq R} e^{i|x|^2}\, dx = |S^{n-1}| \int_0^{\infty} r^{n-1} e^{ir^2}\, dr = \frac{1}{2}|S^{n-1}| \int_0^{\infty} t^{n/2 - 1} e^{it}\, dt$$

diverges for $n \geq 2$.

(e) The above example of $e^{i|x|^2}$ shows that, in general, the Fourier transform of $f \in \mathcal{D}'_{L^1} \cap L^1_{\text{loc}}$ cannot be represented as the limit of the Fourier transforms of radial cut-offs. However, for $T \in \mathcal{D}'_{L^1}$, its Fourier transform $\mathcal{F}T$ can always be represented by spherical *Riesz means of higher order*:

If $T = \sum_{|\alpha| \leq m} \partial^\alpha f_\alpha$, $f_\alpha \in L^1$, and $\phi \in \mathcal{D}_{L^\infty}$, then

$$\langle \phi, Y(R - |x|)\left(1 - \tfrac{|x|^2}{R^2}\right)^m T \rangle$$

$$= \sum_{|\alpha| \leq m} (-1)^\alpha \int_{\mathbf{R}^n} f_\alpha(x) \partial^\alpha \left[Y(R - |x|)\left(1 - \tfrac{|x|^2}{R^2}\right)^m \phi(x) \right] dx$$

$$\to \sum_{|\alpha| \leq m} (-1)^\alpha \int_{\mathbf{R}^n} f_\alpha(x)(\partial^\alpha \phi)(x)\, dx = \langle \phi, T \rangle, \qquad R \to \infty,$$

by Lebesgue's theorem, and hence $\lim_{R \to \infty} Y(R - |x|)\left(1 - \tfrac{|x|^2}{R^2}\right)^m T = T$ holds in $\mathcal{D}'_{L^1}$ and

$$\forall \xi \in \mathbf{R}^n : (\mathcal{F}T)(\xi) = \lim_{R \to \infty} \langle Y(R - |x|)\left(1 - \tfrac{|x|^2}{R^2}\right)^m T_x, e^{-ix\xi} \rangle.$$

Similar spherical Riesz approximations of the Fourier transform can be found in Gonzalez-Vieli [112], Prop. 1, p. 293.

If m is the smallest integer such that a representation $T = \sum_{|\alpha| \leq m} \partial^\alpha f_\alpha$, $f_\alpha \in L^1$, exists, then $\lim_{R \to \infty} Y(R - |x|)\left(1 - \tfrac{|x|^2}{R^2}\right)^l T = T$ holds in $\mathcal{D}'_{L^1}$—according to the above reasoning—for all $l \geq m$. Note that this limit relation can happen to be valid already for $l < m$. This is, e.g., the case for $T = e^{ix^2} \in \mathcal{D}'_{L^1}(\mathbf{R}^1)$, where $m \geq 1$ and $l = 0$ can be taken (see (d)), or $T = e^{i|x|^2} \in \mathcal{D}'_{L^1}(\mathbf{R}^n)$, where $l = [\tfrac{n}{2}]$ works. $\qquad\qquad\qquad\qquad\qquad\qquad\qquad\qquad\qquad\qquad\qquad\qquad\qquad\square$

Example 1.6.10 Let us consider now the Fourier transformation on radially symmetric distributions.

(a) Let us first repeat from Example 1.5.15 that $T \in \mathcal{D}'(\mathbf{R}^n)$ or $T \in \mathcal{D}'(\mathbf{R}^n \setminus \{0\})$ is called *radially symmetric* if and only if $T \circ A = T$ for each A in the orthogonal group $O_n(\mathbf{R})$. We denote by $\mathcal{D}'_r(\mathbf{R}^n)$, $\mathcal{D}'_r(\mathbf{R}^n \setminus \{0\})$ the corresponding closed subspaces of radially symmetric distributions in $\mathcal{D}'(\mathbf{R}^n)$ and in $\mathcal{D}'(\mathbf{R}^n \setminus \{0\})$, respectively.

Since

$$\mathbf{R}^n \setminus \{0\} \longrightarrow (0, \infty) \times \mathbf{S}^{n-1} : x \longmapsto \left(|x|, \tfrac{x}{|x|}\right)$$

is a diffeomorphism, we obtain the following isomorphisms:

$$\mathcal{D}'_r(\mathbf{R}^1 \setminus \{0\}) \overset{\sim}{\to} \mathcal{D}'\big((0, \infty)\big) \overset{\sim}{\to} \mathcal{D}'_r(\mathbf{R}^n \setminus \{0\}) : T \mapsto T|_{(0,\infty)} \mapsto T \circ |x|.$$

Unfortunately, if $\phi \in \mathcal{D}_r(\mathbf{R}^1 \setminus \{0\})$ with $\int_{-\infty}^\infty \phi(t)\, dt = 1$, then $\epsilon^{-1}\phi(\tfrac{t}{\epsilon}) \to \delta$ in $\mathcal{D}'_r(\mathbf{R}^1)$ for $\epsilon \to 0$, whereas $\epsilon^{-1}\phi(\tfrac{|x|}{\epsilon}) \to 0$ in $\mathcal{D}'_r(\mathbf{R}^n)$ for $n \geq 2$, see Example 1.1.11. Hence the above isomorphisms cannot produce an isomorphism from $\mathcal{D}'_r(\mathbf{R}^1)$ to $\mathcal{D}'_r(\mathbf{R}^n)$ by continuous extension.

In the literature (see Schwartz [243], Exp. 7: "Les opérateurs invariants par rotation, l'opérateur Δ"; Treves [271], Lecture 5: "Rotation invariant differential polynomials"; Treves [272], Section 3.10, p. 249), this problem is solved by considering the mapping

$$\mathcal{D}'\big((0,\infty)\big) \xrightarrow{\sim} \mathcal{D}'_r(\mathbf{R}^n \setminus \{0\}) : T \longmapsto |x|^{2-n} \cdot (T \circ |x|^2),$$

which allows an extension to the isomorphism

$$\mathcal{D}'\big([0,\infty)\big) \xrightarrow{\sim} \mathcal{D}'_r(\mathbf{R}^n) : T \longmapsto \Big(\phi \mapsto \tfrac{1}{2}\langle \int_{\mathbf{S}^{n-1}} \phi(\sqrt{t}\,\omega)\,d\omega, T_t\rangle\Big).$$

Here $\mathcal{D}'\big([0,\infty)\big)$ is the dual of $\mathcal{D}\big([0,\infty)\big) = \{\phi : [0,\infty) \to \mathbf{C}\ \mathcal{C}^\infty;$ supp ϕ compact$\}$, and we have an isomorphism

$$\mathcal{D}'\big([0,\infty)\big) \xrightarrow{\sim} \{S \in \mathcal{D}'(\mathbf{R}^1); \text{supp}\,S \subset [0,\infty)\} : T \longmapsto (\phi \mapsto \langle\phi|_{[0,\infty)}, T\rangle).$$

We shall select instead a different route. If we extend the mapping

$$\mathcal{D}'_r(\mathbf{R}^1 \setminus \{0\}) \xrightarrow{\sim} \mathcal{D}'_r(\mathbf{R}^n \setminus \{0\}) : T \longmapsto |x|^{1-n} \cdot (T \circ |x|),$$

by continuity, we obtain the isomorphism

$$\Phi : \mathcal{D}'_r(\mathbf{R}^1) \xrightarrow{\sim} \mathcal{D}'_r(\mathbf{R}^n) : T \longmapsto \Big(\phi \mapsto \tfrac{1}{2}\langle \int_{\mathbf{S}^{n-1}} \phi(t\,\omega)\,d\omega, T_t\rangle\Big), \qquad (1.6.13)$$

which is the transpose of the isomorphism

$$\mathcal{D}_r(\mathbf{R}^n) \xrightarrow{\sim} \mathcal{D}_r(\mathbf{R}^1) : \phi \longmapsto \Big(t \mapsto \tfrac{|\mathbf{S}^{n-1}|}{2}\,\phi(t,0,\ldots,0)\Big)$$

with the inverse

$$\mathcal{D}_r(\mathbf{R}^1) \xrightarrow{\sim} \mathcal{D}_r(\mathbf{R}^n) : \psi \longmapsto \Big(x \mapsto \tfrac{2}{|\mathbf{S}^{n-1}|}\,\psi(|x|)\Big).$$

For example, this implies that

$$\{T \in \mathcal{D}'_r(\mathbf{R}^n); \text{supp}\,T \subset \{0\}\} = \Phi\Big(\Big\{\sum_{k=0}^{m} a_k \delta^{(2k)};\ a_k \in \mathbf{C},\ k=0,\ldots,m \in \mathbf{N}_0\Big\}\Big)$$

$$= \Big\{\sum_{k=0}^{m} a_k \Delta^k \delta;\ a_k \in \mathbf{C},\ k=0,\ldots,m \in \mathbf{N}_0\Big\},$$

since, for $\phi \in \mathcal{D}(\mathbf{R}^n)$,

$$\langle \phi, \Phi(\delta^{(2k)}) \rangle = \frac{1}{2} \left(\frac{d^{2k}}{dt^{2k}} \int_{S^{n-1}} \phi(t\omega)\, d\omega \right)\Big|_{t=0}$$

$$= \frac{1}{2} \int_{S^{n-1}} ((\omega^T \nabla)^{2k} \phi)(0)\, d\omega = \frac{\pi^{(n-1)/2} \Gamma(k + \frac{1}{2})}{\Gamma(k + \frac{n}{2})} (\Delta^k \phi)(0),$$

see Horváth, Ortner and Wagner [147], p. 445, for the last equation.

In other words, a radially symmetric distribution with support at the origin is a linear combination of powers of the Laplacian applied to δ. Similarly, a radially symmetric distribution supported by the sphere $|x| = R$, $R > 0$, is a linear combination of powers of the operator $\sum_{j=1}^{n} x_j \partial_j$ applied to the single layer distribution $S_{RS^{n-1}}(1)$ defined in Example 1.2.14.

(b) Let $\mathcal{S}'_r(\mathbf{R}^n)$ denote the space of temperate, radially symmetric distributions. Because of Proposition 1.6.6, (1), $\mathcal{F} : \mathcal{S}'_r(\mathbf{R}^n) \xrightarrow{\sim} \mathcal{S}'_r(\mathbf{R}^n)$ is well defined and an isomorphism. Combining the Fourier transformation with the isomorphism Φ in (1.6.13) we therefore obtain an isomorphism $\mathcal{H} : \mathcal{S}'_r(\mathbf{R}^1) \xrightarrow{\sim} \mathcal{S}'_r(\mathbf{R}^1)$ (a kind of Hankel transform) such that the following diagram commutes:

$$
\begin{array}{ccc}
\mathcal{S}'_r(\mathbf{R}^1) & \xrightarrow{\ \sim\ } & \mathcal{S}'_r(\mathbf{R}^1) \\
{\scriptstyle \Phi}\downarrow{\scriptstyle \wr} & & {\scriptstyle \Phi}\downarrow{\scriptstyle \wr} \\
\mathcal{S}'_r(\mathbf{R}^n) & \xrightarrow[\ \mathcal{F}\ ]{} & \mathcal{S}'_r(\mathbf{R}^n)
\end{array}
$$

For a radially symmetric *integrable* function $f(x) = g(|x|)$, a classical calculation yields the *Poisson–Bochner formula* for $\mathcal{F}f$, see Bochner [17], Sect. 43.5, (15), p. 235; Schwartz [246], (VII, 7; 22), p. 259:

$$(\mathcal{F}f)(\xi) = \int_{\mathbf{R}^n} e^{-ix\xi} f(x)\, dx = \int_{\mathbf{R}^n} e^{-i|\xi||x_n|} g(|x|)\, dx$$

$$= |S^{n-2}| \int_0^\infty g(r) r^{n-1}\, dr \int_0^\pi e^{-i|\xi|r\cos\theta} \sin^{n-2}\theta\, d\theta$$

$$= (2\pi)^{n/2} |\xi|^{-n/2+1} \int_0^\infty g(r) r^{n/2} J_{n/2-1}(r|\xi|)\, dr, \qquad (1.6.14)$$

see Gröbner and Hofreiter [115], 511.11a, p. 189. Hence, for $g = \check{g} \in L^1(\mathbf{R})$, we have $g \in \mathcal{S}'_r(\mathbf{R}^1)$ and

$$(\mathcal{H}g)(t) = (2\pi|t|)^{n/2} \int_0^\infty g(r) r^{-n/2+1} J_{n/2-1}(r|t|)\, dr.$$

By density, we conclude that $\mathcal{H} : S_r'(\mathbf{R}^1) \xrightarrow{\sim} S_r'(\mathbf{R}^1)$ is the transpose of the map

$$\tilde{\mathcal{H}} : S_r(\mathbf{R}^1) \xrightarrow{\sim} S_r(\mathbf{R}^1) : \phi \longmapsto (2\pi)^{n/2} \int_0^\infty s^{n-1} \phi(s)(s|t|)^{-n/2+1} J_{n/2-1}(s|t|) \, ds.$$

$\mathcal{H}$ is called the *generalized Hankel transformation*, cf. Zemanian [307], Ch. V; Brychkov and Prudnikov [28], 6.4, p. 82.

In particular, if $S \in \mathcal{D}_{L^1}'(\mathbf{R}^n) \cap \mathcal{D}_r'(\mathbf{R}^n)$, and $S = \Phi(T)$, $T \in \mathcal{D}_r'(\mathbf{R}^1)$, then $\mathcal{F}S = \Phi(\mathcal{H}T) \in \mathcal{C}(\mathbf{R}^n)$ is defined pointwise and given by

$$(\mathcal{F}S)(\xi) = \frac{1}{2}(2\pi)^{n/2} \langle |t\xi|^{-n/2+1} J_{n/2-1}(|t\xi|), T_t \rangle.$$

Note that $z^{-n/2+1} J_{n/2-1}(z)$ is an even entire function, which, restricted to $\mathbf{R}$, belongs to $\mathcal{D}_{L^\infty}(\mathbf{R})$, and that $T \in \mathcal{D}_{L^1}'(\mathbf{R}^1)$.

(c) A particular case, which, on the other hand, is equivalent with the Poisson–Bochner formula (1.6.14), arises if $S = \delta(|x| - R) = S_{RS^{n-1}}(1)$, $R > 0$, i.e., S is a uniform mass distribution on the sphere $M = RS^{n-1}$, cf. Example 1.2.14. Then $S = \Phi T$ with $T = R^{n-1}(\delta_R + \delta_{-R}) \in \mathcal{D}_r'(\mathbf{R}^1) \cap \mathcal{E}'(\mathbf{R}^1)$, and

$$\mathcal{F}S = \mathcal{F}\big(\delta(|x| - R)\big) = (2\pi R)^{n/2} |\xi|^{-n/2+1} J_{n/2-1}(R|\xi|). \tag{1.6.15}$$

By integration with respect to R, we obtain

$$\mathcal{F}\big(Y(R - |x|)\big) = \left(\frac{2\pi R}{|\xi|}\right)^{n/2} J_{n/2}(R|\xi|).$$

Note that, for $n \geq 2$, $f(\xi) = |\xi|^{-n/2+1} J_{n/2-1}(R|\xi|)$ yields probably the simplest example of a function in $C_0(\mathbf{R}^n) \cap \bigcap_{q > \frac{2n}{n-1}} L^q(\mathbf{R}^n)$ such that the inverse Fourier transform $\mathcal{F}^{-1}f = (2\pi R)^{-n/2} \delta(|x| - R)$ does not belong to any L^p-space, cf. Example 1.6.8 (c) for $n = 1$. $\qquad\square$

Example 1.6.11

(a) Let us apply the Poisson–Bochner formula (1.6.14) to the radially symmetric functions $f(x) = (|x|^2 + \epsilon^2)^\lambda$, $\epsilon > 0$, $\lambda \in \mathbf{C}$. For $\operatorname{Re} \lambda < -\frac{n}{2}$, they are integrable, and hence

$$\mathcal{F}\big((|x|^2 + \epsilon^2)^\lambda\big) = (2\pi)^{n/2} |\xi|^{-n/2+1} \int_0^\infty (r^2 + \epsilon^2)^\lambda r^{n/2} J_{n/2-1}(r|\xi|) \, dr$$

$$= \frac{2\pi^{n/2}}{\Gamma(-\lambda)} \left(\frac{2\epsilon}{|\xi|}\right)^{n/2+\lambda} K_{n/2+\lambda}(\epsilon|\xi|), \tag{1.6.16}$$

1.6 The Fourier Transformation 99

where K_ν denotes McDonald's function, see Gradshteyn and Ryzhik [113], Eq. 6.565.4.

Since the mapping

$$T : \mathbf{C} \longrightarrow \mathcal{S}'(\mathbf{R}^n) : \lambda \longmapsto T_\lambda = (|x|^2 + \epsilon^2)^\lambda$$

is entire, we conclude that

$$S : \{\lambda \in \mathbf{C}; \operatorname{Re}\lambda < -\tfrac{n}{2}\} \longrightarrow \mathcal{S}'(\mathbf{R}^n) : \lambda \longmapsto \frac{2\pi^{n/2}}{\Gamma(-\lambda)}\left(\frac{2\epsilon}{|\xi|}\right)^{n/2+\lambda} K_{n/2+\lambda}(\epsilon|\xi|)$$

can analytically be continued in $\mathcal{S}'(\mathbf{R}^n)$ to all $\lambda \in \mathbf{C}$. If this continuation is denoted by S as well, then $S_\lambda = \mathcal{F}T_\lambda$ holds for all $\lambda \in \mathbf{C}$. For $\operatorname{Re}\lambda < 0$, S_λ is an integrable function, and (1.6.16) thus persists unchanged for $\operatorname{Re}\lambda < 0$.

In particular, for $\lambda = -k$, $k \in \mathbf{N}$, we obtain anew the unique temperate fundamental solution E of the *iterated metaharmonic operator* $(\Delta_n - \epsilon^2)^k$, $\epsilon > 0$:

$$E = \mathcal{F}^{-1}\big((-|\xi|^2 - \epsilon^2)^{-k}\big) = \frac{(-1)^k \epsilon^{n/2-k}|x|^{-n/2+k}}{2^{n/2+k-1}\pi^{n/2}(k-1)!} K_{n/2-k}(\epsilon|x|),$$

cf. formula (1.4.9).

If we define the *Bessel kernel* by $G_{\lambda,\epsilon} = \mathcal{F}^{-1}\big((|\xi|^2 + \epsilon^2)^{-\lambda/2}\big)$, $\epsilon > 0$, $\lambda \in \mathbf{C}$, cf. Schwartz [246], (VII, 7; 23), p. 260; Donoghue [61], p. 292, then

$$G_{\lambda,\epsilon} = \frac{1}{2^{(\lambda+n)/2-1}\pi^{n/2}\Gamma(\frac{\lambda}{2})}\left(\frac{|x|}{\epsilon}\right)^{(\lambda-n)/2} K_{(n-\lambda)/2}(\epsilon|x|), \qquad x \neq 0, \ \lambda \notin -2\mathbf{N}_0,$$

and hence $G_{\lambda,\epsilon} \in L^1 + \mathcal{E}' \subset \mathcal{D}'_{L^1} \subset \mathcal{D}'_{L^2}$ and the Fourier exchange theorem (see Proposition 1.6.6 (5)) yields

$$\forall \lambda, \mu \in \mathbf{C} : G_{\lambda,\epsilon} * G_{\mu,\epsilon} = G_{\lambda+\mu,\epsilon} \tag{1.6.17}$$

cf. Schwartz [246], (VI, 8; 5), p. 204; Petersen [228], Ch. II, Sect. 9, Ex. 9.1, p. 107. Due to $G_{-2k,\epsilon} = (\epsilon^2 - \Delta_n)^k\delta$, $k \in \mathbf{N}_0$, the *composition law* (1.6.17) also comprises the fact noted above that $G_{2k,\epsilon}$ is the unique fundamental solution in $\mathcal{S}'$ of $(\epsilon^2 - \Delta_n)^k$, $k \in \mathbf{N}_0$.

(b) The limit $\epsilon \searrow 0$ in formula (1.6.16) yields the *elliptic M. Riesz kernels*. For $\operatorname{Re}\lambda > -n$, we have $|x|^\lambda = \lim_{\epsilon\searrow 0}(|x|^2 + \epsilon^2)^{\lambda/2}$ in $\mathcal{S}'(\mathbf{R}^n)$, and hence

$$\mathcal{F}(|x|^\lambda) = \lim_{\epsilon\searrow 0}\left[\frac{2\pi^{n/2}}{\Gamma(-\frac{\lambda}{2})}\left(\frac{2\epsilon}{|\xi|}\right)^{(\lambda+n)/2} K_{(\lambda+n)/2}(\epsilon|\xi|)\right]$$

$$= \frac{2^{\lambda+n}\pi^{n/2}\Gamma(\frac{\lambda+n}{2})}{\Gamma(-\frac{\lambda}{2})}|\xi|^{-\lambda-n}, \qquad -n < \operatorname{Re}\lambda < 0, \tag{1.6.18}$$

see Abramowitz and Stegun [1], 9.6.2, 9.6.10, p. 375.

We now define the M. Riesz kernels by $R_\lambda := \mathcal{F}^{-1}(|\xi|^{-\lambda})$, where

$$\mathbf{C} \setminus (n + 2\mathbf{N}_0) \longrightarrow \mathcal{S}'(\mathbf{R}^n) : \lambda \longmapsto |\xi|^{-\lambda}$$

is holomorphic and has simple poles for $\lambda = n + 2k$, $k \in \mathbf{N}_0$; in these poles, we set $|\xi|^{-n-2k} := \mathrm{Pf}_{\lambda=n+2k} |\xi|^{-\lambda}$, see Example 1.4.9. Then $\lambda \mapsto R_\lambda$ is also holomorphic in $\mathbf{C} \setminus (n + 2\mathbf{N}_0)$, and formula (1.6.18) implies by analytic continuation that

$$R_\lambda = \frac{\Gamma(\frac{n-\lambda}{2})}{2^\lambda \pi^{n/2} \Gamma(\frac{\lambda}{2})} |x|^{\lambda-n}, \qquad \lambda \notin n + 2\mathbf{N}_0, \ \lambda \notin -2\mathbf{N}_0,$$

cf. Riesz [235], (1), p. 16; E.M. Stein [263], (4), p. 117; Horváth [146], p. 180; Wagner [287], Bsp. 1, p. 413.

For $\mathrm{Re}\,\lambda, \mathrm{Re}\,\mu < \frac{n}{2}$, we have $R_\lambda, R_\mu \in L^2 + \mathcal{E}'$, and the Fourier exchange formula in Proposition 1.6.6 (5) then implies the "composition law" $R_\lambda * R_\mu = R_{\lambda+\mu}$. This convolution equation is generally valid for $\mathrm{Re}(\lambda + \mu) < n$, see Ortner [199], pp. 44–46; Ortner and Wagner [219], Ex. 3.3.2, p. 96.

In particular, due to

$$R_{-2k} = \mathcal{F}^{-1}(|\xi|^{2k}) = (-\Delta_n)^k \delta, \qquad k \in \mathbf{N}_0,$$

this composition law also implies that $E = (-1)^k R_{2k}$ is a fundamental solution of Δ_n^k provided $\lambda = 2k$ is not a pole of R_λ, i.e., $2k \notin n + 2\mathbf{N}_0$, or, equivalently, n odd or [n even and $k < \frac{n}{2}$]. Therefore, we obtain as fundamental solution of Δ_n^k in these cases

$$E = (-1)^k R_{2k} = \frac{(-1)^k \Gamma(\frac{n}{2} - k)}{2^{2k}(k-1)! \pi^{n/2}} |x|^{2k-n}, \qquad n \text{ odd or } k < \frac{n}{2}, \qquad (1.6.19)$$

in agreement with the result for $k = 1$ in Example 1.3.14 (a).

Let us remark that, for $k < \frac{n}{2}$, E is the only *homogeneous* fundamental solution of Δ_n^k. In fact, the equation $\Delta_n^k F = \delta$ implies for homogeneous F that F is homogeneous of degree $2k-n$. Furthermore, since homogeneous distributions are temperate, see Donoghue [61], Sect. 32, p. 154, we infer $|x|^{2k} \cdot \mathcal{F}(F-E) = 0$ and thus $\mathrm{supp}\big(\mathcal{F}(F - E)\big) \subset \{0\}$. Therefore $F - E$ is a polynomial of degree $2k - n < 0$ and hence vanishes, cf. also Proposition 2.4.8 (1).

In the excluded cases, where $n = 2l$, $l \in \mathbf{N}$, and $k \geq l$, we have

$$S = \operatorname*{Res}_{\lambda=2k} R_\lambda = \frac{|x|^{2k-n}}{2^{2k}(k-1)! \pi^l} \operatorname*{Res}_{\lambda=2k} \Gamma\Big(\frac{n-\lambda}{2}\Big) = \frac{(-1)^{k-l+1} |x|^{2k-n}}{2^{2k-1}(k-1)!(k-l)! \pi^l}$$

and $\Delta_n^k S = 0$, and hence

$$(-\Delta_n)^k R_{2k} = (-\Delta_n)^k \left(\Pf_{\lambda=2k} R_\lambda \right)$$

$$= (-\Delta_n)^k \lim_{\lambda \to 2k} \left(R_\lambda - \frac{S}{\lambda - 2k} \right) = \lim_{\lambda \to 2k} R_{\lambda-2k} = R_0 = \delta,$$

cf. Ortner [198], pp. 6–9. Thus $(-1)^k R_{2k}$ yields also in these cases a fundamental solution of the polyharmonic operator Δ_n^k. Since

$$R_{2k} = \Pf_{\lambda=2k} \left(\frac{\Gamma(\frac{n-\lambda}{2})}{2^\lambda \pi^{n/2} \Gamma(\frac{\lambda}{2})} |x|^{\lambda-n} \right)$$

$$= |x|^{2(k-l)} \cdot \Pf_{\lambda=2k} \left(\frac{\Gamma(\frac{n-\lambda}{2})}{2^\lambda \pi^{n/2} \Gamma(\frac{\lambda}{2})} \right) + \frac{\partial |x|^{\lambda-2l}}{\partial \lambda} \bigg|_{\lambda=2k} \cdot \Res_{\lambda=2k} \left(\frac{\Gamma(\frac{n-\lambda}{2})}{2^\lambda \pi^{n/2} \Gamma(\frac{\lambda}{2})} \right),$$

and the first summand on the right-hand side is a polynomial solution to $\Delta_{2l}^k u = 0$, we eventually obtain the following fundamental solution E of Δ_{2l}^k:

$$E = \frac{(-1)^{l-1} |x|^{2(k-l)} \log|x|}{2^{2k-1}(k-1)!(k-l)!\pi^l}, \qquad k \geq l \in \mathbb{N}, \tag{1.6.20}$$

cf. Schwartz [246], (VII, 10; 21), p. 288; Ortner and Wagner [219], Ex. 2.7.2, p. 68.   □

Example 1.6.12 In a similar vein, we can investigate the *Poisson kernel*, which appears in the solution of the Dirichlet problem for Δ_n in the half-space. Applying the Poisson–Bochner formula (1.6.14) to the functions $e^{-\lambda|x|}$, $\Re \lambda > 0$, we obtain

$$\mathcal{F}(e^{-\lambda|x|}) = (2\pi)^{n/2} |\xi|^{-n/2+1} \int_0^\infty e^{-\lambda r} r^{n/2} J_{n/2-1}(r|\xi|) \, dr$$

$$= \frac{2^n \pi^{(n-1)/2} \lambda \Gamma(\frac{n+1}{2})}{(\lambda^2 + |\xi|^2)^{(n+1)/2}},$$

see Gradshteyn and Ryzhik [113], Eq. 6.623.2. This yields the Poisson kernels

$$P_\lambda = \mathcal{F}^{-1}(e^{-\lambda|\xi|}) = \frac{\lambda \Gamma(\frac{n+1}{2})}{[\pi(\lambda^2 + |x|^2)]^{(n+1)/2}},$$

which satisfy the composition law $P_\lambda * P_\mu = P_{\lambda+\mu}$ for $\Re \lambda$, $\Re \mu > 0$. The limit relation $\lim_{\epsilon \searrow 0} P_\epsilon = \mathcal{F}^{-1}(1) = \delta$ appeared already in Example 1.1.11 (ii).

Example 1.6.13 Let us next employ the Poisson–Bochner formula in order to calculate $\mathcal{F}(J_0(R|x|))$ in even dimensions, cf. Ibragimov and Mamontov [148], Thm. 2.1, p. 352. In odd dimensions, we shall obtain in the same way $\mathcal{F}\left(\frac{\sin(R|x|)}{|x|}\right)$.

Upon using a partial Fourier transformation (see Definition 1.6.15 below), these two Fourier transforms will furnish the forward fundamental solution of the *wave operator* $\partial_t^2 - \Delta_n$, cf. Example 1.6.17 below.

(a) Let us first deduce—by means of partial integrations—a slight modification of the Poisson–Bochner formula. We shall set $k = [\frac{n}{2} - 1]$ and assume that $f(x) = g(|x|) \in L^1(\mathbf{R}^n)$ is radially symmetric and such that $h(s) = Y(s)g(\sqrt{s}) \in L^1(\mathbf{R})$ and fulfills the condition

$$\lim_{s \to \infty} s^{(n-2j-3)/4} \cdot (h * s_+^j) = 0, \qquad j = 0, \dots, k-1, \text{ where } s_+^j = Y(s)s^j.$$

If we use the substitution $r = \sqrt{s}$ and partial integration in the Poisson–Bochner formula (1.6.14), we obtain

$$
\begin{aligned}
(\mathcal{F}f)(\xi) &= \frac{1}{2}(2\pi)^{n/2}|\xi|^{-n/2+1} \int_0^\infty h(s)s^{n/4-1/2}J_{n/2-1}(\sqrt{s}\,|\xi|)\,\mathrm{d}s \\
&= \frac{(-1)^k(2\pi)^{n/2}}{2(k-1)!|\xi|^{n/2-1}} \int_0^\infty (h * s_+^{k-1})(s)\left(\frac{\mathrm{d}}{\mathrm{d}s}\right)^k\left[s^{n/4-1/2}J_{n/2-1}(\sqrt{s}\,|\xi|)\right]\mathrm{d}s \\
&= \frac{(-1)^k\pi^{n/2}}{(k-1)!}\left(\frac{2}{|\xi|}\right)^{n/2-k-1} \times \\
&\quad \times \int_0^\infty (h * s_+^{k-1})(s)s^{n/4-k/2-1/2}J_{n/2-k-1}(\sqrt{s}\,|\xi|)\,\mathrm{d}s,
\end{aligned}
$$

by Gradshteyn and Ryzhik [113], Eq. 8.472.3. With the abbreviations

$$h_0 = h, \quad h_k = \frac{1}{(k-1)!}\,h * s_+^{k-1} = Y(s)\int_0^s \frac{(s-\sigma)^{k-1}}{(k-1)!}\,h(\sigma)\,\mathrm{d}\sigma,$$

this implies

$$(\mathcal{F}f)(\xi) = (-1)^k\pi^{n/2}\int_0^\infty h_k(s)\left\{\begin{array}{c} J_0(\sqrt{s}\,|\xi|) \\ \frac{2\sin(\sqrt{s}\,|\xi|)}{\sqrt{\pi}\,|\xi|} \end{array}\right\}\,\mathrm{d}s : \quad \left\{\begin{array}{l} \text{if } n \text{ is even,} \\ \text{if } n \text{ is odd.} \end{array}\right. \qquad (1.6.21)$$

By density, this formula can be extended to such $h \in \mathcal{E}'((0,\infty))$ for which also $h_k \in \mathcal{E}'((0,\infty))$.

(b) In particular, if $h = \delta_{R^2}^{(k)}$, $R > 0$, then

$$h_k = h(s) * \frac{s_+^{k-1}}{(k-1)!} = \delta_{R^2} * \left(\frac{s_+^{k-1}}{(k-1)!}\right)^{(k)} = \delta_{R^2},$$

and hence (1.6.21) yields

$$\mathcal{F}(\delta_{R^2}^{(k)} \circ |x|^2) = (-1)^k \pi^{n/2} \begin{cases} J_0(R|\xi|), & \text{if } n \text{ is even,} \\ \frac{2}{\sqrt{\pi}|\xi|} \sin(R|\xi|), & \text{if } n \text{ is odd} \end{cases}, \qquad k = [\tfrac{n}{2} - 1].$$

Conversely, this implies, for $R > 0$, the following:

$$\mathcal{F}(J_0(R|x|)) = 2^n \pi^{n/2} (-1)^{n/2-1} \delta_{R^2}^{(n/2-1)} \circ |\xi|^2 \in \mathcal{S}'(\mathbf{R}_\xi^n), \qquad n \text{ even,}$$
(1.6.22)

and

$$\mathcal{F}\left(\frac{\sin(R|x|)}{|x|}\right) = 2^{n-1} \pi^{(n+1)/2} (-1)^{(n-3)/2} \delta_{R^2}^{\left(\frac{n-3}{2}\right)} \circ |\xi|^2 \in \mathcal{S}'(\mathbf{R}_\xi^n), \qquad n \text{ odd.}$$
(1.6.23)

When evaluated on test functions, the composition $\delta_{R^2}^{(k)} \circ |\xi|^2$ is—according to Definition 1.2.12—given by

$$\langle \phi, \delta_{R^2}^{(k)} \circ |\xi|^2 \rangle = \frac{(-1)^k}{2} \left(\frac{d^k}{ds^k} \int_{S^{n-1}} s^{n/2-1} \phi(\sqrt{s}\,\omega)\, d\omega \right) \bigg|_{s=R^2}, \qquad \phi \in \mathcal{D}(\mathbf{R}^n).$$

The result in Ibragimov and Mamontov [148] is formulated in this way.

Naturally, $\delta_{R^2}^{(k)} \circ |x|^2$ can also be expressed by composition with $|x|$. In fact, if $g : (0, \infty) \to (0, \infty) : t \mapsto t^2$, then

$$\delta_{R^2}^{(k)} \circ |x|^2 = (g^*(\delta_{R^2}^{(k)})) \circ |x| = \sum_{j=0}^{k} c_j \delta_R^{(j)} \circ |x|,$$

and a short calculation yields

$$c_j = \frac{k!}{2R^{2k+1-j}j!} \sum_{i=0}^{j} \binom{j}{i} \binom{(i-1)/2}{k} (-1)^{i+k}.$$

□

Example 1.6.14 Let us next determine the Fourier transform of *Gaussian kernels*. For $n = 1$, Cauchy's integral theorem implies

$$\mathcal{F}(e^{-x^2}) = e^{-\xi^2/4} \int_{\mathbf{R}} \exp\left(-\left(x + \tfrac{i\xi}{2}\right)^2\right) dx = \sqrt{\pi}\, e^{-\xi^2/4}.$$

For positive $\lambda_1, \ldots, \lambda_n$, this yields in $\mathbf{R}^n$

$$\mathcal{F}\left(\exp\left(-\lambda_1 x_1^2 - \cdots - \lambda_n x_n^2\right)\right) = \frac{\pi^{n/2}}{\sqrt{\lambda_1 \cdots \lambda_n}} \exp\left(-\frac{\xi_1^2}{4\lambda_1} - \cdots - \frac{\xi_n^2}{4\lambda_n}\right).$$

For a positive definite, symmetric matrix $A \in \mathbf{R}^{n \times n}$, diagonalization furnishes

$$\mathcal{F}\big(e^{-x^T A x}\big) = \frac{\pi^{n/2}}{\sqrt{\det A}} \, \exp\big(-\tfrac{1}{4} \xi^T A^{-1} \xi\big). \qquad (1.6.24)$$

Finally, by analytic continuation, we see that (1.6.24) persists as long as A belongs to the set U_1, introduced already in (1.4.12), of symmetric complex non-singular matrices with positive semi-definite real part, cf. Hörmander [139], Thm. 7.6.1, p. 206; [140], Lemma 2.4.3, p. 44; Zorich [308], p. 160.

In fact, the mapping

$$U_1 \longrightarrow \mathcal{S}'(\mathbf{R}^n) : A \longmapsto \exp(-x^T A x)$$

is continuous, and it is analytic on

$$U = \{A = A^T \in \mathbf{C}^{n \times n}; \ \mathrm{Re} \, A \text{ is positive definite}\},$$

which is an analytic submanifold of $\mathbf{C}^{n \times n}$. The analytic continuation of $\sqrt{\det A}$ to U is given by

$$\sqrt{\det A} = \pi^{n/2} \bigg(\int_{\mathbf{R}^n} \exp(-x^T A x) \, dx \bigg)^{-1},$$

see Example 1.4.13. Furthermore, the continuity of

$$U_1 \longrightarrow \mathcal{D}'(\mathbf{R}^n) : A \longmapsto \mathcal{F}\big(\exp(-x^T A x)\big) \cdot \exp\big(\tfrac{1}{4} \xi^T A^{-1} \xi\big) = \frac{\pi^{n/2}}{\sqrt{\det A}}$$

implies that $\sqrt{\det A}$ can continuously be extended as a non-vanishing function from U to its closure $U_1 = \overline{U}$. Therefore, both sides of (1.6.24) are well-defined and continuous on U_1, and analytic if restricted to U. Since they coincide for real-valued positive definite symmetric matrices, the equality in (1.6.24) is valid on all of U_1.

In one way or other, this can also be found in Zuily [309], Exercises 66, 78, pp. 141, 145; Vladimirov [280], pp. 114–117; Strichartz [266], pp. 47, 48; Friedlander and Joshi [84], Exercise 8.9, p. 111. Note that the calculation of $\mathcal{F}\big(e^{-x^T A x}\big)$ for $A \in U_1$ by continuity amounts to an approximation with Gauß–Weierstraß kernels, since $A + \epsilon I \in U$ for $A \in U_1$, $\epsilon > 0$, and $e^{-x^T (A + \epsilon I) x} = e^{-\epsilon |x|^2} \cdot e^{-x^T A x}$. □

Definition and Proposition 1.6.15

(1) *For $\phi \in \mathcal{S}(\mathbf{R}_x^m \times \mathbf{R}_y^n)$, the partial Fourier transform defined by*

$$(\mathcal{F}_y \phi)(x, \eta) = \int_{\mathbf{R}^n} \phi(x, y) e^{-i\eta y} \, dy$$

also belongs to $S(\mathbf{R}^{m+n})$, *and* $\mathcal{F}_y : S \to S$ *is continuous. Therefore, the adjoint*

$$\mathcal{F}_y : S'(\mathbf{R}^{m+n}) \longrightarrow S'(\mathbf{R}^{m+n}) : T \longmapsto (\phi \mapsto \langle \mathcal{F}_y \phi, T \rangle)$$

is well defined and continuous. It is again called partial Fourier transform.
(2) *For a subset* $A \subset \mathbf{R}_x^m$, *the mapping*

$$\mathcal{F}_y : C\big(A, S'(\mathbf{R}_y^n)\big) \longrightarrow C\big(A, S'(\mathbf{R}_y^n)\big) : f \longmapsto \big(x \mapsto \mathcal{F}(f(x))\big)$$

is also called partial Fourier transform. The two mappings in (1) *and* (2) *coincide on* $C\big(\mathbf{R}^m, S'(\mathbf{R}^n)\big) \cap S'(\mathbf{R}^{m+n})$.

Proof The statement in (1) is proved analogously as in Proposition 1.6.4, which could be conceived as the special case $m = 0$ of the partial Fourier transform. The consistency statement in (2) is proven by density. $\qquad\square$

Example 1.6.16 Let us determine a fundamental solution of the *heat operator* $\partial_t - \Delta_n$ by partial Fourier transform.

If $E \in S'(\mathbf{R}_{t,x}^{n+1})$ fulfills $(\partial_t - \Delta_n)E = \delta$, then partial Fourier transform with respect to x yields

$$\partial_t(\mathcal{F}_x E) + |\xi|^2 \cdot \mathcal{F}_x E = \delta(t) \otimes 1(\xi).$$

For fixed $\xi \in \mathbf{R}^n$, the only fundamental solution of the ordinary differential operator $\frac{d}{dt} + |\xi|^2$ with support in $[0, \infty)$ is $Y(t) \exp(-t|\xi|^2)$, see Example 1.3.6. The equation

$$(\partial_t + |\xi|^2)\big(Y(t)e^{-t|\xi|^2}\big) = \delta(t) \otimes 1(\xi)$$

then holds in $S'(\mathbf{R}_{t,\xi}^{n+1})$, and hence $E = \mathcal{F}_\xi^{-1}\big(Y(t)e^{-t|\xi|^2}\big)$ is a fundamental solution of the heat operator $\partial_t - \Delta_n$.

In order to calculate this inverse partial Fourier transform, one uses that

$$\mathbf{R} \longrightarrow S'(\mathbf{R}_\xi^n) : t \longmapsto Y(t)e^{-t|\xi|^2}$$

is continuous outside 0 and has limits from both sides at 0. Hence a slight generalization of Proposition 1.6.15 (2) and formula (1.6.24) imply that

$$E = \mathcal{F}_\xi^{-1}\big(Y(t)e^{-t|\xi|^2}\big) = \frac{Y(t)}{(4\pi t)^{n/2}} \exp\left(-\frac{|x|^2}{4t}\right).$$

We have verified this fundamental solution already in Example 1.3.14 (d).

Let us remark that E is the only temperate fundamental solution with support in the half-space $[0, \infty) \times \mathbf{R}^n$. Indeed, if $(\partial_t - \Delta_n)F = 0$, $F \in S'(\mathbf{R}_{t,x}^{n+1})$, and $\operatorname{supp} F \subset [0, \infty) \times \mathbf{R}^n$, then $e^{-\epsilon t}F \in S'$ for $\epsilon > 0$ and $(\partial_t + \epsilon - \Delta_n)(e^{-\epsilon t}F) = 0$, and hence $(i\tau + \epsilon + |\xi|^2)\mathcal{F}(e^{-\epsilon t}F) = 0$, which implies, successively, $\mathcal{F}(e^{-\epsilon t}F) = 0$, $e^{-\epsilon t}F = 0$, and thus $F = 0$. $\qquad\square$

Example 1.6.17 Let us next use the partial Fourier transformation as in Example 1.6.16 to deduce the *forward fundamental solution of the wave operator* $\partial_t^2 - \Delta_n$. For the spatial dimensions $n = 2$ and $n = 3$, respectively, we have already derived this fundamental solution in Example 1.4.12 (b) from that of the Laplacean by analytic continuation with respect to parameters.

If $(\partial_t^2 - \Delta_n)E = \delta$ and $E \in \mathcal{S}'(\mathbf{R}^{n+1})$, then the partial Fourier transform with respect to x fulfills

$$(\partial_t^2 + |\xi|^2)\mathcal{F}_x E = \delta(t) \otimes 1(\xi). \tag{1.6.25}$$

For fixed $\xi \in \mathbf{R}^n$, the only fundamental solution of the ordinary differential operator $\frac{d^2}{dt^2} + |\xi|^2$ with support in $[0, \infty)$ is $Y(t)\frac{\sin(t|\xi|)}{|\xi|}$, see Example 1.3.8 (a), and it satisfies (1.6.25) in $\mathcal{S}'(\mathbf{R}^{n+1})$. Hence $E = \mathcal{F}_\xi^{-1}\big(Y(t)\frac{\sin(t|\xi|)}{|\xi|}\big)$ is a fundamental solution of $\partial_t^2 - \Delta_n$, and it is the only one with support contained in the half-space $t \geq 0$, see Hörmander [138], Thm. 12.5.1, p. 120.

In this case, the function

$$\mathbf{R} \longrightarrow \mathcal{S}'(\mathbf{R}_\xi^n) : t \longmapsto Y(t)\frac{\sin(t|\xi|)}{|\xi|}$$

is continuous, and we can therefore, by Proposition 1.6.15 (2), calculate the inverse Fourier transform with respect to ξ for t fixed. Thus for *odd n*, formula (1.6.23) in Example 1.6.13 yields

$$E = \frac{(-1)^{(n-3)/2}Y(t)}{2\pi^{(n-1)/2}}\delta_{t^2}^{(\frac{n-3}{2})} \circ |x|^2 = \frac{Y(t)}{2\pi^{(n-1)/2}}\delta^{(\frac{n-3}{2})}(t^2 - |x|^2)$$

$$= \frac{Y(t)}{(2\pi)^{(n-1)/2}}\left(\frac{1}{t}\frac{\partial}{\partial t}\right)^{\frac{n-3}{2}}\delta(t^2 - |x|^2)$$

$$= \frac{Y(t)}{(2\pi)^{(n-1)/2}}\left(\frac{1}{t}\frac{\partial}{\partial t}\right)^{\frac{n-3}{2}}\left(\frac{1}{2t}\delta(t - |x|)\right), \qquad n \text{ odd}, n \geq 3. \tag{1.6.26}$$

Note that $E \in \mathcal{S}'(\mathbf{R}^{n+1}) \cap \mathcal{C}(\mathbf{R}_t^1, \mathcal{E}'(\mathbf{R}_x^n))$; the explicit expression for E in (1.6.26) above holds for $t \neq 0$. Formula (1.6.26) is well-known, cf. Shilov [250], 4.7.1, pp. 288–290; Schwartz [241], (17,25), p. 47.

In contrast, for *even n*, we shall represent $Y(t)\frac{\sin(t|\xi|)}{|\xi|}$ as a Bessel transform in order to be able to apply formula (1.6.22). For $t > 0$, we have by Gradshteyn and Ryzhik [113], Eq. 6.554.2, or Oberhettinger [194], 2.7, p. 6, the representation

$$\frac{\sin(t|\xi|)}{|\xi|} = \int_0^t \frac{rJ_0(r|\xi|)}{\sqrt{t^2 - r^2}}\,dr.$$

By means of formula (1.6.22), this yields

$$E = \mathcal{F}_\xi^{-1}\left(Y(t)\frac{\sin(t|\xi|)}{|\xi|}\right) = Y(t)\int_0^t \frac{r}{\sqrt{t^2-r^2}}\mathcal{F}_\xi^{-1}\big(J_0(r|\xi|)\big)\,dr$$

$$= \frac{(-1)^{n/2-1}Y(t)}{\pi^{n/2}}\int_0^t \frac{r}{\sqrt{t^2-r^2}}\delta_{r^2}^{(n/2-1)}\circ|x|^2\,dr$$

$$= \frac{(-1)^{n/2-1}Y(t)}{2\pi^{n/2}}\int_0^{t^2}\delta^{(n/2-1)}(|x|^2-v)\frac{dv}{\sqrt{t^2-v}}$$

$$= \frac{(-1)^{n/2-1}Y(t)}{2\pi^{n/2}}\langle(t^2-|x|^2+s)_+^{-1/2},\delta^{(n/2-1)}(s)\rangle$$

$$= \frac{(-1)^{n/2-1}(n-3)!!}{(2\pi)^{n/2}}Y(t)s_+^{(1-n)/2}\circ(t^2-|x|^2), \qquad n\text{ even}, \qquad (1.6.27)$$

cf. Schwartz [241], (17,31), p. 47. Note that the composition in (1.6.27) is well defined outside the origin since there $\nabla(t^2-|x|^2)\neq 0$, and that $E \in \mathcal{C}\big(\mathbf{R}_t^1,\mathcal{E}'(\mathbf{R}_x^n)\big)$ is already determined by $E|_{\mathbf{R}^{n+1}\setminus\{0\}}$. We observe furthermore that the wave operator $\partial_t^2 - \Delta_n$ is hyperbolic, and thus the fundamental solutions E derived above are the only ones with support in the half-space $t \geq 0$, cf. Proposition 2.4.11 below, or Hörmander [138], Def. 12.3.3, p. 112, and Thm. 12.5.1, p. 120. □

Example 1.6.18 Let us also use partial Fourier transformation in order to determine $H = \mathcal{F}^{-1}T$ if $T = Y(\tau)\delta(\tau^2-|\xi|^2-m^2) \in \mathcal{S}'(\mathbf{R}_{\tau,\xi}^{n+1})$ for fixed $m > 0$. Since $(\tau^2-|\xi|^2-m^2)T = 0$, the distribution H solves the *Klein–Gordon equation*, i.e., $(\partial_t^2-\Delta_n+m^2)H = 0$. Note also that T and H are relativistically invariant, i.e., they are invariant under transformations in the orthochrone Lorentz group. Actually, H yields the kernel for the state space of a spinless elementary particle, a meson, see Schwartz [244], 31, p. 197; [247], Ch. 4, p. 68. In Bogolubov, Logunov and Todorov [18], App. E, (E.4), p. 334, H is called *Pauli–Jordan positive frequency function*.

Before calculating H let us observe that $f(\tau,\xi) = \tau^2-|\xi|^2-m^2$ is submersive in $\mathbf{R}^{n+1}\setminus\{0\}$ and hence $\delta(\tau^2-|\xi|^2-m^2) = f^*\delta \in \mathcal{D}'(\mathbf{R}^{n+1}\setminus\{0\})$ is well defined by Definition 1.2.12. Furthermore, the singular supports of $f^*\delta$ and $Y(\tau)$ are disjoint sets, and hence $T = Y(\tau)\delta(\tau^2-|\xi|^2-m^2) \in \mathcal{S}'(\mathbf{R}^{n+1})$ is also well-defined.

For $\phi \in \mathcal{D}(\mathbf{R}^{n+1})$, Definition 1.2.12 implies

$$\langle\phi,T\rangle = \langle\frac{d}{ds}\int_{\mathbf{R}^n}\left(\int_0^{\sqrt{s+|\xi|^2+m^2}}\phi(\tau,\xi)\,d\tau\right)d\xi,\delta(s)\rangle$$

$$= \int_{\mathbf{R}^n}\frac{\phi(\sqrt{|\xi|^2+m^2},\xi)}{2\sqrt{|\xi|^2+m^2}}\,d\xi.$$

This shows that $T \in \mathcal{C}\big(\mathbf{R}_\xi^n,\mathcal{S}'(\mathbf{R}_\tau^1)\big)$ and that T is given by

$$T: \mathbf{R}_\xi^n \longrightarrow \mathcal{S}'(\mathbf{R}_\tau^1): \xi \longmapsto \frac{1}{2\sqrt{|\xi|^2+m^2}}\delta_{\sqrt{|\xi|^2+m^2}}(\tau).$$

Therefore, Definition 1.6.15 yields

$$H = \mathcal{F}^{-1}T = \mathcal{F}_{\xi}^{-1}(\mathcal{F}_{\tau}^{-1}T) = \mathcal{F}_{\xi}^{-1}\left(\frac{e^{it\sqrt{|\xi|^2+m^2}}}{4\pi\sqrt{|\xi|^2+m^2}}\right).$$

The distribution $S = e^{it\sqrt{|\xi|^2+m^2}}/(4\pi\sqrt{|\xi|^2+m^2})$ continuously depends on t, and hence the partial inverse Fourier transform with respect to ξ can be calculated for t fixed, see Proposition 1.6.15, and $H = \mathcal{F}^{-1}T = \mathcal{F}_{\xi}^{-1}S \in \mathcal{C}(\mathbf{R}_t^1, \mathcal{S}'(\mathbf{R}_x^n))$.

Let us first determine $\mathcal{F}_{\xi}^{-1}S_z$ for

$$S_z := \frac{e^{-z\sqrt{|\xi|^2+m^2}}}{4\pi\sqrt{|\xi|^2+m^2}} \in L^1(\mathbf{R}_{\xi}^n), \qquad z \in \mathbf{C}, \ \mathrm{Re}\, z > 0.$$

Then the Poisson–Bochner formula (1.6.14) and the substitution $\sqrt{r^2+m^2} = ms$ furnish

$$\mathcal{F}_{\xi}^{-1}S_z = \frac{1}{2^{n/2+2}\pi^{n/2+1}|x|^{n/2-1}}\int_0^{\infty}\frac{e^{-z\sqrt{r^2+m^2}}}{\sqrt{r^2+m^2}}r^{n/2}J_{n/2-1}(r|x|)\,\mathrm{d}r$$

$$= \frac{m^{n/2}}{2^{n/2+2}\pi^{n/2+1}|x|^{n/2-1}}\int_1^{\infty}e^{-zms}(s^2-1)^{(n-2)/4}J_{(n-2)/2}(m\sqrt{s^2-1}|x|)\,\mathrm{d}s$$

$$= \frac{m^{(n-1)/2}}{(2\pi)^{(n+3)/2}}(|x|^2+z^2)^{-(n-1)/4}K_{(n-1)/2}(m\sqrt{|x|^2+z^2})$$

by Gradshteyn and Ryzhik [113], Eq. 6.645.2, p. 721. Hence, for fixed $t \in \mathbf{R}$, $H(t)$ is given by the following limit in $\mathcal{S}'(\mathbf{R}_x^n)$:

$$H(t) = \mathcal{F}_{\xi}^{-1}(S_{-it}) = \lim_{\epsilon \searrow 0}\mathcal{F}_{\xi}^{-1}(S_{\epsilon-it})$$

$$= \lim_{\epsilon \searrow 0}\frac{m^{(n-1)/2}}{(2\pi)^{(n+3)/2}}(|x|^2-t^2-2i\epsilon t)^{-(n-1)/4}K_{(n-1)/2}(m\sqrt{|x|^2-t^2-2i\epsilon t}).$$

Let us finally calculate this limit if the space dimension n is at most 3. If $n = 2$, then Gradshteyn and Ryzhik [113], Eq. 8.468, p. 967, yields

$$H(t) = \lim_{\epsilon \searrow 0}\frac{\sqrt{m}}{(2\pi)^{5/2}}(|x|^2-t^2-2i\epsilon t)^{-1/4}K_{1/2}(m\sqrt{|x|^2-t^2-2i\epsilon t})$$

$$= \lim_{\epsilon \searrow 0}\frac{1}{8\pi^2\sqrt{|x|^2-t^2-2i\epsilon t}}e^{-m\sqrt{|x|^2-t^2-2i\epsilon t}}$$

$$= \frac{Y(|x|-|t|)}{8\pi^2\sqrt{|x|^2-t^2}}e^{-m\sqrt{|x|^2-t^2}}$$

$$+ \frac{Y(|t|-|x|)}{8\pi^2\sqrt{t^2-|x|^2}}\left[-\sin(m\sqrt{t^2-|x|^2})+i\,\mathrm{sign}\,t\cdot\cos(m\sqrt{t^2-|x|^2})\right],$$

which is locally integrable in $\mathbf{R}_x^2$. Similarly, for $n = 1$, we use Gradshteyn and Ryzhik [113], Eqs. 8.407.2 and 8.405 to conclude that $H(t)$ is locally integrable and given by

$$H(t) = \frac{Y(|x| - |t|)}{4\pi^2} K_0\big(m\sqrt{x^2 - t^2}\big)$$

$$+ \frac{Y(|t| - |x|)}{8\pi} \big[-N_0\big(m\sqrt{t^2 - x^2}\big) + i\operatorname{sign} t \cdot J_0\big(m\sqrt{t^2 - x^2}\big)\big].$$

$$(1.6.28)$$

We remark that the last formula also yields—after a linear transformation and a differentiation—the Fourier transform of the distribution $x_2\delta(x_1x_2 - 1)$ considered in Example 1.5.11.

The case $n = 3$ is more complicated. For $|t| \neq |x|$, we obtain as above that

$$H(t) = \frac{mY(|x| - |t|)}{8\pi^3 \sqrt{|x|^2 - t^2}} K_1\big(m\sqrt{|x|^2 - t^2}\big)$$

$$+ \frac{mY(|t| - |x|)}{16\pi^2 \sqrt{t^2 - |x|^2}} \big[N_1\big(m\sqrt{t^2 - |x|^2}\big) - i\operatorname{sign} t \cdot J_1\big(m\sqrt{t^2 - |x|^2}\big)\big].$$

Note that—except for $t = 0$, where $H(0) = \frac{m}{8\pi^3|x|} K_1(m|x|) \in L^1_{\text{loc}}(\mathbf{R}^3)$—the distributions $H(t)$ are *not* locally integrable in the case $n = 3$, since $K_1(\epsilon) \sim \epsilon^{-1}$ for $\epsilon \searrow 0$. However, due to $N_1(\epsilon) \sim -\frac{2}{\pi\epsilon}$ for $\epsilon \searrow 0$, the principal value

$$H_1(t) := \operatorname{vp} H(t) = \lim_{\epsilon \searrow 0} Y\big(\big||x| - |t|\big| - \epsilon\big)H(t) \in \mathcal{S}'(\mathbf{R}_x^3)$$

is well defined, and we conclude that the support of $H(t) - H_1(t)$ is contained in the section $|x| = |t|$ of the light cone. In order to determine H on the light cone, we consider the limit

$$H_2(t) := \lim_{\epsilon \searrow 0} \frac{m}{8\pi^3} \left[\frac{K_1\big(m\sqrt{|x|^2 - t^2 - 2i\epsilon t}\big)}{\sqrt{|x|^2 - t^2 - 2i\epsilon t}} - \frac{1}{m(|x|^2 - t^2 - 2i\epsilon t)} \right].$$

Here Lebesgue's theorem can be applied and yields the locally integrable function $H_2(t) = H_1(t) - 1/\big(8\pi^3(|x|^2 - t^2)\big)$. On the other hand, Sokhotski's formula (1.1.2) implies that

$$\lim_{\epsilon \searrow 0} \frac{1}{|x|^2 - t^2 - 2i\epsilon t} = \operatorname{vp}\Big(\frac{1}{|x|^2 - t^2}\Big) + \frac{i\pi}{2t}\delta(|x| - |t|) \in \mathcal{D}'(\mathbf{R}_x^3)$$

for fixed $t \neq 0$. Therefore, we finally obtain for $n = 3$ the following:

$$H(t) = \mathrm{vp}\left(\frac{mY(|x| - |t|)}{8\pi^3 \sqrt{|x|^2 - t^2}} K_1\left(m\sqrt{|x|^2 - t^2}\right) + \frac{mY(|t| - |x|)}{16\pi^2 \sqrt{t^2 - |x|^2}} N_1\left(m\sqrt{t^2 - |x|^2}\right)\right)$$

$$- \mathrm{i}\,\mathrm{sign}\,t \cdot \frac{mY(|t| - |x|)}{16\pi^2 \sqrt{t^2 - |x|^2}} J_1\left(m\sqrt{t^2 - |x|^2}\right)$$

$$+ \frac{\mathrm{i}}{16\pi^2 t} \delta(|t| - |x|), \qquad t \neq 0. \tag{1.6.29}$$

For formula (1.6.29), cf. Schwartz [244], 29.4, p. 186, and 31.5, p. 200; [247], pp. 83, 84; Methée [176, 177]; Bogolubov, Logunov and Todorov [18], App. E, (E.4), p. 334; Ortner and Wagner [223], Cor. 1 (d), p. 139.

Let us yet sketch the connection of H with the forward fundamental solution E (see Definition 2.4.12) of the Klein–Gordon operator $P(\partial) = \partial_t^2 - \Delta_n + m^2$. As we shall expound systematically below (see Props. 2.4.13, 4.4.1, in particular formula (4.4.4)), we have

$$E = \mathcal{F}_{\tau,\xi}^{-1}\left(\lim_{\epsilon \searrow 0}\left[(\mathrm{i}\tau + \epsilon)^2 + |\xi|^2 + m^2\right]^{-1}\right)$$

$$= Y(t)\mathcal{F}_{\tau,\xi}^{-1}\left(\lim_{\epsilon \searrow 0}\left[(\mathrm{i}\tau + \epsilon)^2 + |\xi|^2 + m^2\right]^{-1} - \left[(\mathrm{i}\tau - \epsilon)^2 + |\xi|^2 + m^2\right]^{-1}\right)$$

$$= -2\mathrm{i}\pi Y(t)\mathcal{F}_{\tau,\xi}^{-1}\left(\mathrm{sign}\,\tau \cdot \delta(\tau^2 - |\xi|^2 - m^2)\right) = -2\mathrm{i}\pi Y(t)(H - \check{H}).$$

In particular, for $n = 3$, we obtain

$$E = \frac{\delta(t - |x|)}{4\pi t} - \frac{mY(t - |x|)}{4\pi \sqrt{t^2 - |x|^2}} J_1\left(m\sqrt{t^2 - |x|^2}\right)$$

in accordance with formula (2.3.16) below.

Note that in quantum field theory, instead of E, the fundamental solution

$$\tilde{E} = \mathcal{F}_{\tau,\xi}^{-1}\left(\lim_{\epsilon \searrow 0}\left[-\tau^2 + |\xi|^2 + m^2 - \mathrm{i}\epsilon\right]^{-1}\right),$$

the so-called *Feynman propagator*, is employed, see, e.g., Zeidler [305], Section 14.2.2, p. 776, where $\tilde{E}$ is denoted by $\mathrm{R} = \mathrm{i}\,G_{F,m}$. Since, by Sokhotski's formula,

$$\lim_{\epsilon \searrow 0}\left[-\tau^2 + |\xi|^2 + m^2 - \mathrm{i}\epsilon\right]^{-1} = \mathrm{vp}\left[-\tau^2 + |\xi|^2 + m^2\right]^{-1} + \mathrm{i}\pi\delta\left(-\tau^2 + |\xi|^2 + m^2\right)$$

$$= \lim_{\epsilon \searrow 0}\left[(\mathrm{i}\tau + \epsilon)^2 + |\xi|^2 + m^2\right]^{-1} + 2\mathrm{i}\pi Y(\tau)\delta\left(-\tau^2 + |\xi|^2 + m^2\right)$$

we conclude that

$$\tilde{E} = E + 2i\pi H = 2i\pi\big[Y(-t)H + Y(t)\check{H}\big].$$

For $n = 3$, this equation yields the result in Zeidler [305], Thm. 14.3, p. 780. □

Example 1.6.19 Let us next investigate *Fourier series of periodic distributions.*

(a) $T \in \mathcal{D}'(\mathbf{R}^n)$ is called *periodic* if $T \circ \tau_k = T$ for all $k \in \mathbf{Z}^n$, where $\tau_a : \mathbf{R}^n \to$
    $\mathbf{R}^n : x \mapsto x - a$ denotes the translations. By $\mathbf{T}^n = \mathbf{R}^n/\mathbf{Z}^n$, we denote the
    torus with the induced $\mathcal{C}^\infty$-manifold structure. If, furthermore, $\mathcal{D}'_p(\mathbf{R}^n) = \{T \in$
    $\mathcal{D}'(\mathbf{R}^n); T \text{ periodic}\}$, and $\mathcal{D}'(\mathbf{T}^n)$ is the dual of $\mathcal{C}^\infty(\mathbf{T}^n)$, then

$$\Phi : \mathcal{D}'(\mathbf{T}^n) \xrightarrow{\sim} \mathcal{D}'_p(\mathbf{R}^n) : S \longmapsto \Big(\phi \mapsto \langle \sum_{k \in \mathbf{Z}^n} \tau_k \phi, S \rangle \Big)$$

is an isomorphism.

In analogy with $\mathcal{S}'$, we consider the space s' of *temperate sequences*, which
is defined as

$$s' = s'(\mathbf{Z}^n) = \{(a_k)_{k \in \mathbf{Z}^n} \in \mathbf{C}^{\mathbf{Z}^n}; \exists N \in \mathbf{N} : \forall k \in \mathbf{Z}^n : |a_k| \leq N(1 + |k|_\infty)^N\}.$$

Similar considerations as for the Fourier transformation in Proposition 1.6.5
then show that

$$\mathcal{F}_p : \mathcal{D}'(\mathbf{T}^n) \xrightarrow{\sim} s'(\mathbf{Z}^n) : T \longmapsto \big(\langle e^{-2\pi i k x}, T_x \rangle\big)_{k \in \mathbf{Z}^n}$$

and

$$\mathcal{F}_p^{-1} : s'(\mathbf{Z}^n) \xrightarrow{\sim} \mathcal{D}'(\mathbf{T}^n) : (c_k)_{k \in \mathbf{Z}^n} \longmapsto \sum_{k \in \mathbf{Z}^n} c_k e^{2\pi i k x}$$

are isomorphisms, see Schwartz [246], Ch. VII, Thm. I, p. 225; Vo-Khac Koan
[283], p. 69, 4°.

The following commutative diagram shows that the Fourier transform $\mathcal{F}_p$
induces in particular the Parseval identity, which is the restriction of $\mathcal{F}_p$ to
$L^2(\mathbf{T}^n)$:

$$
\begin{array}{ccccccc}
\mathcal{C}^\infty(\mathbf{T}^n) & \hookrightarrow & L^2(\mathbf{T}^n) & \hookrightarrow & L^1(\mathbf{T}^n) & \hookrightarrow & \mathcal{D}'(\mathbf{T}^n) \\
\downarrow\wr & & \downarrow\wr & & \downarrow & & \mathcal{F}_p\downarrow\wr \\
s(\mathbf{Z}^n) & \hookrightarrow & l^2(\mathbf{Z}^n) & \hookrightarrow & c_0(\mathbf{Z}^n) & \hookrightarrow & s'(\mathbf{Z}^n)
\end{array}
$$

Of course, all the inclusion maps in this diagram are continuous.

(b) A very important periodic distribution is $\delta_{\mathbf{T}^n} := \sum_{k\in\mathbf{Z}^n} \delta_k \in \mathcal{D}'_p(\mathbf{R}^n)$, which is the image of $\delta_0 \in \mathcal{D}'(\mathbf{T}^n)$ under Φ. For $n = 1$, we have

$$\delta_{\mathbf{T}^1} = 1 - \frac{d}{dx}\sum_{k\in\mathbf{Z}} \tau_k\big(xY(x-x^2)\big) \in \mathcal{D}'_{L^\infty}(\mathbf{R}),$$

and hence also $\delta_{\mathbf{T}^n} = \delta_{\mathbf{T}^1} \otimes \cdots \otimes \delta_{\mathbf{T}^1} \in \mathcal{D}'_{L^\infty}(\mathbf{R}^n)$. Let $\chi \in \mathcal{D}(\mathbf{R}^n)$ such that $\sum_{k\in\mathbf{Z}^n} \tau_k\chi = 1$, cf. Vladimirov [280], Sect. 7.1, p. 127. Then, for $T \in \mathcal{D}'_p(\mathbf{R}^n)$,

$$T = \left(\sum_{k\in\mathbf{Z}^n}\tau_k\chi\right)\cdot T = \sum_{k\in\mathbf{Z}^n}\tau_k(\chi\cdot T) = \left(\sum_{k\in\mathbf{Z}^n}\tau_k\delta\right)*(\chi\cdot T)$$

$$= \delta_{\mathbf{T}^n}*(\chi\cdot T) \in \mathcal{D}'_{L^\infty}*\mathcal{E}' = \mathcal{D}'_{L^\infty}\subset\mathcal{S}'(\mathbf{R}^n),$$

see Example 1.5.13. Thus every periodic distribution is temperate, and we obtain $\mathcal{F}T = (\mathcal{F}\delta_{\mathbf{T}^n})\cdot\mathcal{F}(\chi T)$ by Proposition 1.6.6 (5b). For $S \in \mathcal{D}'(\mathbf{T}^n)$ and $k \in \mathbf{Z}^n$, we have

$$\mathcal{F}\big(\chi\cdot\Phi(S)\big)(2\pi k) = \langle e^{-2\pi ikx}, \chi\cdot\Phi(S)\rangle = \langle\chi(x)e^{-2\pi ikx}, \Phi(S)\rangle$$

$$= \langle\sum_{l\in\mathbf{Z}^n}\tau_l\big(\chi(x)e^{-2\pi ikx}\big), S\rangle = \langle e^{-2\pi ikx}, S\rangle = (\mathcal{F}_p S)(k).$$

Furthermore, due to $\tau_k\delta_{\mathbf{T}^n} = \delta_{\mathbf{T}^n}$, the Fourier transform $U := \mathcal{F}(\delta_{\mathbf{T}^n})$ fulfills $U = e^{ikx}\cdot U$ for $k \in \mathbf{Z}^n$, and hence $\operatorname{supp} U \subset 2\pi\mathbf{Z}^n$. This is equivalent to $U = \sum_{k\in\mathbf{Z}^n} P_k(\partial)\delta_{2\pi k}$ for certain polynomials $P_k(x)$. On the other hand, due to $e^{2\pi ikx}\delta_{\mathbf{T}^n} = \delta_{\mathbf{T}^n}$, we have $\tau_{2\pi k}U = U$ and hence $U = P(\partial)\delta_{\mathbf{T}^n}(\frac{x}{2\pi})$ for a single polynomial P. Actually, P must be a constant, since, for $k \in \mathbf{Z}^n$,

$$P(\partial)\delta_{\mathbf{T}^n}(\tfrac{x}{2\pi}) = U = e^{ikx}\cdot U = e^{ikx}\cdot P(\partial)\delta_{\mathbf{T}^n}(\tfrac{x}{2\pi})$$

$$= P(\partial - ik)\big(e^{ikx}\cdot\delta_{\mathbf{T}^n}(\tfrac{x}{2\pi})\big) = P(\partial - ik)\delta_{\mathbf{T}^n}(\tfrac{x}{2\pi}),$$

and thus $P(\partial - ik) = P(\partial)$ for each $k \in \mathbf{Z}^n$. Thus $U = a\delta_{\mathbf{T}^n}(\frac{x}{2\pi})$ for some $a \in \mathbf{C}$. By the Fourier inversion theorem Proposition 1.6.5, we have

$$(2\pi)^n\delta_{\mathbf{T}^n} = \mathcal{F}\mathcal{F}\delta_{\mathbf{T}^n} = \mathcal{F}U = \mathcal{F}\big(a\delta_{\mathbf{T}^n}(\tfrac{x}{2\pi})\big) = a(2\pi)^n(\mathcal{F}\delta_{\mathbf{T}^n})(2\pi x) = a^2(2\pi)^n\delta_{\mathbf{T}^n}.$$

Since $a > 0$ due to

$$0 < \langle e^{-|x|^2}, \delta_{\mathbf{T}^n}\rangle = \langle\mathcal{F}^{-1}(e^{-|x|^2}), a\delta_{\mathbf{T}^n}(\tfrac{x}{2\pi})\rangle,$$

we finally conclude that $a = 1$ and

$$\mathcal{F}(\delta_{\mathbf{T}^n}) = \delta_{\mathbf{T}^n}(\tfrac{x}{2\pi}) = (2\pi)^n\sum_{k\in\mathbf{Z}^n}\delta_{2\pi k},$$

or, in other terms, the equation

$$\sum_{k \in \mathbf{Z}^n} e^{ikx} = (2\pi)^n \sum_{k \in \mathbf{Z}^n} \delta_{2\pi k}. \tag{1.6.30}$$

holds in $\mathcal{S}'(\mathbf{R}^n)$. Formula (1.6.30) is called *Poisson's summation formula*. □

Before showing that (1.6.30) holds more generally, namely in $\mathcal{D}'_{L\infty}(\mathbf{R}^n)$, let us first introduce the standard topologies on the spaces $\mathcal{D}_{L^p}$.

Definition and Proposition 1.6.20 *If the topology on the space $\mathcal{D}_{L^p}$, $1 \leq p \leq \infty$, (see* Definition 1.5.8) *is defined by the seminorms*

$$\mathcal{D}_{L^p}(\mathbf{R}^n) \longrightarrow [0, \infty) : \phi \longmapsto \|\partial^\alpha \phi\|_p, \qquad \alpha \in \mathbf{N}_0^n,$$

then they are Fréchet spaces, i.e., complete metrizable locally convex topological vector spaces. Furthermore, if $\frac{1}{p} + \frac{1}{q} = 1$ and $1 \leq p < \infty$, then the dual of $\mathcal{D}_{L^p}$ coincides with $\mathcal{D}'_{L^q}$ as defined in Definition 1.5.12.

For the **proof**, we refer to Schwartz [246], Ch. VI, Thm. XXV, p. 201. □

In the next proposition, we state the validity of the Poisson summation formula (1.6.30) in $\mathcal{D}'_{L\infty}$.

Proposition 1.6.21 *The equation $\sum_{k \in \mathbf{Z}^n} e^{ikx} = \delta_{\mathbf{T}^n}(\frac{x}{2\pi})$ holds in $\mathcal{D}'_{L\infty}(\mathbf{R}^n)$, i.e., the series $\sum_{k \in \mathbf{Z}^n} (\mathcal{F}\phi)(k)$ converges uniformly for ϕ in bounded subsets of $\mathcal{D}_{L^1}$ to the limit $\langle \phi, \delta_{\mathbf{T}^n}(\frac{x}{2\pi}) \rangle = (2\pi)^n \sum_{k \in \mathbf{Z}^n} \phi(2\pi k)$. (Note, however, that the series $\sum_{k \in \mathbf{Z}^n} \delta_{2\pi k}$ does not converge in $\mathcal{D}'_{L\infty}$, i.e., uniformly on bounded subsets of $\mathcal{D}_{L^1}$.)*

Proof The linear functional

$$U : \mathcal{D}_{L^1} \longrightarrow \mathbf{C} : \phi \longmapsto \sum_{k \in \mathbf{Z}^n} (\mathcal{F}\phi)(k)$$

is well defined and continuous since it is given as the composition of the following four linear and continuous maps:

$$\mathcal{D}_{L^1} \xrightarrow{(1-\Delta_n)^n} \mathcal{D}_{L^1} \hookrightarrow L^1 \xrightarrow{\mathcal{F}} \mathcal{C}_0 \xrightarrow{G} \mathbf{C},$$

where $G(f) = \sum_{k \in \mathbf{Z}^n} f(k) \cdot (1 + |k|^2)^{-n}$.

Since U coincides with $\delta_{\mathbf{T}^n}(\frac{x}{2\pi})$ on the dense subset $\mathcal{S}(\mathbf{R}^n)$ of $\mathcal{D}_{L^1}(\mathbf{R}^n)$ by Example 1.6.19, we conclude that $\sum_{k \in \mathbf{Z}^n} e^{ikx}$ converges to $\delta_{\mathbf{T}^n}(\frac{x}{2\pi})$ in $\mathcal{D}'_{L\infty}$. □

If $\phi \in \mathcal{D}(\mathbf{R}^n)$ with $\phi(0) = 1$ and $\operatorname{supp}\phi \subset \{x \in \mathbf{R}^n; |x| \leq 1\}$, then the set $B = \{\tau_l \phi; l \in \mathbf{Z}^n\}$ is bounded in $\mathcal{D}_{L^1}$, but, obviously,

$$\left\langle \tau_l \phi, \sum_{\substack{k \in \mathbf{Z}^n \\ |k| \leq N}} \delta_k \right\rangle = \begin{cases} 1, \text{ if } |l| \leq N \\ 0, \text{ else} \end{cases}$$

does *not* converge *uniformly* for $l \in \mathbf{Z}^n$ to its limit 1 if $N \to \infty$. Hence, as observed in the proposition, the series $\sum_{k \in \mathbf{Z}^n} \delta_{2\pi k}$ does not converge in $\mathcal{D}'_{L^\infty}$. (However, one can show that $\sum_{k \in \mathbf{Z}^n} \delta_{2\pi k}$ converges uniformly on *compact* subsets of $\mathcal{D}_{L^1}$.)

Example 1.6.22 Here we shall apply the Poisson summation formula to calculate *fundamental solutions on the torus.*

(a) Let us consider a linear partial differential operator $P(\partial) = \sum_{|\alpha| \le m} a_\alpha \partial^\alpha$ with constant coefficients $a_\alpha \in \mathbf{C}$ and acting on $\mathcal{D}'(\mathbf{T}^n)$. We assume that $P(i\xi)$ does not vanish for large real ξ, which is, in particular, the case for elliptic operators. By the Seidenberg–Tarski inequality, see Hörmander [135], Lemma 3, p. 557; [136], App., Lemma 2.1, p. 276, we have

$$\exists m \in \mathbf{N} : \forall k \in \mathbf{Z}^n \text{ with } |k| > m : |P(2\pi i k)| > |k|^{-m}.$$

Therefore, the series

$$E := \sum_{\substack{k \in \mathbf{Z}^n \\ P(2\pi i k) \ne 0}} P(2\pi i k)^{-1} e^{2\pi i k x} \tag{1.6.31}$$

converges in $\mathcal{D}'_{L^\infty}(\mathbf{R}^n)$, cf. the proof of Proposition 1.6.21. Furthermore, $E \in \mathcal{D}'_p(\mathbf{R}^n)$ and it solves the following equation:

$$P(\partial)E = \sum_{\substack{k \in \mathbf{Z}^n \\ P(2\pi i k) \ne 0}} e^{2\pi i k x} = \delta_{\mathbf{T}^n} - \sum_{\substack{k \in \mathbf{Z}^n \\ P(2\pi i k) = 0}} e^{2\pi i k x}.$$

$F := \Phi^{-1} E$ can be conceived of as a "fundamental solution" to $P(\partial)$ on the torus $\mathbf{T}^n$ in the following sense: If we try to solve in $\mathcal{D}'(\mathbf{T}^n)$ the equation $P(\partial)F = \delta + f$ for $f \in C^\infty(\mathbf{T}^n)$ with $\delta = \Phi^{-1}(\delta_{\mathbf{T}^n}) \in \mathcal{D}'(\mathbf{T}^n)$ and $F \in \mathcal{D}'(\mathbf{T}^n)$, then we have to assume $(\mathcal{F}_p f)(k) = -1$ for $k \in \mathbf{Z}^n$ with $P(2\pi i k) = 0$ since

$$P(2\pi i k) \cdot \mathcal{F}_p F = \mathcal{F}_p \big(P(\partial) F \big) = \mathcal{F}_p (\delta + f) = 1 + \mathcal{F}_p f.$$

Therefore,

$$\exists g \in C^\infty(\mathbf{T}^n) : f = g - \sum_{\substack{k \in \mathbf{Z}^n \\ P(2\pi i k) = 0}} e^{2\pi i k x} \text{ and } [(\mathcal{F}_p g)(k) = 0 \text{ for } P(2\pi i k) = 0].$$

If we assume furthermore that $g = 0$, then F is given by (1.6.31) up to a trigonometric polynomial of the form $\sum_{k \in \mathbf{Z}^n, P(2\pi i k) = 0} c_k e^{2\pi i k x}$.

(b) We shall apply now the Poisson summation formula (1.6.30) in order to calculate the fundamental solution $E \in \mathcal{D}'_p(\mathbf{R}^n)$ of the *iterated metaharmonic operator* $P(\partial) = (a^2 - \Delta_n)^l$, $a > 0$, $l \in \mathbf{N}$, on the torus.

If $l > \frac{n}{2}$, then $\phi(\xi) = (a^2 + |\xi|^2)^{-l} \cdot e^{i\xi x} \in \mathcal{D}_{L^1}(\mathbf{R}^n)$ and

$$(\mathcal{F}\phi)(y) = \mathcal{F}_\xi\big((a^2 + |\xi|^2)^{-l}\big)(y - x) = \frac{2\pi^{n/2}}{(l-1)!}\left(\frac{2a}{|y-x|}\right)^{n/2-l} K_{n/2-l}(a|y-x|),$$

see Example 1.6.11 (a). Hence Proposition 1.6.21 yields

$$E = \sum_{k \in \mathbf{Z}^n} \frac{e^{2\pi i k x}}{P(2\pi i k)} = \sum_{k \in \mathbf{Z}^n} \phi(2\pi k) = (2\pi)^{-n} \sum_{k \in \mathbf{Z}^n} (\mathcal{F}\phi)(k)$$

$$= \frac{a^{n/2-l}}{2^{n/2+l-1}(l-1)!\pi^{n/2}} \sum_{k \in \mathbf{Z}^n} \frac{K_{n/2-l}(a|k-x|)}{|k-x|^{n/2-l}}. \qquad (1.6.32)$$

By analytic continuation, formula (1.6.32) yields a fundamental solution of $(a^2 - \Delta_n)^l$ on the torus for all $a \in \mathbf{C}$ with positive real part $\operatorname{Re} a$. Note that the series in (1.6.32) is fast convergent due to the asymptotic expansion

$$K_{n/2-l}(z) = \sqrt{\frac{\pi}{2z}}\, e^{-z}(1 + O(|z|^{-1})), \qquad |z| \to \infty,\ \operatorname{Re} z > 0,$$

see Abramowitz and Stegun [1], 9.7.2, p. 378.

(c) Let us finally consider the *Laplacean on the torus*. According to (a),

$$E_n := -\frac{1}{4\pi^2} \sum_{k \in \mathbf{Z}^n \setminus \{0\}} k^{-2} e^{2\pi i k x} \in \mathcal{D}'_{L^\infty} \cap \mathcal{D}'_p(\mathbf{R}^n)$$

is a fundamental solution of Δ_n on $\mathbf{T}^n$ in the sense that $\Delta_n E_n = \delta_{\mathbf{T}^n} - 1$ in $\mathcal{D}'(\mathbf{R}^n)$ or, equivalently, $\Delta_n \Phi^{-1} E_n = \delta - 1$ in $\mathcal{D}'(\mathbf{T}^n)$.

If $n = 1$, then

$$E_1(x) = -\frac{1}{2\pi^2} \sum_{k=1}^{\infty} k^{-2} \cos(2\pi k x)$$

converges uniformly for $x \in \mathbf{R}$ and hence $E_1 \in \mathcal{BC}(\mathbf{R})$. Since $E_1'' = -1 + \sum_{k \in \mathbf{Z}} \delta_k$ in $\mathcal{D}'_p(\mathbf{R})$, we have $E_1(x) = -\frac{1}{2}x^2 + ax + b$ for $0 \le x \le 1$. The constants a, b are determined by

$$b = E_1(0) = -\frac{1}{2\pi^2} \sum_{k=1}^{\infty} \frac{1}{k^2} = -\frac{1}{12},$$

$$b = E_1(0) = E_1(1) = -\frac{1}{2} + a + b \implies a = \frac{1}{2}.$$

Thus $E_1(x) = -\frac{1}{2}(x^2 - x + \frac{1}{6})$, $0 \le x \le 1$, and this determines E_1 completely by periodicity.

Let us yet calculate E_2, which plays a role in the study of the electronic structure of crystals, see Glasser [110]. By (a), the series

$$E_2(x, y) = -\frac{1}{4\pi^2} \sum_{(k,m)\in\mathbf{Z}^2\backslash\{0\}} \frac{e^{2\pi i(kx+my)}}{k^2 + m^2}$$

converges in $\mathcal{D}'_{L^\infty}(\mathbf{R}^2)$. Apparently, we have

$$E_2(x, y) = E_1(x) - \sum_{m\in\mathbf{Z}\backslash\{0\}} e^{2\pi i m y} \sum_{k\in\mathbf{Z}} \frac{e^{2\pi i k x}}{4\pi^2(k^2 + m^2)}.$$

The sum of the inner series is of course well known. It is also a special case of the fundamental solution on the torus of the (one-dimensional) metaharmonic operator $4\pi^2 m^2 - \frac{d^2}{dx^2}$, which was considered more generally in (b). For $m \in \mathbf{Z} \backslash \{0\}$ and $x \in [0, 1]$, we have

$$\sum_{k\in\mathbf{Z}} \frac{e^{2\pi i k x}}{4\pi^2(k^2 + m^2)} = \frac{1}{2\pi\sqrt{m}} \sum_{k\in\mathbf{Z}} \sqrt{|k - x|}\, K_{-1/2}(2\pi m|k - x|)$$

$$= \frac{1}{4\pi m} \sum_{k\in\mathbf{Z}} e^{-2\pi m|k-x|} = \frac{\cosh\big(m\pi(1 - 2x)\big)}{4\pi m \sinh(m\pi)}.$$

Hence we obtain, for $x \in [0, 1]$ and $y \in \mathbf{R}$, the following:

$$E_2(x, y) = -\frac{1}{2}\Big(x^2 - x + \frac{1}{6}\Big) - \frac{1}{2\pi} \sum_{m=1}^{\infty} \frac{\cos(2\pi m y)}{m \sinh(m\pi)}\, \cosh\big(m\pi(1 - 2x)\big).$$

Let us finally represent E_2 by a Jacobian theta function. We shall employ ϑ_4 in the form

$$\vartheta_4(z, q) = 1 + 2\sum_{m=1}^{\infty}(-1)^m q^{m^2}\cos(2mz), \qquad |q| < 1,$$

see Gradshteyn and Ryzhik [113], Eq. 8.180.1, p. 921; Abramowitz and Stegun [1], 16.27.4, p. 576. Since

$$\frac{\cos(2\pi m y)\cosh\big(m\pi(1 - 2x)\big)}{\sinh(m\pi)} = \sum_{\pm} \frac{\cos\big(2\pi m(y \pm i(x - \frac{1}{2}))\big)}{1 - e^{-2m\pi}}\, e^{-m\pi},$$

we obtain with $q := e^{-\pi}$ and $z = x + iy$ that

$$E_2(x, y) = -\frac{1}{2}\left(x^2 - x + \frac{1}{6}\right) - \frac{1}{2\pi}\sum_{\pm}\sum_{m=1}^{\infty}\frac{\cos\left(2\pi m(y \pm i(x - \frac{1}{2}))\right)}{m(1 - q^{2m})}q^m$$

$$= -\frac{1}{2}\left(x^2 - x + \frac{1}{6}\right) - \frac{1}{2\pi}\log Q_0 + \frac{1}{2\pi}\log\left|\vartheta_4\left(i\pi(z - \frac{1}{2}), e^{-\pi}\right)\right|,$$

$$(1.6.33)$$

where $\log Q_0 = \sum_{m=1}^{\infty}\log(1 - e^{-2m\pi})$, see Oberhettinger [195], (2.26), (2.30), p. 26. Formula (1.6.33) holds for $0 < x < 1$ and $y \in \mathbf{R}$; it seems to have been obtained for the first time in Glasser [110], (20), p. 189.

Since

$$\Delta_2 E_2 = \delta_{\mathbf{T}^2} - 1 = -1 + \sum_{k\in\mathbf{Z}^2}\delta_k \quad \text{and} \quad \Delta_2\log(x^2 + y^2) = 4\pi\delta \text{ in } \mathcal{D}'(\mathbf{R}^2),$$

see Example 1.3.14, we conclude that $f(x, y) := E_2 - \frac{1}{4\pi}\log(x^2 + y^2)$ is an even C^∞ function on the unit disc $x^2 + y^2 < 1$. In particular, $E_2(x, y) = \frac{1}{4\pi}\log(x^2 + y^2) + f(0) + O(x^2 + y^2)$ for $(x, y) \to 0$. Since the constant $f(0)$ plays an important role in statistical mechanics, let us represent it by special functions. We have

$$f(0) = \lim_{(x,y)\to 0}\left[E_2(x, y) - \frac{1}{4\pi}\log(x^2 + y^2)\right]$$

$$= -\frac{1}{12} - \frac{1}{2\pi}\log Q_0 + \frac{1}{2\pi}\log\left(\lim_{z\to 0}\left|z^{-1}\vartheta_4\left(i\pi(z - \frac{1}{2}), e^{-\pi}\right)\right|\right)$$

$$= -\frac{1}{12} - \frac{1}{2\pi}\log Q_0 + \frac{1}{2\pi}\log\left|\pi\vartheta_4'\left(\frac{i\pi}{2}, e^{-\pi}\right)\right|.$$

The numerical evaluation of the two series $\log Q_0 = \sum_{m=1}^{\infty}\log(1 - e^{-2m\pi})$ and

$$\vartheta_4'\left(\frac{i\pi}{2}, e^{-\pi}\right) = -4i\sum_{m=1}^{\infty}(-1)^m m\, e^{-\pi m^2}\sinh(m\pi)$$

yields

$$f(0) = -\frac{1}{12} + \frac{1}{2\pi}\log\left(\frac{4\pi\sum_{m=1}^{\infty}(-1)^{m-1}m\, e^{-\pi m^2}\sinh(m\pi)}{\prod_{m=1}^{\infty}(1 - e^{-2m\pi})}\right) \approx 0.2085777932.$$

$\square$

Chapter 2
General Principles for Fundamental Solutions

The correct definition of a *fundamental solution* of a linear differential operator was anticipated by N. Zeilon in 1911 and finally given in the framework of distribution theory by L. Schwartz in 1950, see Schwartz [246], pp. 135, 136. More generally, L. Schwartz defined *fundamental matrices* $E \in \mathcal{D}'(\mathbf{R}^n)^{l \times l}$ for systems $A(\partial) = \left(A_{ij}(\partial)\right)_{1 \leq i,j \leq l}$ of differential operators by $A(\partial)E = I_l \delta$, or, more explicitly, $\sum_{k=1}^{l} A_{ik}(\partial)E_{kj} = \delta_{ij}\delta$ for $1 \leq i,j \leq l$. The reason for this more general definition lies in the importance of such systems in the natural sciences: Physical phenomena are in general described by vector or tensor fields (as, e.g., displacements, electric and magnetic fields etc.) instead of by single scalar quantities, as e.g., the temperature. Therefore we present three such systems in Examples 2.1.3 and 2.1.4 describing the displacements in isotropic, cubic and hexagonal elastic media, respectively.

The content of the Malgrange–Ehrenpreis Theorem is that every non-trivial linear differential operator with constant coefficients has a fundamental solution. We give a short new constructive proof of this fact in Proposition 2.2.1. Section 2.3 deals with the existence of *temperate* fundamental solutions, a problem which was solved first by S. Łojasiewicz and L. Hörmander.

Apart from the question of existence of fundamental solutions, the search for uniqueness criteria such as support or growth properties in dependence on the operator is essential. This question is investigated in Sect. 2.4. An existence and uniqueness theorem for homogeneous *elliptic* operators (for the definition of ellipticity, see Definition 2.4.7) is given in Proposition 2.4.8. On the other hand, if, for $N \in \mathbf{R}^n \setminus \{0\}$, there exists a tube domain $T = \{i\xi + \sigma N; \xi \in \mathbf{R}^n, \sigma > \sigma_0\} \subset \mathbf{C}^n$ such that $\det A$ does not vanish on T, then a fundamental matrix E of the system $A(\partial)$ is uniquely determined by the condition $\exists \sigma > \sigma_0 : \mathrm{e}^{-\sigma xN}E \in \mathcal{S}'(\mathbf{R}^n)^{l \times l}$, see Definition and Proposition 2.4.13. Furthermore, the support of E is contained in the half-space $H_N = \{x \in \mathbf{R}^n; xN \geq 0\}$. In the literature, such systems are called temperate evolution systems (B. Melrose) or systems correct in the sense of Petrovsky (S.G. Gindikin, L. Hörmander). For shortness and due to the many

© Springer International Publishing Switzerland 2015
N. Ortner, P. Wagner, *Fundamental Solutions of Linear Partial Differential Operators*, DOI 10.1007/978-3-319-20140-5_2

similarities with hyperbolic operators (see Definition 2.4.10), we prefer to call such systems *quasihyperbolic* (cf. Example 2.2.2 and Definition and Proposition 2.4.13).

In Sect. 2.5, the effect on the fundamental matrix of linear transformations of the coordinates is studied since this allows to reduce many operators and systems to simpler ones. Finally, in Sect. 2.6, the construction of fundamental solutions by invariance methods is explained.

2.1   Fundamental Matrices

As we have said above, it is more often *systems* of differential equations than scalar operators which originally occur when one sets up models to describe physical processes. Scalar operators of higher order then appear as the determinants of such systems. Let us therefore introduce the notion of *fundamental matrices* of linear square systems $A(\partial)$ of differential operators, and let us explain the connection with the fundamental solutions of their determinants $P(\partial) = \det A(\partial)$. In the following, $I_l \in \mathrm{Gl}_l(\mathbf{R})$ denotes the $l \times l$ unit matrix.

Definition 2.1.1 Let $A(\partial) = \big(A_{ij}(\partial)\big)_{1 \le i,j \le l}$, where $A_{ij}(\partial) = \sum_{|\alpha| \le m} a_{ij,\alpha} \partial^\alpha$, be an $l \times l$ matrix of linear differential operators in $\mathbf{R}^n$ with constant coefficients $a_{ij,\alpha} \in \mathbf{C}$. A matrix $E \in \mathcal{D}'(\mathbf{R}^n)^{l \times l}$ is called a *right-sided* or a *left-sided fundamental matrix* of $A(\partial)$, respectively, iff the respective equation

$$A(\partial)E = I_l \delta, \text{ i.e., } \forall 1 \le i,j \le l : \sum_{k=1}^{l} A_{ik}(\partial)E_{kj} = \begin{cases} \delta, & \text{if } i = j, \\ 0, & \text{else}, \end{cases}$$

or

$$E * A(\partial)\delta = I_l \delta, \text{ i.e., } \forall 1 \le i,j \le l : \sum_{k=1}^{l} A_{ki}(\partial)E_{jk} = \begin{cases} \delta, & \text{if } i = j, \\ 0, & \text{else} \end{cases}$$

holds in $\mathcal{D}'(\mathbf{R}^n)^{l \times l}$. If E is a right-sided as well as left-sided fundamental matrix of $A(\partial)$, then it is called a *two-sided fundamental matrix* of $A(\partial)$.

Note that the term "fundamental matrix" is not generally in use. Instead, it is also called *Green's matrix, Green's tensor* or simply *fundamental solution.* For Definition 2.1.1, see Malgrange [174], pp. 298, 299; Schwartz [246], Eq. (V, 6; 30), p. 140; Petersen [228], pp. 56, 57; Hörmander [136], p. 94; Jones [152], p. 421.

Let us observe that the transposed matrix E^T of a right-sided fundamental matrix E of $A(\partial)$ is a left-sided fundamental matrix of the transposed system $A(\partial)^T$ and vice versa. As we shall see in Sect. 2.2, a system $A(\partial)$ has a right- or a left-sided fundamental matrix if and only if its determinant operator $\det A(\partial)$ does not vanish identically. In this case, we can construct a two-sided fundamental matrix E of

$A(\partial)$ in the following way: Take a fundamental solution F of $\det A(\partial)$ and set $E := A(\partial)^{\mathrm{ad}}F$, where $A(\partial)^{\mathrm{ad}}$ denotes the adjoint matrix to $A(\partial)$. In fact,

$$P(\partial) = \det A(\partial), \quad P(\partial)F = \delta, \quad E = A(\partial)^{\mathrm{ad}}F$$

$$\Longrightarrow \begin{cases} A(\partial)E = A(\partial)A(\partial)^{\mathrm{ad}}F = I_l P(\partial)F = I_l\delta \\ \text{and } E * A(\partial)\delta = A(\partial)^{\mathrm{ad}}A(\partial)F = I_l\delta. \end{cases} \qquad (2.1.1)$$

(This procedure is a classical one: see Weierstrass [300], pp. 287–288; Hörmander [136], Section 3.8, p. 94.) However, in general, a right-sided fundamental matrix is not necessarily a left-sided one, and conversely, see Example 2.1.2.

Example 2.1.2 For $A(\partial) = \begin{pmatrix} d/dx & 1 \\ 0 & 1 \end{pmatrix}$, a right-sided fundamental matrix is of the form $\begin{pmatrix} Y(x) + C_1 & -Y(x) + C_2 \\ 0 & \delta \end{pmatrix}$, whereas a left-sided fundamental matrix is given by $\begin{pmatrix} Y(x) + C_3 & -Y(x) - C_3 \\ C_4 & \delta - C_4 \end{pmatrix}$, $C_1, \ldots, C_4 \in \mathbf{C}$ being arbitrary.

The two-sided fundamental matrices are of the form $E = \begin{pmatrix} Y(x)+C & -Y(x)-C \\ 0 & \delta \end{pmatrix}$, $C \in \mathbf{C}$. In fact, $E = A(\partial)^{\mathrm{ad}}F$, where $F = Y(x) + C$ is a fundamental solution of $\det A(\partial) = \frac{d}{dx}$. $\qquad \square$

Example 2.1.3 Let us use formula (2.1.1) in order to calculate the fundamental matrix of the Lamé system $A(\partial)$ governing elastodynamics inside a homogeneous isotropic medium.

If $u = (u_1, u_2, u_3)^T$ denotes the displacement in an elastic medium, ρ, f the densities of mass and force, respectively, then $A(\partial)u = \rho f$ where $\partial = (\partial_t, \partial_1, \partial_2, \partial_3)$, $\nabla = (\partial_1, \partial_2, \partial_3)^T$,

$$A(\partial) := \rho I_3 \partial_t^2 - B(\nabla), \qquad B(\nabla) := \mu \Delta_3 I_3 + (\lambda + \mu)\nabla \cdot \nabla^T, \qquad (2.1.2)$$

and $\lambda, \mu > 0$ denote Lamé's constants.

Generally, the matrix $A = \alpha I_l + \beta \xi \cdot \xi^T \in \mathbf{C}^{l \times l}$ (with $\alpha, \beta \in \mathbf{C}$ and $\xi \in \mathbf{R}^l$) has the eigenvalues $\alpha + \beta|\xi|^2$ with multiplicity 1 and α with multiplicity $l-1$ and hence $\det A = \alpha^{l-1}(\alpha + \beta|\xi|^2)$. Similarly, the ansatz $A^{\mathrm{ad}} = \gamma I_l + \epsilon \xi \cdot \xi^T$ yields

$$A^{\mathrm{ad}} = \alpha^{l-2}\big[(\alpha + \beta|\xi|^2)I_l - \beta \xi \cdot \xi^T\big].$$

Therefore,

$$\begin{aligned} P(\partial) = \det A(\partial) &= \det\big((\rho\partial_t^2 - \mu\Delta_3)I_3 - (\lambda + \mu)\nabla \cdot \nabla^T\big) \\ &= (\rho\partial_t^2 - \mu\Delta_3)^2\big(\rho\partial_t^2 - (\lambda + 2\mu)\Delta_3\big) \end{aligned} \qquad (2.1.3)$$

and

$$A(\partial)^{\text{ad}} = (\rho\partial_t^2 - \mu\Delta_3)\left[(\rho\partial_t^2 - (\lambda + 2\mu)\Delta_3)I_3 + (\lambda + \mu)\nabla \cdot \nabla^T\right].$$

For the two different irreducible factors of $P(\partial)$ in (2.1.3), let us introduce the abbreviations

$$W_s(\partial) = \rho\partial_t^2 - \mu\Delta_3, \qquad W_p(\partial) = \rho\partial_t^2 - (\lambda + 2\mu)\Delta_3.$$

These two wave operators account for the propagation of shear and of pressure waves, respectively, in the medium, see Achenbach [2], 4.1, pp. 122–124. By formula (1.4.11) and Proposition 1.3.19, their forward fundamental solutions F_s, F_p are given by

$$F_s = \frac{\delta\left(t - \frac{|x|}{c_s}\right)}{4\pi\mu|x|}, \qquad F_p = \frac{\delta\left(t - \frac{|x|}{c_p}\right)}{4\pi(\lambda + 2\mu)|x|},$$

where $c_s = \sqrt{\frac{\mu}{\rho}}$, $c_p = \sqrt{\frac{\lambda + 2\mu}{\rho}}$ are the velocities of the shear and pressure waves, respectively.

From the uniqueness (see Proposition 2.4.11 below) of the forward fundamental solution F of the hyperbolic operator $P(\partial) = W_s(\partial)^2 W_p(\partial)$ and the convolvability of F_s, F_p, we obtain that $F = F_s * F_s * F_p$. Hence we infer from (2.1.1) that the unique fundamental matrix E of $A(\partial)$ with support in the half-space $t \geq 0$ is given by

$$
\begin{aligned}
E = A(\partial)^{\text{ad}}F &= W_s(\partial)\left[W_p(\partial)I_3 + (\lambda + \mu)\nabla \cdot \nabla^T\right](F_s * F_s * F_p) \\
&= \left[W_p(\partial)I_3 + (\lambda + \mu)\nabla \cdot \nabla^T\right](F_s * F_p) \\
&= I_3 F_s + (\lambda + \mu)\nabla \cdot \nabla^T (F_s * F_p).
\end{aligned}
\tag{2.1.4}
$$

From formula (2.1.4), we conclude that the fundamental matrix E can be expressed by means of the fundamental solution $F_s * F_p$ of the fourth-order operator $W_s(\partial)W_p(\partial)$; in particular, there is no need to calculate the fundamental solution of the sixth-order operator $P(\partial) = \det A(\partial) = W_s(\partial)^2 W_p(\partial)$, as was done in Piskorek [229], p. 95.

In order to derive Stokes's representation of the fundamental matrix E from the year 1849, see Stokes [265], let us apply the "difference device", which shall be developed in more generality in Sect. 3.3. Due to

$$
\begin{aligned}
W_s(\partial)W_p(\partial)\left[(\lambda + 2\mu)F_p - \mu F_s\right] &= \left[(\lambda + 2\mu)W_s(\partial) - \mu W_p(\partial)\right]\delta \\
&= (\lambda + \mu)\rho\partial_t^2\delta,
\end{aligned}
$$

we conclude, by convolution with the fundamental solution $tY(t) \otimes \delta(x)$ of the operator ∂_t^2, that the forward fundamental solution $F_s * F_p$ of $W_s(\partial)W_p(\partial)$ has the following representation:

$$F_s * F_p = \frac{1}{(\lambda + \mu)\rho} \left[tY(t) \otimes \delta(x) \right] * \left[(\lambda + 2\mu)F_p - \mu F_s \right]. \tag{2.1.5}$$

Since $F_p \in C\big(\mathbf{R}_x^3 \setminus \{0\}, \mathcal{E}'(\mathbf{R}_t^1)\big)$ is given by $(\lambda+2\mu)F_p(x) = \frac{1}{4\pi|x|}\delta_{|x|/c_p}(t)$ for $x \neq 0$, we obtain

$$\left[tY(t) \otimes \delta(x) \right] * (\lambda + 2\mu)F_p = \frac{\left(t - \frac{|x|}{c_p}\right)Y\left(t - \frac{|x|}{c_p}\right)}{4\pi|x|} \tag{2.1.6}$$

for $x \neq 0$, and (2.1.6) then holds in $\mathcal{D}'(\mathbf{R}^4)$ by homogeneity. If we insert (2.1.6) and the analogous equation for $\left[tY(t) \otimes \delta(x) \right] * \mu F_s$ into (2.1.5), we obtain

$$(\lambda + \mu)F_s * F_p = \frac{1}{4\pi\rho|x|} \left[\left(t - \frac{|x|}{c_p} \right) Y\left(t - \frac{|x|}{c_p} \right) - \left(t - \frac{|x|}{c_s} \right) Y\left(t - \frac{|x|}{c_s} \right) \right]$$

$$= \frac{1}{4\pi\rho} \left[\left(\frac{1}{c_s} - \frac{1}{c_p} \right) + \left(\frac{t}{|x|} - \frac{1}{c_s} \right) Y(|x| - c_s t) - \left(\frac{t}{|x|} - \frac{1}{c_p} \right) Y(|x| - c_p t) \right].$$

For the differentiation of $F_s * F_p$, we then employ the many-dimensional jump formula (1.3.13):

$$(\lambda + \mu)\nabla \cdot \nabla^T(F_s * F_p) = \frac{1}{4\pi\rho} \left\{ \frac{3xx^T - I_3|x|^2}{|x|^5} t\left[Y(|x| - c_s t) - Y(|x| - c_p t) \right] \right.$$

$$\left. - xx^T \left[\frac{1}{c_s^4 t^3} \delta(|x| - c_s t) - \frac{1}{c_p^4 t^3} \delta(|x| - c_p t) \right] \right\}$$

Inserting this into (2.1.4) finally yields Stokes's formula for the forward fundamental matrix E of Lamé's system $A(\partial)$ defined in (2.1.2):

$$E = \frac{I_3|x|^2 - xx^T}{4\pi\mu|x|^3} \delta\left(t - \frac{|x|}{c_s} \right) + \frac{xx^T}{4\pi(\lambda + 2\mu)|x|^3} \delta\left(t - \frac{|x|}{c_p} \right)$$

$$+ \frac{t}{4\pi\rho|x|^3} \left(I_3 - \frac{3xx^T}{|x|^2} \right) \left[Y\left(t - \frac{|x|}{c_s} \right) - Y\left(t - \frac{|x|}{c_p} \right) \right], \tag{2.1.7}$$

cf. Achenbach [2], (3.95/96/98), pp. 99f.; Achenbach and Wang [3], (8.15), p. 282; Duff [62], pp. 270f.; [64], p. 79; Eringen and Şuhubi [69], (5.10.30), p. 400; Love [171], (36), p. 305; Mura [185], (9.34), p. 63; Willis [302], (34), p. 387; Wagner [293], p. 406.   □

Example 2.1.4 Let us consider now the equations of *anisotropic* elastodynamics. The investigation of this 3 by 3 system will also show the importance of higher order partial differential operators in mathematical physics. We shall develop here only the algebraic part of the construction of the fundamental matrix, and we shall postpone the application of the Herglotz–Petrovsky formula to Chap. 4, where hyperbolic operators and systems will be analyzed.

(a) In a homogeneous anisotropic medium, the displacements u_p, the stresses σ_{pq} and the strains v_{pq} satisfy the following equations (where we use Einstein's summation convention):

$$\rho \partial_t^2 u_p = \partial_q \sigma_{pq} + \rho f_p, \qquad \sigma_{pq} = \sigma_{qp};$$

$$v_{pq} = \frac{1}{2}(\partial_q u_p + \partial_p u_q); \tag{2.1.8}$$

$$\sigma_{pq} = c_{pqrs} v_{rs} \text{ (Hooke's law)}, \qquad c_{pqrs} = c_{qprs} = c_{rspq}.$$

Herein $p, q, r, s \in \{1, 2, 3\}$, c_{pqrs} are the elastic constants, and ρ, f denote the densities of mass and of force, respectively, cf. Achenbach and Wang [3], (2.1/2), (2.4), p. 274; Buchwald [31], (2.1–5), p. 564; Duff [62], (1.1), p. 249; Eringen and Şuhubi [69], (5.2.19), p. 346; Herglotz [127], (3.48), p. 75, and (6.9), p. 156; Musgrave [186], (6.1.4/6), p. 67; (3.11.1/2), p. 28; Payton [226], (1.1.1–5), p. 1; Poruchikov [231], (2.1.1–6), p. 4.

Abbreviating the symmetric matrix $(c_{pqrs}\partial_q\partial_s)_{r,s}$ by $B(\nabla)$ we derive the system

$$A(\partial)u = (\rho I_3 \partial_t^2 - B(\nabla))u = \rho f$$

(cf. Duff [62], (1.2), p. 250; Musgrave [186], (6.1.7), p. 68) by elimination of σ_{pq} and v_{pq}. In the sequel, we put $\rho = 1$.

The dimension of the linear space of tensors (c_{pqrs}) of rank 4 fulfilling the symmetry relations stated in (2.1.8) equals 21. This fact is exploited when the "contracted index notation" is used (cf. Musgrave [186], (3.13.4–6), p. 33; Payton [226], p. 3): The indices $11, 22, 33, 12, 13, 23$ are replaced by $1, 2, 3, 6, 5, 4$, respectively. Consequently, (2.1.8) takes the form $\sigma = C\tilde{v}$, where $\sigma = (\sigma_{11}, \sigma_{22}, \sigma_{33}, \sigma_{23}, \sigma_{13}, \sigma_{12})^T \in \mathbf{R}^6$, $\tilde{v} = (v_{11}, v_{22}, v_{33}, 2v_{23}, 2v_{13}, 2v_{12})^T \in \mathbf{R}^6$, and $C \in \mathbf{R}^{6\times6}$.

Let us consider such particular cases of the elastic constants for which the 6×6-matrix C has the form $C = \begin{pmatrix} H & 0 \\ 0 & L \end{pmatrix}$ with two symmetric 3×3-matrices H, L. This implies the equations

$$\begin{pmatrix} \sigma_{11} \\ \sigma_{22} \\ \sigma_{33} \end{pmatrix} = H \begin{pmatrix} v_{11} \\ v_{22} \\ v_{33} \end{pmatrix}, \quad \begin{pmatrix} \sigma_{23} \\ \sigma_{13} \\ \sigma_{12} \end{pmatrix} = 2L \begin{pmatrix} v_{23} \\ v_{13} \\ v_{12} \end{pmatrix}, \quad c_{pqrs} = \begin{cases} h_{jk} & : p = q, r = s, \\ l_{j-3,k-3} & : p \neq q, r \neq s, \\ 0 & : \text{else}, \end{cases}$$

if j corresponds to pq or qp, and k to rs or sr, respectively ($1 \leq p, q, r, s \leq 3$, $1 \leq j, k \leq 6$). Hence the matrix $B(\xi)$ assumes the form

$$B(\xi) = \left(h_{jk}\xi_j\xi_k\right)_{j,k=1,2,3}$$

$$+ \begin{pmatrix} l_{22}\xi_3^2 + 2l_{23}\xi_2\xi_3 + l_{33}\xi_2^2 & l_{33}\xi_1\xi_2 + \xi_3\,\Xi & l_{22}\xi_1\xi_3 + \xi_2\,\Xi \\ l_{33}\xi_1\xi_2 + \xi_3\,\Xi & l_{11}\xi_3^2 + 2l_{13}\xi_1\xi_3 + l_{33}\xi_1^2 & l_{11}\xi_2\xi_3 + \xi_1\,\Xi \\ l_{22}\xi_1\xi_3 + \xi_2\,\Xi & l_{11}\xi_2\xi_3 + \xi_1\,\Xi & l_{11}\xi_2^2 + 2l_{12}\xi_1\xi_2 + l_{22}\xi_1^2 \end{pmatrix}$$

$$\tag{2.1.9}$$

with $\Xi := l_{23}\xi_1 + l_{13}\xi_2 + l_{12}\xi_3$.

(b) Let us specify the above for *isotropic media.* In such media, the tensor (c_{pqrs}) is determined by *two* independent constants, the Lamé constants λ, μ, whereby

$$c_{pqrs} = \lambda\delta_{pq}\delta_{rs} + \mu(\delta_{pr}\delta_{qs} + \delta_{ps}\delta_{qr})$$

or

$$C = \begin{pmatrix} H & 0 \\ 0 & L \end{pmatrix}, \qquad H = \begin{pmatrix} \lambda + 2\mu & \lambda & \lambda \\ \lambda & \lambda + 2\mu & \lambda \\ \lambda & \lambda & \lambda + 2\mu \end{pmatrix}, \qquad L = \mu I_3,$$

cf. Eringen and Şuhubi [69], (5.2.20), p. 346; Payton [226], (1.1.7), p. 2; Sommerfeld [259], p. 272.

Then $B(\xi) = \mu|\xi|^2 I_3 + (\lambda + \mu)\xi \cdot \xi^T$ as in (2.1.2) and the determinant operator of the system degenerates:

$$P(\partial) = \det\left(I_3\partial_t^2 - B(\nabla)\right) = (\partial_t^2 - \mu\Delta_3)^2(\partial_t^2 - (\lambda + 2\mu)\Delta_3),$$

cf. (2.1.3).

(c) *Cubic media* are characterized by the *three* independent constants $a = c_{11} - c_{44}$, $b = c_{12} + c_{44}$, $c = c_{44}$, whereby the tensor (c_{pqrs}) is given as

$$c_{pqrs} = (b - c)\delta_{pq}\delta_{rs} + c(\delta_{pr}\delta_{qs} + \delta_{ps}\delta_{qr}) + (a - b)\delta_{pj}\delta_{qj}\delta_{rj}\delta_{sj}$$

or

$$C = \begin{pmatrix} H & 0 \\ 0 & L \end{pmatrix}, \qquad H = \begin{pmatrix} a + c & b - c & b - c \\ b - c & a + c & b - c \\ b - c & b - c & a + c \end{pmatrix}, \qquad L = c I_3,$$

$$\tag{2.1.10}$$

cf. Chadwick and Smith [46], (6.1), p. 60; Dederichs and Leibfried [55], (13), p. 1176. If $a = b$, then the cubic medium is isotropic with $\lambda = b - c$, $\mu = c$. Thus the difference $b - a$ is a measure of the anisotropy of the cubic material, cf. Liess [165], p. 274.

From (2.1.9) and (2.1.10), we infer $\Xi = 0$ and

$$B(\xi) = c|\xi|^2 I_3 + b\xi \cdot \xi^T - (b-a)\begin{pmatrix} \xi_1^2 & 0 & 0 \\ 0 & \xi_2^2 & 0 \\ 0 & 0 & \xi_3^2 \end{pmatrix}$$

$$= \begin{pmatrix} c|\xi|^2 + a\xi_1^2 & b\xi_1\xi_2 & b\xi_1\xi_3 \\ b\xi_1\xi_2 & c|\xi|^2 + a\xi_2^2 & b\xi_2\xi_3 \\ b\xi_1\xi_3 & b\xi_2\xi_3 & c|\xi|^2 + a\xi_3^2 \end{pmatrix},$$

cf. Duff [62], p. 271; Liess [165], p. 274; Sommerfeld [259], p. 272.

If, as before, $A(\tau, \xi) = \tau^2 I_3 - B(\xi)$, then the determinant operator $P(\partial) = \det A(\partial)$ is of degree 6 and, in general, irreducible. In fact, $A(\tau, \xi) = M - b\xi \cdot \xi^T$, where M is the diagonal matrix with the elements $\tau^2 - c|\xi|^2 + (b-a)\xi_j^2$, $j = 1, 2, 3$. For the determinant of the difference $M - b\xi \cdot \eta^T$ of the diagonal matrix M and the rank-one matrix $b\xi \cdot \eta^T$, we have the general formula

$$\det(M - b\xi \cdot \eta^T) = \det M - b\xi^T (M^{\mathrm{ad}})^T \eta = \det M - b\eta^T M^{\mathrm{ad}}\xi,$$

and this implies

$$P(\partial) = \det A(\partial) = \prod_{j=1}^{3} W_j(\partial) - b\sum_{j=1}^{3} \partial_j^2 W_{j+1}(\partial) W_{j+2}(\partial), \qquad (2.1.11)$$

where

$$W_j(\partial) = \partial_t^2 - c\Delta_3 + (b-a)\partial_j^2, \ j = 1, 2, 3, \ \text{and} \ W_4(\partial) = W_1(\partial), W_5(\partial) = W_2(\partial).$$

According to (2.1.11), the *slowness surface* $\{(\tau, \xi) \in \mathbf{R}^4; P(\tau, \xi) = 0\}$ is then given by $\displaystyle\sum_{j=1}^{3} \frac{b\xi_j^2}{\tau^2 - c|\xi|^2 + (b-a)\xi_j^2} = 1$ cf. Duff [62], p. 271; Mura [185], (3.35), p. 14; Liess [165], p. 274.

A similar calculation yields for the adjoint matrix of $A(\partial)$ the following:

$$A(\partial)_{jj}^{\mathrm{ad}} = W_{j+1}(\partial) W_{j+2}(\partial) - b\partial_{j+1}^2 W_{j+2}(\partial) - b\partial_{j+2}^2 W_{j+1}(\partial),$$

and $A(\partial)_{j,j+1}^{\mathrm{ad}} = b\partial_j \partial_{j+1} W_{j+2}(\partial), \ j = 1, 2, 3,$

Therefore, the forward fundamental matrix E of $A(\partial)$ is given by $E = A^{\mathrm{ad}}(\partial) F$ where F is the forward fundamental solution of $P(\partial) = \det A(\partial)$.

Let us finally determine for which values of a, b, c the determinant operator $\det A(\partial)$ is reducible. Due to the apparent symmetry in ξ_1, ξ_2, ξ_3, the polynomial

$P(\tau, \xi)$ splits into factors either if it has the form $\prod_{j=1}^{3} (\tau^2 - \alpha|\xi|^2 + \beta\xi_j^2)$, i.e., if $b = 0$, or if there exists a factor $\tau^2 - \alpha|\xi|^2$, which is symmetric in ξ_1, ξ_2, ξ_3. This second assumption implies that either $a = b$, i.e., the medium is isotropic, or else $a = -2b$. In this last case,

$$P(\tau, \xi) = (\tau^2 - c|\xi|^2)\big[(\tau^2 + (b-c)|\xi|^2)^2 - b^2(\xi_1^4 + \xi_2^4 + \xi_3^4 - \xi_1^2\xi_2^2 - \xi_1^2\xi_3^2 - \xi_2^2\xi_3^2)\big].$$

(Concerning the case $b = 0$ cf. Chadwick and Norris [45] (4.2), p. 601; (1.3), p. 590: "For cubic media there is just one constraint on the elastic moduli, i.e., $b = 0$, under which the slowness surface is composed of three spheroids." Cf. also Chadwick and Smith [46], (8.10), p. 74.)

(d) For media of *hexagonal symmetry*, the elastic constants fulfill

$$C = \begin{pmatrix} H & 0 \\ 0 & L \end{pmatrix}, \qquad H = \begin{pmatrix} c_{11} & c_{12} & c_{13} \\ c_{12} & c_{11} & c_{13} \\ c_{13} & c_{13} & c_{33} \end{pmatrix}, \qquad L = \begin{pmatrix} c_{44} & 0 & 0 \\ 0 & c_{44} & 0 \\ 0 & 0 & \frac{1}{2}(c_{11} - c_{12}) \end{pmatrix},$$

cf. Fedorov [72], (9.22), p. 31; Musgrave [186], p. 94; Payton [226], (1.3.2), p. 3. In the tensor (c_{pqrs}), there thus remain 5 independent constants, which we will choose, in accordance with Buchwald [31] (6.7), (6.10), pp. 572, 573, as

$$a_1 = c_{11}, \quad a_2 = c_{33}, \quad a_3 = c_{13} + c_{44}, \quad a_4 = \frac{1}{2}(c_{11} - c_{12}), \quad a_5 = c_{44}.$$
$$(2.1.12)$$

With this notation, we obtain

$$B(\xi) = \begin{pmatrix} a_1\xi_1^2 + a_4\xi_2^2 + a_5\xi_3^2 & (a_1 - a_4)\xi_1\xi_2 & a_3\xi_1\xi_3 \\ (a_1 - a_4)\xi_1\xi_2 & a_4\xi_1^2 + a_1\xi_2^2 + a_5\xi_3^2 & a_3\xi_2\xi_3 \\ a_3\xi_1\xi_3 & a_3\xi_2\xi_3 & a_5(\xi_1^2 + \xi_2^2) + a_2\xi_3^2 \end{pmatrix},$$
$$(2.1.13)$$

cf. Kröner [158], (4), p. 404; Payton [226], (1.5.10), p. 6.

As observed already by Christoffel in 1877 (see Payton [226], p. 7), the determinant operator $P(\partial) = \det A(\partial) = \det(I_3\partial_t^2 - B(\nabla))$ splits. In fact, $\tau^2 I_3 - B(\xi) = M - \eta \cdot \eta^T$ with $\eta = (\sqrt{a_1 - a_4}\,\xi_1, \sqrt{a_1 - a_4}\,\xi_2, a_3\xi_3/\sqrt{a_1 - a_4})^T$ and M the diagonal matrix with the elements

$$m_{11} = m_{22} = \tau^2 - a_4(\xi_1^2 + \xi_2^2) - a_5\xi_3^2, \quad m_{33} = \tau^2 - a_5(\xi_1^2 + \xi_2^2) - \Big(a_2 - \frac{a_3^2}{a_1 - a_4}\Big)\xi_3^2.$$

Putting $\rho^2 = \xi_1^2 + \xi_2^2$, we obtain

$$P(\tau, \xi) = \det M - \eta^T M^{\mathrm{ad}} \eta = m_{11}\left[m_{11}m_{33} - m_{33}(a_1 - a_4)\rho^2 - m_{11}\frac{a_3^2\xi_3^2}{a_1 - a_4} \right]$$

$$= (\tau^2 - a_4\rho^2 - a_5\xi_3^2)\left[(\tau^2 - a_4\rho^2 - a_5\xi_3^2)(\tau^2 - a_5\rho^2 - a_2\xi_3^2) \right.$$

$$\left. - (a_1 - a_4)\rho^2\left(\tau^2 - a_5\rho^2 - \left(a_2 - \frac{a_3^2}{a_1 - a_4}\right)\xi_3^2\right) \right]$$

Hence $P(\partial)$ is the product of the wave operator $\partial_t^2 - a_4\Delta_2 - a_5\partial_3^2$ and of the quartic operator

$$R(\partial) := \partial_t^4 - \partial_t^2(a_1\Delta_2 + a_2\partial_3^2 + a_5\Delta_3) + a_1a_5\Delta_2^2 + (a_1a_2 - a_3^2 + a_5^2)\Delta_2\partial_3^2 + a_2a_5\partial_3^4, \tag{2.1.14}$$

cf. Mura [185], (3.38), p. 14; Payton [226], (1.5.13), p. 6.

There are exactly two cases in which the operator $R(\partial)$ in (2.1.14) is a product of two wave operators:

$$R(\tau, \xi) = 0 \iff 2\tau^2 = (a_1 + a_5)\rho^2 + (a_2 + a_5)\xi_3^2 \pm \sqrt{D},$$

$$D := \left[(a_1 - a_5)\rho^2 - (a_2 - a_5)\xi_3^2 \right]^2 + 4a_3^2\rho^2\xi_3^2.$$

$\sqrt{D}$ is a polynomial in ξ if and only if either $a_3 = 0$ or $a_3^2 = (a_1 - a_5)(a_2 - a_5)$, cf. Chadwick and Norris [45], (1.2), p. 589; Payton [226], p. 96. In these cases only, $P(\partial)$ splits into 3 wave operators. Explicitly, in the case of $a_3 = 0$, we have

$$P(\partial) = (\partial_t^2 - a_4\Delta_2 - a_5\partial_3^2)(\partial_t^2 - a_1\Delta_2 - a_5\partial_3^2)(\partial_t^2 - a_5\Delta_2 - a_2\partial_3^2),$$

and in the case of $a_3^2 = (a_1 - a_5)(a_2 - a_5)$, we obtain

$$P(\partial) = (\partial_t^2 - a_4\Delta_2 - a_5\partial_3^2)(\partial_t^2 - a_1\Delta_2 - a_2\partial_3^2)(\partial_t^2 - a_5\Delta_3).$$

$\square$

2.2 The Malgrange–Ehrenpreis Theorem

The Malgrange–Ehrenpreis theorem states that every (not identically vanishing) partial differential operator with constant coefficients possesses a fundamental solution in the space of distributions, i.e.,

$$\forall P(\partial) \in \mathbf{C}[\partial_1, \ldots, \partial_n] \setminus \{0\} : \exists E \in \mathcal{D}'(\mathbf{R}^n) : P(\partial)E = \delta, \tag{2.2.1}$$

see Malgrange [174], Thm. 1, p. 288; Ehrenpreis [68], Thm. 6, p. 892.

Let us first give a short historical account concerning the development of the concept of fundamental solution. Before 1950, in which year the first edition of the first part of Schwartz [246] appeared, not even the *question* about the existence of a fundamental solution did make sense, since there did not at all exist a generally adopted definition of a fundamental solution. The definitions before L. Schwartz usually referred to special types of operators and, correspondingly, to a special kind of the singularity of the fundamental solution, see, e.g., Courant and Hilbert [51], pp. 351, 363–365, 370; Levi [164], p. 276; Bureau [33], p. 15; Somigliana [258]; Fredholm [82]. Let us also mention the different definition of J. Hadamard, later used by F. Bureau, of a fundamental solution of a hyperbolic second order operator, which is not equivalent to Schwartz's definition, see Lützen [173], p. 103; Leray [163], p. 66; Hadamard [123]; Bureau [36, 37].

In 1950, L. Schwartz wrote: "Les définitions habituelles d'une solution élémentaire comme solution *usuelle* du système homogène ayant en un point une singularité d'un certain type, doivent, à notre avis, être totalement rejetées" (Schwartz [246], p. 135, 136).

In particular, the earlier definitions determined fundamental solutions only up to multiplicative constants. E.g., before 1950, both functions $E = -\frac{1}{4\pi|x|}$ and $F = \frac{1}{|x|}$ served as fundamental solutions for the three-dimensional Laplacean Δ_3. L. Schwartz's definition (i.e., that in Definition 1.3.5) excludes F, since $\Delta_3 F = -4\pi\delta$. Hence "... Schwartz clarifie la notion de solution élémentaire en la définissant comme une solution d'une équation ayant la mesure de Dirac δ pour second membre" (Malgrange [175], p. 29; cf. also Horváth [142], p. 236, 237; Dieudonné [60], p. 255).

Let us remark that L. Schwartz's definition was, for locally integrable fundamental solutions, anticipated by N. Zeilon in 1911 (see Schwartz [246], first ed., Vol. 1, (V, 6; 25), p. 135 and footnote (1)) : "*Es soll:*

jede Funktion $F(x, y, z)$ ein Fundamentalintegral der linearen Differentialgleichung

$$f\left(\frac{\partial}{\partial x}, \frac{\partial}{\partial y}, \frac{\partial}{\partial z}\right)u = 0$$

genannt werden, die der Bedingung genügt, dass

$$f\left(\frac{\partial}{\partial x}, \frac{\partial}{\partial y}, \frac{\partial}{\partial z}\right)\int_D F(x - \lambda, y - \mu, z - \nu)\,d\lambda\,d\mu\,d\nu$$

gleich 1 ist, wenn das Integrationsgebiet D den Punkt x,y,z einschliesst, und gleich 0, wenn dieser Punkt ausserhalb D liegt. Oder, was auf dasselbe herauskommt: Wenn $\phi(x, y, z)$ eine willkürliche Funktion ist, so soll:

$$u = \int_D F(x - \lambda, y - \mu, z - \nu)\phi(\lambda, \mu, \nu)\,d\lambda\,d\mu\,d\nu$$

im Gebiete D eine Lösung geben der Gleichung:

$$f\left(\frac{\partial}{\partial x}, \frac{\partial}{\partial y}, \frac{\partial}{\partial z}\right)u = \phi.$$

Dabei ist D als ganz willkürlich vorausgesetzt, namentlich muss es gestattet sein, es beliebig klein zu machen." (Zeilon [306], pp. 1, 2; Lützen [173], p. 103.)

Already in 1948, L. Schwartz posed the problem to show that every not identically vanishing linear partial operator with constant coefficients has a fundamental solution (see Treves, Pisier, and Yor [276], p. 1078; Gårding [93], p. 80). This problem was solved, independently, in Malgrange [174], Thm. 1, p. 288, and in Ehrenpreis [68], Thm. 6, p. 892. Both used the Hahn–Banach theorem in order to extend a certain linear functional. The key step in their proofs consisted in showing the continuity of this functional on a suitable subspace of the space of all test functions.

Immediately thereafter, the search for explicit general formulae yielding fundamental solutions began; in particular, since such formulae were known for several special classes of differential operators (e.g., for hyperbolic operators and, more generally, for operators correct in the sense of Petrovsky, see Hörmander [138], p. 120, (12.5.3) and p. 143; for elliptic and, more generally, hypoelliptic operators, see Hörmander [133], p. 223; Mizohata [182], p. 142–144). We owe the first explicit general formula to L. Hörmander, who generalized the procedure used for hypoelliptic operators in his thesis (Hörmander [133], p. 223). F. Trèves adapted this method (dubbed "Hörmander's staircase", see Gel'fand and Shilov [106], Ch. II, Section 3.3, p. 103) in order to obtain a fundamental solution depending continuously on the coefficients of the differential operator (see Treves [270, 272]). A detailed description is contained in Ortner and Wagner [215].

An inconvenience of the "staircase" construction consists in its use of partitions of unity based on the location of the zeroes of $P(z)$. In König [155], a new method of proof of the Malgrange–Ehrenpreis theorem (2.2.1) was given. It avoided the use of partitions of unity, but involved n parametric integrations over inverse Fourier transforms of modulus one functions. In Ortner and Wagner [213], a formula involving only *one* parametric integration was given; in the still simpler proof below, which is due to Wagner [296], we represent a fundamental solution by *sums* of inverse Fourier transforms of modulus one functions.

Proposition 2.2.1 *Let $P(\xi) = \sum_{|\alpha| \le m} c_\alpha \xi^\alpha \in \mathbf{C}[\xi] \setminus \{0\}$ be a not identically vanishing polynomial on $\mathbf{R}^n$ of degree m. If $P_m(\xi) = \sum_{|\alpha|=m} c_\alpha \xi^\alpha$ and $\eta \in \mathbf{R}^n$ with $P_m(\eta) \ne 0$, the real numbers $\lambda_0, \ldots, \lambda_m$ are pairwise different, and $a_j = \prod_{k=0, k \ne j}^m (\lambda_j - \lambda_k)^{-1}$, then*

$$E = \frac{1}{P_m(2\eta)} \sum_{j=0}^m a_j e^{\lambda_j \eta x} \mathcal{F}_\xi^{-1}\left(\frac{\overline{P(i\xi + \lambda_j \eta)}}{P(i\xi + \lambda_j \eta)}\right) \tag{2.2.2}$$

is a fundamental solution of $P(\partial)$, i.e., $P(\partial)E = \delta$.

Proof

(1) Let us first observe that, for $\lambda \in \mathbf{R}$ fixed, $N = \{\xi \in \mathbf{R}^n;\ P(i\xi + \lambda\eta) = 0\}$ is a set of Lebesgue measure zero. In fact, after a linear change of the coordinates, we can assume that $P_m(1, 0, \ldots, 0) \neq 0$, and then $\int_N d\xi = \int_{\mathbf{R}^{n-1}} \left(\int_{N_{\xi'}} d\xi_1 \right) d\xi' = 0$ by Fubini's theorem and since the sets $N_{\xi'} = \{\xi_1 \in \mathbf{R};\ P(i(\xi_1, \xi') + \lambda\eta) = 0\}$ are finite for $\xi' = (\xi_2, \ldots, \xi_n) \in \mathbf{R}^{n-1}$. Hence

$$S(\xi) = \frac{\overline{P(i\xi + \lambda\eta)}}{P(i\xi + \lambda\eta)} \in L^\infty(\mathbf{R}^n) \subset \mathcal{S}'(\mathbf{R}^n)$$

and formula (2.2.2) is meaningful.

(2) For $S \in \mathcal{S}'(\mathbf{R}^n)$ and $\zeta \in \mathbf{C}^n$, we have

$$P(\partial)(e^{\zeta x} \mathcal{F}^{-1} S) = e^{\zeta x} P(\partial + \zeta) \mathcal{F}^{-1} S = e^{\zeta x} \mathcal{F}_\xi^{-1}(P(i\xi + \zeta)S).$$

Taking $S = \overline{P(i\xi + \lambda\eta)} / P(i\xi + \lambda\eta)$ with $\lambda \in \mathbf{R}$, this implies

$$P(\partial)\left(e^{\lambda\eta x} \mathcal{F}^{-1}\left(\frac{\overline{P(i\xi + \lambda\eta)}}{P(i\xi + \lambda\eta)} \right) \right) = e^{\lambda\eta x} \mathcal{F}_\xi^{-1}\left(\overline{P(i\xi + \lambda\eta)} \right).$$

Furthermore,

$$\mathcal{F}_\xi^{-1}\left(\overline{P(i\xi + \lambda\eta)} \right) = \mathcal{F}_\xi^{-1}\left(\overline{P}(-i\xi + \lambda\eta) \right) = \overline{P}(-\partial + \lambda\eta)\delta,$$

and hence

$$P(\partial)\left(e^{\lambda\eta x} \mathcal{F}^{-1}\left(\frac{\overline{P(i\xi + \lambda\eta)}}{P(i\xi + \lambda\eta)} \right) \right) = e^{\lambda\eta x} \overline{P}(-\partial + \lambda\eta)\delta = \overline{P}(-\partial + 2\lambda\eta)(e^{\lambda\eta x}\delta)$$

$$= \overline{P}(-\partial + 2\lambda\eta)\delta = \lambda^m \overline{P_m(2\eta)}\delta + \sum_{k=0}^{m-1} \lambda^k T_k,$$

for certain distributions $T_k \in \mathcal{E}'(\mathbf{R}^n)$. (Note that $e^{\lambda\eta x}\delta = \delta$.)

Since $a_0, \ldots, a_m$ fulfill the system of linear equations

$$\sum_{j=0}^m a_j \lambda_j^k = \begin{cases} 0, & \text{if } k = 0, \ldots, m-1, \\ 1, & \text{if } k = m, \end{cases}$$

cf. Proposition 1.3.7, we obtain

$$P(\partial)E = \frac{1}{\overline{P_m(2\eta)}} \sum_{j=0}^{m} a_j \big[\lambda_j^m \overline{P_m(2\eta)}\, \delta + \sum_{k=0}^{m-1} \lambda_j^k T_k\big] = \delta,$$

i.e., E is a fundamental solution of the operator $P(\partial)$. This completes the proof. $\square$

Example 2.2.2 Let us illustrate the construction formula (2.2.2) for fundamental solutions in the case of *quasihyperbolic operators*, which will be studied more thoroughly in Chap. 4.

An operator $P(\partial)$ in $\mathbf{R}^n$ is called *quasihyperbolic* in the direction $N \in \mathbf{R}^n \setminus \{0\}$ iff the condition

$$\exists \sigma_0 \in \mathbf{R} : \forall \sigma > \sigma_0 : \forall \xi \in \mathbf{R}^n : P(i\xi + \sigma N) \neq 0 \qquad (2.2.3)$$

holds. Hence $P(\partial)$ is quasihyperbolic iff $P(z)$, $z \in \mathbf{C}^n$, has no zeroes for large $\operatorname{Re} z$ in direction N, cf. Ortner and Wagner [207], Def. 2, p. 442. For quasihyperbolic operators, there exists one and only one fundamental solution F satisfying

$$\exists \sigma > \sigma_0 : e^{-\sigma x N} F \in \mathcal{S}'(\mathbf{R}^n) \qquad (2.2.4)$$

if σ_0 is as in (2.2.3). Furthermore, $\operatorname{supp} F \subset H_N := \{x \in \mathbf{R}^n; Nx \geq 0\}$, and $e^{-\sigma x N} F \in \mathcal{S}'(\mathbf{R}^n)$ and the equation $F = e^{\sigma x N} \mathcal{F}^{-1}\big(P(i\xi + \sigma N)^{-1}\big)$ hold for each $\sigma > \sigma_0$ if σ_0 is as in (2.2.3), see Proposition 2.4.13 below or Ortner and Wagner [209], Prop. 1, p. 530.

Let us show now that the fundamental solution E in (2.2.2) coincides with F fulfilling (2.2.4) if $\eta = N$ and the real numbers $\lambda_0, \ldots, \lambda_m$ in Proposition 2.2.1 are chosen larger than σ_0. In fact, with these choices, we obtain

$$e^{\lambda_j N x} \mathcal{F}_\xi^{-1}\left(\frac{\overline{P(i\xi + \lambda_j N)}}{P(i\xi + \lambda_j N)}\right) = e^{\lambda_j N x} \bar{P}(-\partial + \lambda_j N) \mathcal{F}_\xi^{-1}\left(\frac{1}{P(i\xi + \lambda_j N)}\right)$$

$$= \bar{P}(-\partial + 2\lambda_j N) e^{\lambda_j N x} \mathcal{F}_\xi^{-1}\left(\frac{1}{P(i\xi + \lambda_j N)}\right)$$

$$= \bar{P}(-\partial + 2\lambda_j N) F = \left[\overline{P_m(2N)}\lambda_j^m + \sum_{k=0}^{m-1} Q_k(\partial)\lambda_j^k\right] F$$

and hence

$$E = \frac{1}{\overline{P_m(2N)}} \sum_{j=0}^{m} a_j\, e^{\lambda_j N x} \mathcal{F}_\xi^{-1}\left(\frac{\overline{P(i\xi + \lambda_j N)}}{P(i\xi + \lambda_j N)}\right)$$

$$= \frac{1}{\overline{P_m(2N)}} \sum_{j=0}^{m} a_j \left[\overline{P_m(2N)}\lambda_j^m + \sum_{k=0}^{m-1} Q_k(\partial)\lambda_j^k\right] F = F. \qquad \square$$

By means of formula (2.1.1), we can infer from the Malgrange–Ehrenpreis theorem for scalar operators (i.e., (2.2.1)) the existence of fundamental matrices for square systems of linear partial differential operators with constant coefficients, cf. also Malgrange [174], Prop. 6, p. 299; Agranovich [4], pp. 37, 38; Hörmander [136], pp. 94, 95.

Proposition 2.2.3 *For the system $A(\partial) \in \mathbf{C}[\partial]^{l \times l}$, the following four assertions are equivalent:*

(1) $A(\partial)$ *has a right-sided fundamental matrix;*
(2) $A(\partial)$ *has a left-sided fundamental matrix;*
(3) $A(\partial)$ *has a two-sided fundamental matrix;*
(4) $\det A(\partial)$ *does not vanish identically.*

Proof Trivially, (3) implies (1) and (2). Furthermore, if (4) is satisfied, then the Malgrange–Ehrenpreis theorem in the scalar case (see (2.2.1) or (2.2.2)) implies the existence of a fundamental solution F of $P(\partial) = \det A(\partial)$, and formula (2.1.1) then yields the two-sided fundamental matrix $E = A(\partial)^{\mathrm{ad}} F$ of the system $A(\partial)$. Hence (4) implies (3). It thus remains to show only that (1) implies (4). (Then, by symmetry, i.e., using transposition, also (2) will imply (4).)

If E is a right-sided fundamental matrix of $A(\partial)$, i.e., if $A(\partial)E = I_l \delta$, then, evidently, $A(\partial)$ cannot be the zero matrix. Let $k > 0$ denote the rank of the matrix $A(\partial)$ and assume, contrary to (4), that $k < l$. After a possible renumbering of the coordinates, we can suppose that the operator $Q(\partial) = \det(A_{ij}(\partial)_{i,j=2,\dots,k+1})$ does not vanish identically. If $C(\partial) \in \mathbf{C}[\partial]^{l \times l}$ contains the adjoint matrix of $A_{ij}(\partial)_{i,j=1,\dots,k+1}$ in the rows and columns corresponding to $i, j = 1, \dots, k + 1$, and consists of zeroes in the remaining places, then $C_{11}(\partial) = Q(\partial) \not\equiv 0$. On the other hand, $C(\partial)\delta = C(\partial)A(\partial)E$ must have a zero in the upper left corner since

$$(C_{ij}(\partial))_{i,j=1,\dots,k+1} \cdot (A_{ij}(\partial))_{i,j=1,\dots,k+1} = I_{k+1} \det((A_{ij}(\partial))_{i,j=1,\dots,k+1}) = 0,$$

A having rank k. This contradicts the assumption $k < l$ and thus shows that (1) implies (4). Hence the proof is complete.   □

2.3 Temperate Fundamental Solutions

We next investigate the problem of the existence of a *temperate* fundamental solution E of an operator $P(\partial)$. After Fourier transformation, this problem is equivalent to the *division problem* $P(i\xi) \cdot \mathcal{F}E = 1$ in $\mathcal{S}'(\mathbf{R}^n)$ formulated by L. Schwartz in 1950, cf. Gårding [93], p. 80; Schwartz [248], p. 9. Obviously, in the dense open subset $\mathbf{R}^n \setminus Z$, $Z := \{\xi \in \mathbf{R}^n; P(i\xi) = 0\}$ of $\mathbf{R}^n$, the distribution $\mathcal{F}E$ must coincide with $P(i\xi)^{-1}$. Hence the division problem consists in extending $P(i\xi)^{-1} \in \mathcal{D}'(\mathbf{R}^n \setminus Z)$ to a distribution in $\mathcal{S}'(\mathbf{R}^n)$. The historically first solutions in Hörmander [135] (see Thm. 3, p. 567) and in Łojasiewicz [168, 169] were based on an estimate of $|P(\mathrm{i}\xi)|$ from below by powers of the distance from ξ to Z. These methods of proof rely on the so-called Hörmander–Łojasiewicz inequalities, Whitney's extension theorem and partitions of unity and hence do not produce explicit formulae for fundamental solutions, cf. the elaborated presentations in Treves [271], pp. 221–242; Krantz and Parks [156], pp. 115–135.

In 1969, Bernstein and Gel'fand [13] presented a new method of proof of the division problem relying on the analytic continuation of the function $\lambda \mapsto P^\lambda$, which continuation was posed as problem by I.M. Gel'fand at the International Congress of Mathematicians in 1954 (see Gel'fand [102], p. 262):

"...the following two problems are of interest:

I. Let $P(x_1, x_2, \ldots, x_n)$ be a polynomial. Consider the area in which $P > 0$. Let $\varphi(x_1, \ldots, x_n)$ be an infinitely many differentiable function equal to zero outside a certain finite area. We shall examine the functional

$$(P^\lambda \cdot \varphi) = \int_{P>0} P^\lambda(x_1, \ldots, x_n)\varphi(x_1, \ldots, x_n)dx_1 \ldots dx_n.$$

It is necessary to prove that this is a meromorphic function of λ (it would be natural to call it a ζ-function of the given polynomial), whose poles are located in points forming several arithmetic progressions, as well as to calculate the residues of this function."

Whereas the proof in Bernstein and Gel'fand [13] is based on Hironaka's theorem on the resolution of singularities, I.N. Bernstein succeeded later to perform the analytic continuation of P^λ by means of a functional equation, similarly as for the gamma function where one uses the equation $\Gamma(\lambda) = \frac{\Gamma(\lambda+1)}{\lambda}$, see Bernstein [11, 12]. We shall employ this approach to prove the existence of temperate fundamental solutions.

Proposition 2.3.1 *Let* $P(\partial) = \sum_{|\alpha| \leq m} a_\alpha \partial^\alpha \in \mathbf{C}[\partial_1, \ldots, \partial_n] \setminus \{0\}$ *be a not identically vanishing linear differential operator with constant coefficients. Then* $P(\partial)$ *possesses a temperate fundamental solution, i.e.,* $\exists E \in \mathcal{S}'(\mathbf{R}^n) : P(\partial)E = \delta$.

Proof

(1) Let us first assume that $P(\mathrm{i}\xi)$ is real-valued and non-negative, i.e., $\forall \xi \in \mathbf{R}^n$: $P(\mathrm{i}\xi) \geq 0$. Then, obviously, the mapping

$$F : \{\lambda \in \mathbf{C};\ \mathrm{Re}\,\lambda > 0\} \longrightarrow \mathcal{S}'(\mathbf{R}^n) : \lambda \longmapsto P(\mathrm{i}\xi)^\lambda \qquad (2.3.1)$$

is well-defined and holomorphic. Bernstein's functional equation (see Bernstein [12], p. 273, Thm. 1'; Björk [15], Ch. 1, 5.7, 5.8) stipulates the existence of a differential operator $Q(\lambda, \xi, \partial) \in \mathbf{C}[\lambda, \xi_1, \ldots, \xi_n, \partial_1, \ldots, \partial_n]$ with polynomial coefficients and of a polynomial $b(\lambda) \in \mathbf{C}[\lambda]$ such that

$$Q(\lambda, \xi, \partial)P(\mathrm{i}\xi)^{\lambda+1} = b(\lambda)P(\mathrm{i}\xi)^\lambda \qquad (2.3.2)$$

holds for all complex λ with $\operatorname{Re}\lambda > 0$. (The normalized polynomial $b(\lambda)$ of minimal degree such that Eq. (2.3.2) is fulfilled with a suitable Q is called the *Bernstein–Sato polynomial* of $P(i\xi)$. A systematic study of Bernstein–Sato polynomials is contained in Yano [304].)

By means of the functional equation (2.3.2), we can holomorphically continue the function F in (2.3.1) to the whole complex plane $\mathbf{C}$ with the exception of the points in the arithmetic progressions $\{\lambda - k;\ b(\lambda) = 0,\ k \in \mathbf{N}_0\}$. In these exceptional points λ_0, $F(\lambda)$ is meromorphic, and we set $F(\lambda_0) := \operatorname{Pf}_{\lambda=\lambda_0} F(\lambda)$, cf. Definition 1.4.6, Proposition 1.4.7, which holds for every Hausdorff, quasicomplete locally convex space, and in particular for $\mathcal{S}'(\mathbf{R}^n)$. Therefore, for $\operatorname{Re}\lambda > -k,\ k \in \mathbf{N}_0$, we have

$$F(\lambda) = \operatorname{Pf}\left[\frac{Q(\lambda,\xi,\partial)}{b(\lambda)}\frac{Q(\lambda+1,\xi,\partial)}{b(\lambda+1)}\cdots\frac{Q(\lambda+k-1,\xi,\partial)}{b(\lambda+k-1)}P(i\xi)^{\lambda+k}\right] \in \mathcal{S}'(\mathbf{R}^n).$$

By analytic continuation, the equation $P(i\xi)F(\lambda) = F(\lambda+1)$ holds for each $\lambda \in \mathbf{C}$. In particular, $F(-1)$ solves the division problem $P(i\xi)F(-1) = 1$, and $E = \mathcal{F}^{-1}(F(-1))$ is a temperate fundamental solution of $P(\partial)$. (It is sometimes called *Bernstein's fundamental solution*.)

(2) For general $P(\partial)$, we set $E = \overline{P}(-\partial)E_1$ where E_1 is the temperate fundamental solution constructed in (1) of the operator $Q(\partial) = P(\partial)\overline{P}(-\partial)$. Note that $Q(\partial)$ has the symbol

$$Q(i\xi) = P(i\xi)\overline{P}(-i\xi) = |P(i\xi)|^2,$$

which is real-valued and non-negative. The proof is complete. □

Let us explain Bernstein's method of construction of temperate fundamental solutions by several explicit examples.

Example 2.3.2 Let us first consider the negative *Laplace operator* $P(\partial) = -\Delta_n$, similarly as in Dieudonné [59], 17.9.2; Horváth [146], Ex. 1, p. 176; Wagner [287], Bsp. 1, p. 413.

(a) In this case, $P(i\xi) = |\xi|^2$ is non-negative, and $F(\lambda) = P(i\xi)^\lambda = |\xi|^{2\lambda}$ is holomorphic in $\mathbf{C}\setminus\{-\frac{n}{2}-j;\ j \in \mathbf{N}_0\}$ and has simple poles in $\lambda = -\frac{n}{2}-j,\ j \in \mathbf{N}_0$. In fact, from Example 1.4.9, we obtain, for $k \in \mathbf{N}_0$,

$$\operatorname*{Res}_{\lambda=-(n+k)/2} |\xi|^{2\lambda} = \frac{1}{2}\operatorname*{Res}_{\lambda=-n-k}|\xi|^\lambda = \frac{1}{2}(-1)^k \sum_{|\alpha|=k}\frac{\langle\omega^\alpha,1\rangle}{\alpha!}\partial^\alpha\delta$$

$$= \begin{cases} 0 & : k \text{ odd}, \\ \frac{\pi^{n/2}\Delta_n^j\delta}{2^{2j}j!\,\Gamma(\frac{n}{2}+j)} & : k \text{ even},\ k=2j. \end{cases}$$

For the last equation see Gel'fand and Shilov [104], Ch. I, 3.9, (5'), p. 73; Ortner and Wagner [219], Ex. 2.3.1, p. 41.

(b) In this case, the functional equation (2.3.2) is simple and reads as

$$\frac{1}{4}\Delta_n |\xi|^{2\lambda+2} = (\lambda + 1)(\lambda + \tfrac{n}{2})|\xi|^{2\lambda}, \qquad (2.3.3)$$

cf. Yano [304], (2), p. 112. Hence $Q(\lambda, \xi, \partial) = \frac{1}{4}\Delta_n$ is here independent of ξ and λ, and $b(\lambda) = (\lambda + 1)(\lambda + \tfrac{n}{2})$.

As in the proof of Proposition 2.3.1, we set $F(\lambda_0) = \mathrm{Pf}_{\lambda=\lambda_0} F(\lambda)$ in the poles $\lambda_0 = -\tfrac{n}{2} - j$, $j \in \mathbf{N}_0$. Then the equation $P(i\xi)F(\lambda) = F(\lambda + 1)$ holds for each $\lambda \in \mathbf{C}$, and, therefore, $E(\lambda) = \mathcal{F}^{-1}\big(F(\lambda)\big)$ fulfills $-\Delta_n E(\lambda) = E(\lambda + 1)$. On the other hand, if we apply the inverse Fourier transform to (2.3.2), then we obtain the *Bernstein–Sato recursion formula*

$$Q(\lambda, -i\partial, -ix)E(\lambda + 1) = b(\lambda)E(\lambda). \qquad (2.3.4)$$

(Equations (2.3.2) and (2.3.4) hold in all points $\lambda \in \mathbf{C}$ where $F(\lambda)$, the analytic continuation of $P(i\xi)^\lambda$, is holomorphic.) In our case, (2.3.4) reads

$$-\tfrac{1}{4}|x|^2 \cdot E(\lambda + 1) = (\lambda + 1)(\lambda + \tfrac{n}{2})E(\lambda), \qquad \lambda \in \mathbf{C} \setminus \{-\tfrac{n}{2} - j; j \in \mathbf{N}_0\}. \qquad (2.3.5)$$

(c) The distributions $E(\lambda)$ coincide with the elliptic M. Riesz kernels introduced and investigated in Example 1.6.11(b), i.e., $E(\lambda) = R_{-2\lambda}$. Note that Eq. (2.3.5) furnishes a new method to derive the fundamental solutions $E(-k)$ of the iterated operator $(-\Delta_n)^k$, $k \in \mathbf{N}$, from the fundamental solution $E(-1)$ of $-\Delta_n$ if $b(-2), \dots, b(-k)$ do not vanish, namely

$$E(-k) = \frac{Q(-k, -i\partial, -ix)}{b(-k)} \cdots \frac{Q(-2, -i\partial, -ix)}{b(-2)} E(-1). \qquad (2.3.6)$$

Hence, if n is odd, or $k \in \mathbf{N}$, $k < \tfrac{n}{2}$, then Example 1.3.14(a) and (2.3.6) imply that

$$E(-k) = \frac{\Gamma(\tfrac{n}{2} - k)}{2^{2k}(k - 1)!\pi^{n/2}} |x|^{2k-n}$$

is a fundamental solution of $(-\Delta_n)^k$. This result agrees with (1.6.19).

Similarly, if the numbers $b(-j)$ do not vanish for $l + 1 \le j \le k$, then

$$E(-k) = \frac{Q(-k, -i\partial, -ix)}{b(-k)} \cdots \frac{Q(-l - 1, -i\partial, -ix)}{b(-l - 1)} E(-l),$$

and this yields for $P(\partial) = -\Delta_n$, n even, $l = \tfrac{n}{2}$, $k > l$,

$$E(-k) = \frac{(\tfrac{n}{2} - 1)!}{(k - \tfrac{n}{2})!(k - 1)!} \left(-\tfrac{1}{4}|x|^2\right)^{-n/2+k} E(-\tfrac{n}{2})$$

in agreement with (1.6.20). Note that—in both cases—formula (2.3.5) allows to derive the fundamental solutions $E(-k)$ of the iterated operators $P(\partial)^k$ from $E(-1)$ respectively from $E(-\frac{n}{2})$ simply by multiplications with powers of $|x|$. □

Let us consider now, more generally as in Proposition 2.3.1 and Example 2.3.2, powers of *complex-valued* polynomials.

Proposition 2.3.3 *Given a polynomial* $P(\xi) = \sum_{|\alpha| \le m} a_\alpha \xi^\alpha$, $a_\alpha \in \mathbf{C}$, $\xi \in \mathbf{R}^n$, *of degree* m *and a measurable bounded function* $k : \{\xi \in \mathbf{R}^n; P(\xi) \ne 0\} \longrightarrow \mathbf{Z}$, *we set*

$$P(\xi)^\lambda := \begin{cases} 0 & : P(\xi) = 0, \\ \exp\big(\lambda[2\pi i k(\xi) + i \arg(P(\xi)) + \log|P(\xi)|]\big) & : P(\xi) \ne 0, \end{cases}$$

where $z = |z| \cdot e^{i \arg z}$, $z \in \mathbf{C}$, *with* $\arg z \in (-\pi, \pi]$.
  Then $P(\xi)^\lambda \in L^1_{\text{loc}}(\mathbf{R}^n)$ *for all* $\lambda \in \mathbf{C}$ *with* $\operatorname{Re} \lambda > -\frac{1}{m}$, *and the mapping*

$$\{\lambda \in \mathbf{C}; \operatorname{Re} \lambda > -\tfrac{1}{m}\} \longrightarrow \mathcal{S}'(\mathbf{R}^n) : \lambda \longmapsto P(\xi)^\lambda$$

is well-defined and holomorphic.

Proof Obviously, for $\operatorname{Re} \lambda > 0$, the function

$$|P(\xi)^\lambda| = \exp\big(-\operatorname{Im} \lambda[2\pi k(\xi) + \arg P(\xi)]\big) \cdot |P(\xi)|^{\operatorname{Re}\lambda}$$

is polynomially bounded and thus yields a temperate distribution. Furthermore, if $\phi \in \mathcal{S}(\mathbf{R}^n)$, then

$$\langle \phi, P(\xi)^\lambda \rangle = \int_{\mathbf{R}^n} \phi(\xi) \, P(\xi)^\lambda \, d\xi$$

analytically depends on $\lambda \in \mathbf{C}$ with $\operatorname{Re} \lambda > 0$.
  For negative real values of λ, we use the estimate

$$\exists C > 0 : \forall N > 0 : \forall \epsilon > 0 : \int_{|\xi| < N} Y(\epsilon - |P(\xi)|) \, d\xi \le C N^{n-1} \epsilon^{1/m}$$

to conclude that $\int_{|\xi| < N} |P(\xi)^\lambda| \, d\xi$ is finite for $\operatorname{Re} \lambda > -\frac{1}{m}$ and grows at most polynomially if $N \to \infty$, cf. also Ricci and E.M. Stein [233], Prop., p. 182. □

Example 2.3.4 Let us first observe that Bernstein [12], p. 273, Thm. 1'; Björk [15], Ch. 1, 5.7, 5.8, prove the existence of polynomials Q, b for arbitrary complex-valued polynomials $P(i\xi)$ such that the functional equation

$$Q(\lambda, \xi, \partial) P(i\xi)^{\lambda+1} = b(\lambda) P(i\xi)^\lambda \qquad (2.3.2)$$

holds in the *algebraic sense*, i.e., if $P(i\xi)^\lambda$ is considered as a symbol which is subject to the relations $P(i\xi)P(i\xi)^\lambda = P(i\xi)^{\lambda+1}$ and $\partial_j P(i\xi)^{\lambda+1} = (\lambda + 1)P(i\xi)^\lambda \cdot \frac{\partial P(i\xi)}{\partial \xi_j}$. For *non-negative* polynomials $P(i\xi)$, we set $k(\xi) = 0$ in Proposition 2.3.3. Then the algebraic validity of (2.3.2) implies that (2.3.2) also holds in $S'(\mathbf{R}^n)$, first for large $\mathrm{Re}\,\lambda$ since there $P(i\xi)^\lambda$ is sufficiently often differentiable, and then, by analytic continuation, for all complex λ unless λ is one of the poles of $F(\lambda) = P(i\xi)^\lambda$. For *complex-valued* polynomials, the relation between the algebraic and the distributional validity of (2.3.2) is more complicated. Let us illustrate this fact in the simple case of the Cauchy–Riemann operator $P(\partial) = \partial_1 + i\partial_2$.

For $P(i\xi) = i\xi_1 - \xi_2$, the equations

$$\partial_1(i\xi_1 - \xi_2)^{\lambda+1} = i(\lambda + 1)(i\xi_1 - \xi_2)^\lambda, \qquad \partial_2(i\xi_1 - \xi_2)^{\lambda+1} = -(\lambda + 1)(i\xi_1 - \xi_2)^\lambda$$

hold in the algebraic sense. Let us define the distribution-valued function

$$F : \{\lambda \in \mathbf{C};\ \mathrm{Re}\,\lambda > -1\} \longrightarrow S'(\mathbf{R}^2) : \lambda \longmapsto (i\xi_1 - \xi_2)^\lambda = e^{\lambda \log(i\xi_1 - \xi_2)},$$

where $\log(i\xi_1 - \xi_2) = \log|\xi| + i\arg(i\xi_1 - \xi_2)$, i.e., we choose $k \equiv 0$ in Proposition 2.3.3. Then, due to the discontinuity of $F(\lambda)$ along the half-line $\xi_1 = 0,\ \xi_2 > 0$, the jump formula (1.3.9) yields

$$\partial_1(F(\lambda + 1)) = i(\lambda + 1)F(\lambda) - 2i\sin(\lambda\pi)\delta(\xi_1) \otimes Y(\xi_2)\xi_2^{\lambda+1},$$
$$\text{and } \partial_2(F(\lambda + 1)) = -(\lambda + 1)F(\lambda) \tag{2.3.7}$$

for $\mathrm{Re}\,\lambda > -1$. Since $F(0) = 1$ and thus $\partial_2(F(0)) = 0$, the second equation in (2.3.7) shows that $F(\lambda)$ can analytically be extended to the whole complex plane. Therefore (2.3.7) remains valid for all complex λ if $Y(\xi_2)\xi_2^{\lambda+1}$ is replaced by $\xi_{2+}^{\lambda+1}$. (Note that $\sin(\lambda\pi)\xi_{2+}^{\lambda+1}$ also depends holomorphically on λ.)

In particular, for $\lambda = -2$, we obtain

$$\partial_1(F(-1)) = -iF(-2) - 2\pi i\delta, \qquad \partial_2(F(-1)) = F(-2),$$

and hence $(\partial_1 + i\partial_2)(F(-1)) = -2\pi i\delta$, i.e.,

$$(\partial_1 + i\partial_2)\frac{1}{2\pi(\xi_1 + i\xi_2)} = \delta,$$

which is in accordance with Example 1.3.14(b).   □

We consider next quasihyperbolic operators, for which the continuation of $\lambda \mapsto P(i\xi)^\lambda$ to the whole complex plane can be achieved without poles, and without the use of Bernstein's equation (2.3.2).

Proposition 2.3.5 *For $P(\partial) \in \mathbf{C}[\partial_1, \ldots, \partial_n]$ and $N \in \mathbf{R}^n \setminus \{0\}$ suppose that*

$$\forall \sigma > 0 : \forall \xi \in \mathbf{R}^n : P(\mathrm{i}\xi + \sigma N) \neq 0, \tag{2.3.8}$$

i.e., $P(\partial)$ is quasihyperbolic in the direction N with $\sigma_0 = 0$, see (2.2.3). Furthermore, let the function

$$X := \mathbf{R}^n \times (0, \infty) \longrightarrow \mathbf{C} : (\xi, \sigma) \longmapsto \log(P(\mathrm{i}\xi + \sigma N))$$

be determined by the choice of a value at $(\xi, \sigma) = (0, 1)$ and continuous extension in the simply connected space $\dot{X}$.

Then the distribution-valued function

$$F : \mathbf{C} \longrightarrow \mathcal{S}'(\mathbf{R}^n) : \lambda \longmapsto F(\lambda) = \lim_{\sigma \searrow 0} P(\mathrm{i}\xi + \sigma N)^\lambda$$

is well-defined and entire. Furthermore, $E(\lambda) := \mathcal{F}^{-1}(F(\lambda)) \in \mathcal{S}'(\mathbf{R}^n)$ fulfills

$$\forall \lambda \in \mathbf{C} : \operatorname{supp} E(\lambda) \subset \{x \in \mathbf{R}^n; \ x \cdot N \geq 0\}$$

and $\forall k \in \mathbf{N}_0 : P(\partial)^k E(-k) = \delta$, i.e., $E(-k)$ is a temperate fundamental solution of $P(\partial)^k$.

Proof

(a) An appeal to the Seidenberg–Tarski lemma, i.e., Lemma 2 in Hörmander [135], p. 557, furnishes that

$$|P(\mathrm{i}\xi + \sigma N)| \geq c\sigma^k (1 + |\xi|^2 + \sigma^2)^{-k} \tag{2.3.9}$$

for some positive constants c, k and all $\xi \in \mathbf{R}^n$ and $\sigma > 0$. This implies that $P(\mathrm{i}\xi + \sigma N)^\lambda \in \mathcal{S}'(\mathbf{R}^n_\xi)$ for all $\lambda \in \mathbf{C}$ and $\sigma > 0$. Furthermore, also the boundary value for $\sigma \searrow 0$ exists in $\mathcal{S}'(\mathbf{R}^n_\xi)$ due to Atiyah, Bott and Gårding [5], pp. 121–122; Hörmander [139], Thm. 3.1.15; Zuily [309], Exercise 52, p. 93. Hence F is well-defined.

Let U be the open complex right half-plane $U = \{z \in \mathbf{C}; \operatorname{Re} z > 0\}$ and consider

$$T : U \times \mathbf{C} \longrightarrow \mathcal{S}'(\mathbf{R}^n) : (z, \lambda) \longmapsto P(\mathrm{i}\xi + zN)^\lambda.$$

Then T is holomorphic since this holds for the integrals

$$\int \phi(\xi) P(\mathrm{i}\xi + zN)^\lambda \, \mathrm{d}\xi, \qquad \phi \in \mathcal{S},$$

due to the estimate (2.3.9) and Lebesgue's theorem. By Morera's theorem, we conclude that $F(\lambda) = \lim_{\sigma \searrow 0} T(\sigma, \lambda)$ is entire.

(b) Let us next show that $E(\lambda)|_H = 0$ if $H := \{x \in \mathbf{R}^n;\, Nx < 0\}$. For that reason, let us define

$$G(z, \lambda) := e^{zNx} \mathcal{F}_\xi^{-1}\big(P(i\xi + zN)^\lambda\big),$$

which is a holomorphic function on $U \times \mathbf{C}$ with values in $\mathcal{D}'(\mathbf{R}^n)$. Since $G(z, \lambda) = G(\mathrm{Re}\, z, \lambda)$, this implies that G is independent of z. For $\lambda \in \mathbf{C}$ fixed, we then obtain

$$E(\lambda) = \mathcal{F}^{-1}(F(\lambda)) = \lim_{\sigma \searrow 0} \mathcal{F}_\xi^{-1}\big(P(i\xi + \sigma N)^\lambda\big)$$

$$= \lim_{\sigma \searrow 0} e^{\sigma Nx} \mathcal{F}_\xi^{-1}\big(P(i\xi + \sigma N)^\lambda\big) = G(1, \lambda).$$

This implies

$$E(\lambda) = \lim_{\sigma \to \infty} G(\sigma, \lambda) = \lim_{\sigma \to \infty}\Big[e^{\sigma Nx} \sigma^l \cdot \sigma^{-l} \mathcal{F}^{-1}\big(P(i\xi + \sigma N)^\lambda\big)\Big] = 0 \text{ in } \mathcal{D}'(H)$$

since $e^{\sigma Nx} \sigma^l$ converges to 0 in $\mathcal{E}(H)$ for each $l \in \mathbf{N}$ and the set $\{\sigma^{-l} P(i\xi + \sigma N)^\lambda;\, \sigma \geq 1\}$ is bounded in $\mathcal{S}'(\mathbf{R}_\xi^n)$ for suitable $l \in \mathbf{N}$. (More precisely, if $\mathrm{Re}\, \lambda \geq 0$, then we can choose any $l \geq m \cdot \mathrm{Re}\, \lambda$ if $m = \deg P$; for $\mathrm{Re}\, \lambda < 0$, one can either use the estimate (2.3.9) or argue by analytic continuation from the case $\mathrm{Re}\, \lambda > 0$.)

Finally, $P(i\xi)^k \cdot F(-k) = 1$ implies $P(\partial)^k E(-k) = \delta$, and hence the proof is complete. $\qquad\qquad\square$

Example 2.3.6 Let us investigate the entire function $\lambda \mapsto E(\lambda) = \mathcal{F}^{-1}(F(\lambda))$ in Proposition 2.3.5 in the particular case of the *wave operator* $P(\partial) = \partial_t^2 - \Delta_n$. (Note that $E(-1)$ has already been calculated in Example 1.4.12 for $n = 2, 3$ and in Example 1.6.17 for general n.)

$P(\partial)$ is (quasi)hyperbolic in the direction $N = (1, 0)$ since

$$P\big(i(\tau, \xi) + \sigma N\big) = P(i\tau + \sigma, i\xi) = (i\tau + \sigma)^2 + |\xi|^2 \neq 0$$

for $\sigma > 0$ and $(\tau, \xi) \in \mathbf{R}^{n+1}$. We observe that $(i\tau + \sigma)^2 + |\xi|^2 \in \mathbf{C} \setminus (-\infty, 0]$ for $\sigma > 0$ and therefore

$$\log P\big(i(\tau, \xi) + \sigma N\big) = \log \big|P\big(i(\tau, \xi) + \sigma N\big)\big| + i \arg P\big(i(\tau, \xi) + \sigma N\big)$$

is a continuous function on $X = \mathbf{R}_{\tau,\xi}^{n+1} \times (0, \infty)$ if we take the argument of $P\big(i(\tau, \xi) + \sigma N\big)$ in the interval $(-\pi, \pi)$. This yields

$$\lim_{\sigma \searrow 0} \log\big[(i\tau + \sigma)^2 + |\xi|^2\big] = \log\big|\tau^2 - |\xi|^2\big| + i\pi Y(\tau^2 - |\xi|^2)\, \mathrm{sign}\, \tau.$$

Hence, for $\text{Re }\lambda > -1$, the distributions $F(\lambda) = \lim_{\sigma \searrow 0} P(\mathrm{i}(\tau, \xi) + \sigma N)^{\lambda}$ are locally integrable and given by

$$F(\lambda) = |\tau^2 - |\xi|^2|^{\lambda} \cdot \left[Y(|\xi|^2 - \tau^2) + Y(\tau^2 - |\xi|^2) \cdot e^{\mathrm{i}\pi\lambda \operatorname{sign}\tau} \right].$$

(a) Let us calculate $E(\lambda) = \mathcal{F}^{-1}F(\lambda)$ first by partial Fourier transform, see Definition 1.6.15.

For $\sigma > 0$, $\text{Re }\lambda < -\frac{1}{2}$ and fixed $\xi \in \mathbf{R}^n$, the function $\left[(\mathrm{i}\tau + \sigma)^2 + |\xi|^2 \right]^{\lambda}$ is absolutely integrable with respect to τ, and its inverse Fourier transform is given by

$$\mathcal{F}_{\tau}^{-1}\left(\left[(\mathrm{i}\tau + \sigma)^2 + |\xi|^2 \right]^{\lambda} \right) = \frac{1}{2\pi} \int_{-\infty}^{\infty} e^{\mathrm{i}\tau t} \left[(\mathrm{i}\tau + \sigma)^2 + |\xi|^2 \right]^{\lambda} \mathrm{d}\tau$$

$$= \frac{e^{-\sigma t}}{2\pi \mathrm{i}} \int_{\sigma - \mathrm{i}\infty}^{\sigma + \mathrm{i}\infty} e^{pt} (p^2 + |\xi|^2)^{\lambda} \, \mathrm{d}p.$$

The last integral is a well-known inverse Laplace transform, see Badii and Oberhettinger [7], Part II, Eq. 4.27, p. 240, which gives

$$\mathcal{F}_{\tau}^{-1}\left(\left[(\mathrm{i}\tau + \sigma)^2 + |\xi|^2 \right]^{\lambda} \right) = e^{-\sigma t} Y(t) \frac{\sqrt{\pi}}{\Gamma(-\lambda)} \left(\frac{2|\xi|}{t} \right)^{1/2 + \lambda} J_{-1/2 - \lambda}(|\xi|t).$$

Hence

$$E(\lambda) = \frac{\sqrt{\pi}}{\Gamma(-\lambda)} \mathcal{F}_{\xi}^{-1}\left[Y(t) \left(\frac{2|\xi|}{t} \right)^{1/2 + \lambda} J_{-1/2 - \lambda}(|\xi|t) \right]. \qquad (2.3.10)$$

For $\text{Re }\lambda < -\frac{1}{2}$, the right-hand side in (2.3.10) continuously depends on t, and we can fix t in order to perform the inverse Fourier transform with respect to ξ by means of the Poisson–Bochner formula (1.6.14). This yields, for $\text{Re }\lambda < -n$,

$$E(\lambda) = \frac{\sqrt{\pi}\, 2^{1/2 + \lambda} Y(t) |x|^{-n/2 + 1}}{\Gamma(-\lambda)(2\pi)^{n/2} t^{1/2 + \lambda}} \int_0^{\infty} \rho^{\lambda + (n+1)/2} J_{n/2 - 1}(\rho|x|) \cdot J_{-1/2 - \lambda}(\rho t) \, \mathrm{d}\rho$$

$$= \frac{2^{2\lambda + 1} Y(t - |x|)(t^2 - |x|^2)^{-\lambda - (n+1)/2}}{\pi^{(n-1)/2} \Gamma(-\lambda) \Gamma(-\lambda - \frac{n-1}{2})}$$

$$\qquad (2.3.11)$$

by Gradshteyn and Ryzhik [113], Eq. 6.575.1, p. 692. Note that the right-hand side in (2.3.11) is locally integrable for $\text{Re }\lambda < -\frac{n-1}{2}$, and hence the same is true for $E(\lambda)$ and (2.3.11) holds for $\text{Re }\lambda < -\frac{n-1}{2}$ by analytic continuation.

Let us remark that $Z_{\lambda} = E(-\frac{\lambda}{2})$ is traditionally called *hyperbolic Marcel Riesz kernel*, and that (2.3.11) is also given in Schwartz [246], Eq. (I, 3; 31),

p. 50; Atiyah, Bott and Gårding [5], (4.20), p. 147; Dieudonné [59], (17.9.4.5), p. 267; Riesz [234], p. 156; Riesz [235], p. 4.

As observed already by M. Riesz, the "composition law" $Z_\lambda * Z_\mu = Z_{\lambda+\mu}$ holds for all $\lambda, \mu \in \mathbf{C}$. In fact, $Z_\lambda, Z_\mu \in \mathcal{D}'_\Gamma = \{T \in \mathcal{D}'(\mathbf{R}^{n+1}); \operatorname{supp} T \subset \Gamma\}$ if Γ is the forward wave cone $\Gamma = \{(t, x) \in \mathbf{R}^{n+1}; t \geq |x|\}$, and hence Z_λ, Z_μ are convolvable by support, see Example 1.5.11. Furthermore, for $\operatorname{Re}\lambda < 1$, we have $Z_\lambda \in \mathcal{D}'_{L^2}$ (since $F(-\frac{\lambda}{2})$ is a polynomial times an L^2-function), and hence $Z_\lambda * Z_\mu = Z_{\lambda+\mu}$ holds by the exchange theorem Proposition 1.6.6 (5). This relation then persists for all complex λ, μ by analytic continuation.

Note that $E(-k)$, $k \in \mathbf{N}$, is the only fundamental solution of the hyperbolic operator $(\partial_t^2 - \Delta_n)^k$ with support in the half-space $t \geq 0$, see Hörmander [138], Thm. 12.5.1, p. 120. If $k > \frac{n-1}{2}$, then $E(-k)$ is locally integrable and, according to (2.3.11), given by

$$E(-k) = \frac{2^{1-2k}Y(t - |x|)(t^2 - |x|^2)^{k-(n+1)/2}}{(k-1)!\, \pi^{(n-1)/2}\Gamma(k - \frac{n-1}{2})}. \tag{2.3.12}$$

For $k \leq \frac{n-1}{2}$, we obtain the fundamental solution $E(-k)$ by analytically continuing $E(\lambda)$, $\operatorname{Re}\lambda < -\frac{n-1}{2}$, in (2.3.11) since $\lambda \mapsto E(\lambda)$ is entire. For $t \neq |x|$ and $k = 1$, the result coincides with the formulas in (1.6.26), (1.6.27).

(b) A second evaluation of the inverse Fourier transform of $F(\lambda)$ employs the Lorentz invariance and the homogeneity of $F(\lambda)$, which properties are passed on to $E(\lambda)$. Since, furthermore, $\operatorname{supp} E(\lambda)$ is contained in the half-space $\{(t, x) \in \mathbf{R}^{n+1}; t \geq 0\}$ by Proposition 2.3.5, we conclude that

$$\forall \lambda \in \mathbf{C} \text{ with } \operatorname{Re}\lambda < -\tfrac{n-1}{2} : \exists c \in \mathbf{C} : E(\lambda) = cY(t - |x|)(t^2 - |x|^2)^{-\lambda-(n+1)/2}.$$

Finally, in order to determine the constant $c = c(\lambda)$, we use formula (2.3.10) for $\operatorname{Re}\lambda < -n$, and we obtain

$$c = E(\lambda)(1, 0) = \frac{\sqrt{\pi}}{\Gamma(-\lambda)(2\pi)^n} \int_{\mathbf{R}^n} (2|\xi|)^{1/2+\lambda} J_{-1/2-\lambda}(|\xi|)\, d\xi$$

$$= \frac{2^{3/2+\lambda-n}}{\pi^{(n-1)/2}\Gamma(-\lambda)\Gamma(\frac{n}{2})} \int_0^\infty r^{-1/2+\lambda+n} J_{-1/2-\lambda}(r)\, dr$$

$$= \frac{2^{2\lambda+1}}{\pi^{(n-1)/2}\Gamma(-\lambda)\Gamma(-\lambda - \frac{n-1}{2})}.$$

Therefrom (2.3.11) follows for $\operatorname{Re}\lambda < -\frac{n-1}{2}$ by analytic continuation.

(c) Let us eventually calculate $E(\lambda)$ by means of an improved version of the so-called *Cagniard–de Hoop method,* see de Hoop [132], Achenbach [2].

If $\operatorname{Re}\lambda < -\frac{n+1}{2}$ and $\sigma > 0$, then $\left[(i\tau + \sigma)^2 + |\xi|^2\right]^\lambda \in L^1(\mathbf{R}^{n+1}_{\tau,\xi})$ and hence $E(\lambda)$ is given by the following absolutely convergent Fourier integral (see the proof of Proposition 2.3.5):

$$E(\lambda) = G(\sigma, \lambda) = \frac{e^{\sigma t}}{(2\pi)^{n+1}} \int_{-\infty}^{\infty} e^{i t \tau} \left(\int_{\mathbf{R}^n} e^{i x \xi} \left[(i\tau + \sigma)^2 + |\xi|^2 \right]^{\lambda} d\xi \right) d\tau.$$

The substitution $p = i\tau + \sigma$ then yields

$$E(\lambda) = \frac{1}{2\pi i} \int_{\operatorname{Re} p = \sigma} e^{pt} U(p) \, dp,$$

where U denotes the analytic function

$$U : \{ p \in \mathbf{C}; \operatorname{Re} p > 0 \} \longrightarrow \mathcal{S}'(\mathbf{R}_x^n) : p \longmapsto \mathcal{F}_{\xi}^{-1} \big((p^2 + |\xi|^2)^{\lambda} \big).$$

Let us represent $U(p)$ by a one-fold integral. If $p > 0$, then the scale transformations $\xi = p \cdot \eta$, $\eta = (s, \eta') \in \mathbf{R}^n$ and $\eta' = \sqrt{1 + s^2} \, \zeta \in \mathbf{R}^{n-1}$ yield, due to the rotational symmetry of $U(p)$, the following:

$$U(p) = \frac{p^{2\lambda + n}}{(2\pi)^n} \int_{\mathbf{R}^n} e^{ipxn} [1 + |\eta|^2]^{\lambda} \, d\eta$$

$$= \frac{p^{2\lambda + n}}{(2\pi)^n} \int_{-\infty}^{\infty} e^{ip|x|s} (1 + s^2)^{\lambda + (n-1)/2} \, ds \cdot \int_{\mathbf{R}^{n-1}} (1 + |\zeta|^2)^{\lambda} \, d\zeta.$$

The inner integral can easily be evaluated:

$$\int_{\mathbf{R}^{n-1}} (1 + |\zeta|^2)^{\lambda} \, d\zeta = \frac{2\pi^{(n-1)/2}}{\Gamma(\frac{n-1}{2})} \int_0^{\infty} (1 + t^2)^{\lambda} t^{n-2} \, dt = \frac{\pi^{(n-1)/2} \Gamma(-\lambda - \frac{n-1}{2})}{\Gamma(-\lambda)}.$$

Applying Cauchy's integral theorem we next deform the integration contour for s from the real axis to one along $s = iv \pm 0$, $1 \leq v < \infty$. This yields

$$\int_{-\infty}^{\infty} e^{ip|x|s} (1 + s^2)^{\lambda + (n-1)/2} \, ds$$

$$= -2 \sin\big(\pi(\lambda + \tfrac{n-1}{2})\big) \int_1^{\infty} e^{-p|x|v} (v^2 - 1)^{\lambda + (n-1)/2} \, dv$$

$$= -\frac{2}{|x|} \sin\big(\pi(\lambda + \tfrac{n-1}{2})\big) \int_{|x|}^{\infty} e^{-pt} \Big(\frac{t^2}{|x|^2} - 1 \Big)^{\lambda + (n-1)/2} \, dt.$$

Making use of the complement formula for the gamma function we obtain

$$U(p) = \frac{p^{2\lambda + n}}{2^{n-1} \pi^{(n-1)/2} \Gamma(-\lambda) \Gamma(\lambda + \frac{n+1}{2}) |x|} \cdot \mathcal{L}_t \left(Y(t - |x|) \Big(\frac{t^2}{|x|^2} - 1 \Big)^{\lambda + (n-1)/2} \right)$$

where $\mathcal{L}_t(f) = \int_0^\infty f(t)e^{-pt}\,\mathrm{d}t$ denotes the Laplace transform of f. Due to

$$p^{2\lambda+n} = \frac{1}{\Gamma(-2\lambda-n)}\,\mathcal{L}_t\big(Y(t)t^{-2\lambda-n-1}\big),$$

the convolution theorem for the Laplace transformation then yields

$$U(p) = \frac{1}{2^{n-1}\pi^{(n-1)/2}\Gamma(-\lambda)|x|}\times$$

$$\times \mathcal{L}_t\bigg(\frac{Y(t-|x|)}{\Gamma(-2\lambda-n)\Gamma(\lambda+\frac{n+1}{2})}\int_{|x|}^t (t-s)^{-2\lambda-n-1}\Big(\frac{s^2}{|x|^2}-1\Big)^{\lambda+(n-1)/2}\,\mathrm{d}s\bigg).$$

The definite integral therein is evaluated by means of Gröbner and Hofreiter [115], 421.4, p. 175:

$$\frac{Y(t-|x|)}{\Gamma(-2\lambda-n)\Gamma(\lambda+\frac{n+1}{2})}\int_{|x|}^t (t-s)^{-2\lambda-n-1}\Big(\frac{s^2}{|x|^2}-1\Big)^{\lambda+(n-1)/2}\,\mathrm{d}s$$

$$= \frac{2^{2\lambda+n}Y(t-|x|)|x|}{\Gamma(-\lambda-\frac{n-1}{2})}\,(t^2-|x|^2)^{-\lambda-(n+1)/2};$$

Hence

$$U(p) = \frac{2^{2\lambda+1}\mathcal{L}_t\big(Y(t-|x|)(t^2-|x|^2)^{-\lambda-(n+1)/2}\big)}{\pi^{(n-1)/2}\Gamma(-\lambda)\Gamma(-\lambda-\frac{n-1}{2})},$$

and, due to $U = \mathcal{L}\big(E(\lambda)\big)$, we infer that

$$E(\lambda) = \frac{2^{2\lambda+1}Y(t-|x|)(t^2-|x|^2)^{-\lambda-(n+1)/2}}{\pi^{(n-1)/2}\Gamma(-\lambda)\Gamma(-\lambda-\frac{n-1}{2})}$$

for $\mathrm{Re}\,\lambda < -\frac{n+1}{2}$ in accordance with (2.3.11).   □

Example 2.3.7 Let us generalize the last example so as also to cover the convolution group $E(\lambda)$ of the *Klein–Gordon operator* $P(\partial) = \partial_t^2 - \Delta_n + m^2$, $m > 0$.

Similarly as in Example 2.3.6 (a), we use partial Fourier transformation with respect to t and x. As above, we have $E(\lambda) = \mathcal{F}^{-1}\big(F(\lambda)\big)$ where

$$F(\lambda) = \big|\tau^2-|\xi|^2+m^2\big|^\lambda \cdot \Big[Y(|\xi|^2-\tau^2+m^2)+Y(\tau^2-|\xi|^2-m^2)\cdot e^{i\pi\lambda\,\mathrm{sign}\,\tau}\Big]$$

and

$$\mathcal{F}_\tau^{-1}\big(F(\lambda)\big) = \frac{\sqrt{\pi}\,Y(t)}{\Gamma(-\lambda)}\left(\frac{2\sqrt{|\xi|^2 + m^2}}{t}\right)^{1/2+\lambda} J_{-1/2-\lambda}\big(t\sqrt{|\xi|^2 + m^2}\big), \quad \operatorname{Re}\lambda < 0.$$
(2.3.13)

Hence, applying the Poisson–Bochner formula (1.6.14) yields, for $\operatorname{Re}\lambda < -n$,

$$E(\lambda) = \frac{\sqrt{\pi}\,2^{1/2+\lambda}Y(t)|x|^{-n/2+1}}{\Gamma(-\lambda)(2\pi)^{n/2}t^{1/2+\lambda}} \times$$

$$\times \int_0^\infty \rho^{n/2}(\rho^2 + m^2)^{\lambda/2+1/4} J_{n/2-1}(\rho|x|) \cdot J_{-1/2-\lambda}\big(t\sqrt{\rho^2 + m^2}\big)\,d\rho.$$

Finally, Gradshteyn and Ryzhik [113], Eq. 6.596.6, p. 706, furnishes

$$E(\lambda) = \frac{2^{\lambda-(n-1)/2}Y(t - |x|)m^{\lambda+(n+1)/2}}{\pi^{(n-1)/2}\Gamma(-\lambda)}\,(t^2 - |x|^2)^{-\lambda/2-(n+1)/4}\times$$

$$\times J_{-\lambda-(n+1)/2}\big(m\sqrt{t^2 - |x|^2}\big).$$
(2.3.14)

As in Example 2.3.6, formula (2.3.14) holds for $\operatorname{Re}\lambda < -\frac{n-1}{2}$ by analytic continuation.

In particular, for $n = 2$ and $\lambda = -1$, we obtain the fundamental solution of the Klein–Gordon operator $\partial_t^2 - \Delta_2 + m^2$ in two space dimensions:

$$E(-1) = \frac{Y(t - |x|)}{2\pi\sqrt{t^2 - |x|^2}}\cos\big(m\sqrt{t^2 - |x|^2}\big) \in L_{\text{loc}}^1(\mathbf{R}_{t,x}^3).$$
(2.3.15)

(As always for hyperbolic operators, this is the only fundamental solution with support in the half-space $t \geq 0$.)

For $n = 3$, the calculation of $E(-1)$ is slightly more difficult since the assumption $\operatorname{Re}\lambda < -\frac{n-1}{2}$ is not satisfied for $\lambda = -1$. From the equations

$$\partial_\tau F(\lambda + 1) = -2(\lambda + 1)\tau F(\lambda) \quad \text{and} \quad E(\lambda) = \mathcal{F}^{-1}\big(F(\lambda)\big),$$

we deduce $tE(\lambda + 1) = -2(\lambda + 1)\partial_t E(\lambda)$. This implies, for $n = 3$ and $t \neq 0$,

$$E(-1) = \frac{2}{t}\frac{\partial}{\partial t}E(-2) = \frac{2}{t}\frac{\partial}{\partial t}\left[\frac{Y(t - |x|)}{8\pi}J_0\big(m\sqrt{t^2 - |x|^2}\big)\right]$$

$$= \frac{\delta(t - |x|)}{4\pi t} - \frac{mY(t - |x|)}{4\pi\sqrt{t^2 - |x|^2}}J_1\big(m\sqrt{t^2 - |x|^2}\big).$$
(2.3.16)

(For the definition of $\frac{1}{t}\delta(t-|x|)$, see Example 1.4.12.) Since $E(\lambda) \in C(\mathbf{R}_t^1, \mathcal{D}'(\mathbf{R}^n))$ for $\text{Re}\,\lambda < -\frac{1}{2}$ by Eq. (2.3.13), formula (2.3.16) yields a representation of the fundamental solution $E(-1)$ of the Klein–Gordon operator $\partial_t^2 - \Delta_3 + m^2$ which is valid in $\mathcal{D}'(\mathbf{R}^4)$.

In general, $E(-k)$ is the unique fundamental solution of the iterated Klein–Gordon operator $(\partial_t^2 - \Delta_3 + m^2)^k$, $k \in \mathbf{N}$, with support in the half-space $t \geq 0$ according to Proposition 2.3.5.   □

Example 2.3.8 Let us next consider quasihyperbolic operators in $\mathbf{R}^{n+1}$ of the form $P(\partial) = \partial_t + R(\partial_1, \dots, \partial_n)$, which contain as particular cases the *heat operator* and the *Schrödinger operator*.

(a) If $N = (1, 0, \dots, 0)$, then the condition (2.3.8) of quasihyperbolicity takes the form

$$\forall \sigma > 0 : \forall (\tau, \xi) \in \mathbf{R}^{n+1} : \sigma + i\tau + R(i\xi) \neq 0,$$

which is equivalent to

$$\inf\{\text{Re}\,R(i\xi); \xi \in \mathbf{R}^n\} \geq 0. \tag{2.3.17}$$

If condition (2.3.17) is satisfied, then the numbers $z = \sigma + i\tau + R(i\xi)$ belong to the complex half-plane $\text{Re}\,z > 0$ for $\sigma > 0$, $(\tau, \xi) \in \mathbf{R}^{n+1}$, and we can take the usual determination of z^λ for $\lambda \in \mathbf{C}$. Thus, by Proposition 2.3.5, the convolution group of $P(\partial)$ is defined by

$$E(\lambda) = \mathcal{F}^{-1}(F(\lambda)) = \lim_{\sigma \searrow 0} \mathcal{F}^{-1}[(\sigma + i\tau + R(i\xi))^\lambda].$$

For $\text{Re}\,\lambda < 0$ and $z \in \mathbf{C}$ with $\text{Re}\,z > 0$, the function $t \mapsto Y(t)e^{-zt}t^{-\lambda-1} = e^{-zt}t_+^{-\lambda-1}$ is integrable and its Fourier transform is

$$\mathcal{F}_t(e^{-zt}t_+^{-\lambda-1}) = \int_0^\infty e^{-(z+i\tau)t}t^{-\lambda-1}\,dt = \Gamma(-\lambda)(z+i\tau)^\lambda.$$

Hence

$$\mathcal{F}_\tau^{-1}((z+i\tau)^\lambda) = \frac{e^{-zt}}{\Gamma(-\lambda)}t_+^{-\lambda-1}. \tag{2.3.18}$$

Note that (2.3.18) holds for each $\lambda \in \mathbf{C}$ since the left-hand side is obviously entire in λ, and so is the right-hand side if one takes into account that $t_+^{-\lambda-1}$ and $\Gamma(-\lambda)$ have both simple poles at $\lambda \in \mathbf{N}_0$, cf. Example 1.4.8.

By partial Fourier transformation, we conclude that

$$E(\lambda) = \mathcal{F}_\xi^{-1}\mathcal{F}_\tau^{-1}(F(\lambda)) = \mathcal{F}_\xi^{-1}\left(\lim_{\sigma \searrow 0}\left[\frac{e^{-(\sigma+R(i\xi))t}}{\Gamma(-\lambda)}\, t_+^{-\lambda-1}\right]\right)$$
$$= \mathcal{F}_\xi^{-1}\left(e^{-R(i\xi)t}\right) \cdot \frac{t_+^{-\lambda-1}}{\Gamma(-\lambda)}. \qquad (2.3.19)$$

We observe that the function

$$\mathbf{R} \longrightarrow \mathcal{S}'(\mathbf{R}_\xi^n) : t \longmapsto \mathcal{F}_\xi^{-1}\left(e^{-R(i\xi)t}\right)$$

is infinitely differentiable and hence can be multiplied with $t_+^{-\lambda-1}/\Gamma(-\lambda)$.

Let us mention that the convolution equation $E(\lambda) * E(\mu) = E(\lambda + \mu)$ holds generally for all complex λ, μ if $P(\partial)$ is quasihyperbolic, see Ortner and Wagner [218], Prop., p. 147. However, in the non-hyperbolic case, the convolvability of $E(\lambda), E(\mu)$ is more difficult to establish; it relies on the fact that $E(\lambda) \in \mathcal{D}'_{[0,\infty)} \hat\otimes \mathcal{O}'_C(\mathbf{R}_x^n)$, where $\mathcal{D}'_{[0,\infty)} = \{T \in \mathcal{D}'(\mathbf{R}^1); \operatorname{supp} T \subset [0,\infty)\}$, cf. Ortner and Wagner [219], Section 3.7, p. 114, and $\mathcal{O}'_C(\mathbf{R}^n) = \cap_{k\in\mathbf{N}}(1 + |x|^2)^{-k}\mathcal{D}'_{L^\infty}(\mathbf{R}^n)$, see Schwartz [246], p. 244.

In particular, the equation $E(k) * E(-k) = \delta$, $k \in \mathbf{N}$, shows that

$$E(-k) = \mathcal{F}_\xi^{-1}\left(e^{-R(i\xi)t}\right) \cdot \frac{t_+^{k-1}}{(k-1)!}$$

is a fundamental solution of $(\partial_t + R(\partial))^k$. As we will see in Proposition 2.4.13 below, this fundamental solution is the only one which is temperate and vanishes for $t < 0$.

(b) Let us specialize the above now to the heat and the Schrödinger operator, respectively.

If we set $R(\xi) = -|\xi|^2$, then we obtain the convolution group of the *heat operator* $\partial_t - \Delta_n$. From (2.3.19) and (1.6.24), we obtain

$$E(\lambda) = \mathcal{F}_\xi^{-1}\left(e^{-|\xi|^2 t}\right) \cdot \frac{t_+^{-\lambda-1}}{\Gamma(-\lambda)} = \frac{e^{-|x|^2/(4t)} \cdot t_+^{-n/2-1-\lambda}}{\Gamma(-\lambda)(4\pi)^{n/2}}.$$

Hence the locally integrable functions

$$E(-k) = \frac{Y(t)t^{-n/2-1+k}e^{-|x|^2/(4t)}}{(k-1)!(4\pi)^{n/2}}$$

are the fundamental solutions of $(\partial_t - \Delta_n)^k$, $k \in \mathbf{N}$, cf. (1.3.14) and Example 1.6.16 for the case $k = 1$.

For the *Schrödinger operator* $\partial_t - \mathrm{i}\Delta_n$, we have $R(\mathrm{i}\xi) = \mathrm{i}|\xi|^2$ and

$$E(\lambda) = \mathcal{F}_\xi^{-1}\big(\mathrm{e}^{-\mathrm{i}|\xi|^2 t}\big) \cdot \frac{t_+^{-\lambda-1}}{\Gamma(-\lambda)} = \frac{\mathrm{e}^{\mathrm{i}|x|^2/(4t)-\mathrm{i}n\pi/4} \cdot t_+^{-n/2-1-\lambda}}{\Gamma(-\lambda)(4\pi)^{n/2}}, \qquad (2.3.20)$$

see Example 1.6.14.

The multiplication in these products is understood as explained in (a), i.e., for $\phi \in \mathcal{D}(\mathbf{R}_{t,x}^{n+1})$, we set

$$\langle \phi, E(\lambda) \rangle = (4\pi)^{-n/2}\mathrm{e}^{-\mathrm{i}n\pi/4} \cdot \langle t^{-n/2} \int \phi(t,x)\mathrm{e}^{\mathrm{i}|x|^2/(4t)}\,\mathrm{d}x, \frac{t_+^{-\lambda-1}}{\Gamma(-\lambda)} \rangle.$$

Note that $E(\lambda)$ is not locally integrable for $\operatorname{Re}\lambda \geq -\frac{n}{2}$. In particular, the fundamental solution $E(-1)$ of the Schrödinger operator $\partial_t - \mathrm{i}\Delta_n$ is locally integrable only for $n = 1$ and else is given by the iterated, not absolutely convergent integral

$$\langle \phi, E(-1) \rangle = \frac{\mathrm{e}^{-\mathrm{i}n\pi/4}}{(4\pi)^{n/2}} \int_0^\infty \bigg(\int_{\mathbf{R}^n} \phi(t,x)\mathrm{e}^{\mathrm{i}|x|^2/(4t)}\,\mathrm{d}x \bigg) \frac{\mathrm{d}t}{t^{n/2}},$$

cf. Treves [274], 6.2, p. 45.

(c) Following S.L. Sobolev, let us finally consider the quasihyperbolic operator $\partial_t - \partial_1\partial_2\partial_3$ and construct its fundamental solution E. By (2.3.19), we have

$$E = \mathcal{F}_\xi^{-1}\big(\mathrm{e}^{-\mathrm{i}\xi_1\xi_2\xi_3 t}\big) \cdot Y(t).$$

Since the mapping

$$\mathbf{R}_{\xi_1,\xi_2}^2 \longrightarrow \mathcal{S}'(\mathbf{R}_{\xi_3}^1) : (\xi_1,\xi_2) \longmapsto \mathrm{e}^{-\mathrm{i}\xi_1\xi_2\xi_3 t}$$

is continuous, we can apply the partial Fourier transform (see Definition 1.6.15), and we conclude that

$$\mathcal{F}_\xi^{-1}\big(\mathrm{e}^{-\mathrm{i}\xi_1\xi_2\xi_3 t}\big) = \mathcal{F}_{\xi_1,\xi_2}^{-1}\mathcal{F}_{\xi_3}^{-1}\big(\mathrm{e}^{-\mathrm{i}\xi_1\xi_2\xi_3 t}\big) = \mathcal{F}_{\xi_1,\xi_2}^{-1}\big[\delta(x_3 - t\xi_1\xi_2)\big].$$

Note that, for fixed $t > 0$,

$$\delta(x_3 - t\xi_1\xi_2) \in \mathcal{C}\big(\mathbf{R}_{\xi_1,\xi_2}^2, \mathcal{S}'(\mathbf{R}_{x_3}^1)\big) \cap \mathcal{S}'(\mathbf{R}_{\xi_1,\xi_2,x_3}^3)$$

and also

$$\delta(x_3 - t\xi_1\xi_2) \in \mathcal{C}\big(\mathbf{R}_{x_3}^1, \mathcal{S}'(\mathbf{R}_{\xi_1,\xi_2}^2)\big) \cap \mathcal{S}'(\mathbf{R}_{\xi_1,\xi_2,x_3}^3).$$

Therefore, we can fix now x_3 in order to evaluate the partial Fourier transform $\mathcal{F}^{-1}_{\xi_1,\xi_2}\big[\delta(x_3 - t\xi_1\xi_2)\big]$. This inverse Fourier transform of a delta distribution along the hyperbola $\xi_1\xi_2 = \frac{x_3}{t}$ has already been calculated in Example 1.6.18 up to a linear transformation. There we have shown that

$$\mathcal{F}^{-1}_{\eta}\big(Y(\eta_1)\delta(\eta_1^2 - \eta_2^2 - m^2)\big) = \frac{Y(|y_2| - |y_1|)}{4\pi^2} K_0\big(m\sqrt{y_2^2 - y_1^2}\big)$$

$$+ \frac{Y(|y_1| - |y_2|)}{8\pi}\Big[-N_0\big(m\sqrt{y_1^2 - y_2^2}\big) + \mathrm{i}\operatorname{sign}y_1 \cdot J_0\big(m\sqrt{y_1^2 - y_2^2}\big)\Big] \in L^1_{\mathrm{loc}}(\mathbf{R}_y^2),$$

see (1.6.28).

By Proposition 1.6.6 (1), $\mathcal{F}^{-1}(T \circ A) = \frac{1}{|\det A|}(\mathcal{F}^{-1}T) \circ A^{-1T}$ for $T \in \mathcal{S}'(\mathbf{R}^n)$, $A \in \mathrm{Gl}_n(\mathbf{R})$. Hence, if we set

$$T = Y(\eta_1)\delta(\eta_1^2 - \eta_2^2 - m^2) \qquad \text{and} \qquad A\xi = \frac{1}{2}(\xi_1 + \xi_2, \xi_1 - \xi_2)^T = \eta,$$

we infer

$$\mathcal{F}^{-1}\big(Y(\xi_1 + \xi_2)\delta(\xi_1\xi_2 - m^2)\big) = \frac{Y(-x_1x_2)}{2\pi^2} K_0\big(2m\sqrt{-x_1x_2}\big)$$

$$+ \frac{Y(x_1x_2)}{4\pi}\Big[-N_0\big(2m\sqrt{x_1x_2}\big) + \mathrm{i}\operatorname{sign}(x_1 + x_2)\cdot J_0\big(2m\sqrt{x_1x_2}\big)\Big].$$

Adding this with the distribution reflected at the origin yields

$$\mathcal{F}^{-1}\big(\delta(\xi_1\xi_2 - m^2)\big) = \frac{Y(-x_1x_2)}{\pi^2} K_0\big(2m\sqrt{-x_1x_2}\big) - \frac{Y(x_1x_2)}{2\pi} N_0\big(2m\sqrt{x_1x_2}\big).$$

Finally, upon distinguishing the cases $x_3 > 0$ and $x_3 < 0$, we arrive at

$$E = Y(t)\mathcal{F}^{-1}_{\xi_1,\xi_2}\big[\delta(x_3 - t\xi_1\xi_2)\big]$$

$$= Y(t)\bigg[\frac{Y(-x_1x_2x_3)}{\pi^2 t} K_0\big(2\sqrt{-x_1x_2x_3/t}\big) - \frac{Y(x_1x_2x_3)}{2\pi t} N_0\big(2\sqrt{x_1x_2x_3/t}\big)\bigg].$$

$$(2.3.21)$$

Note that—in contrast to the Schrödinger operator—the fundamental solution E in (2.3.21) is locally integrable. This was observed already in Sobolev [255], p. 1247, where E is derived by introducing the similarity variable $\frac{x_1x_2x_3}{t}$ and by performing the "ansatz" $E = \frac{Y(t)}{t}\Lambda(\frac{x_1x_2x_3}{t})$. The ensuing third-order ordinary differential equation for Λ splits and yields Bessel functions, see Example 2.6.4 below. Note that two errors with respect to signs should be corrected in Sobolev's final result, see Sobolev [255], (8), p. 1247. A further derivation of the formula in (2.3.21) is given in Ortner [205], see Prop. 6, p. 158.

We also remark that Sobolev's operator $\partial_t - \partial_1\partial_2\partial_3$ serves as a prototype of q-hyperbolic operators (here $q = \frac{3}{2}$) introduced and studied in Gindikin [107], pp. 6, 71.   □

Let us next investigate fundamental solutions of *homogeneous* differential operators. For the particular case of elliptic homogeneous operators, we refer to Hörmander [139], Thm. 7.1.20, p. 169.

Proposition 2.3.9 *If $P(\partial) = \sum_{|\alpha|=m} a_\alpha \partial^\alpha$ is a linear differential operator which is homogeneous of degree $m \in \mathbf{N}$, then there exists a fundamental solution E which is associated homogeneous of degree $m - n$. More precisely, if $m < n$, then $E = F \cdot |x|^{m-n}$, and if $m \geq n$, then $E = F \cdot |x|^{m-n} + Q(x) \log |x|$, where $F \in \mathcal{D}'(\mathbf{S}^{n-1})$ and Q is a homogeneous polynomial of degree $m - n$. (Recall that $\langle \phi, F \cdot |x|^\lambda \rangle = \langle \langle \phi(t\omega), F(\omega) \rangle, t_+^{\lambda+n-1} \rangle$, $\lambda \in \mathbf{C}$, $\phi \in \mathcal{D}(\mathbf{R}^n)$, see Example 1.4.9.)*

Proof

(a) Let us first consider the following division problem on the sphere:

$$P(i\omega) \cdot U = 1, \quad U \in \mathcal{D}'(\mathbf{S}^{n-1}),$$

cf. Gårding [89], p. 407.

Without restriction, we may assume that $P(N) \neq 0$ for $N = (0,\ldots,0,1)$. We employ the stereographic projection

$$p : \mathbf{R}^{n-1} \longrightarrow \mathbf{S}^{n-1} \setminus \{N\} : \eta \longmapsto \frac{1}{1+|\eta|^2}(2\eta, |\eta|^2 - 1)$$

in order to transform the equation $P(i\omega) \cdot U = 1$ into

$$P(2\eta, |\eta|^2 - 1) \cdot i^m (1 + |\eta|^2)^{-m} p^*(U) = 1.$$

By Proposition 2.3.1 and by the identification of temperate distributions on $\mathbf{R}^{n-1}$ with distributions on $\mathbf{S}^{n-1}$ (see Schwartz [246], Ch. VII, Thm. V, p. 238), we obtain $U \in \mathcal{D}'(\mathbf{S}^{n-1})$ which solves $P(i\omega) \cdot U = 1$ on $\mathbf{S}^{n-1} \setminus \{N\}$. Finally, near N, U is uniquely determined by $P(i\omega) \cdot U = 1$ due to $P(N) \neq 0$.

(b) If we define, as in (1.4.3), $V = U \cdot |\xi|^{-m} \in \mathcal{S}'(\mathbf{R}^n)$ by

$$\langle \phi, V \rangle = \langle \langle \phi(t\omega), U(\omega) \rangle, t_+^{-m+n-1} \rangle, \qquad \phi \in \mathcal{S}'(\mathbf{R}^n),$$

then $P(i\xi) \cdot V = 1$ holds in $\mathcal{S}'(\mathbf{R}^n)$. Hence $E = \mathcal{F}^{-1}V$ is a fundamental solution of $P(\partial)$.

(c) If $m < n$, then V is homogeneous in $\mathbf{R}^n$ of degree $-m$ and hence E is homogeneous of degree $m - n$. By Gårding [89], Lemmes 1.5, 4.1, pp. 393, 400, or Ortner and Wagner [219], Thm. 2.5.1, p. 58, $E = \mathcal{F}^{-1}V$ can be cast in the form $E = F \cdot |x|^{m-n}$ for some $F \in \mathcal{D}'(\mathbf{S}^{n-1})$.

(d) If $m \geq n$, then V is still homogeneous in $\mathbf{R}^n \setminus \{0\}$, but can cease to be homogeneous in $\mathbf{R}^n$, cf. Examples 1.2.10, 1.4.10 for the case of $m = n$.

Generally, by analytic continuation, $U \cdot |\xi|^\lambda$ is homogeneous of degree λ where this function of λ is analytic, i.e. for $\lambda \in \mathbf{C} \setminus \{-n, -n-1, \dots\}$. In the possible poles $\lambda = -m$, $m \geq n$, we set

$$R := \operatorname*{Res}_{\lambda = -m} (U \cdot |\xi|^\lambda) = (-1)^{m-n} \sum_{\substack{\alpha \in \mathbf{N}_0^n \\ |\alpha| = m-n}} \frac{1}{\alpha!} \langle \omega^\alpha, U \rangle \, \partial^\alpha \delta, \qquad (2.3.22)$$

see Example 1.4.9, and we have

$$V = \operatorname*{Pf}_{\lambda = -m} (U \cdot |\xi|^\lambda) = \lim_{\lambda \to -m} \left[U \cdot |\xi|^\lambda - \frac{R}{\lambda + m} \right].$$

From this we conclude that

$$\begin{aligned}
V(c\xi) &= \lim_{\lambda \to -m} \left[(U \cdot |\xi|^\lambda)(c\xi) - c^{-m} \frac{R}{\lambda + m} \right] \\
&= c^{-m} \lim_{\lambda \to -m} \left[U \cdot |\xi|^\lambda - \frac{R}{\lambda + m} \right] + \lim_{\lambda \to -m} (c^\lambda - c^{-m}) U \cdot |\xi|^\lambda \\
&= c^{-m} V + c^{-m} (\log c) R
\end{aligned}$$

for $c > 0$, cf. Ortner and Wagner [219], (2.5.1), p. 59.

Hence V is *associated homogeneous* of order m in $\mathbf{R}^n$, i.e., $\forall c > 0 : V(c\xi) = c^{-m} V + c^{-m} (\log c) R$ with R homogeneous. Due to Proposition 1.6.6 (1), this implies

$$E(cx) = c^{m-n} E - c^{m-n} (\log c) \cdot \mathcal{F}^{-1} R, \qquad c > 0, \qquad (2.3.23)$$

where $Q := -\mathcal{F}^{-1} R$ is a homogeneous polynomial of degree $m - n$. By the structure theorem Prop. 2.5.3 in Ortner and Wagner [219] (cf. also Grudzinski [119], Thm. 4.25', p. 178), we conclude that E has the representation $E = F \cdot |x|^{m-n} + Q(x) \log |x|$ for some $F \in \mathcal{D}'(\mathbf{S}^{n-1})$. This completes the proof. $\qquad \square$

Example 2.3.10

(a) Reconsidering the case of $P(\partial) = (-\Delta_n)^k$, let us comment on the structure of the fundamental solution E given explicitly in (1.6.19) and (1.6.20). Indeed, E is homogeneous if the degree $m = 2k$ of $P(\partial)$ satisfies $m < n$. On the other hand, if $m \geq n$ and n is even, then E is equal to a polynomial times a logarithm, see (1.6.20). However, if n is odd, then there is no logarithmic term present since the residue R in (2.3.22) vanishes due to $\langle \omega^\alpha, 1 \rangle = 0$ for $\alpha \in \mathbf{N}_0^n$ of degree $|\alpha| = m - n = 2k - n$, this degree being odd.

(b) Similarly, for a homogeneous *quasihyperbolic* operator, the logarithmic term disappears if the solution U of the division problem $P(i\omega) \cdot U = 1$ is chosen as in Proposition 2.3.5, i.e., $U = \lim_{\sigma \searrow 0} P(i\omega + \sigma N)^{-1}$. Then $V = U \cdot |\xi|^{-m}$ must be homogeneous. In fact, as we have seen in the proof of Proposition 2.3.5,

$\lim_{\sigma \searrow 0} P(\mathrm{i}\xi + \sigma N)^\lambda$ is entire and coincides with $U \cdot |\xi|^\lambda$ for $\mathrm{Re}\,\lambda > 0$. Thus also $\lambda \mapsto U \cdot |\xi|^\lambda$ is entire and $\mathrm{Res}_{\lambda=-m}\, U \cdot |\xi|^\lambda = 0$.   □

2.4 Uniqueness and Representations of Fundamental Solutions

In the following, we investigate properties of distributions which imply uniqueness for fundamental solutions, namely

(i) growth and decay properties,

(ii) support properties.

We first consider systems with non-vanishing symbol, cf. Petersen [228], Lemma 8.6 B, p. 296; Szmydt and Ziemian [267], Prop. 2, p. 219; Gindikin and Volevich [108], p. 16, [109], p. 58, for the scalar case.

Proposition 2.4.1 *Let $A(\partial) \in \mathbf{C}[\partial]^{l \times l}$ be a quadratic system of linear partial differential operators in $\mathbf{R}^n$. Then the following conditions are equivalent:*

(1) $A(\partial)$ *has one and only one two-sided fundamental matrix in $\mathcal{S}'(\mathbf{R}^n)^{l \times l}$;*
(2) $A(\partial)$ *has a two-sided fundamental matrix in $\mathcal{O}'_C(\mathbf{R}^n)^{l \times l}$;*
(3) $A(\mathrm{i}\xi)$ *is invertible for each $\xi \in \mathbf{R}^n$, i.e., $\forall \xi \in \mathbf{R}^n : \det A(\mathrm{i}\xi) \neq 0$.*

Proof

(1) $\Rightarrow$ (3) : This follows from the fact that

$$T := B \cdot \mathrm{e}^{\mathrm{i}x\xi_0} \in \mathcal{S}'(\mathbf{R}^n)^{l \times l} \setminus \{0\}$$

solves the homogeneous equation $A(\partial)T = 0$ if $\det A(\mathrm{i}\xi_0) = 0$ and $B \in \mathbf{C}^{l \times l} \setminus \{0\}$ satisfies $A(\mathrm{i}\xi_0)B = 0$.

(3) $\Rightarrow$ (2) : Due to the Hörmander–Łojasiewicz inequality (see Hörmander [135], Lemma 2, (2.5), p. 557; Hörmander [138], Ex. A.2.7, (A.2.6), p. 368), the assumption in (3) implies that $A(\mathrm{i}\xi)^{-1}$ and its derivatives have at most polynomial growth for $|\xi| \to \infty$, i.e.,

$$A(\mathrm{i}\xi)^{-1} \in \mathcal{O}_M(\mathbf{R}^n)^{l \times l},$$

cf. Schwartz [246], p. 243, for the definition of the spaces $\mathcal{O}_M(\mathbf{R}^n), \mathcal{O}'_C(\mathbf{R}^n)$ and the equation $\mathcal{O}_M(\mathbf{R}^n) = \mathcal{F}\mathcal{O}'_C(\mathbf{R}^n)$. Hence

$$E = A(\partial)^{\mathrm{ad}} \mathcal{F}^{-1}\big(\det(A(\mathrm{i}\xi))^{-1}\big) \in \mathcal{O}'_C(\mathbf{R}^n)^{l \times l}$$

is a two-sided fundamental matrix of $A(\partial)$.

$(2) \Rightarrow (1):$ If $E \in \mathcal{O}'_C(\mathbf{R}^n)^{l \times l}$ and $F \in \mathcal{S}'(\mathbf{R}^n)^{l \times l}$ are two-sided fundamental matrices of $A(\partial)$, then they are convolvable and hence

$$E = E * A(\partial)F = \big(E * A(\partial)\delta\big) * F = F.$$

The proof is complete.   □

Example 2.4.2 As in Example 1.4.11, let us consider the metaharmonic operator $P(\partial) = \Delta_n + \lambda$, $\lambda \in \mathbf{C} \setminus [0, \infty)$. Then condition (3) in Proposition 2.4.1 is satisfied since $P(\mathrm{i}\xi) = \lambda - |\xi|^2 \neq 0$ for $\xi \in \mathbf{R}^n$. The fast decreasing fundamental solution E of $P(\partial)$ is then given by

$$E = -\mathrm{i}d_n(\lambda)|x|^{-n/2+1}H^{(1)}_{n/2-1}\big(\sqrt{\lambda}|x|\big), \qquad d_n(\lambda) = \frac{\lambda^{n/4-1/2}}{2^{n/2+1}\pi^{n/2-1}},$$

where $\sqrt{\lambda}$ is defined in the slit plane $\mathbf{C} \setminus [0, \infty)$ by $0 < \arg\sqrt{\lambda} < \pi$, see Example 1.4.11. (Note that $H^{(1)}_\nu(z)$ decreases exponentially if $|z| \to \infty$ with $\operatorname{Im} z > 0$.)

In particular, for $\lambda = -\mu$, $\mu > 0$, the unique temperate fundamental solution of $(\Delta_n - \mu)^k$, $k \in \mathbf{N}$, is given in terms of MacDonald's function, see (1.4.9) and Example 1.6.11(a).   □

Example 2.4.3 As another application of Proposition 2.4.1, we consider the *time-harmonic Lamé system*

$$A(\nabla) = -\rho\tau^2 I_3 - B(\nabla), \qquad B(\nabla) := \mu\Delta_3 I_3 + (\lambda + \mu)\nabla \cdot \nabla^T.$$

As in Example 2.1.3, $\lambda, \mu > 0$ denote Lamé's constants. Then the system $A(\nabla)$ arises from the one in formula (2.1.2) by partial Fourier transform with respect to the time variable t. Hence we obtain, for fixed $\tau > 0$, the following locally integrable fundamental matrix F of $A(\nabla)$ by partial Fourier transform applied to the fundamental matrix E in Stokes's formula (2.1.7):

$$F = \mathcal{F}_t E = \frac{I_3|x|^2 - xx^T}{4\pi\mu|x|^3}\,\mathrm{e}^{-\mathrm{i}\tau|x|/c_s} + \frac{xx^T}{4\pi(\lambda+2\mu)|x|^3}\,\mathrm{e}^{-\mathrm{i}\tau|x|/c_p}$$

$$+ \frac{1}{4\pi\rho|x|^3\tau^2}\Big(I_3 - \frac{3xx^T}{|x|^2}\Big)\Big[\mathrm{e}^{-\mathrm{i}\tau|x|/c_p}\Big(1 + \frac{\mathrm{i}\tau|x|}{c_p}\Big) - \mathrm{e}^{-\mathrm{i}\tau|x|/c_s}\Big(1 + \frac{\mathrm{i}\tau|x|}{c_s}\Big)\Big].$$

$$(2.4.1)$$

Herein, $c_s = \sqrt{\frac{\mu}{\rho}}$, $c_p = \sqrt{\frac{\lambda+2\mu}{\rho}}$ are the velocities of the shear and pressure waves, respectively.

By analytic continuation, with respect to ρ, (2.4.1) yields a fundamental matrix of $A(\nabla)$ for all $\rho \in \mathbf{C} \setminus \{0\}$. For $\rho \in \mathbf{C} \setminus [0, \infty)$, let us choose $\sqrt{\rho}$ such that $\operatorname{Im}\sqrt{\rho} < 0$, i.e., $\operatorname{Im}(c_s^{-1})$, $\operatorname{Im}(c_p^{-1}) < 0$. This implies that $F \in \mathcal{O}'_C(\mathbf{R}^3)^{3 \times 3}$ and, by Proposition 2.4.1, F is the only temperate fundamental matrix if $\rho \in \mathbf{C} \setminus [0, \infty)$ and $\sqrt{\rho}$ as above.

For $\rho > 0$, which is the physically relevant case, there exist many temperate fundamental matrices. If we take the real part in (2.4.1), we obtain

$$
\operatorname{Re} F(x) = \frac{I_3 |x|^2 - xx^T}{4\pi\mu |x|^3} \cos\Big(\frac{\tau |x|}{c_s}\Big) + \frac{xx^T \cos(\tau |x|/c_p)}{4\pi(\lambda + 2\mu)|x|^3}
$$

$$
+ \frac{1}{4\pi\rho |x|^3 \tau^2}\Big(I_3 - \frac{3xx^T}{|x|^2}\Big)\Big[\cos\Big(\frac{\tau |x|}{c_p}\Big) - \cos\Big(\frac{\tau |x|}{c_s}\Big) \tag{2.4.2}
$$

$$
+ \tau |x|\Big(c_p^{-1}\sin\Big(\frac{\tau |x|}{c_p}\Big) - c_s^{-1}\sin\Big(\frac{\tau |x|}{c_s}\Big)\Big)\Big].
$$

For (2.4.2), see Mura [185], (9.40), p. 65; Norris [190], (B3), p. 187; Ortner and Wagner [217], p. 331. □

Let us generalize now Proposition 2.4.1 to symbols with finitely many real zeroes, cf. Zuily [309], Ex. 82, p. 147; Gel'fand and Shilov [106], Ch. III, Section 2.4, p. 135; Friedman [85], p. 98.

Proposition 2.4.4 *Let $P(\partial)$ be a linear differential operator such that the set $Z = \{\xi \in \mathbf{R}^n; P(i\xi) = 0\}$ is finite. Then two temperate fundamental solutions of $P(\partial)$ differ only by an exponential polynomial $\sum_{\xi \in Z} Q_\xi(x)e^{ix\xi}$, where $Q_\xi(x) = \sum_{|\alpha| \le m} a_{\xi\alpha}x^\alpha$ with $m \in \mathbf{N}_0$ and $a_{\xi\alpha} \in \mathbf{C}$.*

Proof This is evident by Fourier transformation: If $T \in \mathcal{S}'$ and $P(\partial)T = 0$, then $P(i\xi)(\mathcal{F}T) = 0$ and thus $\operatorname{supp}(\mathcal{F}T) \subset Z$. Therefore, by Proposition 1.3.15,

$$
\mathcal{F}T = (2\pi)^n \sum_{\xi \in Z} Q_\xi(i\partial)\delta_\xi,
$$

and hence $T = \sum_{\xi \in Z} Q_\xi(x)e^{ix\xi}$. □

As an example, let us first investigate temperate fundamental solutions of ordinary differential operators, cf. Proposition 1.3.7, Example 1.3.8.

Proposition 2.4.5 *Let $m \in \mathbf{N}$, $\alpha \in \mathbf{N}_0^m$ and $\lambda_1, \dots, \lambda_m \in \mathbf{C}$ be pairwise different. Let us set* $\operatorname{sign} t = \left\{ \begin{array}{ll} 1 & : t \ge 0, \\ -1 & : t < 0 \end{array} \right\}$. *Then the ordinary differential operator*

$$
P_{\lambda,\alpha}\big(\tfrac{d}{dx}\big) = \prod_{j=1}^m \Big(\frac{d}{dx} - \lambda_j\Big)^{\alpha_j + 1}
$$

has the following temperate fundamental solution:

$$
E = \sum_{j=1}^m Y(-x\operatorname{sign}(\operatorname{Re}\lambda_j))\frac{\operatorname{sign}(-\operatorname{Re}\lambda_j)}{\alpha_j!}\Big(\frac{\partial}{\partial\lambda_j}\Big)^{\alpha_j}\Big(e^{\lambda_j x}\prod_{k \neq j}(\lambda_j - \lambda_k)^{-\alpha_k - 1}\Big).
$$

$$\tag{2.4.3}$$

E is uniquely determined in $S'(\mathbf{R}^1)$ up to an exponential polynomial of the form

$$\sum_{\operatorname{Re}\lambda_j=0}\sum_{k=0}^{\alpha_j} c_{jk}x^k e^{\lambda_j x}, \qquad c_{jk}\in\mathbf{C}.$$

Proof Similarly as in the proof of Proposition 1.3.7, we first assume that $\alpha=0$ and set

$$E = \sum_{j=1}^{m} Y\big(-x\operatorname{sign}(\operatorname{Re}\lambda_j)\big)a_j e^{\lambda_j x}, \qquad a_j\in\mathbf{C}.$$

Considering the jump conditions for E as in the proof of Proposition 1.3.7 yields

$$\sum_{j=1}^{m} a_j\lambda_j^k\operatorname{sign}(-\operatorname{Re}\lambda_j) = \begin{cases} 0 & : k=0,\dots,m-2, \\ 1 & : k=m-1, \end{cases}$$

and hence $a_j = \operatorname{sign}(-\operatorname{Re}\lambda_j)\prod_{k\neq j}(\lambda_j-\lambda_k)^{-1}$.

If $\alpha\neq 0$ and $\forall j=1,\dots,m : \operatorname{Re}\lambda_j \neq 0$, then differentiation with respect to λ as in the second part of the proof of Proposition 1.3.7 furnishes formula (2.4.3). Finally, if the real part of some of the roots λ_j vanishes, then one uses a limit process. The uniqueness statement of Proposition 2.4.5 is a consequence of Proposition 2.4.4. This completes the proof. $\qquad\square$

Example 2.4.6 In the simple case of $P(\frac{d}{dx}) = \frac{d^2}{dx^2} - \lambda^2$, $\lambda\in\mathbf{C}\setminus\{0\}$, formula (2.4.3) yields the temperate fundamental solution

$$E = -\frac{\operatorname{sign}(\operatorname{Re}\lambda)}{2\lambda}\big[Y(-x\operatorname{sign}(\operatorname{Re}\lambda))e^{\lambda x} + Y(x\operatorname{sign}(\operatorname{Re}\lambda))e^{-\lambda x}\big],$$

which is unique if $\operatorname{Re}\lambda\neq 0$ and else is unique up to $c_1 e^{\lambda x} + c_2 e^{-\lambda x}$. $\qquad\square$

Definition 2.4.7 The operator $P(\partial) = \sum_{|\alpha|\leq m} a_\alpha\partial^\alpha$ of order m with the principal part $P_m(\partial) = \sum_{|\alpha|=m} a_\alpha\partial^\alpha$ is called *elliptic* if and only if $\forall\xi\in\mathbf{R}^n\setminus\{0\} : P_m(\xi)\neq 0$.

For this definition, cf. Hörmander [139], Def. 7.1.19, p. 169. Note that elliptic operators are "hypoelliptic", i.e., each solution $u\in\mathcal{D}'(\Omega)$ of $P(\partial)u = 0$ in some open set $\Omega\subset\mathbf{R}^n$ is necessarily C^∞ in Ω, see Hörmander [138], Thms. 11.1.1, 11.1.10, pp. 61, 67; Zuily [309], Exercise 97, p. 187. Hence each fundamental solution E of an elliptic operator $P(\partial)$ is C^∞ outside the origin.

Proposition 2.4.8 *Let $P(\partial)$ be an elliptic operator in $\mathbf{R}^n$ which is homogeneous of degree m.*

(1) *$P(\partial)$ has a fundamental solution E such that $E(x)\cdot|x|^{-m+n}/\log|x|$ is bounded for $|x|\to\infty$, i.e.,*

$$\exists C > 0 : \forall x\in\mathbf{R}^n \text{ with } |x|\geq 2 : |E(x)|\leq C|x|^{m-n}\log|x|. \qquad (2.4.4)$$

If $m < n$ then there exists precisely one fundamental solution E satisfying (2.4.4); if $m \geq n$, then E is uniquely determined by condition (2.4.4) up to polynomials of degree $m - n$.

(2) *For odd dimensions $n \geq 3$, there exists one and only one fundamental solution E which is homogeneous and even.*

Proof

(1) We have already shown in Proposition 2.3.9 that $P(\partial)$ has an associated homogeneous fundamental solution of the form

$$E(x) = F \cdot |x|^{m-n} + Q(x) \log |x|$$

for some $F \in \mathcal{D}'(\mathbf{S}^{n-1})$ and a homogeneous polynomial Q of degree m. Due to the ellipticity of $P(\partial)$, the distribution F is C^∞, and hence the estimate (2.4.4) follows.

On the other hand, a fundamental solution E_1 satisfying (2.4.4) is necessarily temperate, and, because of $P(\partial)(E - E_1) = 0$, we conclude that $P(i\xi) \cdot \mathcal{F}(E - E_1) = 0$ and that the support of $\mathcal{F}(E - E_1)$ is contained in $\{0\}$. Therefore, $\mathcal{F}(E - E_1)$ is a sum of derivatives of δ (see Proposition 1.3.15) and $E - E_1 = R$ for a polynomial R. Due to (2.4.4), R must vanish if $m < n$, and R is of degree at most $m - n$ if $m \geq n$.

(2) If $P(\partial)$ is elliptic in $\mathbf{R}^n$ and $n \geq 3$, then the order m of the symbol $P(i\xi)$ is necessarily even, see Lions and Magenes [167], Prop. 1.1, p. 121. Let us recall now some steps in the construction of the fundamental solution E in the course of the proof of Proposition 2.3.9.

First, the division problem $P(i\omega) \cdot U = 1$ is solved on the sphere $\mathbf{S}^{n-1}$. Due to the ellipticity of $P(\partial)$, the solution $U = P(i\omega)^{-1} \in C^\infty(\mathbf{S}^{n-1})$ is uniquely determined. Note that U is an even function on $\mathbf{S}^{n-1}$ since m is even. This implies that the distribution $V = U \cdot |\xi|^{-m} \in \mathcal{D}'(\mathbf{R}^n)$ is even and homogeneous of degree $-m$. In fact, $U \cdot |\xi|^\lambda$ is analytic in λ and yields homogeneous distributions for $\lambda \in \mathbf{C} \setminus \{-n - k; k \in \mathbf{N}_0\}$, see Example 1.4.9. Hence V is clearly homogenous if $m < n$. In the case $m \geq n$, this is also true since then $R = \mathrm{Res}_{\lambda=-m} U \cdot |\xi|^\lambda$ vanishes due to the fact that U is even:

$$\operatorname*{Res}_{\lambda=-m} U \cdot |\xi|^\lambda = \frac{(-1)^{m-n}}{(m-n)!} \langle (\omega^T \cdot \nabla)^{m-n}, U(\omega) \rangle \delta = 0$$

for $m \geq n$, m even, n odd. Thus $E = \mathcal{F}^{-1}V$ is an even and homogeneous fundamental solution of $P(\partial)$.

If E_1, E_2 are two even, homogeneous fundamental solutions of $P(\partial)$, then E_i, $i = 1, 2$, are both homogeneous of degree $m - n$ and $E_1 - E_2$ is a polynomial by part (1). Hence $E_1 - E_2$ is an even polynomial of the odd degree $m - n$ and consequently vanishes. This shows that an even, homogeneous fundamental solution of $P(\partial)$ is uniquely determined and completes the proof. $\square$

The second part of Proposition 2.4.8 goes back to Wagner [292], Prop. 1, p. 1193.

Example 2.4.9 Let us illustrate the uniqueness assertions in Proposition 2.4.8 for homogeneous elliptic operators of degree 2, i.e.,

$$P(\partial) = \sum_{i=1}^{n}\sum_{j=1}^{n} a_{ij}\partial_i\partial_j = \nabla^T A \nabla, \qquad A = A^T \in \mathbf{C}^{n \times n},$$

by calculating explicitly the associated homogeneous fundamental solutions of $P(\partial)$.

(a) Let us consider first the two-dimensional case $n = 2$. Then the set

$$M_2 = \{A \in \mathbf{C}^{2\times 2};\ A = A^T,\ \nabla^T A \nabla \text{ is elliptic}\}$$
$$= \{A \in \mathbf{C}^{2\times 2};\ A = A^T \text{ and } \forall x \in \mathbf{R}^2 \setminus \{0\} : x^T A x \neq 0\}$$

consists of three connectivity components. In fact, $x^T A x$ does not vanish for $x \neq 0$ if and only if $x^T A x = a(x_1 - \lambda_1 x_2)(x_1 - \lambda_2 x_2)$ with $a \in \mathbf{C}\setminus\{0\}$, $\lambda_1, \lambda_2 \in \mathbf{C}\setminus\mathbf{R}$. Therefore, M_2 is the disjoint union of the components M_2^0, M_2^+, M_2^- given by the sets of matrices of the form

$$a \begin{pmatrix} 1 & -(\lambda_1 + \lambda_2)/2 \\ -(\lambda_1 + \lambda_2)/2 & \lambda_1 \lambda_2 \end{pmatrix}, \qquad a \in \mathbf{C}\setminus\{0\},\ \lambda_1, \lambda_2 \in \mathbf{C}\setminus\mathbf{R},$$

where either $\operatorname{Im}\lambda_1 \cdot \operatorname{Im}\lambda_2 < 0$ or $\operatorname{Im}\lambda_i > 0$ $(i = 1, 2)$ or $\operatorname{Im}\lambda_i < 0$ $(i = 1, 2)$, respectively. Hence M_2^0, M_2^+, M_2^- are the components which contain the matrices corresponding to the operators $\Delta_2, (\partial_1 - i\partial_2)^2, (\partial_1 + i\partial_2)^2$, respectively. (Note that the set Γ_2 in Example 1.4.12 corresponds to the diagonal matrices in M_2, and that $\Gamma_2 \subset M_2^0$.)

According to Proposition 2.3.9, we obtain a fundamental solution E of $P(\partial) = \nabla^T A \nabla$, $A \in M_2$, in the form

$$E = \mathcal{F}^{-1}V, \quad V = U(\omega) \cdot |\xi|^{-2} \in \mathcal{S}'(\mathbf{R}^2), \quad U(\omega) = -\frac{1}{\omega^T A \omega} \in C^\infty(\mathbf{S}^1).$$

Hence

$$\langle \phi, V \rangle = -\langle \langle \phi(t\omega), (\omega^T A \omega)^{-1} \rangle, t_+^{-1} \rangle, \qquad \phi \in \mathcal{S}(\mathbf{R}^2).$$

Furthermore, according to Proposition 2.4.8, E is uniquely determined up to a constant by the condition

$$\exists C > 0 : \forall \in \mathbf{R}^2 \text{ with } |x| \geq 2 : |E(x)| \leq C \log |x|.$$

As in the proof of Proposition 2.3.9, we observe that E and V are homogeneous if and only if

$$R = \operatorname*{Res}_{\lambda=-2} U \cdot |\xi|^{\lambda} = \left(\int_{S^1} U(\omega)\, d\sigma(\omega) \right) \delta$$

vanishes. In order to evaluate this integral, let us use the *gnomonian projection* $\omega = \pm \frac{1}{\sqrt{1+t^2}}\binom{t}{1}$, $d\sigma(\omega) = \frac{dt}{1+t^2}$, which projects both the upper and the lower semicircle onto the real axis $\mathbf{R}_t^1$. This yields

$$\int_{S^1} \frac{d\sigma(\omega)}{(\omega_1 - \lambda_1\omega_2)(\omega_1 - \lambda_2\omega_2)} = 2 \int_{-\infty}^{\infty} \frac{dt}{(t - \lambda_1)(t - \lambda_2)}$$

$$= \begin{cases} 0 & : \operatorname{Im}\lambda_1 \cdot \operatorname{Im}\lambda_2 > 0, \\ \frac{4\pi i\, \operatorname{sign}(\operatorname{Im}\lambda_1)}{\lambda_1 - \lambda_2} & : \operatorname{Im}\lambda_1 \cdot \operatorname{Im}\lambda_2 < 0. \end{cases}$$

$$(2.4.5)$$

Therefore, E and V are homogeneous iff $A \in M_2^+ \cup M_2^-$.

Let us finally calculate E. For $A \in M_2^0$, we can find E by analytic continuation from the real-valued case. Starting from the fundamental solution $E = \frac{1}{4\pi} \log |x|^2$ of Δ_2, see Example 1.3.14(a), the linear transformation formula in Proposition 1.3.19 yields, upon addition of a constant,

$$E = \frac{1}{4\pi \sqrt{\det A}} \log(x^T A^{\mathrm{ad}} x) \qquad (2.4.6)$$

as fundamental solution of $\nabla^T A \nabla$ for positive definite $A \in \mathbf{R}^{2\times 2}$. (Note that $x^T A x = |\sqrt{A}x|^2$.) Formula (2.4.6) is then generally valid for $A \in M_2^0$ if we take into account that $x^T A^{\mathrm{ad}} x \neq t a_{11}$ for $t < 0$ and $x \in \mathbf{R}^2$. (On $\mathbf{C} \setminus a_{11} \cdot (-\infty, 0]$, the logarithm can be defined continuously.) Furthermore, $\sqrt{\det A}$ is uniquely determined if it is chosen positive for positive definite A and continuously extended for $A \in M_2^0$.

If $P(\partial)$ is expressed in the form $P(\partial) = (\partial_1 - \lambda_1\partial_2)(\partial_1 - \lambda_2\partial_2)$, $\operatorname{Im}\lambda_1 \cdot \operatorname{Im}\lambda_2 < 0$, (and hence $A \in M_2^0$), then we obtain

$$E = -\frac{\operatorname{sign}(\operatorname{Im}\lambda_1)}{2\pi i(\lambda_1 - \lambda_2)} \log[(x_2 + \lambda_1 x_1)(x_2 + \lambda_2 x_1)]. \qquad (2.4.7)$$

We observe that the multiplicative constant $d(\lambda_1, \lambda_2)$ in (2.4.7) preceding the logarithm is connected with the residue $R = \operatorname{Res}_{\lambda=-2}(U \cdot |\xi|^{\lambda})$ in the proof of Proposition 2.3.9. In fact, (2.3.23) yields $\mathcal{F}^{-1}R = -2d(\lambda_1, \lambda_2)$ in accordance with formula (2.4.5) which furnished

$$R = -\frac{4\pi i\, \operatorname{sign}(\operatorname{Im}\lambda_1)}{\lambda_1 - \lambda_2} \delta, \qquad \operatorname{Im}\lambda_1 \cdot \operatorname{Im}\lambda_2 < 0.$$

E.g., if $P(\partial) = \partial_1^2 + \partial_2^2 + 2i\partial_1\partial_2$, then we obtain for a fundamental solution of $P(\partial)$

$$E = \frac{1}{4\sqrt{2}\,\pi}\,\log(|x|^2 - 2ix_1x_2) = \frac{1}{4\sqrt{2}\,\pi}\Big[\frac{1}{2}\log(|x|^4 + 4x_1^2x_2^2) - i\arctan\Big(\frac{2x_1x_2}{|x|^2}\Big)\Big].$$

Let us yet give an explicit formula for a homogeneous fundamental solution E of $P(\partial) = \nabla^T A\nabla$ if $A \in M_2^{\pm}$. We set, without loss of generality, $P(\partial) = (\partial_1 - \lambda_1\partial_2)(\partial_1 - \lambda_2\partial_2)$ with $\epsilon = \text{sign}(\text{Im}\,\lambda_1) = \text{sign}(\text{Im}\,\lambda_2) \in \{\pm 1\}$ and $\lambda_1 \neq \lambda_2$, and

$$E = -\frac{\epsilon}{2\pi i(\lambda_1 - \lambda_2)}\cdot\log\Big(\frac{x_2 + \lambda_1x_1}{x_2 + \lambda_2x_1}\Big). \tag{2.4.8}$$

Note that $\frac{x_2+\lambda_1x_1}{x_2+\lambda_2x_1} \in \mathbf{C}\setminus(-\infty, 0]$ for $x \in \mathbf{R}^2\setminus\{0\}$. We define the logarithm in the usual way in the slit plane $\mathbf{C}\setminus(-\infty, 0]$. In order to verify (2.4.8), we use the jump formula (1.3.9) and obtain

$$(\partial_1 - \lambda_j\partial_2)\log(x_2 + \lambda_jx_1) = 2\pi i\epsilon\delta(x_1)\otimes Y(-x_2),\ j = 1, 2,$$

and hence

$$P(\partial)E = -\frac{\epsilon}{2\pi i(\lambda_1 - \lambda_2)}\cdot\big[(\partial_1 - \lambda_2\partial_2) - (\partial_1 - \lambda_1\partial_2)\big]2\pi i\epsilon\delta(x_1)\otimes Y(-x_2) = \delta.$$

The case of $\lambda_1 = \lambda_2 \in \mathbf{C}\setminus\mathbf{R}$ follows by a limit procedure. We obtain

$$(\partial_1 - \lambda\partial_2)^2\Big[-\frac{\text{sign}(\text{Im}\,\lambda)}{2\pi i}\cdot\frac{x_1}{x_2 + \lambda x_1}\Big] = \delta,\qquad \lambda \in \mathbf{C}\setminus\mathbf{R}.$$

The formulas (2.4.6–2.4.8) will be generalized in Proposition 3.3.2 below, where we shall consider $R(\partial) = \prod_{j=1}^{l}(\partial_1 - \lambda_j\partial_2)^{\alpha_j+1}$, see also Somigliana [258], Wagner [285], Ch. III, Satz 4, p. 40.

(b) In contrast to the two-dimensional case $n = 2$, the set

$$M_n = \{A \in \mathbf{C}^{n\times n};\ A = A^T,\ \nabla^T A\nabla \text{ is elliptic}\}$$

$$= \{A \in \mathbf{C}^{n\times n};\ A = A^T \text{ and } \forall x \in \mathbf{R}^n\setminus\{0\} : x^T Ax \neq 0\}$$

is *connected* if $n \geq 3$. Let us prove this fact analogously to the proof of Prop. 1.1 in Ch. II of Lions and Magenes [167], p. 121.

For $\xi, \eta \in \mathbf{R}^n$ linearly independent, we first show that

$$B := \begin{pmatrix} \xi^T A\xi & \xi^T A\eta \\ \xi^T A\eta & \eta^T A\eta \end{pmatrix} \in M_2^0.$$

In fact, if $\eta \in \mathbf{R}^n \setminus \{0\}$ is fixed and $\xi \in \mathbf{R}^n \setminus \mathbf{R}\eta$, then the zeroes λ_1, λ_2 of the parabola $\tau \mapsto (\xi + \tau\eta)^T A(\xi + \tau\eta)$ belong to $\mathbf{C} \setminus \mathbf{R}$ and depend continuously on ξ. If ξ is replaced by $-\xi$, then λ_1, λ_2 change their signs, and hence, since $\mathbf{R}^n \setminus \mathbf{R}\eta$ is connected for $n \geq 3$, we must have Im $\lambda_1 \cdot$ Im $\lambda_2 < 0$, i.e., $B \in M_2^0$.

If $A \in M_n$ and, without loss of generality, $a_{11} = 1$, then this implies $\xi^T A \xi \in \mathbf{C} \setminus (-\infty, 0]$ by part (a). Hence A can be joined in M_n to I_n by means of the path $tA + (1-t)I_n$, $0 \leq t \leq 1$, and thus M_n is connected.

If $A = I_n$, then the unique homogeneous fundamental solution of $P(\partial) = \nabla^T A \nabla = \Delta_n$ is given by

$$E = \frac{\Gamma(\frac{n}{2})}{(2-n)2\pi^{n/2}} |x|^{2-n}$$

according to Example 1.3.14(a). A linear transformation in $\mathrm{Gl}_n(\mathbf{R})$ then yields, by Proposition 1.3.19, the homogeneous fundamental solution E of $P(\partial) = \nabla^T A \nabla$ for real-valued positive definite A:

$$E = \frac{\Gamma(\frac{n}{2})}{(2-n)2\pi^{n/2}} (\det A)^{(n-3)/2} (x^T A^{\mathrm{ad}} x)^{-n/2+1}. \tag{2.4.9}$$

Finally, (2.4.9) remains valid by analytic continuation for each $A \in M_n$ if the values of $\sqrt{\det A}$ and of $\sqrt{x^T A^{\mathrm{ad}} x}$ are determined by continuity.

To illustrate formula (2.4.9) for a concrete example, let us consider the operator

$$P(\partial) = \partial_1^2 + \partial_2^2 + 2i\partial_1\partial_2 + \partial_3^2 = \nabla^T A \nabla, \qquad A = \begin{pmatrix} 1 & i & 0 \\ i & 1 & 0 \\ 0 & 0 & 1 \end{pmatrix}.$$

Then (2.4.9) yields

$$E = -\frac{1}{4\pi \sqrt{x_1^2 + x_2^2 - 2ix_1x_2 + 2x_3^2}}$$

in agreement with Hörmander [139], Exercise 7.1.39, p. 388.

(c) In order to deduce the associated homogeneous fundamental solutions E of Proposition 2.4.8 for the powers $(\nabla^T A \nabla)^k$, $k \in \mathbf{N}$, we can apply the same procedure as above. Passing from the fundamental solutions of Δ_n^k in (1.6.19) and (1.6.20) by a linear transformation to $(\nabla^T A \nabla)^k$ for A real-valued and positive definite, we obtain

$$E = \begin{cases} \dfrac{(-1)^k \Gamma(\frac{n}{2}-k)(\det A)^{(n-1)/2-k}}{2^{2k}(k-1)!\, \pi^{n/2}} (x^T A^{\mathrm{ad}} x)^{-n/2+k} & : n \text{ odd or } k < \frac{n}{2}, \\[4mm] \dfrac{(-1)^{n/2-1}(\det A)^{(n-1)/2-k}}{2^{2k}(k-1)!(k-\frac{n}{2})!\, \pi^{n/2}} (x^T A^{\mathrm{ad}} x)^{-n/2+k} \log(x^T A^{\mathrm{ad}} x) & : n \text{ even and } k \geq \frac{n}{2}. \end{cases} \tag{2.4.10}$$

Again, (2.4.10) remains valid by analytic continuation if $n = 2$ and $A \in M_2^0$ or if $n \geq 3$ and $A \in M_n$. (Therein, the functions $A \mapsto (\det A)^{(n-1)/2-k}$ and $A \mapsto \log(x^T A^{\mathrm{ad}} x)$ have to be determined by continuity.) Furthermore, E is the only homogeneous fundamental solution if n is odd or $k < \frac{n}{2}$ and else is uniquely determined by the condition (2.4.4) up to polynomials of degree $2k - n$. $\qquad\square$

We next consider hyperbolic systems of differential operators. For these, the uniqueness of the fundamental matrix is implied by support properties.

Definition 2.4.10 Let $N \in \mathbf{R}^n \setminus \{0\}$.

(1) A linear differential operator $P(\partial) = \sum_{|\alpha| \leq m} a_\alpha \partial^\alpha$, $a_\alpha \in \mathbf{C}$, in $\mathbf{R}^n$ with principal part $P_m(\partial) = \sum_{|\alpha|=m} a_\alpha \partial^\alpha$ is called *hyperbolic* in the direction N if and only if

$$\text{(i) } P_m(N) \neq 0 \text{ and (ii) } \exists \sigma_0 \in \mathbf{R} : \forall \sigma > \sigma_0 : \forall \xi \in \mathbf{R}^n : P(i\xi + \sigma N) \neq 0.$$

(2) A quadratic system $A(\partial) \in \mathbf{C}[\partial]^{l \times l}$ of linear differential operators in $\mathbf{R}^n$ is called *hyperbolic* in the direction N if and only if $P(\partial) = \det A(\partial)$ is hyperbolic in the direction N.

These definitions go back to Gårding [88]; see also Hörmander [138], Def. 12.3.3, p. 112; Atiyah, Bott and Gårding [5], Section 3, p. 126; Gårding [93], Section 8, p. 55. In the cited literature, it is also shown that $A(\partial)$ is hyperbolic in the direction N if and only if $A(\partial)$ possesses a two-sided fundamental matrix E with support in a cone contained in $\{x \in \mathbf{R}^n; xN > 0\} \cup \{0\}$, see also Ch. IV.

Let us next show that, if $A(\partial)$ is hyperbolic in the direction N, then there exists only one fundamental matrix with support in the closed half-space $H_N = \{x \in \mathbf{R}^n; xN \geq 0\}$.

Proposition 2.4.11 *Let $A(\partial)$ be a system which is hyperbolic in the direction N and let E be a two-sided fundamental matrix with support in a cone K contained in $\{x \in \mathbf{R}^n; xN > 0\} \cup \{0\}$. Then each right-sided and each left-sided fundamental matrix of $A(\partial)$ with support in $H_N = \{x \in \mathbf{R}^n; xN \geq 0\}$ coincides with E.*

Proof If F is a fundamental matrix satisfying $\operatorname{supp} F \subset H_N$, then E and F are convolvable by support (see Example 1.5.11), and we conclude that

$$E = E * A(\partial) F = \big(E * A(\partial)\delta\big) * F = (I_l \delta) * F = F.$$

$\qquad\square$

In particular, Proposition 2.4.11 applies to the system of elastodynamics considered in Examples 2.1.3, 2.1.4.

Definition 2.4.12 Let $P(\partial)$ be an operator in $\mathbf{R}^n$ which is hyperbolic in the direction N. Then the only fundamental solution E of $P(\partial)$ with support in $H_N = \{x \in \mathbf{R}^n; xN \geq 0\}$ is called the *forward fundamental solution of $P(\partial)$ with respect* *to* N. Similarly, if $A(\partial)$ is a system which is hyperbolic in the direction N, then the only fundamental matrix with support in H_N is called the *forward fundamental matrix of* $A(\partial)$ *with respect to* N. In particular, if $P(\partial) = P(\partial_t, \partial_1, \dots, \partial_n)$ and $N = (1, 0, \dots, 0)$, we will just speak of the *forward fundamental solution* without mentioning N, and similarly for a matrix A.

If an operator is hyperbolic with respect to N, then it is also hyperbolic with respect to the direction $-N$ (see, e.g., Hörmander [136], Thm. 5.5.1, p. 132), and the forward fundamental solution with respect to $-N$ is called the *backward fundamental solution with respect to* N. We also note that, in a physical context, in particular in connection with the wave equation or the Klein–Gordon equation, the forward and backward fundamental solutions are also called "retarded" and "advanced fundamental solutions", respectively, or "retarded and advanced potentials", see, e.g., Friedlander [83], p. 117; Zeidler [305], 12.5.3, p. 715; Komech [154], Ch. V, 6.2. (Let us mention that the forward fundamental solution of the wave equation is called "advanced" in Hörmander [138], p. 195.)

If $A(\partial)$ is only *quasihyperbolic* instead of hyperbolic, i.e., $P(\partial) = \det A(\partial)$ fulfills condition (2.2.3), then support conditions alone do not suffice to ensure the uniqueness of the fundamental matrix.

Definition and Proposition 2.4.13 $A(\partial) \in \mathbf{C}[\partial]^{l \times l}$ *is called quasihyperbolic in the direction* $N \in \mathbf{R}^n \setminus \{0\}$ *iff* $P(\partial) = \det A(\partial)$ *fulfills the condition*

$$\exists \sigma_0 \in \mathbf{R} : \forall \sigma > \sigma_0 : \forall \xi \in \mathbf{R}^n : P(\mathrm{i}\xi + \sigma N) \neq 0. \tag{2.4.11}$$

If $A(\partial)$ *is quasihyperbolic and* σ_0 *is as in* (2.4.11), *then there exists a two-sided fundamental matrix* E *satisfying*

$$\operatorname{supp} E \subset H_N = \{x \in \mathbf{R}^n; xN \geq 0\} \text{ and } \forall \sigma \geq \sigma_0 : \mathrm{e}^{-\sigma xN} E \in \mathcal{S}'(\mathbf{R}^n)^{l \times l}. \tag{2.4.12}$$

Furthermore, each right-sided and each left-sided fundamental matrix F *of* $A(\partial)$ *satisfying* $\exists \sigma > \sigma_0 : \mathrm{e}^{-\sigma xN} F \in \mathcal{S}'(\mathbf{R}^n)^{l \times l}$ *coincides with* E. *For* $\sigma > \sigma_0$, σ_0 *as in* (2.4.11), E *has the representation*

$$E = \mathrm{e}^{\sigma N x} \cdot \mathcal{F}^{-1}\big(A(\mathrm{i}\xi + \sigma N)^{-1}\big). \tag{2.4.13}$$

If E_1 *is the fundamental solution of* $P(\partial)$ *satisfying* $\exists \sigma > \sigma_0 : \mathrm{e}^{-\sigma xN} E_1 \in \mathcal{S}'(\mathbf{R}^n)$, *then* $E = A^{\mathrm{ad}}(\partial) E_1$.

Proof In order to reduce the assertion to Proposition 2.3.5, we set $Q(\partial) = P(\partial + \sigma_0 N)$. Then $Q(\partial)$ satisfies condition (2.3.8) in Proposition 2.3.5, i.e.,

$$\forall \sigma > 0 : \forall \xi \in \mathbf{R}^n : Q(\mathrm{i}\xi + \sigma N) = P\big(\mathrm{i}\xi + (\sigma + \sigma_0)N\big) \neq 0.$$

Hence, by Proposition 2.3.5, $Q(\partial)$ possesses a fundamental solution $G \in \mathcal{S}'(\mathbf{R}^n)$ such that $\operatorname{supp} G \subset H_N$. Then $E_1 = \mathrm{e}^{\sigma_0 N x} G$ is a fundamental solution of $P(\partial)$

satisfying $e^{-\sigma_0 Nx}E_1 \in \mathcal{S}'(\mathbf{R}^n)$ and $\operatorname{supp} E_1 \subset H_N$. For $\sigma \geq \sigma_0$, this implies

$$e^{-\sigma Nx}E_1 = \chi(Nx)e^{-(\sigma-\sigma_0)Nx}e^{-\sigma_0 Nx}E_1 \in \mathcal{S}'(\mathbf{R}^n)$$

if we take $\chi \in \mathcal{E}(\mathbf{R}_t^1)$, $\chi(t) = 1$ for $t > 0$ and $\chi(t) = 0$ for $t < -1$, and observe that $\chi(Nx)e^{-(\sigma-\sigma_0)Nx} \in \mathcal{D}_{L^\infty}(\mathbf{R}^n)$. Therefore, if $E = A^{\mathrm{ad}}(\partial)E_1$, then E is a two-sided fundamental matrix of $P(\partial)$ fulfilling $\operatorname{supp} E \subset H_N$ and $e^{-\sigma Nx}E \in \mathcal{S}'(\mathbf{R}^n)^{l\times l}$ for each $\sigma \geq \sigma_0$.

For a further right-sided fundamental matrix F of $A(\partial)$ which fulfills $G = e^{-\sigma Nx}F \in \mathcal{S}'(\mathbf{R}^n)^{l\times l}$ for some $\sigma > \sigma_0$, we conclude that

$$A(\partial + \sigma N)(G - e^{-\sigma Nx}E) = e^{-\sigma Nx}A(\partial)(F - E) = 0$$

and hence $A(i\xi + \sigma N)\mathcal{F}(G - e^{-\sigma Nx}E) = 0$. Because of $\det A(i\xi + \sigma N) = P(i\xi + \sigma N) \neq 0$ for $\xi \in \mathbf{R}^n$, we have $G = e^{-\sigma Nx}E$, i.e., $F = E$. An analogous reasoning applies if F is a left-sided fundamental solution. This completes the proof. $\square$

For scalar hyperbolic operators, the formula in (2.4.13) coincides with (12.5.3) in Hörmander [138], p. 120; for the generalization to quasihyperbolic operators and systems, see Ortner and Wagner [207], Prop. 1, p. 442, Ortner and Wagner [209], Prop. 1, p. 530, and Ortner and Wagner [218], Prop. 9, p. 147.

Example 2.4.14 Let us apply Proposition 2.4.13 to a less known quasihyperbolic but non-hyperbolic system arising in elasticity, namely *Rayleigh's system*.

According to S. Timoshenko, the transverse vibrations in a homogeneous bar are governed by the following 2×2 system of linear partial differential equations (cf. Graff [114], pp. 181–183, in particular Eqs. (3.4.11/12); Timoshenko and Young [269], pp. 330, 331):

$$\rho A \partial_t^2 u - \kappa AG(\partial_x^2 u - \partial_x \psi) = q, \qquad (2.4.14)$$

$$EI\partial_x^2 \psi + \kappa AG(\partial_x u - \psi) - \rho I \partial_t^2 \psi = 0. \qquad (2.4.15)$$

In the above equations, $u(t,x)$ denotes the displacement of the bar at the co-ordinate x and at time t, and $\psi(t,x)$ is the slope of the deflection curve diminished by the angle of shear at the neutral axis. The positive parameters A, I, ρ, E, G and κ stand for the cross-section area, the moment of inertia, the mass density, Young's modulus, the shear modulus, and Timoshenko's shear coefficient, respectively.

If we neglect the inertia term $-\rho I \partial_t^2 \psi$ in (2.4.15), this equation becomes

$$EI\partial_x^2 \psi + \kappa AG(\partial_x u - \psi) = 0,$$

cf. Flügge [79], (6a/b), p. 313; Love [171], Ch. XX, § 280, (7), p. 431, "Rayleigh's equation". With the abbreviations $\alpha = \sqrt{\frac{\rho}{\kappa G}}$ and $\beta = \sqrt{\frac{\kappa AG}{EI}}$, we can rewrite this system as

$$B(\partial)\begin{pmatrix} u \\ \psi \end{pmatrix} = \begin{pmatrix} q/(\kappa AG) \\ 0 \end{pmatrix}, \qquad B(\partial) = \begin{pmatrix} \alpha^2\partial_t^2 - \partial_x^2 & \partial_x \\ \beta^2\partial_x & \partial_x^2 - \beta^2 \end{pmatrix}.$$

Let us show that $B(\partial)$ is quasihyperbolic in the direction $N = (1,0)$. In fact, for $\sigma > 0 = \sigma_0$ and for $(\tau, \xi) \in \mathbf{R}^2$, the determinant

$$P(\partial) = \det B(\partial) = (\alpha^2\partial_t^2 - \partial_x^2)\partial_x^2 - \alpha^2\beta^2\partial_t^2, \qquad \alpha, \beta > 0, \qquad (2.4.16)$$

fulfills

$$P(\sigma + i\tau, i\xi) = -\alpha^2(\sigma + i\tau)^2(\xi^2 + \beta^2) - \xi^4 \neq 0.$$

On the other hand, $\deg P = 4$ and $P_4(N) = 0$ imply that $B(\partial)$ and $P(\partial)$ are not hyperbolic in the direction N.

By Proposition 2.4.13, there exists a temperate two-sided fundamental matrix E of $B(\partial)$ with support in $\{(t,x) \in \mathbf{R}^2; \ t \geq 0\}$, and E is uniquely determined by the condition $e^{-\sigma t}E \in S'(\mathbf{R}^2)^{2\times 2}$ for some $\sigma > 0$. As explained in Sect. 2.1, E has the representation $E = B(\partial)^{\mathrm{ad}}F$ where F is the fundamental solution of $P(\partial) = \det B(\partial)$ (see (2.4.16)) with $e^{-\sigma t}F \in S'(\mathbf{R}^2)$, $\sigma \geq 0$. An explicit expression for F was given in Ortner and Wagner [214], Prop. 2, p. 226. In Example 4.1.9, we shall come back to Rayleigh's operator $P(\partial)$ in (2.4.16).   □

2.5   Linear Transformations

We recall that Proposition 1.3.19 describes the effect of linear transformations of the independent variables x on the fundamental solution of a scalar operator $P(\partial)$, i.e., $|\det B|^{-1} \cdot E \circ B^{-1T}$ is a fundamental solution of $(P \circ B)(\partial)$ if $P(\partial)E = \delta$ and $B \in \mathrm{Gl}_n(\mathbf{R})$. Let us slightly generalize this formula and then provide some examples.

Proposition 2.5.1 *Let* $A(\partial) = \big(A_{ij}(\partial)\big)_{1 \leq i,j \leq l} \in \mathbf{C}[\partial]^{l \times l}$ *be an* $l \times l$ *matrix of linear differential operators in* $\mathbf{R}^n$ *with constant coefficients,* $S, T \in \mathrm{Gl}_l(\mathbf{C})$, $B \in \mathrm{Gl}_n(\mathbf{R})$, $c \in \mathbf{C}^n$, *and denote by* $\tilde{A}(\partial)$ *the following transformed system of constant coefficient operators:*

$$\tilde{A}(\partial) = S \cdot A(B\partial + c) \cdot T,$$

i.e.,

$$\tilde{A}_{ij}(\partial) = \sum_{k=1}^{l}\sum_{m=1}^{l} s_{ik}t_{mj}A_{km}\left(c_1 + \sum_{s=1}^{n} b_{1s}\partial_s, \ldots, c_n + \sum_{s=1}^{n} b_{ns}\partial_s\right).$$

Then

$$\tilde{E} = |\det B|^{-1} \cdot T^{-1} \cdot \left((e^{-cx}E) \circ B^{-1T} \right) \cdot S^{-1} \qquad (2.5.1)$$

yields a right-sided fundamental matrix of $\tilde{A}(\partial)$ if E is a right-sided fundamental matrix of $A(\partial)$.

Proof For $U \in \mathcal{D}'(\mathbf{R}^n)$ and $c \in \mathbf{R}^n$, we generally have $P(\partial + c)(e^{-cx}U) = e^{-cx}P(\partial)U$ if $P(\partial)$ is a linear differential operator with constant coefficients. This implies that $A(\partial + c)(e^{-cx}E) = I_l\delta$ if $A(\partial)E = I_l\delta$, i.e., if E is a right-sided fundamental matrix of $A(\partial)$, see Definition 2.1.1. Obviously, we then obtain $A^1(\partial)E^1 = I_l\delta$ for $A^1(\partial) = S \cdot A(\partial + c) \cdot T$ and $E^1 = e^{-cx}T^{-1} \cdot E \cdot S^{-1}$. Finally, analogously to the proof of Proposition 1.3.19,

$$\left(\tilde{A}(\partial)\tilde{E} \right)_{ij} = |\det B|^{-1} \cdot \left[(A^1 \circ B)(\partial)(E^1 \circ B^{-1T}) \right]_{ij}$$

$$= |\det B|^{-1} \sum_{k=1}^{l} (A^1_{ik} \circ B)(\partial)(E^1_{kj} \circ B^{-1T})$$

$$= |\det B|^{-1}\delta_{ij} \cdot \delta \circ B^{-1T} = \delta_{ij} \cdot \delta(x),$$

and hence $\tilde{A}(\partial)\tilde{E} = I_l\delta$. $\square$

Example 2.5.2

(a) The *iterated transport operator* in $\mathbf{R}^n$ has the form $\tilde{P}(\partial) = \left(\lambda + \sum_{j=1}^{n} a_j\partial_j \right)^m$, $a \in \mathbf{R}^n \setminus \{0\}$, $\lambda \in \mathbf{C}$, $m \in \mathbf{N}$. Since $P(\partial) = \partial_1^m$ has the fundamental solution

$$E = \frac{Y(x_1)x_1^{m-1}}{(m-1)!} \otimes \delta(x'), \qquad x' = (x_2, \ldots, x_n),$$

and

$$\tilde{P}(\partial) = P(B\partial + c) = (B\partial + c)_1^m, \quad B = \begin{pmatrix} a_1 & a_2 & \ldots & \ldots & a_n \\ 0 & 1 & 0 & \ldots & 0 \\ \vdots & & & & \vdots \\ 0 & 0 & \ldots & 0 & 1 \end{pmatrix}, \quad c = \begin{pmatrix} \lambda \\ 0 \\ \vdots \\ 0 \end{pmatrix},$$

(where we suppose without restriction of generality that $a_1 \neq 0$), we infer that $\tilde{E} = \frac{1}{|a_1|}(e^{-cx}E) \circ B^{-1T}$ is a fundamental solution of $\tilde{P}(\partial)$.

For a test function $\phi \in \mathcal{D}(\mathbf{R}^n)$, we obtain

$$\langle \phi, \tilde{E} \rangle = \langle \phi \circ B^T, e^{-cx}E \rangle = \frac{1}{(m-1)!} \langle \phi \circ B^T, Y(x_1)e^{-\lambda x_1}x_1^{m-1} \otimes \delta(x') \rangle$$

$$= \frac{1}{(m-1)!} \int_0^\infty \phi(at)e^{-\lambda t}t^{m-1} \, dt.$$

$$(2.5.2)$$

For formula (2.5.2), cf. Garnir [97], Vladimirov [279], § 10, Section 11, p. 154.

(b) More generally, let us consider now the powers of a linear first-order operator with *complex* constant coefficients, i.e., $\tilde{P}(\partial) = \left(\lambda + \sum_{j=1}^{n} a_j \partial_j\right)^m$, $\lambda \in \mathbf{C}$, $a \in \mathbf{C}^n$, $m \in \mathbf{N}$, with $\alpha = \operatorname{Re} a, \beta = \operatorname{Im} a$ linearly independent in $\mathbf{R}^n$. Without restriction of generality, we can assume that $\det \begin{pmatrix} \alpha_1 & \alpha_2 \\ \beta_1 & \beta_2 \end{pmatrix} \neq 0$. We then have

$\tilde{P}(\partial) = P(B\partial + c)$ where $P(\partial) = (\partial_1 + i\partial_2)^m$ is the iterated Cauchy–Riemann operator and

$$B = \begin{pmatrix} \alpha_1 & \alpha_2 & \cdots\cdots\cdots & \alpha_n \\ \beta_1 & \beta_2 & \cdots\cdots\cdots & \beta_n \\ 0 & 0 & 1 & 0 & \cdots & 0 \\ \vdots & \vdots & & & & \vdots \\ 0 & 0 & \cdots\cdots & 0 & 1 \end{pmatrix}, \qquad c = \begin{pmatrix} \lambda \\ 0 \\ \vdots \\ 0 \end{pmatrix}.$$

By Example 1.3.14(b) and Lemma 2.5.3 below,

$$E = \frac{x_1^{m-1} \otimes \delta(x'')}{(m-1)! \cdot 2\pi(x_1 + ix_2)}, \qquad x'' = (x_3, \dots, x_n),$$

is a fundamental solution of $P(\partial)$ in $\mathbf{R}^n$. Hence Proposition 2.5.1 yields the following representation of a fundamental solution $\tilde{E}$ of $\tilde{P}(\partial) = \left(\lambda + \sum_{j=1}^{n} a_j \partial_j\right)^m$:

$$\phi \in \mathcal{D}(\mathbf{R}^n) \Longrightarrow \langle \phi, \tilde{E} \rangle = \langle \phi \circ B^T, e^{-cx} E \rangle$$

$$= \frac{1}{2\pi(m-1)!} \langle \phi \circ B^T, \frac{e^{-\lambda x_1} x_1^{m-1}}{x_1 + ix_2} \otimes \delta(x'') \rangle$$

$$= \frac{1}{2\pi(m-1)!} \int_{\mathbf{R}^2} \phi(s\alpha + t\beta) e^{-\lambda s} \frac{s^{m-1}}{s + it} \, ds\, dt.$$

$\square$

Lemma 2.5.3 *If E_1 is a fundamental solution of the operator $P_1(\partial) = \partial_1 + R(\partial')$, $\partial' = (\partial_2, \dots, \partial_n)$, then the iterated operators $P_m(\partial) = P_1(\partial)^m = \left(\partial_1 + R(\partial')\right)^m$, $m \in \mathbf{N}$, have the fundamental solutions*

$$E_m = \frac{x_1^{m-1}}{(m-1)!} E_1.$$

Proof In fact, for $m \geq 2$, we have

$$
P_1(\partial)E_m = \frac{x_1^{m-2}}{(m-2)!} E_1 + \frac{x_1^{m-1}}{(m-1)!} P_1(\partial)E_1
$$

$$
= E_{m-1} + \frac{x_1^{m-1}}{(m-1)!} \delta = E_{m-1}.
$$

□

Let us point out that the formula in Lemma 2.5.3 can be conceived as a special case of the Bernstein–Sato recursion formula $Q(\lambda, -i\partial, -ix)E(\lambda + 1) = b(\lambda)E(\lambda)$, see (2.3.4), when taking $Q(\lambda, \xi, \partial) = \partial_1$, $b(\lambda) = i(\lambda + 1)$ and $E_m = E(-m)$, i.e., $-ix_1E_m = -imE_{m+1}$.

Example 2.5.4 The general *iterated anisotropic metaharmonic operator* has the form

$$
\tilde{P}(\partial) = (\nabla^T A \nabla + b^T \nabla - \lambda)^m = \left(\sum_{j=1}^{n} \sum_{k=1}^{n} a_{jk} \partial_j \partial_k + \sum_{j=1}^{n} b_j \partial_j - \lambda \right)^m,
$$

where $A = (a_{jk})_{1 \leq j,k \leq n}$ is a real, symmetric, positive definite matrix, $b \in \mathbf{C}^n$, $\lambda \in \mathbf{C}$, $\mu = \lambda + \frac{1}{4}b^T A^{-1} b \in \mathbf{C} \setminus (-\infty, 0]$, $m \in \mathbf{N}$.

For $\mu > 0$, the uniquely determined temperate fundamental solution E of $P(\partial) = (\Delta_n - \mu)^m$ was derived in Example 1.4.11, see (1.4.9), and also in Example 1.6.11 (a):

$$
E = \frac{(-1)^m |x|^{m-n/2} \mu^{n/4-m/2}}{2^{n/2+m-1} \pi^{n/2} (m-1)!} K_{n/2-m}(\sqrt{\mu} \, |x|).
$$

By analytic continuation, this expression continues to yield the only temperate fundamental solution as long as $\mu \in \mathbf{C} \setminus (-\infty, 0]$, cf. also Example 2.4.2.

If we define the matrix B as the square root of the positive definite real matrix A, i.e., $B = \sqrt{A}$, and set $c = \frac{1}{2}B^{-1}b$, then

$$
P(B\partial + c) = \left((c^T + \nabla^T B^T)(B\nabla + c) - \mu \right)^m
$$

$$
= \left(\nabla^T B^T B \nabla + 2c^T B \nabla + c^T c - \mu \right)^m = (\nabla^T A \nabla + b^T \nabla - \lambda)^m = \tilde{P}(\partial)
$$

because of $\mu = \lambda + \frac{1}{4}b^T A^{-1}b$. Hence the temperate fundamental solution of $\tilde{P}(\partial) = (\nabla^T A \nabla + b^T \nabla - \lambda)^m$ is given by

$$
\tilde{E} = |\det B|^{-1} (e^{-cx}E) \circ B^{-1T} = \frac{\exp(-\frac{1}{2}b^T A^{-1} x)}{\sqrt{\det A}} (E \circ A^{-1/2})
$$

$$
= \frac{(-1)^m \exp(-\frac{1}{2}b^T A^{-1} x)}{2^{n/2+m-1} \pi^{n/2} (m-1)! \sqrt{\det A}} \cdot \left(\frac{x^T A^{-1} x}{\mu} \right)^{m/2-n/4} \cdot K_{n/2-m}\left(\sqrt{\mu \cdot x^T A^{-1} x} \right).
$$

$$
(2.5.3)
$$

A direct verification of this fundamental solution by Fourier transformation is given in Lorenzi [170], pp 841–844.   □

Example 2.5.5 Similarly as in the last example, let us deduce now a fundamental solution $\tilde{E}$ of the *iterated anisotropic heat and Schrödinger operators*

$$\tilde{P}(\partial) = (\partial_t - \nabla^T A \nabla + b^T \nabla - \lambda)^m = \left(\partial_t - \sum_{j=1}^{n} \sum_{k=1}^{n} a_{jk} \partial_j \partial_k + \sum_{j=1}^{n} b_j \partial_j - \lambda \right)^m,$$

where $A = A^T \in \mathbf{C}^{n \times n}$ is non-singular (i.e., $\det A \neq 0$) and with positive semi-definite real part, $b \in \mathbf{C}^n$, $\lambda \in \mathbf{C}$.

By Example 1.4.13 and Lemma 2.5.3, a fundamental solution E of $P(\partial) = (\partial_t - \nabla^T A \nabla)^m$ is given by

$$E = \frac{Y(t) t^{m-1}}{(4\pi t)^{n/2} \sqrt{\det A}(m-1)!} \, e^{-x^T A^{-1} x/(4t)} \in \mathcal{C}\big([0, \infty), \mathcal{D}'(\mathbf{R}_x^n)\big),$$

see (1.4.14). Setting $\mu = \lambda - \frac{1}{4} b^T A^{-1} b$, $c = (-\mu, -\frac{1}{2} b^T A^{-1})^T \in \mathbf{C}^{n+1}$, we obtain

$$P(\partial + c) = \left[\partial_t - \mu - (\nabla^T - \tfrac{1}{2} b^T A^{-1}) A (\nabla - \tfrac{1}{2} A^{-1} b) \right]^m$$
$$= (\partial_t - \nabla^T A \nabla + b^T \nabla - \lambda)^m = \tilde{P}(\partial)$$

and hence

$$\tilde{E} = e^{\mu t} \cdot e^{b^T A^{-1} x/2} \cdot E$$

$$= \frac{Y(t) t^{m-1} e^{\mu t}}{(4\pi t)^{n/2} \sqrt{\det A}(m-1)!} \, e^{-x^T A^{-1} x/(4t) + b^T A^{-1} x/2} \in \mathcal{C}\big([0, \infty), \mathcal{D}'(\mathbf{R}_x^n)\big).$$

□

Example 2.5.6 As our final example, we consider an iterated *anisotropic Klein–Gordon operator* of the form

$$\tilde{P}(\partial) = (\partial_t^2 + \beta \partial_t - \nabla^T A \nabla + b^T \nabla - \lambda)^m = \left(\partial_t^2 + \beta \partial_t - \sum_{j=1}^{n} \sum_{k=1}^{n} a_{jk} \partial_j \partial_k + \sum_{j=1}^{n} b_j \partial_j - \lambda \right)^m,$$

where $A = (a_{jk})_{1 \leq j,k \leq n}$ is a real, symmetric, positive definite matrix, $b \in \mathbf{C}^n$, $\beta, \lambda \in \mathbf{C}$, $\mu^2 = \frac{1}{4}(b^T A^{-1} b - \beta^2) - \lambda \neq 0$, $m \in \mathbf{N}$.

By formula (2.3.14) in Example 2.3.7, the forward fundamental solution of the operator $P(\partial) = (\partial_t^2 - \Delta_n + \mu^2)^m$ is locally integrable if $m > \frac{n-1}{2}$ and given by

$$E = \frac{2^{-m-(n-1)/2} Y(t - |x|) \mu^{-m+(n+1)/2}}{\pi^{(n-1)/2}(m-1)!} (t^2 - |x|^2)^{m/2-(n+1)/4} \times$$

$$\times J_{m-(n+1)/2}\left(\mu\sqrt{t^2 - |x|^2}\right).$$

As in Example 2.5.4, we set $B = \begin{pmatrix} 1 & 0 \\ 0 & \sqrt{A} \end{pmatrix} \in \mathrm{Gl}_{n+1}(\mathbf{R})$, $c = \frac{1}{2}\begin{pmatrix} \beta \\ -A^{-1/2}b \end{pmatrix}$ and conclude that

$$P(B\partial + c) = (\partial_t^2 + \beta\partial_t + \tfrac{\beta^2}{4} - \nabla^T A\nabla + b^T\nabla - \tfrac{1}{4}b^T A^{-1}b + \mu^2)^m = \tilde{P}(\partial).$$

Hence the forward fundamental solution of $\tilde{P}(\partial)$ is

$$\tilde{E} = |\det B|^{-1}(e^{-\beta t/2 + b^T A^{-1/2}x/2}E) \circ B^{-1}$$

$$= \frac{2^{-m-(n-1)/2} Y(t - \sqrt{x^T A^{-1}x}) \mu^{-m+(n+1)/2}}{\pi^{(n-1)/2}(m-1)!\sqrt{\det A}} e^{-\beta t/2 + b^T A^{-1}x/2} \times$$

$$\times (t^2 - x^T A^{-1}x)^{m/2-(n+1)/4} J_{m-(n+1)/2}\left(\mu\sqrt{t^2 - x^T A^{-1}x}\right). \qquad \square$$

2.6   Invariance with Respect to Transformation Groups

In some cases, the invariance of a differential operator $P(\partial)$ under a group of linear transformations offers a means for the calculation of a fundamental solution. More precisely, such an invariance leads from the constant coefficient operator $P(\partial)$ in n dimensions to a linear operator in fewer variables having however, in general, non-constant coefficients. In order to exploit the invariance of $P(\partial)$, we have to employ some uniqueness class for the fundamental solution (see Sect. 2.4), and we use the following proposition, cf. Wagner [285], Satz 6, p. 11.

Proposition 2.6.1 *Let* $P(\partial) = \sum_{|\alpha|\le m} a_\alpha \partial^\alpha$ *be a linear constant coefficient operator in* $\mathbf{R}^n$ *and* $A \in \mathrm{Gl}_n(\mathbf{R})$ *such that* $P \circ A = \lambda \cdot P$ *for some* $\lambda \in \mathbf{C} \setminus \{0\}$. *Furthermore, assume that the subspace* $\mathcal{H} \subset \mathcal{D}'(\mathbf{R}^n)$ *is an* A^T-*invariant "set of uniqueness" for* $P(\partial)$, *i.e.,*

$$(i) \forall T \in \mathcal{H} : T \circ A^T \in \mathcal{H}; \qquad (ii) \exists_1 E \in \mathcal{H} : P(\partial)E = \delta.$$

Then the unique fundamental solution $E \in \mathcal{H}$ *of* $P(\partial)$ *fulfills*

$$E \circ A^T = \frac{\lambda}{|\det A|} E. \qquad (2.6.1)$$

Proof By Proposition 1.3.19, we have

$$(P \circ A^{-1})(\partial)(|\det A| \cdot E \circ A^T) = \delta.$$

Due to $P \circ A^{-1} = \lambda^{-1} P$, we infer that $\lambda^{-1} |\det A| \cdot E \circ A^T$ is a fundamental solution of $P(\partial)$ belonging to $\mathcal{H}$ by assumption (i). From the uniqueness property (ii), we then conclude the equation in (2.6.1). □

Example 2.6.2 Let us first consider the *iterated Laplacean* $P(\partial) = \Delta_n^m$, $m \in \mathbf{N}$. Then $P \circ A = P$ if A belongs to the orthogonal group $O_n(\mathbf{R})$, and $P \circ A = c^{2m} P$ if $A = cI$, $c \in \mathbf{R}$. On the other hand, the subspace $\mathcal{H} = \mathcal{E}' + \mathcal{C}_0 \subset \mathcal{S}'$ is invariant under orthogonal linear transformations and under dilatations. Furthermore, there exists at most one fundamental solution of $P(\partial)$ in $\mathcal{H}$. In fact, if $T \in \mathcal{H}$ and $P(\partial)T = 0$, then $|x|^{2m} \mathcal{F} T = 0$ and hence $\mathcal{F} T = \sum_{|\alpha| \le 2m} c_\alpha \partial^\alpha \delta$, $c_\alpha \in \mathbf{C}$, see Proposition 1.3.15. Therefore, $T = (2\pi)^{-n} \sum_{|\alpha| \le 2m} c_\alpha (-ix)^\alpha$ is a polynomial, which must vanish due to $T \in \mathcal{H}$.

If there exists at all a fundamental solution E of $P(\partial)$ in $\mathcal{H}$, then Proposition 2.6.1 implies

$$\forall A \in O_n(\mathbf{R}) : E \circ A^T = E \quad \text{and} \quad \forall c \in \mathbf{R} : E(cx) = c^{2m-n} E,$$

i.e., E is radially symmetric and homogeneous of degree $2m-n$. By Example 1.6.10, E then must have the form $E = c|x|^{2m-n} \in L^1_{\mathrm{loc}}(\mathbf{R}^n)$, $c \in \mathbf{C}$. Note that such distributions belong to $\mathcal{H}$ only if $m < \frac{n}{2}$.

If $m < \frac{n}{2}$, then there exists a fundamental solution in $\mathcal{H}$, which necessarily has the form $E = c|x|^{2m-n}$, see formula (1.6.19), and E is unique in $\mathcal{H}$ by the above reasoning. In contrast, for $m \le \frac{n}{2}$ and n even, $\Delta_n^m |x|^{2m-n} = 0$. □

Example 2.6.3 Let us next consider the *heat operator* $P(\partial) = \partial_t - \Delta_n$. This operator has the following invariance properties: $P \circ A = P$ for $A = \begin{pmatrix} 1 & 0 \\ 0 & B \end{pmatrix}$, $B \in O_n(\mathbf{R})$,

and $P \circ A = c^2 P$ for $A = \begin{pmatrix} c^2 & 0 \\ 0 & cI_n \end{pmatrix}$, $c \in \mathbf{R}$. By Proposition 2.4.13, there exists a unique fundamental solution E in

$$\mathcal{H} = \{T \in \mathcal{S}'(\mathbf{R}^{n+1}); \ T = 0 \text{ for } t < 0\}.$$

Furthermore, $\mathcal{H}$ is invariant under the linear transformations A considered above, and hence Proposition 2.6.1 implies that $E(t, Bx) = E$ for each $B \in O_n(\mathbf{R})$ and $E(c^2 t, cx) = c^{-n} E$ for $c > 0$.

Since the heat operator $\partial_t - \Delta_n$ is hypoelliptic, we know that E is infinitely differentiable outside the origin. Therefore, $f(s) := E(1, \sqrt{s}, 0, \ldots, 0) \in \mathcal{C}^\infty((0, \infty))$

and E can be represented by f in $\mathbf{R}^{n+1} \setminus \{0\}$:

$$t > 0, \ c = \frac{1}{\sqrt{t}} \implies E(t,x) = c^n E(c^2 t, cx) = t^{-n/2} E\left(1, \frac{x}{\sqrt{t}}\right) = t^{-n/2} f\left(\frac{|x|^2}{t}\right),$$

i.e., $E|_{\mathbf{R}^{n+1}\setminus\{0\}} = Y(t) t^{-n/2} f(|x|^2/t)$.

In order to determine the explicit form of the function f, we use the differential equation $(\partial_t - \Delta_n)E = 0$ in $\mathbf{R}^{n+1} \setminus \{0\}$. Setting $s = |x|^2/t$ and $r = |x|$ we obtain, for $t > 0$,

$$(\partial_t - \Delta_n)E = \left(\partial_t - \partial_r^2 - \tfrac{n-1}{r}\,\partial_r\right)E = -t^{-n/2-1}\left[4sf'' + (2n + s)f' + \tfrac{n}{2}f\right]$$

$$= -t^{-n/2-1}\left(s\frac{\mathrm{d}}{\mathrm{d}s} + \frac{n}{2}\right)\left(4\frac{\mathrm{d}}{\mathrm{d}s} + 1\right)f.$$

Hence f must fulfill

$$\left(s\frac{\mathrm{d}}{\mathrm{d}s} + \frac{n}{2}\right)\left(4\frac{\mathrm{d}}{\mathrm{d}s} + 1\right)f(s) = 0, \qquad s > 0.$$

If $g = 4f' + f$, then $sg' + \tfrac{n}{2}g = 0$ yields $g(s) = Ds^{-n/2}$, and $4f' + f = Ds^{-n/2}$ implies

$$f(s) = e^{-s/4}\left(D\int_1^s \sigma^{-n/2} e^{\sigma/4}\,\mathrm{d}\sigma + C\right), \quad C, D \in \mathbf{C}.$$

Thus we obtain that

$$E(t,x) = Y(t) t^{-n/2} e^{-|x|^2/(4t)}\left(D\int_1^{|x|^2/t} \sigma^{-n/2} e^{\sigma/4}\,\mathrm{d}\sigma + C\right).$$

Since $E(t,x)$ is defined and regular for $t = 1$ and $x = 0$, we conclude that the constant D must vanish, i.e.,

$$E_n = C_n Y(t) t^{-n/2} e^{-|x|^2/(4t)} \in L^1_{\mathrm{loc}}(\mathbf{R}^{n+1}), \qquad C_n \in \mathbf{R},$$

where we indicate the dependence on the space dimension n now by an additional index.

Let us yet determine the value of C_n by Hadamard's method of descent, cf. Delache and Leray [56], p. 317. The distribution

$$W = E_n * (\delta_{(t,x')} \otimes 1_{x_n}) - E_{n-1} \otimes 1_{x_n} \in \mathcal{S}'(\mathbf{R}^{n+1}), \qquad x' = (x_1, \ldots, x_{n-1}),$$

is well-defined and satisfies $(\partial_t - \Delta_n)W = 0$ and $W = 0$ for $t < 0$. By Proposition 2.4.13, we have $W = 0$, i.e., $E_n * (\delta_{(t,x')} \otimes 1_{x_n}) = E_{n-1} \otimes 1_{x_n}$. For

$t = 1, x' = 0$, this implies

$$C_{n-1} = E_{n-1}(1,0) = \int_{\mathbf{R}} E_n(1,0,x_n)\,dx_n = C_n \int_{-\infty}^{\infty} e^{-x_n^2/4}\,dx_n = 2\sqrt{\pi}\,C_n.$$

Because of $E_0 = Y(t)$, i.e., $C_0 = 1$, we finally obtain

$$E_n = \frac{Y(t)}{(4\pi t)^{n/2}}\, e^{-|x|^2/(4t)}$$

in accordance with (1.3.14) in Example 1.3.14 (d), see also Examples 1.6.16 and 2.3.8 (b).   □

Example 2.6.4 We shall apply Proposition 2.6.1 also in the case of the *Sobolev operator* $P(\partial) = \partial_t - \partial_1\partial_2\partial_3$ in $\mathbf{R}^4$, cf. Example 2.3.8 (c). This operator is quasi-hyperbolic with respect to t, and hence there exists one and only one fundamental solution E in the subspace

$$\mathcal{H} = \{T \in \mathcal{S}'(\mathbf{R}^4);\ T = 0 \text{ for } t < 0\}.$$

The relations $P \circ A = cP$ for $A = \begin{pmatrix} c & 0 \\ 0 & B \end{pmatrix} \in \mathrm{Gl}_4(\mathbf{R})$, $B = \begin{pmatrix} c_1 & 0 & 0 \\ 0 & c_2 & 0 \\ 0 & 0 & c_3 \end{pmatrix}$, $c_j \in \mathbf{R}$

with $c = c_1 c_2 c_3 > 0$ then yield $E(ct, c_1 x_1, c_2 x_2, c_3 x_3) = c^{-1} E$. This implies that E can be represented by composition with the function $h(t, x) = x_1 x_2 x_3/t$. More precisely, if

$$U = \{(t, x) \in \mathbf{R}^4;\ t \neq 0, (x_1 x_2, x_1 x_3, x_2 x_3) \neq 0\}$$
$$\text{and } h : U \longrightarrow \mathbf{R} : (t, x) \longmapsto \frac{x_1 x_2 x_3}{t},$$

then h is C^∞ and submersive, and

$$E|_U = \frac{Y(t)}{t}\, h^*(\Lambda) = \frac{Y(t)}{t}\, \Lambda\!\left(\frac{x_1 x_2 x_3}{t}\right) \in \mathcal{D}'(U)$$

holds for some distribution in one variable $\Lambda \in \mathcal{D}'(\mathbf{R})$.

If we express $(\partial_t - \partial_1\partial_2\partial_3)E$ by the function $\Lambda(s)$ of the variable $s = \frac{x_1 x_2 x_3}{t}$, we obtain that

$$(\partial_t - \partial_1\partial_2\partial_3)E = -\frac{Y(t)}{t^2}\big(s^2\Lambda''' + 3s\Lambda'' + (s+1)\Lambda' + \Lambda\big) \circ h$$

$$= -\frac{Y(t)}{t^2}\left[\left(s\frac{d}{ds} + 1\right)\left(s\frac{d^2}{ds^2} + \frac{d}{ds} + 1\right)\Lambda\right] \circ h$$

holds in $\mathcal{D}'(U)$. From this we conclude that $(s\frac{d}{ds}+1)(s\frac{d^2}{ds^2}+\frac{d}{ds}+1)\Lambda = 0$ in $\mathcal{D}'(\mathbf{R})$.

The solutions of $s \cdot T' + T = 0$ in $\mathcal{D}'(\mathbf{R}_s^1)$ are given by $T = A\delta + B\,\mathrm{vp}(\frac{1}{s})$, $A, B \in \mathbf{C}$. Hence Λ fulfills in $\mathcal{D}'(\mathbf{R})$ the ordinary differential equation

$$s\Lambda'' + \Lambda' + \Lambda = A\delta + B\,\mathrm{vp}(\tfrac{1}{s}).$$

For $s > 0$, we substitute $u = 2\sqrt{s}$ and write $M(u) = \Lambda(s) = \Lambda(u^2/4)$. Then $\frac{d}{ds} = \frac{2}{u}\frac{d}{du}$ and thus M fulfills $M'' + \frac{1}{u}M' + M = 4Bu^{-2}$. From Gradshteyn and Ryzhik [113], Eqs. 8.577, 8.571, we then obtain

$$\Lambda(s) = C_1 J_0(2\sqrt{s}) + C_2 N_0(2\sqrt{s})$$

$$+ C_3\left[J_0(2\sqrt{s})\int_{2\sqrt{s}}^{\infty} N_0(u)\,\frac{du}{u} - N_0(2\sqrt{s})\int_{2\sqrt{s}}^{\infty} J_0(u)\,\frac{du}{u}\right], \quad s > 0.$$

Similarly, for $s < 0$, we substitute $u = 2\sqrt{-s}$ and obtain

$$\Lambda(s) = C_4 I_0(2\sqrt{-s}) + C_5 K_0(2\sqrt{-s})$$

$$+ C_6\left[I_0(2\sqrt{-s})\int_{1}^{2\sqrt{-s}} K_0(u)\,\frac{du}{u} - K_0(2\sqrt{-s})\int_{1}^{2\sqrt{-s}} I_0(u)\,\frac{du}{u}\right], \quad s < 0,$$

cf. Sobolev [255], Eq. (7), p. 1247.

Let us consider now $(\partial_t - \partial_1\partial_2\partial_3)Y(t)t^{-1}\Lambda(x_1 x_2 x_3/t)$ in $\mathcal{D}'(\mathbf{R}^4)$. If we specify Λ by

$$\Lambda(s) = Y(s)N_0(2\sqrt{s}) - \tfrac{2}{\pi}Y(-s)K_0(2\sqrt{-s})$$

then the asymptotic expansions of N_0 and K_0 at 0 (see Gradshteyn and Ryzhik [113], Eqs. 8.444.1, 8.447.3) imply that

$$\Lambda(s) = \frac{1}{\pi}(2\gamma + \log|s|) + O(s^2 \log|s|), \quad s \to 0, \qquad (2.6.2)$$

and hence $f(t,x) = Y(t)t^{-1}\Lambda(x_1 x_2 x_3/t)$ fulfills $(\partial_t - \partial_1\partial_2\partial_3)f = 0$ for $t \neq 0$. Therefore, by the jump formula,

$$(\partial_t - \partial_1\partial_2\partial_3)f = \lim_{\epsilon\searrow 0}(\partial_t - \partial_1\partial_2\partial_3)Y(t-\epsilon)f = T\otimes\delta(t), \quad T = \lim_{\epsilon\searrow 0} f(\epsilon, x) \in \mathcal{D}'(\mathbf{R}^3).$$

Because of (2.6.2), the equation $(s\Lambda')' + \Lambda = 0$ holds in $\mathcal{D}'(\mathbf{R})$. The functions $g_\epsilon(s) = \Lambda(s/\epsilon)$ then satisfy $sg_\epsilon'' + g_\epsilon' + \epsilon^{-1}g_\epsilon = 0$, and $\lim_{\epsilon\searrow 0} g_\epsilon = 0$ in $\mathcal{D}'(\mathbf{R})$ inductively yields that $\lim_{\epsilon\searrow 0}\epsilon^\alpha g_\epsilon = 0$ holds in $\mathcal{D}'(\mathbf{R})$ for each $\alpha \in \mathbf{R}$.

In particular, for $\alpha = -1$, we conclude by composition with $h_1(x) = x_1 x_2 x_3$ that

$$T = \lim_{\epsilon \searrow 0} f(\epsilon, x) = \lim_{\epsilon \searrow 0} \frac{1}{\epsilon} \Lambda\left(\frac{x_1 x_2 x_3}{\epsilon}\right) = h_1^*\left(\lim_{\epsilon \searrow 0} \frac{1}{\epsilon} \Lambda\left(\frac{s}{\epsilon}\right)\right) = 0$$

holds in $U_1 = \{x \in \mathbf{R}^3; (x_1 x_2, x_1 x_3, x_2 x_3) \neq 0\}$, i.e., outside the three coordinate axes.

If $c_1, c_2, c_3 \in \mathbf{R}$ with $c = c_1 c_2 c_3 > 0$, then $f(ct, c_1 x_1, c_2 x_2, c_3 x_3) = c^{-1} f$ and hence $T(c_1 x_1, c_2 x_2, c_3 x_3) = c^{-1} T$. Since $T|_{U_1} = 0$, this implies that T is a multiple of δ. Thus $(\partial_t - \partial_1 \partial_2 \partial_3) f = C\delta$, and $C \neq 0$ due to $f \in S'(\mathbf{R}^4)$ and the uniqueness statement in Proposition 2.4.13. As a consequence,

$$E = C_1 \frac{Y(t)}{t}\left[Y(x_1 x_2 x_3) N_0\left(2\sqrt{x_1 x_2 x_3/t}\right) - \frac{2}{\pi} Y(-x_1 x_2 x_3) K_0\left(2\sqrt{-x_1 x_2 x_3/t}\right)\right]$$

is, for appropriate C_1, the unique fundamental solution of $\partial_t - \partial_1 \partial_2 \partial_3$ satisfying $E \in S'(\mathbf{R}^4)$ and $E = 0$ for $t < 0$. In fact, (2.3.21) shows that we must choose $C_1 = -\frac{1}{2\pi}$.   □

Example 2.6.5

(a) In a very similar way, one could consider the operators $\partial_t - i\prod_{j=1}^{2m} \partial_j$ and $\partial_t - \prod_{j=1}^{2m+1} \partial_j$, which are quasihyperbolic with respect to t and hence have one and only one temperate fundamental solution E with support in the half-space where $t \geq 0$.

   For $P(\partial) = \partial_t - i\partial_1 \partial_2$, formula (1.4.14) yields

$$E = \frac{Y(t)}{2\pi t} e^{i x_1 x_2 / t} \in C([0, \infty), S'(\mathbf{R}^2)).$$

On the other hand, the invariance method as in Example 2.6.4 leads to the representation

$$E = \frac{Y(t)}{t} \Lambda\left(\frac{x_1 x_2}{t}\right) = \frac{Y(t)}{t} \Lambda(s), \qquad s = \frac{x_1 x_2}{t},$$

and to the ordinary differential equation $(\frac{d}{ds} s)(1 + i\frac{d}{ds})\Lambda = 0$, which has the correct solution Ce^{is}, $c \in \mathbf{C}$.

(b) By partial Fourier transformation as in Example 2.3.8 (c), the fundamental solutions of $\partial_t - i\partial_1 \partial_2 \partial_3 \partial_4$ and of $\partial_t - \partial_1 \partial_2 \partial_3 \partial_4 \partial_5$, respectively, can be expressed as simple definite integrals involving Bessel functions. E.g., for $P(\partial) = \partial_t - \partial_1 \partial_2 \partial_3 \partial_4 \partial_5$, we obtain the following representation of the fundamental solution

E by a simple definite integral:

$$E = \frac{2Y(t)}{\pi^4 t} \int_0^\infty \left\{ K_0(u) \left[Y(X) K_0\left(\frac{4}{u}\sqrt{\frac{X}{t}}\right) - \frac{\pi}{2} Y(-X) N_0\left(\frac{4}{u}\sqrt{-\frac{X}{t}}\right) \right] \right.$$

$$\left. - \frac{\pi}{2} N_0(u) \left[Y(-X) K_0\left(\frac{4}{u}\sqrt{-\frac{X}{t}}\right) - \frac{\pi}{2} Y(X) N_0\left(\frac{4}{u}\sqrt{\frac{X}{t}}\right) \right] \right\} \frac{du}{u}$$

where $X = \prod_{j=1}^5 x_j$. In contrast, the invariance method leads to the fourth-order ordinary differential equation $\left[1 + \frac{d}{ds}(s\frac{d}{ds})^3 \right] \Lambda(s) = 0$. □

Example 2.6.6 Let us finally apply the invariance method to the *Klein–Gordon operator* $P(\partial) = \partial_t^2 - \Delta_n + c^2$, $c \in \mathbf{C} \setminus (-\infty, 0]$, which is invariant under Lorentz transformations. For the method in general, see Szmydt and Ziemian [267, 268].

Since the operator $P(\partial)$ is hyperbolic, there exists one and only one fundamental solution E in

$$\mathcal{H} = \{ T \in \mathcal{D}'(\mathbf{R}^{n+1}); \ T = 0 \text{ for } t < 0 \},$$

see Proposition 2.4.11. If we define the Lorentz product by

$$\left[\begin{pmatrix} t \\ x \end{pmatrix}, \begin{pmatrix} t \\ x \end{pmatrix} \right] = t^2 - |x|^2, \qquad \begin{pmatrix} t \\ x \end{pmatrix} \in \mathbf{R}^{n+1},$$

then $A = (a_{ij})_{0 \le i,j \le n} \in \mathrm{Gl}_{n+1}(\mathbf{R})$ is called a *proper Lorentz transformation* if $a_{00} > 0$ and $\left[A\begin{pmatrix} t \\ x \end{pmatrix}, A\begin{pmatrix} t \\ x \end{pmatrix} \right] = t^2 - |x|^2$ for each $\begin{pmatrix} t \\ x \end{pmatrix} \in \mathbf{R}^{n+1}$. Since $P(\partial)$ and $\mathcal{H}$ are invariant under proper Lorentz transformations A, Proposition 2.6.1 yields $E = E \circ A$.

Therefore, outside the origin, E is the pull-back of a one-dimensional distribution T by the submersive function

$$h : \mathbf{R}^{n+1} \setminus \{0\} \longrightarrow \mathbf{R} : \begin{pmatrix} t \\ x \end{pmatrix} \longmapsto t^2 - |x|^2,$$

cf. Proposition 1.2.13 and Gårding and Lions [96], i.e.,

$$E|_{\mathbf{R}^{n+1}\setminus\{0\}} = Y(t) h^*(T), \qquad T \in \mathcal{D}'(\mathbf{R}^1), \qquad \mathrm{supp}\, T \subset [0, \infty).$$

Note that $\mathrm{supp}\, h^*(T)$ is contained in the union $\{ \begin{pmatrix} t \\ x \end{pmatrix} \in \mathbf{R}^{n+1}; |t| \ge |x| \}$ of the forward and the backward propagation cones, and that therefore $Y(t)$ and $h^*(T)$ can be multiplied in $\mathbf{R}^{n+1} \setminus \{0\}$ without difficulties.

The equation $(\partial_t^2 - \Delta_n + c^2)E|_{\mathbf{R}^{n+1}\setminus\{0\}} = 0$ then furnishes for $T \in \mathcal{D}'(\mathbf{R}_u^1)$ the ordinary differential equation

$$u \cdot T'' + \frac{n+1}{2}T' + \frac{c^2}{4}T = 0.$$

From Gradshteyn and Ryzhik [113], Eq. 8.491.3 (corrected), p. 971, we infer that

$$T|_{(0,\infty)} = u^{(1-n)/4}\big[C_1 J_{(1-n)/2}(c\sqrt{u}) + C_2 N_{(1-n)/2}(c\sqrt{u})\big] \in \mathcal{C}^\infty((0,\infty)).$$

By the recursion formula

$$\frac{\mathrm{d}}{\mathrm{d}u}\big(u^{\nu/2}Z_\nu(c\sqrt{u})\big) = -\frac{c}{2}u^{(\nu-1)/2}Z_{\nu-1}(c\sqrt{u}),$$

where $u > 0$ and $Z_\nu = J_\nu$ or $Z_\nu = N_\nu$ (cf. Gradshteyn and Ryzhik [113], Eq. 8.472.3, p. 968), we see that T coincides on $\mathbf{R} \setminus \{0\}$ with the distribution

$$S = \frac{\mathrm{d}^k}{\mathrm{d}u^k}\big[Y(u)u^{\nu/2}\big(D_1 J_\nu(c\sqrt{u}) + D_2 N_\nu(c\sqrt{u})\big)\big] \in \mathcal{D}'(\mathbf{R}),$$

if $\nu = \left\{ \begin{array}{l} 0 : \ n \text{ odd,} \\ \frac{1}{2} : \ n \text{ even} \end{array} \right\}$ and $k = \nu + (n-1)/2 = [\frac{n}{2}]$, $D_1, D_2 \in \mathbf{C}$. From Proposition 1.3.15, we conclude that

$$T = S + \sum_{j=0}^{m} a_j \delta^{(j)}, \qquad a_j \in \mathbf{C}.$$

Since, by Leibniz' formula,

$$u \cdot \frac{\mathrm{d}^{k+2}U}{\mathrm{d}u^{k+2}} = \frac{\mathrm{d}^{k+2}}{\mathrm{d}u^{k+2}}(u \cdot U) - (k+2)\frac{\mathrm{d}^{k+1}U}{\mathrm{d}u^{k+1}}$$

holds for $U \in \mathcal{D}'(\mathbf{R}_u^1)$, we obtain, due to $\frac{n+1}{2} - (k+2) = -(\nu+1)$,

$$u \cdot S'' + \frac{n+1}{2}S' + \frac{c^2}{4}S = D_1 \frac{\mathrm{d}^k}{\mathrm{d}u^k}\Big[u\big(Y(u)u^{\nu/2}J_\nu(c\sqrt{u})\big)''$$

$$+ (1-\nu)\big(Y(u)u^{\nu/2}J_\nu(c\sqrt{u})\big)' + \frac{c^2}{4}Y(u)u^{\nu/2}J_\nu(c\sqrt{u})\Big]$$

$$D_2 \frac{\mathrm{d}^k}{\mathrm{d}u^k}\Big[u\big(Y(u)u^{\nu/2}N_\nu(c\sqrt{u})\big)''$$

$$+ (1-\nu)\big(Y(u)u^{\nu/2}N_\nu(c\sqrt{u})\big)' + \frac{c^2}{4}Y(u)u^{\nu/2}N_\nu(c\sqrt{u})\Big].$$

$$(2.6.3)$$

The expressions in the brackets on the right-hand side of (2.6.3) must vanish for $u \neq 0$ since they are classical solutions of $uU'' + (1 - v)U' + \frac{c^2}{4}U = 0$, $u > 0$. Moreover, this equation holds also in $\mathcal{D}'(\mathbf{R}_u^1)$ for $U = Y(u)u^{v/2}J_v(c\sqrt{u})$. On the other hand, for $U = Y(u)N_0(c\sqrt{u})$, the asymptotic expansion

$$N_0(z) = \frac{2}{\pi}\left[\log\frac{z}{2} + \gamma\right] + O(z^2 \log z), \qquad z \searrow 0,$$

yields

$$uU'' + U' + \frac{c^2}{4}U = (uU')' + \frac{c^2}{4}U = \frac{1}{\pi}\delta.$$

Similarly, for $v = \frac{1}{2}$, we set

$$U = Y(u)u^{1/4}N_{1/2}(c\sqrt{u}) = -\sqrt{\frac{2}{\pi c}}\,Y(u)\cos(c\sqrt{u})$$

and obtain

$$uU'' + \frac{1}{2}U' + \frac{c^2}{4}U = (uU')' - \frac{1}{2}U' + \frac{c^2}{4}U = \frac{1}{\sqrt{2\pi c}}\delta.$$

Therefore, eventually,

$$u \cdot S'' + \frac{n+1}{2}S' + \frac{c^2}{4}S = CD_2\delta^{(k)}, \qquad C = \begin{cases} \frac{1}{\pi}: & n \text{ odd,} \\ \frac{1}{\sqrt{2\pi c}}: & n \text{ even.} \end{cases}$$

Since a non-trivial linear combination of derivatives of δ, i.e., $U = \sum_{j=0}^m a_j\delta^{(j)} \in \mathcal{D}'(\mathbf{R}) \setminus \{0\}$, cannot fulfill an equation of the type

$$uU'' + \frac{n+1}{2}U' + \frac{c^2}{4}U = \alpha\delta^{([n/2])}, \qquad \alpha \in \mathbf{C},$$

we conclude that $D_2 = 0$ and $E|_{\mathbf{R}^{n+1}\setminus\{0\}} = Y(t)h^*(T)$ where T is a multiple of $\frac{d^k}{du^k}\left[Y(u)u^{v/2}J_v(c\sqrt{u})\right] \in \mathcal{D}'(\mathbf{R})$, $v = \begin{cases} 0: & n \text{ odd,} \\ \frac{1}{2}: & n \text{ even} \end{cases}$ and $k = [\frac{n}{2}]$.

By the recursion formula in Gradshteyn and Ryzhik [113], Eq. 8.472.3, p. 968, the analytic distribution-valued function

$$V: \{v \in \mathbf{C}; \operatorname{Re} v > -1\} \longrightarrow \mathcal{D}'(\mathbf{R}_u^1) : v \longmapsto Y(u)u^{v/2}J_v(c\sqrt{u})$$

satisfies $\frac{d}{du}V_\nu = -\frac{c}{2}V_{\nu-1}$, and hence it can be extended to an entire function. Therefore, if E_n denotes the fundamental solution of $\partial_t^2 - \Delta_n + c^2$, then

$$E_n|_{\mathbf{R}^{n+1}\setminus\{0\}} = C_n Y(t)h^*(V_{(1-n)/2}) = C_n Y(t)V_{(1-n)/2}(t^2 - |x|^2)$$

for some $C_n \in \mathbf{R}$.

In order to determine the values of the constants C_n, we employ—as in Example 2.6.3—Hadamard's method of descent in the form used in Delache and Leray [56], p. 317. Since

$$W = E_n * (\delta_{(t,x')} \otimes 1_{x_n}) - E_{n-1} \otimes 1_{x_n} \in \mathcal{D}'(\mathbf{R}^{n+1}), \qquad x' = (x_1,\dots,x_{n-1}),$$

is well-defined and satisfies $(\partial_t^2 - \Delta_n + c^2)W = 0$ and $W = 0$ for $t < 0$, we conclude that W vanishes, i.e., $E_n * (\delta_{(t,x')} \otimes 1_{x_n}) = E_{n-1} \otimes 1_{x_n}$. This yields the equation

$$2C_n Y(t - |x'|) \int_0^{\sqrt{t^2-|x'|^2}} (t^2 - |x'|^2 - x_n^2)^{\nu/2} J_\nu\big(c\sqrt{t^2 - |x'|^2 - x_n^2}\big)\, dx_n \bigg|_{\nu=(1-n)/2}$$

$$= C_{n-1} Y(t)V_{(2-n)/2}(t^2 - |x'|^2).$$

Upon inserting $t = 1$ and $x' = 0$, this implies

$$C_{n-1} J_{(n-2)/2}(c) = 2C_n \int_0^1 (1 - x_n^2)^{\nu/2} J_\nu\big(c\sqrt{1 - x_n^2}\big)\, dx_n \bigg|_{\nu=(1-n)/2}$$

$$= 2C_n \int_0^1 u^{\nu+1} J_\nu(cu) \frac{du}{\sqrt{1 - u^2}} \bigg|_{\nu=(1-n)/2} = \sqrt{\frac{2\pi}{c}}\, C_n J_{(2-n)/2}(c),$$

see Oberhettinger [194], Eq. 4.38, p. 39. Hence $C_n = \sqrt{\frac{c}{2\pi}}\, C_{n-1}$.

Because of

$$E_0 = C_0 Y(t)V_{1/2}(t^2) = C_0 Y(t)\sqrt{t}\, J_{1/2}(ct) = C_0 Y(t)\sqrt{\frac{2}{\pi c}}\,\sin(ct)$$

and $E_0 = Y(t)c^{-1}\sin(ct)$, see Example 1.3.8(a), we have $C_0 = \sqrt{\frac{\pi}{2c}}$ and, consequently,

$$C_n = \Big(\frac{c}{2\pi}\Big)^{n/2} C_0 = \frac{c^{(n-1)/2}}{2^{(n+1)/2}\pi^{(n-1)/2}}.$$

The final formula

$$E_n = \frac{c^{(n-1)/2}Y(t-|x|)}{2^{(n+1)/2}\pi^{(n-1)/2}} \left. (t^2-|x|^2)^{\nu/2}J_\nu(c\sqrt{t^2-|x|^2})\right|_{\nu=(1-n)/2}$$

agrees with the result in (2.3.14). For the explicit result for E_n, see also Schwartz [246], (VI, 5; 30), p. 179; Linés [166], 12.9, 12.11, pp. 49, 50; Courant and Hilbert [52], Ch. VI, § 12.6, pp. 693–695; Léonard [162], p. 36; Ortner and Wagner [207], Ex. 5, p. 457. □

Chapter 3
Parameter Integration

In its simplest form, the method of parameter integration yields a fundamental solution E of a product $P_1(\partial)P_2(\partial)$ of differential operators as a simple integral with respect to λ over fundamental solutions E_λ of the squared convex sums $\left(\lambda P_1(\partial) + (1-\lambda)P_2(\partial)\right)^2$. Heuristically, this relies on the representations of E and of E_λ as inverse Fourier transforms, i.e.,

$$\mathcal{F}E = \frac{1}{P_1(\mathrm{i}\xi)P_2(\mathrm{i}\xi)} \overset{(F)}{=} \int_0^1 \frac{\mathrm{d}\lambda}{\left(\lambda P_1(\xi) + (1-\lambda)P_2(\xi)\right)^2} = \int_0^1 \mathcal{F}E_\lambda \, \mathrm{d}\lambda$$

where the equation (F) is Feynman's first formula, see (3.1.1) below (for the name cf. Schwartz [245], Ex. I-8, p. 72). Note that Eq. (3.1.1) boils down to the formula $a^{-1} - b^{-1} = \int_a^b x^{-2}\mathrm{d}x$, $0 < a < b$, from elementary calculus.

By generalizing the integration over $[0, 1]$ to one over the simplex

$$\Sigma_{l-1} = \left\{(\lambda_1, \ldots, \lambda_l) \in \mathbf{R}^l;\ \lambda_1 \geq 0, \ldots, \lambda_l \geq 0, \sum_{j=1}^l \lambda_j = 1\right\}$$

one can represent a fundamental solution of the product $\prod_{j=1}^l P_j(\partial)^{\alpha_j+1}$, $\alpha \in \mathbf{N}_0^l$, by a *parameter integral* with respect to λ over fundamental solutions E_λ of the iterated operator $\left(\sum_{j=1}^l \lambda_j P_j(\partial)\right)^{|\alpha|+l}$. These representations are applied to constructing fundamental solutions of products of wave and of Laplace operators, respectively, i.e., of

$$\prod_{j=1}^l (\partial_t^2 - \lambda_j \Delta_n)^{\alpha_j+1} \quad \text{and of} \quad \prod_{j=1}^l (\Delta_{n-1} + \lambda_j \partial_n^2)^{\alpha_j+1}$$

for positive, pairwise different $\lambda_1, \ldots, \lambda_l$, see Sects. 3.2 and 3.3.

© Springer International Publishing Switzerland 2015
N. Ortner, P. Wagner, *Fundamental Solutions of Linear Partial Differential Operators*, DOI 10.1007/978-3-319-20140-5_3

By means of more sophisticated parameter integration formulas, it is also possible to represent fundamental solutions of indecomposable operators as parameter integrals over fundamental solutions of simpler operators. For example, starting from formula

$$\frac{1}{a^2 - b^2 - c^2} = \frac{1}{2\pi} \int_{\lambda^2 + \mu^2 \le 1} \frac{d\lambda d\mu}{(a + \lambda b + \mu c)^2 \sqrt{1 - \lambda^2 - \mu^2}}$$

for $a, b, c \in \mathbf{R}$ with $a^2 > b^2 + c^2$, we represent a fundamental solution of Timoshenko's beam operator

$$(\partial_t^2 - a\partial_x^2 + b)^2 - (c\partial_x^2 - d)^2 - e^2, \quad a, b, c, d \in \mathbf{R}, \ a \ge |c|, \ e \in \mathbf{C},$$

as a double integral over fundamental solutions $E_{\lambda,\mu}$ of the squared one-dimensional Klein–Gordon operators $(\partial_t^2 - a\partial_x^2 + b + \lambda(c\partial_x^2 - d) + \mu e)^2$. This double integral can be reduced to a simple one by known integral formulas, see Example 3.5.4.

An important application of the method of parameter integration is the explicit representation of the forward fundamental solution of the product $(\partial_t^2 - \Delta_3)(\partial_t^2 - a\Delta_2 - b\partial_3^2)$ of wave operators in Example 4.2.7. For proper choices of the parameters $a, b > 0$, this operator exhibits the phenomenon of conical refraction in a most illustrative way. Obviously, the qualitative discussion of conical refraction could be performed similarly in the case of the operator $(\partial_t^2 - \Delta_n)(\partial_t^2 - a_1\partial_1^2 - \cdots - a_n\partial_n^2)$, $a_j > 0, j = 1, \ldots, n$.

3.1 Parameter Integration for Decomposable Operators

By employing the so-called *Feynman formula*

$$\frac{1}{ab} = \int_0^1 \frac{d\lambda}{[\lambda a + (1 - \lambda)b]^2} \tag{3.1.1}$$

for $a, b \in \mathbf{C}$ with $\lambda a + (1 - \lambda)b \ne 0$ for $\lambda \in [0, 1]$ (cf. Feynman [73], Appendix, (14a), p. 785), D.W. Bresters derived the forward fundamental solution E of the product of two Klein–Gordon operators $R(\partial) = (\alpha_1\partial_t^2 - \Delta_n + \beta_1)(\alpha_2\partial_t^2 - \Delta_n + \beta_2)$ from the forward fundamental solutions E_λ of the iterated Klein–Gordon operators $Q_\lambda(\partial)^2$ where

$$Q_\lambda(\partial) = (\lambda\alpha_1 + (1 - \lambda)\alpha_2)\partial_t^2 - \Delta_n + \lambda\beta_1 + (1 - \lambda)\beta_2, \quad \lambda \in [0, 1],$$

cf. Bresters [25, 26].

Let us explain this construction method in the slightly more general case where $R(\partial) = P_1(\partial)P_2(\partial)$ is quasihyperbolic in the direction N, and the "intermediate" operators

$$Q_\lambda(\partial) = \lambda P_1(\partial) + (1-\lambda)P_2(\partial), \qquad \lambda \in [0,1],$$

are "uniformly" quasihyperbolic in this direction (see Definition 3.1.1 below).

Then $R(\partial)$ and $Q_\lambda(\partial)^2$ have fundamental solutions E and E_λ, respectively, given by

$$E = e^{\sigma x N} \mathcal{F}^{-1}\left(R(i\xi + \sigma N)^{-1}\right),$$

$$E_\lambda = e^{\sigma x N} \mathcal{F}^{-1}\left([\lambda P_1(i\xi + \sigma N) + (1-\lambda)P_2(i\xi + \sigma N)]^{-2}\right).$$

Therefore, setting $\eta = i\xi + \sigma N$, formula (3.1.1) yields the following:

$$R(\eta)^{-1} = \left[P_1(\eta)P_2(\eta)\right]^{-1} = \int_0^1 \frac{d\lambda}{\left[\lambda P_1(\eta) + (1-\lambda)P_2(\eta)\right]^2}$$

and hence $E = \int_0^1 E_\lambda \, d\lambda$.

In Proposition 3.1.2 below, we shall generalize this to the case that R is a product of l factors. Instead of (3.1.1), which refers to two factors, we shall make use of "Feynman's first formula" referring to l factors, i.e.,

$$\frac{1}{a_1 \dots a_l} = (l-1)! \int_{\Sigma_{l-1}} \frac{d\sigma(\lambda)}{(\lambda_1 a_1 + \dots + \lambda_l a_l)^l}, \qquad a_1 > 0, \dots, a_l > 0, \qquad (3.1.2)$$

see Schwartz [245], p. 72; Bresters [25], Eq. 4.17, p. 129; Brychkov, Marichev and Prudnikov [29], Eq. 3.3.4.3, p. 590; Folland [78], Eq. (7.6), p. 200. Herein, Σ_{l-1} denotes the $(l-1)$-dimensional standard simplex, i.e.,

$$\Sigma_{l-1} = \left\{(\lambda_1, \dots, \lambda_l) \in \mathbf{R}^l; \lambda_1 \geq 0, \dots, \lambda_l \geq 0, \sum_{j=1}^l \lambda_j = 1\right\}$$

and $d\sigma(\lambda)$ is the measure $d\lambda_1 \dots d\lambda_{l-1}$ on Σ_{l-1}.

For the operators involved, we shall assume *uniform* quasihyperbolicity, i.e., that the constant σ_0 in (2.2.3) can be chosen independently of the parameter λ. More precisely:

Definition 3.1.1 A set of differential operators $\{Q_\lambda(\partial); \lambda \in \Lambda\}$ is called *uniformly quasihyperbolic* in the direction $N \in \mathbf{R}^n \setminus \{0\}$ if and only if the condition

$$\exists \sigma_0 \in \mathbf{R} : \forall \sigma > \sigma_0 : \forall \xi \in \mathbf{R}^n : \forall \lambda \in \Lambda : Q_\lambda(i\xi + \sigma N) \neq 0 \qquad (3.1.3)$$

is satisfied.

For a product of factors from a set of uniformly quasihyperbolic operators, Feynman's formula (3.1.2) can be applied, cf. Ortner and Wagner [211], Prop. 1, p. 307.

Proposition 3.1.2 *Suppose that $P_j(\partial)$, $j = 1, \ldots, l$, are differential operators in $\mathbf{R}^n$ such that the set $\{Q_\lambda(\partial) := \sum_{j=1}^{l} \lambda_j P_j(\partial); \ \lambda \in \Sigma_{l-1}\}$ is uniformly quasihyperbolic in the direction $N \in \mathbf{R}^n \setminus \{0\}$. Let σ_0 be as in (3.1.3), $\alpha \in \mathbf{N}_0^l$, and denote by E and by E_λ, respectively, the fundamental solutions of $\prod_{j=1}^{l} P_j(\partial)^{\alpha_j+1}$ and of $Q_\lambda(\partial)^{|\alpha|+l}$ which satisfy $e^{-\sigma x N} E, e^{-\sigma x N} E_\lambda \in \mathcal{S}'(\mathbf{R}^n)$ for $\sigma > \sigma_0$, see Proposition 2.4.13. Then E_λ continuously depends on $\lambda \in \Sigma_{l-1}$ and we have*

$$E = \frac{(|\alpha| + l - 1)!}{\alpha!} \int_{\Sigma_{l-1}} E_\lambda \, \lambda^\alpha d\sigma(\lambda). \qquad (3.1.4)$$

Proof By differentiation with respect to a_j, Feynman's formula (3.1.2) implies

$$\prod_{j=1}^{l} a_j^{-\alpha_j-1} = \frac{(|\alpha| + l - 1)!}{\alpha!} \int_{\Sigma_{l-1}} \frac{\lambda^\alpha \, d\sigma(\lambda)}{(\lambda_1 a_1 + \ldots \lambda_l a_l)^{|\alpha|+l}} \qquad (3.1.5)$$

for $a_1 > 0, \ldots, a_l > 0$ and $\alpha \in \mathbf{N}_0^l$.

We next observe that (3.1.5) holds more generally for $a \in \mathbf{C}^l$ under the assumption that the convex hull of $a_1, \ldots, a_l \in \mathbf{C}$ does not contain 0, i.e.,

$$\forall \lambda \in \Sigma_{l-1} : \lambda_1 a_1 + \ldots \lambda_l a_l \neq 0.$$

In fact, this results from analytic continuation since the set

$$\{a \in \mathbf{C}^l; \ \forall \lambda \in \Sigma_{l-1} : \lambda_1 a_1 + \ldots \lambda_l a_l \neq 0\}$$

is arcwise connected in $\mathbf{C}^l$.

Due to the uniform quasihyperbolicity of $Q_\lambda(\partial)$, $\lambda \in \Sigma_{l-1}$, we may insert $a_j = P_j(i\xi + \sigma N)$, $\sigma > \sigma_0$, into (3.1.5), and this yields

$$\prod_{j=1}^{l} P_j(i\xi + \sigma N)^{-\alpha_j-1} = \frac{(|\alpha| + l - 1)!}{\alpha!} \int_{\Sigma_{l-1}} Q_\lambda(i\xi + \sigma N)^{-|\alpha|-l} \lambda^\alpha \, d\sigma(\lambda)$$

$$(3.1.6)$$

According to formula (2.4.13) in Proposition 2.4.13, we have

$$E_\lambda = e^{\sigma N x} \cdot \mathcal{F}^{-1}\big(Q_\lambda(i\xi + \sigma N)^{-|\alpha|-l}\big).$$

As in the proof of Proposition 2.3.5, see in particular (2.3.9), the Seidenberg–Tarski lemma implies that

$$\Sigma_{l-1} \longrightarrow \mathcal{S}'(\mathbf{R}^n) : \lambda \longmapsto Q_\lambda(\mathrm{i}\xi + \sigma N)^{-|\alpha|-l}$$

is continuous, and hence the same holds for $\lambda \mapsto E_\lambda$. This finally implies, for $\phi \in \mathcal{D}$,

$$\langle \phi, E \rangle = \langle \mathcal{F}^{-1}(e^{\sigma N x} \cdot \phi), \prod_{j=1}^{l} P_j(\mathrm{i}\xi + \sigma N)^{-\alpha_j - 1} \rangle$$

$$= \langle \mathcal{F}^{-1}(e^{\sigma N x} \cdot \phi), \frac{(|\alpha| + l - 1)!}{\alpha!} \int_{\Sigma_{l-1}} Q_\lambda(\mathrm{i}\xi + \sigma N)^{-|\alpha|-l} \lambda^\alpha \mathrm{d}\sigma(\lambda) \rangle$$

$$= \frac{(|\alpha| + l - 1)!}{\alpha!} \int_{\Sigma_{l-1}} \langle \mathcal{F}^{-1}(e^{\sigma N x} \cdot \phi), Q_\lambda(\mathrm{i}\xi + \sigma N)^{-|\alpha|-l} \rangle \lambda^\alpha \mathrm{d}\sigma(\lambda)$$

$$= \frac{(|\alpha| + l - 1)!}{\alpha!} \int_{\Sigma_{l-1}} \langle \phi, E_\lambda \rangle \lambda^\alpha \mathrm{d}\sigma(\lambda)$$

$$= \langle \phi, \frac{(|\alpha| + l - 1)!}{\alpha!} \int_{\Sigma_{l-1}} E_\lambda \lambda^\alpha \mathrm{d}\sigma(\lambda) \rangle.$$

$\square$

Example 3.1.3 Generalizing Example 2.5.2 we now consider *products of transport operators*, i.e.,

$$R(\partial) = \prod_{j=1}^{l} \left(d_j + \sum_{i=1}^{n} a_{ij} \partial_i \right)^{\alpha_j + 1}$$

for $A = (a_{ij}) \in \mathbf{R}^{n \times l}$, $d \in \mathbf{C}^l$, $\alpha \in \mathbf{N}_0^l$.

(a) Without loss of generality, we may suppose that

$$\exists N \in \mathbf{R}^n \setminus \{0\} : \forall j = 1, \dots, l : \sum_{i=1}^{n} a_{ij} N_i > 0 \qquad (3.1.7)$$

holds. (Otherwise, just multiply some of the vectors $(a_{1j}, \dots, a_{nj})^T \in \mathbf{R}^n$, $j = 1, \dots, l$, by -1.) Assumption (3.1.7) implies that the operators

$$Q_\lambda(\partial) = \sum_{j=1}^{l} \lambda_j \left(d_j + \sum_{i=1}^{n} a_{ij} \partial_i \right), \quad \lambda \in \Sigma_{l-1},$$

are uniformly quasihyperbolic (and even hyperbolic) with respect to N. Hence we can apply Proposition 3.1.2.

The fundamental solution E_λ of the operator $Q_\lambda(\partial)^{|\alpha|+l}$ is given by the distribution

$$\langle \phi, E_\lambda \rangle = \frac{1}{(|\alpha|+l-1)!} \int_0^\infty \phi(A\lambda t)e^{-t\lambda \cdot d}t^{|\alpha|+l-1}\, dt, \qquad \phi \in \mathcal{D}(\mathbf{R}^n),$$

see (2.5.2). Therefore, formula (3.1.4) in Proposition 3.1.2 yields the following representation for the fundamental solution E of $R(\partial)$:

$$\langle \phi, E \rangle = \frac{1}{\alpha!} \int_{\Sigma_{l-1}} \lambda^\alpha \int_0^\infty t^{|\alpha|+l-1}e^{-t\lambda \cdot d}\phi(A\lambda t)\, dt d\sigma(\lambda)$$

$$= \frac{1}{\alpha!}\int_0^\infty \cdots \int_0^\infty y^\alpha e^{-y \cdot d}\phi(Ay)\, dy_1 \ldots dy_l. \qquad (3.1.8)$$

(In (3.1.8), the substitution $y_1 = t\lambda_1, \ldots, y_{l-1} = t\lambda_{l-1}, y_l = t(1 - \lambda_1 - \cdots - \lambda_{l-1})$, $dy = t^{l-1}dt d\sigma(\lambda)$ was used.) For the case of $\alpha = 0$, $l \le n$, cf. Garnir [97], p. 97.

(b) Let us consider in particular the case of $l \le n$, rank $A = l$, and describe E in a more explicit manner.

If V is the subspace of $\mathbf{R}^n$ spanned by the l columns $A_j = (a_{1j}, \ldots, a_{nj})^T$, $j = 1, \ldots, l$, of A, and $V^\perp$ denotes the orthogonal complement of V in $\mathbf{R}^n$, if C is the cone in V spanned by $A_1, \ldots, A_l$, $d\tau(x)$ the Euclidean measure on V induced from $\mathbf{R}^n$ and h is the vector space isomorphism

$$h : \mathbf{R}^l \longrightarrow V : y \longmapsto Ay,$$

then

$$\langle \phi, E \rangle = \frac{1}{\sqrt{\det(A^TA)} \cdot \alpha!} \int_C h^{-1}(x)^\alpha\, e^{-d \cdot h^{-1}(x)}\phi(x)\, d\tau(x),$$

or, equivalently,

$$E = \frac{\chi_C(x)h^{-1}(x)^\alpha\, e^{-d \cdot h^{-1}(x)}}{\sqrt{\det(A^TA)} \cdot \alpha!} \otimes \delta_{V^\perp},$$

where χ_C is the characteristic function of C and E is represented as the tensor product of a locally integrable function with the Dirac measure at 0 in $V^\perp$.

In particular, for $l = n$, we obtain

$$E(x) = \frac{\chi_C(x)h^{-1}(x)^\alpha\, e^{-d \cdot h^{-1}(x)}}{|\det A| \cdot \alpha!} \in L^1_{\mathrm{loc}}(\mathbf{R}^n)$$

as fundamental solution of $R(\partial) = \prod_{j=1}^{n}(d_j + \sum_{i=1}^{n} a_{ij}\partial_i)^{\alpha_j+1}$ for $A \in$ $\mathrm{Gl}_n(\mathbf{R})$, $d \in \mathbf{C}^n$. The two-dimensional iterated wave operator $(\partial_1^2 - \partial_2^2)^m$ considered in Example 1.5.5 is contained herein as the special case of $d = 0$, $A = \begin{pmatrix} 1 & 1 \\ 1 & -1 \end{pmatrix}$, $\alpha = (m-1, m-1)$, and $C = \{x \in \mathbf{R}^2; x_1 \geq |x_2|\}$.

We shall consider the remaining case $l > n$ in Example 3.4.5 below.   □

Example 3.1.4 Let us generalize now Example 2.5.5 and consider *products of heat or Schrödinger operators*, i.e.,

$$R(\partial) = \prod_{j=1}^{l}\left(\partial_t - \nabla^T A_j \nabla - d_j + \sum_{k=1}^{n} b_{jk}\partial_k\right)^{\alpha_j+1},$$

where the matrices $A_j \in \mathbf{C}^{n\times n}$, $j = 1,\dots,n$, are symmetric and have positive semi-definite real parts, $B = (b_{jk}) \in \mathbf{C}^{l\times n}$, $d = (d_j) \in \mathbf{C}^l$, and the condition

$$\forall \lambda \in \Sigma_{l-1} : \det\left(\sum_{j=1}^{l}\lambda_j A_j\right) \neq 0 \tag{3.1.9}$$

is satisfied.

Then $R(\partial)$ is quasihyperbolic with respect to t and its uniquely determined fundamental solution E with supp $E = 0$ for $t < 0$ and at most exponential growth (see Proposition 2.4.13) can be represented by formula (3.1.4) in Proposition 3.1.2. In fact, if

$$P_j(\partial) = \partial_t - \nabla^T A_j \nabla - d_j + \sum_{k=1}^{n} b_{jk}\partial_k,$$

then the set $\{Q_\lambda(\partial) := \sum_{j=1}^{l}\lambda_j P_j(\partial); \lambda \in \Sigma_{l-1}\}$ is uniformly quasihyperbolic since

$$Q_\lambda(\partial) = \sum_{j=1}^{l}\lambda_j P_j(\partial) = \partial_t - \nabla^T A(\lambda)\nabla - d(\lambda) + \sum_{k=1}^{n} b_k(\lambda)\partial_k,$$

where $A(\lambda) = \sum_{j=1}^{l}\lambda_j A_j$, $b_k(\lambda) = \sum_{j=1}^{l}\lambda_j b_{jk}$, $d(\lambda) = \sum_{j=1}^{l}\lambda_j d_j$, and $\operatorname{Re} A(\lambda)$ is positive semi-definite.

Hence E is given by formula (3.1.4):

$$E = \frac{(|\alpha| + l - 1)!}{\alpha!}\int_{\Sigma_{l-1}} E_\lambda \lambda^\alpha \mathrm{d}\sigma(\lambda), \tag{3.1.4}$$

where E_λ is the fundamental solution of $Q_\lambda(\partial)^{|\alpha|+l}$, i.e.,

$$E_\lambda = \frac{Y(t)t^{|\alpha|+l-n/2-1}\exp([d(\lambda)-\frac{1}{4}b(\lambda)^T A(\lambda)^{-1}b(\lambda)]t)}{(4\pi)^{n/2}\sqrt{\det A(\lambda)}(|\alpha|+l-1)!}$$

$$\times \exp\Big(-\frac{1}{4t}x^T A(\lambda)^{-1}x + \frac{1}{2}b(\lambda)^T A(\lambda)^{-1}x\Big) \in \mathcal{C}\big([0,\infty),\mathcal{D}'(\mathbf{R}_x^n)\big),$$

see Examples 2.5.5, 1.4.13 and 1.6.14 for the definition of $\sqrt{\det A(\lambda)}$. Note that E_λ is locally integrable if $|\alpha|+l > \frac{n}{2}$ or $\operatorname{Re} A(\lambda)$ is positive definite. Similarly, $E \in L^1_{\text{loc}}(\mathbf{R}^n)$ holds if one of the conditions

$$\text{(i)}\ |\alpha|+l > \frac{n}{2} \quad \text{or (ii)}\ \forall j=1,\dots,l : \operatorname{Re} A_j \text{ is positive definite} \qquad (3.1.10)$$

is satisfied.

The representation of E in (3.1.4) was given first in Ortner and Wagner [211], Prop. 6, p. 318. It also exhibits the support, the singular support and the analytic singular support of E, namely $\operatorname{supp} E = \{(t,x);\ t \geq 0\}$,

$$\operatorname{sing\,supp} E = \begin{cases} \{0\},\ \text{i.e.,}\ P(\partial)\ \text{hypoelliptic} & : \text{(ii) in (3.1.10) holds,} \\ \{(0,x);\ x \in \mathbf{R}^n\} & : \text{else,} \end{cases}$$

and $\operatorname{sing\,supp}_A E = \{(0,x);\ x \in \mathbf{R}^n\}$.

For $l = 2$, the definite integral in (3.1.4) is a simple one. Hence the fundamental solution of a product of *two* anisotropic heat or Schrödinger operators can be represented as a simple integral over elementary functions. The fundamental solution of the more general operator

$$\Big(\partial_t - \nabla^T A_1 \nabla - d_1 + \sum_{k=1}^n b_{1k}\partial_k\Big)\Big(\partial_t - \nabla^T A_2 \nabla - d_2 + \sum_{k=1}^n b_{2k}\partial_k\Big) - h^2$$

is investigated in Ortner and Wagner [207], Prop. 4, p. 450, and Remark 4, p. 452. We will come back to operators of this kind in Sect. 3.5.   □

Let us now apply the method of parameter integration to products of operators which are not necessarily quasihyperbolic.

Proposition 3.1.5 *Suppose that* $P_j(\partial)$, $j = 1,\dots,l$, *are differential operators in* $\mathbf{R}^n$ *such that the symbols of* $Q_\lambda(\partial) := \sum_{j=1}^l \lambda_j P_j(\partial)$, $\lambda \in \Sigma_{l-1}$, *do not vanish, i.e.,*

$$\forall \lambda \in \Sigma_{l-1} : \forall \xi \in \mathbf{R}^n : Q_\lambda(i\xi) \neq 0.$$

Let E and E_λ, respectively, denote the uniquely determined temperate fundamental solutions of $\prod_{j=1}^{l} P_j(\partial)^{\alpha_j+1}$ and of $Q_\lambda(\partial)^{|\alpha|+l}$. (By Proposition 2.4.1, E and E_λ even belong to the space $\mathcal{O}'_C(\mathbf{R}^n)$.)

Then the parameter integration formula holds:

$$E = \frac{(|\alpha| + l - 1)!}{\alpha!} \int_{\Sigma_{l-1}} E_\lambda \, \lambda^\alpha \mathrm{d}\sigma(\lambda). \tag{3.1.4}$$

Proof The proof proceeds in literally the same way as the proof of Proposition 3.1.2 if we set therein $\sigma = 0$. □

Example 3.1.6 Let us consider the following *product of anisotropic metaharmonic operators*:

$$R(\partial) = \prod_{j=1}^{l} \left(\nabla^T A_j \nabla - d_j + \sum_{k=1}^{n} b_{jk} \partial_k \right)^{\alpha_j+1}. \tag{3.1.11}$$

Here we suppose that $\alpha \in \mathbf{N}_0^l$ and $A_j \in \mathbf{R}^{n \times n}$ are real-valued symmetric positive definite matrices. This implies that their convex combinations $A(\lambda) = \sum_{j=1}^{l} \lambda_j A_j$, $\lambda \in \Sigma_{l-1}$, are also positive definite. Furthermore, $b_{jk}, d_j \in \mathbf{C}$ must fulfill the condition

$$\forall \lambda \in \Sigma_{l-1} : \mu(\lambda) := d(\lambda) + \frac{1}{4} b(\lambda)^T A(\lambda)^{-1} b(\lambda) \in \mathbf{C} \setminus (-\infty, 0]$$

if $b_k(\lambda) = \sum_{j=1}^{l} \lambda_j b_{jk}$ and $d(\lambda) = \sum_{j=1}^{l} \lambda_j d_j$.

Then the temperate fundamental solutions E_λ of

$$Q_\lambda(\partial)^{|\alpha|+l} = \left(\nabla^T A(\lambda) \nabla - d(\lambda) + \sum_{k=1}^{n} b_k(\lambda) \partial_k \right)^{|\alpha|+l}, \quad \lambda \in \Sigma_{l-1},$$

are given by formula (2.5.3) in Example 2.5.4. Therefore, Proposition 3.1.5 yields for the temperate fundamental solution E of $R(\partial)$ in (3.1.11) the integral representation

$$E = \frac{(-1)^{|\alpha|+l}}{2^{n/2+|\alpha|+l-1} \alpha! \pi^{n/2}} \int_{\Sigma_{l-1}} \frac{\lambda^\alpha \, \exp(-\frac{1}{2} b(\lambda)^T A(\lambda)^{-1} x)}{\sqrt{\det A(\lambda)}}$$

$$\times \left(\frac{x^T A(\lambda)^{-1} x}{\mu(\lambda)} \right)^{(|\alpha|+l)/2 - n/4} \cdot K_{n/2-|\alpha|-l}\left(\sqrt{\mu(\lambda) \cdot x^T A(\lambda)^{-1} x} \right) \mathrm{d}\sigma(\lambda).$$

Let us specialize this formula for the case of two factors, i.e., $l = 2$. Here we shall assume that

$$R(\partial) = (\Delta_n - d_1)\Big(\sum_{k=1}^{n} a_k\partial_k^2 - d_2\Big), \qquad a_k > 0, \ d_1, d_2 \in \mathbf{C},$$

with $d(\lambda) = \lambda d_1 + (1 - \lambda)d_2 \in \mathbf{C} \setminus (-\infty, 0]$ for $0 \le \lambda \le 1$. Then the temperate fundamental solution of $R(\partial)$ is a locally integrable function given by

$$E = \frac{1}{2(2\pi)^{n/2}} \int_0^1 \Big(\frac{1}{\lambda d_1 + (1 - \lambda)d_2} \sum_{k=1}^{n} \frac{x_k^2}{\lambda + (1 - \lambda)a_k}\Big)^{-n/4+1}$$

$$K_{n/2-2}\Big(\sqrt{\big(\lambda d_1 + (1 - \lambda)d_2\big) \sum_{k=1}^{n} \frac{x_k^2}{\lambda + (1-\lambda)a_k}}\Big) \prod_{k=1}^{n}\big(\lambda + (1 - \lambda)a_k\big)^{-1/2} \, \mathrm{d}\lambda.$$

$$(3.1.12)$$

Formula (3.1.12) coincides with the one given in Garnir [100], 4., p. 1132, if the substitution $\mu = \lambda/(1 - \lambda)$ is performed. In Garnir [100], this formula is verified by differentiation.

In particular, if $1 \le n \le 3$, then E is continuous at 0 and

$$E(0) = \frac{\Gamma(2 - \frac{n}{2})}{(4\pi)^{n/2}} \int_0^1 \big(\lambda d_1 + (1 - \lambda)d_2\big)^{n/2-2} \prod_{k=1}^{n}\big(\lambda + (1 - \lambda)a_k\big)^{-1/2} \, \mathrm{d}\lambda.$$

Hence $E(0)$ can be expressed by elementary functions for $n = 1, 2$, whereas, for $n = 3$, it is an elliptic integral in the generic case of pairwise different a_1, a_2, a_3.

Let us also mention that, by limit considerations, (3.1.12) remains valid and yields a locally integrable fundamental solution in the case $d_1 = 0$ and $d_2 \in \mathbf{C} \setminus (-\infty, 0]$.   □

Let us next apply the method of parameter integration to products of *homogeneous elliptic operators*.

Proposition 3.1.7 *Suppose that $P_j(\partial)$, $j = 1, \dots, l$, are differential operators in $\mathbf{R}^n$ which are homogeneous of degree m and such that each operator*

$$Q_\lambda(\partial) = \sum_{j=1}^{l} \lambda_j P_j(\partial), \qquad \lambda \in \Sigma_{l-1},$$

is elliptic. Let $\alpha \in \mathbf{N}_0^l$ and assume that E_λ are fundamental solutions of $Q_\lambda(\partial)^{|\alpha|+l}$ which depend continuously on $\lambda \in \Sigma_{l-1}$ and satisfy the estimate

$$\forall \lambda \in \Sigma_{l-1} : \exists C > 0 : \forall x \in \mathbf{R}^n \text{ with } |x| \ge 2 : |E_\lambda(x)| \le C|x|^{m(|\alpha|+l)-n} \log |x|.$$

$$(3.1.13)$$

Then the parameter integration formula

$$E = \frac{(|\alpha| + l - 1)!}{\alpha!} \int_{\Sigma_{l-1}} E_\lambda \, \lambda^\alpha \mathrm{d}\sigma(\lambda). \tag{3.1.4}$$

yields a fundamental solution E of $R(\partial) = \prod_{j=1}^{l} P_j(\partial)^{\alpha_j + 1}$ which grows at most as a constant multiple of $|x|^{m(|\alpha|+l)-n} \log |x|$ for $|x| \to \infty$.

Proof Upon applying the Fourier transform to the equation $Q_\lambda(\partial)^{|\alpha|+l} E_\lambda = \delta$ we obtain $Q_\lambda(\mathrm{i}\xi)^{|\alpha|+l} \mathcal{F}E_\lambda = 1$ and hence $\mathcal{F}E_\lambda$ coincides with $Q_\lambda(\mathrm{i}\xi)^{-|\alpha|-l}$ for $\xi \neq 0$.

We note that $F_\lambda(\omega) := Q_\lambda(\mathrm{i}\omega)^{-|\alpha|-l} \in C^\infty(S^{n-1})$ and that, by Example 1.4.9, the map $\mu \mapsto F_\lambda \cdot |\xi|^\mu \in \mathcal{S}'(\mathbf{R}^n)$ is meromorphic with at most simple poles in $-n - \mathbf{N}_0$. Therefore,

$$\tilde{E}_\lambda := \mathcal{F}^{-1}\left(\Pf_{\mu=-m(|\alpha|+l)} F_\lambda \cdot |\xi|^\mu \right)$$

is a fundamental solution of $Q_\lambda(\partial)^{|\alpha|+l}$ which is associated homogeneous of degree $m(|\alpha| + l) - n$. Hence, due to Proposition 2.4.8 and the estimate (3.1.13), $E_\lambda = \tilde{E}_\lambda + q_\lambda$ where q_λ is a polynomial of degree at most $m(|\alpha| + l) - n$.

Since E_λ and $\tilde{E}_\lambda$ continuously depend on $\lambda \in \Sigma_{l-1}$, the same holds for the polynomials q_λ. If E is defined by (3.1.4), we thus obtain, with another polynomial q of degree $\leq m(|\alpha| + l) - n$, the following:

$$
\begin{aligned}
\mathcal{F}(E - q) &= \frac{(|\alpha| + l - 1)!}{\alpha!} \int_{\Sigma_{l-1}} \mathcal{F}(\tilde{E}_\lambda) \, \lambda^\alpha \mathrm{d}\sigma(\lambda) \\
&= \frac{(|\alpha| + l - 1)!}{\alpha!} \int_{\Sigma_{l-1}} \Pf_{\mu=-m(|\alpha|+l)} \left(F_\lambda \cdot |\xi|^\mu \right) \lambda^\alpha \mathrm{d}\sigma(\lambda) \\
&= \frac{(|\alpha| + l - 1)!}{\alpha!} \Pf_{\mu=-m(|\alpha|+l)} \left(\left(\int_{\Sigma_{l-1}} Q_\lambda(\mathrm{i}\omega)^{-|\alpha|-l} \lambda^\alpha \mathrm{d}\sigma(\lambda) \right) \cdot |\xi|^\mu \right) \\
&= \Pf_{\mu=-m(|\alpha|+l)} \left(R(\mathrm{i}\omega)^{-1} \cdot |\xi|^\mu \right)
\end{aligned}
$$

by Feynman's formula (3.1.5). This implies $R(\partial)E = \delta$. Furthermore, E has the required growth by construction. $\qquad\square$

Example 3.1.8

(a) Similarly as in Example 3.1.6, let us consider here *products of anisotropic Laplaceans*, i.e., $R(\partial) = \prod_{j=1}^{l} (\nabla^T A_j \nabla)^{1+\alpha_j}$. As before, $\alpha \in \mathbf{N}_0^l$ and $A_j \in \mathbf{R}^{n \times n}$ are real-valued symmetric positive definite matrices, and we set

$$Q_\lambda = \nabla^T A(\lambda) \nabla, \qquad A(\lambda) = \sum_{j=1}^{l} \lambda_j A_j.$$

For the fundamental solutions E_λ of $Q_\lambda(\partial)^{|\alpha|+l}$, we obtain from Example 1.6.11 (b) and Proposition 1.3.19

$$E_\lambda = \frac{(x^T A(\lambda)^{-1}x)^{|\alpha|+l-n/2}}{2^{2(|\alpha|+l)}(|\alpha|+l-1)!\pi^{n/2}\sqrt{\det A(\lambda)}}$$

$$\times \begin{cases} (-1)^{|\alpha|+l}\Gamma(\frac{n}{2}-|\alpha|-l) & : n \text{ odd or } |\alpha|+l < \frac{n}{2}, \\[2mm] \dfrac{(-1)^{n/2-1}\log(x^T A(\lambda)^{-1}x)}{(|\alpha|+l-\frac{n}{2})!} & : n \text{ even and } |\alpha|+l \geq \frac{n}{2}. \end{cases}$$

Herewith, Proposition 3.1.7 furnishes

$$E = \frac{(|\alpha|+l-1)!}{\alpha!}\int_{\Sigma_{l-1}} E_\lambda \lambda^\alpha d\sigma(\lambda)$$

$$= \frac{(-1)^{|\alpha|+l}\Gamma(\frac{n}{2}-|\alpha|-l)}{2^{2(|\alpha|+l)}\alpha!\pi^{n/2}}\int_{\Sigma_{l-1}}\frac{\lambda^\alpha(x^T A(\lambda)^{-1}x)^{|\alpha|+l-n/2}}{\sqrt{\det A(\lambda)}} d\sigma(\lambda)$$

if n is odd or $|\alpha|+l < \frac{n}{2}$, and

$$E = \frac{(-1)^{n/2-1}}{2^{2(|\alpha|+l)}\alpha!(|\alpha|+l-\frac{n}{2})!\pi^{n/2}}\int_{\Sigma_{l-1}}\frac{\lambda^\alpha(x^T A(\lambda)^{-1}x)^{|\alpha|+l-n/2}}{\sqrt{\det A(\lambda)}}$$

$$\times \log(x^T A(\lambda)^{-1}x)\, d\sigma(\lambda)$$

if n is even and $|\alpha|+l \geq \frac{n}{2}$.

In particular, for $l = 2$ and $\alpha = 0$, we obtain the following fundamental solution E of the operator $R(\partial) = (\nabla^T A_1 \nabla)(\nabla^T A_2 \nabla)$:

$$E = \begin{cases} \dfrac{\Gamma(\frac{n}{2}-2)}{16\pi^{n/2}}\displaystyle\int_0^1 \frac{(x^T A(\lambda)^{-1}x)^{2-n/2}}{\sqrt{\det A(\lambda)}} d\lambda & : n \neq 2,4, \\[4mm] \dfrac{(-1)^{n/2-1}}{16\pi^{n/2}}\displaystyle\int_0^1 \frac{(x^T A(\lambda)^{-1}x)^{2-n/2}}{\sqrt{\det A(\lambda)}} \log(x^T A(\lambda)^{-1}x)\, d\lambda & : n = 2,4, \end{cases}$$

(3.1.14)

where $A(\lambda) = \lambda A_1 + (1-\lambda)A_2$.

Finally, let us evaluate formula (3.1.14) in the cases $n = 2$ and $n = 3$, which can be found in the literature.

(b) For $n = 2$, we shall calculate a fundamental solution of $R(\partial) = \Delta_2(a_1\partial_1^2 + a_2\partial_2^2)$, $a_1 > 0$, $a_2 > 0$, $a_1 \neq a_2$. Up to linear transformations, $R(\partial)$ represents the general case of a product of two anisotropic Laplaceans in $\mathbf{R}^2$.

Omitting a quadratic polynomial in $x = (x_1, x_2)$ (which is a solution of the homogeneous equation), formula (3.1.14) yields with $A_1 = I_2$ and $A_2 =$

$\begin{pmatrix} a_1 & 0 \\ 0 & a_2 \end{pmatrix}$ the following representation for a fundamental solution of $R(\partial)$:
$E = E_1 + E_2$, where

$$E_1 = \frac{x_1^2}{16\pi} \int_0^1 \frac{\log(x_1^2[\lambda + (1-\lambda)a_2] + x_2^2[\lambda + (1-\lambda)a_1])}{[\lambda + (1-\lambda)a_2]^{1/2}[\lambda + (1-\lambda)a_1]^{3/2}} \, d\lambda$$

and E_2 is derived from E_1 by interchanging the roles of x_1, x_2 and of a_1, a_2, respectively.

Let us evaluate the last integral by partial integration and employing the formula

$$\int \frac{d\lambda}{[\alpha\lambda + \beta]^{1/2}[\gamma\lambda + \delta]^{3/2}} = \frac{2}{\alpha\delta - \beta\gamma} \sqrt{\frac{\alpha\lambda + \beta}{\gamma\lambda + \delta}} + C.$$

This implies

$$E_1 = \frac{x_1^2}{8\pi(a_1 - a_2)} \sqrt{\frac{\lambda + (1-\lambda)a_2}{\lambda + (1-\lambda)a_1}}$$

$$\times \log \left(x_1^2[\lambda + (1-\lambda)a_2] + x_2^2[\lambda + (1-\lambda)a_1] \right)\Big|_{\lambda=0}^1$$

$$- \frac{x_1^2}{8\pi(a_1 - a_2)} \int_0^1 \sqrt{\frac{\lambda + (1-\lambda)a_2}{\lambda + (1-\lambda)a_1}}$$

$$\times \frac{x_1^2(1 - a_2) + x_2^2(1 - a_1)}{x_1^2[\lambda + (1-\lambda)a_2] + x_2^2[\lambda + (1-\lambda)a_1]} \, d\lambda$$

$$= \frac{x_1^2}{8\pi(a_1 - a_2)} \left\{ \log|x|^2 - \sqrt{\frac{a_2}{a_1}} \log(a_2 x_1^2 + a_1 x_2^2) \right\}$$

$$+ \frac{x_1^2[x_1^2(1 - a_2) + x_2^2(1 - a_1)]}{8\pi(a_2 - a_1)} \cdot I_1,$$

where

$$I_1 = \int_0^1 \frac{1}{\sqrt{[\lambda + (1-\lambda)a_1][\lambda + (1-\lambda)a_2]}} \cdot \frac{d\lambda}{x_1^2 + u^2 x_2^2},$$

and $u^2 = \dfrac{\lambda + (1-\lambda)a_1}{\lambda + (1-\lambda)a_2}$, $\lambda = \dfrac{a_1 - a_2 u^2}{(1 - a_2)u^2 - (1 - a_1)}$.

Upon substituting λ by u we obtain

$$I_1 = -2 \int_{\sqrt{a_1/a_2}}^{1} \frac{du}{(x_1^2 + u^2 x_2^2)[(1-a_2)u^2 - (1-a_1)]}$$

$$= \frac{2}{x_1^2(1-a_2) + x_2^2(1-a_1)} \int_{\sqrt{a_1/a_2}}^{1} \left[\frac{x_2^2}{x_1^2 + u^2 x_2^2} - \frac{1-a_2}{(1-a_2)u^2 - (1-a_1)} \right] du.$$

Hence we have, up to second-order polynomials,

$$E = F + \frac{x_1^2 - x_2^2}{8\pi(a_1 - a_2)} \log |x|^2 - \frac{a_2 x_1^2 - a_1 x_2^2}{8\pi \sqrt{a_1 a_2}(a_1 - a_2)} \log(a_2 x_1^2 + a_1 x_2^2),$$

where

$$F = \frac{x_1^2 x_2^2}{4\pi(a_2 - a_1)} \left[\int_{\sqrt{a_1/a_2}}^{1} \frac{du}{x_1^2 + u^2 x_2^2} - \int_{\sqrt{a_2/a_1}}^{1} \frac{du}{x_2^2 + u^2 x_1^2} \right]$$

$$= \frac{x_1^2 x_2^2}{2\pi(a_2 - a_1)} \int_{\sqrt{a_1/a_2}}^{1} \frac{du}{x_1^2 + u^2 x_2^2}.$$

(In the second integral, we have used the substitution $v = \frac{1}{u}$.) By means of the addition theorem of the arctangent, we finally obtain

$$F = \frac{x_1 x_2}{2\pi(a_1 - a_2)} \arctan\left(\frac{(\sqrt{a_1} - \sqrt{a_2})x_1 x_2}{\sqrt{a_2}x_1^2 + \sqrt{a_1}x_2^2} \right).$$

Hence we conclude that the operator $\Delta_2(a_1\partial_1^2 + a_2\partial_2^2)$, $a_1 > 0$, $a_2 > 0$, $a_1 \neq a_2$, has the fundamental solution

$$E = \frac{x_1^2 - x_2^2}{8\pi(a_1 - a_2)} \log |x|^2 - \frac{a_2 x_1^2 - a_1 x_2^2}{8\pi \sqrt{a_1 a_2}(a_1 - a_2)} \log(a_2 x_1^2 + a_1 x_2^2)$$

$$+ \frac{x_1 x_2}{2\pi(a_1 - a_2)} \arctan\left(\frac{(\sqrt{a_1} - \sqrt{a_2})x_1 x_2}{\sqrt{a_2}x_1^2 + \sqrt{a_1}x_2^2} \right). \qquad (3.1.15)$$

By linear transformations (see Sect. 2.5), the result in (3.1.15) is equivalent to the formula for a fundamental solution of the operator $(a^2\partial_1^2 + \partial_2^2)(b^2\partial_1^2 + \partial_2^2)$ given in Galler [86], p. 49. Similarly, $\Delta_2(a_1\partial_1^2 + a_2\partial_2^2)$ is also linearly equivalent to the operator $\partial_1^4 + \partial_2^4 + \gamma\partial_1^2\partial_2^2$, which describes deflections of elastic orthotropic plates. Its fundamental solution can be found in P. Stein [262], Eq. (B9), p. 11; Ortner [200], p. 140; Wagner [285], p. 44. Below, we shall also deduce (3.1.15) as a special case of Somigliana's formula, which refers to arbitrary homogeneous operators in the plane, i.e., to $\prod_{j=1}^{l}(\partial_1 - \lambda_j \partial_2)^{\alpha_j+1}$, $\lambda_j \in$ $\mathbf{C}$, $\alpha \in \mathbf{N}_0^l$, see Proposition 3.3.2 and Example 3.3.3 below.

(c) Let us yet evaluate formula (3.1.14) for $n = 3$. Without restriction of generality, we can suppose that A_1, A_2 are diagonal matrices with the positive entries a_j and b_j, respectively, in the diagonal. We assume, furthermore, that $\frac{b_1}{a_1} > \frac{b_2}{a_2} > \frac{b_3}{a_3}$. Upon setting $c_j(\lambda) = a_j\lambda + b_j(1 - \lambda)$, the fundamental solution E in (3.1.14) of the operator

$$R(\partial) = (a_1\partial_1^2 + a_2\partial_2^2 + a_3\partial_3^2)(b_1\partial_1^2 + b_2\partial_2^2 + b_3\partial_3^2)$$

assumes the form

$$E = -\frac{1}{8\pi}\int_0^1 \sqrt{\sum_{j=1}^{3}\frac{x_j^2}{c_j(\lambda)}} \cdot \frac{d\lambda}{\sqrt{c_1(\lambda)c_2(\lambda)c_3(\lambda)}}$$

$$= -\frac{1}{8\pi}\sum_{j=1}^{3}x_j^2\int_0^1 \frac{d\lambda}{c_j(\lambda)\sqrt{x_1^2 c_2(\lambda)c_3(\lambda) + x_2^2 c_1(\lambda)c_3(\lambda) + x_3^2 c_1(\lambda)c_2(\lambda)}}$$

$$= -\frac{1}{8\pi}\sum_{j=1}^{3}x_j^2\int_0^\infty \frac{d\mu}{(a_j\mu + b_j)\sqrt{\mu^2 f + \mu h + g}}, \qquad (3.1.16)$$

where we have used the substitution $\mu = \frac{\lambda}{1-\lambda}$ and the notations

$$f = a_1 a_2 a_3 \sum_{j=1}^{3}\frac{x_j^2}{a_j}, \quad g = b_1 b_2 b_3 \sum_{j=1}^{3}\frac{x_j^2}{b_j},$$

$$h = x_1^2(a_2 b_3 + a_3 b_2) + x_2^2(a_1 b_3 + a_3 b_1) + x_3^2(a_1 b_2 + a_2 b_1),$$

cf. Bureau [35], p. 474.

By Gröbner and Hofreiter [115], Eq. 213.6a, the integrals in (3.1.16) yield logarithms or arctangents in dependence on the signs of $b_j^2 f - a_j b_j h + a_j^2 g$, $j = 1, 2, 3$, respectively. With the abbreviation

$$\gamma_j = (a_j b_i - b_j a_i)(a_j b_k - b_j a_k), \qquad \{i, j, k\} = \{1, 2, 3\},$$

we obtain

$$b_j^2 f - a_j b_j h + a_j^2 g = \gamma_j x_j^2. \qquad (3.1.17)$$

From the inequalities $\frac{b_1}{a_1} > \frac{b_2}{a_2} > \frac{b_3}{a_3}$, we infer $\gamma_1 > 0$, $\gamma_2 < 0$, $\gamma_3 > 0$, and this furnishes $E = \sum_{j=1}^{3} E_j$ with

$$
E_j = \begin{cases}
-\dfrac{|x_j|}{8\pi\sqrt{\gamma_j}} \log\left(\dfrac{a_j[|x_j|\sqrt{\gamma_j g} + a_j g - \frac{1}{2}b_j h]}{b_j[|x_j|\sqrt{\gamma_j f} + \frac{1}{2}a_j h - b_j f]}\right) & : j = 1, 3, \\[3mm]
-\dfrac{x_2}{8\pi\sqrt{-\gamma_2}} \arctan\left(\dfrac{x_2\sqrt{-\gamma_2}(b_2\sqrt{f} + a_2\sqrt{g})}{a_2 b_2(\sqrt{fg} - \frac{1}{2}h) + b_2^2 f + a_2^2 g}\right) & : j = 2.
\end{cases}
$$

$$(3.1.18)$$

Formula (3.1.18) coincides, up to some signs, with the formulas given in Garnir [100], 8., p. 1139, and in Bureau [35], (2), p. 474.

A more symmetric representation of E can be reached by employing Eq. (3.1.17) in (3.1.18). This yields

$$
E = -\frac{1}{8\pi}\left[\sum_{j=1,3}\frac{x_j}{\sqrt{\gamma_j}}\log\left(\frac{a_j\sqrt{g} + b_j\sqrt{f} + \sqrt{\gamma_j}x_j}{a_j\sqrt{g} + b_j\sqrt{f} - \sqrt{\gamma_j}x_j}\right)\right.
$$
$$
\left. + \frac{2x_2}{\sqrt{-\gamma_2}}\arctan\left(\frac{\sqrt{-\gamma_2}x_2}{a_2\sqrt{g} + b_2\sqrt{f}}\right)\right]. \qquad (3.1.19)
$$

Formula (3.1.19) was given first in Herglotz [125], III, (40), p. 81 (where $e_\alpha = a_\alpha = \frac{1}{b_\alpha}$, $c_\alpha = \sqrt{-\gamma_\alpha}$), and it is a particular case of the fundamental solution derived for certain homogeneous elliptic quartic operators in Wagner [292], see p. 1205.   □

3.2 Products of Operators Belonging to a One-Dimensional Affine Subspace

We shall next show that the parameter integration formula (3.1.4) in Proposition 3.1.2 reduces to a formula containing only *one-dimensional* parameter integrals for operators of the form

$$
R(\partial) = \prod_{j=1}^{l}\left(P_1(\partial) + \lambda_j P_2(\partial)\right)^{\alpha_j+1}, \qquad \alpha \in \mathbf{N}_0^l, \ \lambda \in \mathbf{R}^l,
$$

i.e., when all the factors $P_1(\partial) + \lambda_j P_2(\partial)$ of $R(\partial)$ are contained in an affine line of the vector space $\mathbf{C}[\partial]$ of all linear constant coefficient operators. For $\alpha = 0$, this formula was derived in Wagner [285], p. 18; for general α, see Ortner and Wagner [211], Prop. 2, p. 308; Ortner [202], Prop. 5, p. 92.

Proposition 3.2.1 *Suppose that the operators* $Q_\lambda(\partial) = P_1(\partial) + \lambda P_2(\partial)$, $\lambda \in$ *$[c, d] \subset \mathbf{R}$, are uniformly quasihyperbolic with respect to $N \in \mathbf{R}^n \setminus \{0\}$. Let* $\lambda_0, \ldots, \lambda_l \in [c, d]$ *and such that* $\lambda_1, \ldots, \lambda_l$ *are pairwise different. Let* $\sigma_0 \in \mathbf{R}$ *be as in* (3.1.3) *with* $\Lambda = [c, d]$, $\alpha \in \mathbf{N}_0^l$, *and denote by E and by E_λ, respectively, the fundamental solutions of*

$$R(\partial) = \prod_{j=1}^{l} (P_1(\partial) + \lambda_j P_2(\partial))^{\alpha_j + 1} \quad \text{and of} \quad Q_\lambda(\partial)^{|\alpha| + l},$$

respectively, which satisfy $e^{-\sigma x N} E \in \mathcal{S}'(\mathbf{R}^n)$ *and* $e^{-\sigma x N} E_\lambda \in \mathcal{S}'(\mathbf{R}^n)$ *for* $\sigma > \sigma_0$. *Then E_λ continuously depends on* $\lambda \in [c, d]$ *and we have*

$$E = (|\alpha| + l - 1) \sum_{j=1}^{l} \frac{1}{\alpha_j!} \left(\frac{\partial}{\partial \lambda_j} \right)^{\alpha_j} \left[c_j \int_{\lambda_0}^{\lambda_j} (\lambda_j - \lambda)^{|\alpha| + l - 2} E_\lambda \, d\lambda \right]$$

where $c_j := \prod_{\substack{k=1 \\ k \neq j}}^{l} (\lambda_j - \lambda_k)^{-\alpha_k - 1}$.

Proof

(1) Let us first show that

$$\prod_{j=1}^{l} (s + \lambda_j t)^{-\alpha_j - 1} = (|\alpha| + l - 1) \sum_{j=1}^{l} \frac{1}{\alpha_j!} \left(\frac{\partial}{\partial \lambda_j} \right)^{\alpha_j} \left[c_j \int_{\lambda_0}^{\lambda_j} \frac{(\lambda_j - \lambda)^{|\alpha| + l - 2}}{(s + \lambda t)^{|\alpha| + l}} \, d\lambda \right]$$
(3.2.1)

if $s, t \in \mathbf{C}$ such that $s + \lambda t \neq 0$ for $\lambda \in [c, d]$.

In fact, by Gröbner and Hofreiter [115], Eq. 421.4, p. 175,

$$\int_{\lambda_0}^{\lambda_j} \frac{(\lambda_j - \lambda)^{|\alpha| + l - 2}}{(s + \lambda t)^{|\alpha| + l}} \, d\lambda = \frac{(\lambda_j - \lambda_0)^{|\alpha| + l - 1}}{(|\alpha| + l - 1)(s + \lambda_j t)(s + \lambda_0 t)^{|\alpha| + l - 1}}$$

and hence Eq. (3.2.1) is equivalent to

$$(s + \lambda_0 t)^{|\alpha| + l - 1} \prod_{j=1}^{l} (s + \lambda_j t)^{-\alpha_j - 1} = \sum_{j=1}^{l} \frac{1}{\alpha_j!} \left(\frac{\partial}{\partial \lambda_j} \right)^{\alpha_j} \left[\frac{c_j (\lambda_j - \lambda_0)^{|\alpha| + l - 1}}{s + \lambda_j t} \right].$$
(3.2.2)

Setting $d_j = \prod_{\substack{k=1 \\ k \neq j}}^{l} (\lambda_j - \lambda_k)^{-1}$, we have $c_j = \left(\prod_{\substack{k=1 \\ k \neq j}}^{l} \alpha_k!^{-1} \partial_{\lambda_k}^{\alpha_k} \right) d_j$ and hence (3.2.2) is equivalent to

$$\partial_\lambda^\alpha \left[(s + \lambda_0 t)^{|\alpha| + l - 1} \prod_{j=1}^{l} (s + \lambda_j t)^{-1} \right] = (-t)^{|\alpha|} \partial_\lambda^\alpha \left[\sum_{j=1}^{l} \frac{d_j (\lambda_j - \lambda_0)^{|\alpha| + l - 1}}{s + \lambda_j t} \right].$$
(3.2.3)

We therefore consider the rational function $f(s)$ of s given by

$$f(s) := (s + \lambda_0 t)^{|\alpha|+l-1} \prod_{j=1}^{l} (s + \lambda_j t)^{-1} - (-t)^{|\alpha|} \sum_{j=1}^{l} \frac{d_j (\lambda_j - \lambda_0)^{|\alpha|+l-1}}{s + \lambda_j t}.$$

By the definition of d_j, all its residues $\mathrm{Res}_{s=-\lambda_j t} f(s)$ vanish, and thus f is a polynomial in s of degree at most $|\alpha| - 1$. Furthermore, $\partial_\lambda^\alpha f = 0$ because the coefficients of $f(s)$ are polynomials in λ of degree at most $|\alpha| - 1$.

(2) Due to the uniform quasihyperbolicity of the family of operators $Q_\lambda(\partial) = P_1(\partial) + \lambda P_2(\partial)$, $\lambda \in [c, d]$, we can insert $s = P_1(\sigma N + i\xi)$, $t = P_2(\sigma N + i\xi)$, $\sigma > \sigma_0$, $\xi \in \mathbf{R}^n$, into (3.2.1), and this yields

$$R(\sigma N + i\xi)^{-1} = \prod_{j=1}^{l} [P_1(\sigma N + i\xi) + \lambda_j P_2(\sigma N + i\xi)]^{-\alpha_j - 1}$$

$$= (|\alpha| + l - 1) \sum_{j=1}^{l} \frac{1}{\alpha_j!} \left(\frac{\partial}{\partial \lambda_j} \right)^{\alpha_j} \left[c_j \int_{\lambda_0}^{\lambda_j} \frac{(\lambda_j - \lambda)^{|\alpha|+l-2} \, d\lambda}{[P_1(\sigma N + i\xi) + \lambda P_2(\sigma N + i\xi)]^{|\alpha|+l}} \right].$$

$$(3.2.4)$$

Furthermore, by formula (2.4.13) in Proposition 2.4.13 applied to the fundamental solutions E of $R(\partial)$ and E_λ of $Q_\lambda(\partial)^{|\alpha|+l}$, respectively, we obtain

$$E = e^{\sigma N x} \mathcal{F}_\xi^{-1} \big(R(\sigma N + i\xi)^{-1} \big)$$

$$= (|\alpha| + l - 1) \sum_{j=1}^{l} \frac{1}{\alpha_j!} \left(\frac{\partial}{\partial \lambda_j} \right)^{\alpha_j} \left[c_j \int_{\lambda_0}^{\lambda_j} (\lambda_j - \lambda)^{|\alpha|+l-2} \right.$$

$$\left. \cdot e^{\sigma N x} \mathcal{F}_\xi^{-1} \big([P_1(\sigma N + i\xi) + \lambda P_2(\sigma N + i\xi)]^{-|\alpha|-l} \big) \, d\lambda \right]$$

$$= (|\alpha| + l - 1) \sum_{j=1}^{l} \frac{1}{\alpha_j!} \left(\frac{\partial}{\partial \lambda_j} \right)^{\alpha_j} \left[c_j \int_{\lambda_0}^{\lambda_j} (\lambda_j - \lambda)^{|\alpha|+l-2} E_\lambda \, d\lambda \right].$$

Here, as before, the interchange of the Fourier transform with the parametric differentiations and with the integration is justified by means of the Seidenberg–Tarski Lemma.   □

Example 3.2.2 We shall apply Proposition 3.2.1 to the construction of the forward fundamental solution E of the following *product $R(\partial)$ of wave operators*:

$$R(\partial) = \prod_{j=1}^{l} (\partial_t^2 - \lambda_j \Delta_n)^{\alpha_j + 1}, \quad \alpha \in \mathbf{N}_0^l, \ \lambda_j > 0, \ \lambda_j \neq \lambda_k \text{ for } j \neq k. \quad (3.2.5)$$

This basic operator was investigated for $\alpha = 0$ already in Herglotz [125], II, see (171), (175), p. 313, and later, for arbitrary α, in Gal'pern and Kondrashov [87] and in Bresters [25] with the goal of deducing regularity results for the fundamental solution, $R(\partial)$ being considered as a prototype of general hyperbolic operators.

If G denotes the fundamental solution of $(\partial_t^2 - \Delta_n)^{|\alpha|+l}$ with support in $t \geq 0$, then $(\partial_t^2 - \lambda \Delta_n)^{|\alpha|+l}$ has for $\lambda > 0$ the fundamental solution

$$E_\lambda = \lambda^{-n/2} G\left(t, \frac{x}{\sqrt{\lambda}}\right),$$

see Proposition 1.3.19. Therefore, Proposition 3.2.1 entails the formula

$$E = (|\alpha|+l-1) \sum_{j=1}^{l} \frac{1}{\alpha_j!} \left(\frac{\partial}{\partial \lambda_j}\right)^{\alpha_j} \left[c_j \int_0^{\lambda_j} (\lambda_j - \lambda)^{|\alpha|+l-2} G\left(t, \frac{x}{\sqrt{\lambda}}\right) \frac{d\lambda}{\lambda^{n/2}} \right] \quad (3.2.6)$$

for the fundamental solution E of $R(\partial)$ in (3.2.5). (Here, as before, we have set $c_j := \prod_{\substack{k=1 \\ k \neq j}}^{l} (\lambda_j - \lambda_k)^{-\alpha_k-1}$.)

Let us suppose now additionally that $|\alpha|+l > \frac{n-1}{2}$. Then $G \in L^1_{\text{loc}}(\mathbf{R}^{n+1})$ is given by formula (2.3.12) and E is locally integrable as well. From (3.2.6), we obtain the following representation of E :

$$E(t,x) = \frac{2^{1-2|\alpha|-2l}}{(|\alpha|+l-2)! \pi^{(n-1)/2} \Gamma(|\alpha|+l-\frac{n-1}{2})} \sum_{j=1}^{l} \frac{1}{\alpha_j!} \left(\frac{\partial}{\partial \lambda_j}\right)^{\alpha_j} (c_j F_j)$$

$$(3.2.7)$$

where

$$F_j(t,x) = Y\left(t - \frac{|x|}{\sqrt{\lambda_j}}\right) \int_{|x|^2/t^2}^{\lambda_j} (\lambda_j - \lambda)^{|\alpha|+l-2} \left(t^2 - \frac{|x|^2}{\lambda}\right)^{|\alpha|+l-(n+1)/2} \cdot \frac{d\lambda}{\lambda^{n/2}}$$

$$= Y\left(t - \frac{|x|}{\sqrt{\lambda_j}}\right) \frac{t^{2-2l-2|\alpha|}}{\lambda_j^{l+|\alpha|-1/2}} (\lambda_j t^2 - |x|^2)^{2l+2|\alpha|-(n+3)/2}$$

$$\times \int_0^1 u^{|\alpha|+l-2} (1-u)^{|\alpha|+l-(n+1)/2} \left(1 - \left(1 - \frac{|x|^2}{\lambda_j t^2}\right) u\right)^{-|\alpha|-l+1/2} du.$$

$$(3.2.8)$$

For the last equation, we employed the linear substitution $\lambda = (|x|^2/t^2) u + \lambda_j (1-u)$.

The integral in (3.2.8) is a particular case of Euler's integral representation of Gauß' hypergeometric function $_2F_1$. This yields

$$F_j(t, x) = Y\left(t - \frac{|x|}{\sqrt{\lambda_j}}\right) \frac{(\lambda_j t^2 - |x|^2)^{2l+2|\alpha|-(n+3)/2}}{\lambda_j^{l+|\alpha|-1/2} t^{2l+2|\alpha|-2}} B(|\alpha| + l - 1, |\alpha| + l - \tfrac{n-1}{2})$$

$$\times {}_2F_1\left(|\alpha| + l - 1, |\alpha| + l - \tfrac{1}{2}; 2|\alpha| + 2l - \tfrac{n+1}{2}; 1 - \frac{|x|^2}{\lambda_j t^2}\right).$$

$$(3.2.9)$$

In order to represent F_j by elementary functions, we shall distinguish two cases according to the parity of n.   □

Example 3.2.3 Let us now first investigate in detail the product $R(\partial)$ of wave operators in (3.2.5) if the space dimension n is odd, i.e., $n = 2k + 1$, $k \in \mathbf{N}_0$.

We then can evaluate formula (3.2.9) by using Eqs. 7.2.1.12 and 7.3.1.105 in Brychkov, Marichev and Prudnikov [30] (see pp. 431, 461):

$$\frac{d^k}{dz^k}\left(z^{c-1} {}_2F_1(a, b; c; z)\right) = (-1)^k (1 - c)_k z^{c-k-1} {}_2F_1(a, b; c - k; z)$$

and

$$_2F_1(a, a + \tfrac{1}{2}; 2a + 1; z) = \left(\frac{2}{1 + \sqrt{1 - z}}\right)^{2a}.$$

This yields

$$_2F_1(a, a + \tfrac{1}{2}; 2a + 1 - k; z) = \frac{z^{k-2a}}{2a(2a - 1)\cdots(2a - k + 1)} \left(\frac{d}{dz}\right)^k \left(\frac{2z}{1 + \sqrt{1 - z}}\right)^{2a}$$

$$= \frac{z^{k-2a} 2^{2a}}{2a(2a - 1)\cdots(2a - k + 1)} \left(\frac{d}{dz}\right)^k (1 - \sqrt{1 - z})^{2a}. \qquad (3.2.10)$$

Inserting (3.2.10) into (3.2.9) and setting $r = |x|$ furnishes

$$F_j = Y\left(t - \frac{r}{\sqrt{\lambda_j}}\right) \frac{\Gamma(|\alpha| + l - 1)\Gamma(|\alpha| + l - k) 2^{2|\alpha|+2l-2}}{\Gamma(2|\alpha| + 2l - 1)\sqrt{\lambda_j}} \left(-\frac{\partial}{2r\partial r}\right)^k (\sqrt{\lambda_j} t - r)^{2|\alpha|+2l-2}.$$

Combining the last formula with (3.2.7) finally yields for $|\alpha| + l > k$ the following:

$$E = \frac{1}{2^{k+1}\pi^k (2|\alpha| + 2l - 2)!} \times$$

$$\times \sum_{j=1}^{l} \frac{Y\left(t - \frac{r}{\sqrt{\lambda_j}}\right)}{\alpha_j!} \left(-\frac{\partial}{r\partial r}\right)^k \left(\frac{\partial}{\partial \lambda_j}\right)^{\alpha_j} \left[\frac{c_j}{\sqrt{\lambda_j}}(\sqrt{\lambda_j} t - r)^{2|\alpha|+2l-2}\right], \qquad (3.2.11)$$

where, as always, $c_j := \prod_{\substack{m=1 \\ m \neq j}}^{l} (\lambda_j - \lambda_m)^{-\alpha_m - 1}$. In the case $\alpha = 0$, formula (3.2.11)

coincides with Wagner [285], Satz 4, p. 20; Galler [86], Satz 16.1, p. 61. Formulas in terms of F. John's spherical means for the fundamental solutions of products of wave operators in odd space dimensions were deduced in Gal'pern and Kondrashov [87], see (56), p. 133.

Let us yet specify (3.2.11) for the space dimensions $n = 1$ and $n = 3$, respectively. For the operator $\prod_{j=1}^{l} (\partial_t^2 - \lambda_j \partial_x^2)^{\alpha_j + 1}$ with λ_j positive and pairwise different and $\alpha \in \mathbf{N}_0^l$, we obtain the following forward fundamental solution $E \in L^1_{\mathrm{loc}}(\mathbf{R}^2)$ by setting $k = 0$ in (3.2.11):

$$E = \frac{1}{2 \cdot (2|\alpha| + 2l - 2)!} \sum_{j=1}^{l} \frac{Y\left(t - \frac{|x|}{\sqrt{\lambda_j}}\right)}{\alpha_j!} \left(\frac{\partial}{\partial \lambda_j}\right)^{\alpha_j} \left[\frac{c_j}{\sqrt{\lambda_j}} (\sqrt{\lambda_j} t - |x|)^{2|\alpha| + 2l - 2}\right].$$

For $n = 3$, we obtain the forward fundamental solution $E \in L^1_{\mathrm{loc}}(\mathbf{R}^4)$ of the operator $\prod_{j=1}^{l} (\partial_t^2 - \lambda_j \Delta_3)^{\alpha_j + 1}$ (for pairwise different positive λ_j, $\alpha \neq 0$ or $l \geq 2$) by setting $k = 1$ in (3.2.11):

$$E = \frac{1}{4\pi (2|\alpha| + 2l - 3)!} \sum_{j=1}^{l} \frac{Y\left(t - \frac{r}{\sqrt{\lambda_j}}\right)}{\alpha_j! \, r} \left(\frac{\partial}{\partial \lambda_j}\right)^{\alpha_j} \left[\frac{c_j}{\sqrt{\lambda_j}} (\sqrt{\lambda_j} t - r)^{2|\alpha| + 2l - 3}\right].$$

$\square$

Example 3.2.4 Let us now consider the product $R(\partial)$ of wave operators in (3.2.5) for even space dimensions $n = 2k$, $k \in \mathbf{N}$.

We then evaluate formula (3.2.9) by employing the differentiation formulas 7.2.1.10, 7.2.1.12, 7.2.1.14 and the representation 7.3.2.157 in Brychkov, Marichev and Prudnikov [30] (see pp. 431, 432, 477). This yields, for $a \in \mathbf{N}$:

$$_2F_1(a, a + \tfrac{1}{2}; 2a + \tfrac{3}{2} - k; z) =$$

$$= \frac{\Gamma(2a - k + \tfrac{3}{2}) z^{k - 2a - 1/2}}{\Gamma(2a + \tfrac{1}{2})} \left(\frac{\mathrm{d}}{\mathrm{d}z}\right)^{k-1} \left[z^{2a - 1/2} \, _2F_1(a, a + \tfrac{1}{2}; 2a + \tfrac{1}{2}; z)\right]$$

$$= \frac{\Gamma(2a - k + \tfrac{3}{2}) \tfrac{1}{2} \sqrt{\pi} z^{k - 2a - 1/2}}{\Gamma(a) \Gamma(a + \tfrac{1}{2}) \Gamma(a + \tfrac{3}{2})} \left(\frac{\mathrm{d}}{\mathrm{d}z}\right)^{k-1} \left[z^{2a - 1/2} \left(\frac{\mathrm{d}}{\mathrm{d}z}\right)^{a-1} \, _2F_1(1, \tfrac{3}{2}; a + \tfrac{3}{2}; z)\right]$$

$$= \frac{\Gamma(2a - k + \tfrac{3}{2}) 2^{4a-2} z^{k - 2a - 1/2}}{3\sqrt{\pi} \, \Gamma(2a)^2} \left(\frac{\mathrm{d}}{\mathrm{d}z}\right)^{k-1} \left[z^{2a - 1/2} \left(\frac{\mathrm{d}}{\mathrm{d}z}\right)^{a-1} \times \right.$$

$$\left. \times (1 - z)^{a-1} \left(\frac{\mathrm{d}}{\mathrm{d}z}\right)^{a-1} \, _2F_1(1, \tfrac{3}{2}; \tfrac{5}{2}; z)\right]$$

$$= \frac{\Gamma(2a-k+\frac{3}{2})2^{4a-3}z^{k-2a-1/2}}{\sqrt{\pi}\,\Gamma(2a)^2}\left(\frac{d}{dz}\right)^{k-1}\left[z^{2a-1/2}\left(\frac{d}{dz}\right)^{a-1}\times\right.$$

$$\left.\times\,(1-z)^{a-1}\left(\frac{d}{dz}\right)^{a-1}\frac{1}{z^{3/2}}\left[\log\left(\frac{1+\sqrt{z}}{1-\sqrt{z}}\right)-2\sqrt{z}\right]\right].$$

Inserting $a=|\alpha|+l-1$ we obtain from (3.2.7) and (3.2.9)

$$E=\frac{2^{2|\alpha|+2l-6}}{\pi^k\Gamma(2|\alpha|+2l-2)^2}\sum_{j=1}^{l}\frac{1}{\alpha_j!}\left(\frac{\partial}{\partial\lambda_j}\right)^{\alpha_j}\left[\left\{c_j z^{k-2|\alpha|-2l+3/2}\left(\frac{d}{dz}\right)^{k-1}z^{2|\alpha|+2l-5/2}\times\right.\right.$$

$$\times\left(\frac{d}{dz}\right)^{|\alpha|+l-2}(1-z)^{|\alpha|+l-2}\left(\frac{d}{dz}\right)^{|\alpha|+l-2}\frac{1}{z^{3/2}}\left[\log\left(\frac{1+\sqrt{z}}{1-\sqrt{z}}\right)-2\sqrt{z}\right]\right\}\Big|_{z=1-|x|^2/(\lambda_j t^2)}$$

$$\times\,Y\left(t-\frac{|x|}{\sqrt{\lambda_j}}\right)\cdot\frac{(\lambda_j t^2-|x|^2)^{2l+2|\alpha|-k-3/2}}{\lambda_j^{l+|\alpha|-1/2}t^{2l+2|\alpha|-2}}\right].$$

By means of

$$r=|x|,\quad 1-z=\frac{r^2}{\lambda_j t^2},\quad \frac{d}{dz}=-\frac{\lambda_j t^2}{2}\frac{\partial}{r\partial r},\quad \frac{1+\sqrt{z}}{1-\sqrt{z}}=\left(\frac{\sqrt{\lambda_j}t}{r}+\sqrt{\frac{\lambda_j t^2}{r^2}-1}\right)^2,$$

it follows that

$$E=\frac{(-1)^{k-1}}{(2\pi)^k\Gamma(2|\alpha|+2l-2)^2}\sum_{j=1}^{l}\frac{Y\left(t-\frac{r}{\sqrt{\lambda_j}}\right)}{\alpha_j!}\left(\frac{\partial}{\partial\lambda_j}\right)^{\alpha_j}\left[\frac{c_j}{\sqrt{\lambda_j}}\left(\frac{\partial}{r\partial r}\right)^{k-1}\times\right.$$

$$\times\,(\lambda_j t^2-r^2)^{2|\alpha|+2l-5/2}\left(\frac{\partial}{r\partial r}\right)^{|\alpha|+l-2}r^{2|\alpha|+2l-4}\left(\frac{\partial}{r\partial r}\right)^{|\alpha|+l-2}\times$$

$$\left.\left\{\sqrt{\lambda_j}\,t(\lambda_j t^2-r^2)^{-3/2}\log\left(\frac{\sqrt{\lambda_j}t}{r}+\sqrt{\frac{\lambda_j t^2}{r^2}-1}\right)-(\lambda_j t^2-r^2)^{-1}\right\}\right]$$

$$(3.2.12)$$

is the fundamental solution of $R(\partial)=\prod_{j=1}^{l}(\partial_t^2-\lambda_j\Delta_{2k})^{\alpha_j+1}$ for $\alpha\in\mathbf{N}_0^l$ with $|\alpha|>k-l-\frac{1}{2}$, for pairwise different $\lambda_1,\ldots,\lambda_l>0$ and $c_j=\prod_{\substack{m=1\\m\neq j}}^{l}(\lambda_j-\lambda_m)^{-\alpha_m-1}$.

In the case $\alpha=0$ and $k=1$, we obtain a relatively simple formula for the fundamental solution of $\prod_{j=1}^{l}(\partial_t^2-\lambda_j\Delta_2)$, i.e.,

$$E=\frac{1}{2\pi(2l-3)!^2}\sum_{j=1}^{l}Y\left(t-\frac{r}{\sqrt{\lambda_j}}\right)\lambda_j^{-1/2}\left(\prod_{\substack{k=1\\k\neq j}}^{l}(\lambda_j-\lambda_k)^{-1}\right)\times$$

$$\times (\lambda_j t^2 - r^2)^{2l-5/2} \Big(\frac{\partial}{r \partial r} \Big)^{l-2} r^{2l-4} \Big(\frac{\partial}{r \partial r} \Big)^{l-2} \times$$

$$\Big\{ \sqrt{\lambda_j}\, t (\lambda_j t^2 - r^2)^{-3/2} \log \Big(\frac{\sqrt{\lambda_j}\, t}{r} + \sqrt{\frac{\lambda_j t^2}{r^2} - 1} \Big) - (\lambda_j t^2 - r^2)^{-1} \Big\}.$$

(3.2.13)

In different form, this fundamental solution was presented in Wagner [285], Satz 5, p. 21.

In particular, for $l = 2$, the forward fundamental solution of the operator $(\partial_t^2 - \lambda_1 \Delta_2)(\partial_t^2 - \lambda_2 \Delta_2)$ is given by

$$E = \frac{1}{2\pi(\lambda_1 - \lambda_2)} \Bigg[Y\Big(t - \frac{r}{\sqrt{\lambda_1}}\Big)\Big(t \operatorname{arcosh}\Big(\frac{\sqrt{\lambda_1}\, t}{r}\Big) - \sqrt{t^2 - \frac{r^2}{\lambda_1}}\Big)$$

$$- Y\Big(t - \frac{r}{\sqrt{\lambda_2}}\Big)\Big(t \operatorname{arcosh}\Big(\frac{\sqrt{\lambda_2}\, t}{r}\Big) - \sqrt{t^2 - \frac{r^2}{\lambda_2}}\Big) \Bigg],$$

cf. Wagner [285], Ch. II, Bsp. 7, p. 29. □

Example 3.2.5 Let us finally derive *Herglotz' integral representation* of the forward fundamental solution of the product $R(\partial)$ of wave operators in (3.2.5), cf. Herglotz [125], II.

Let us assume first that $2|\alpha| + 2l > n$. By means of one of Kummer's transformation formulas for the hypergeometric functions, i.e.,

$$_2F_1(a, 1 - a; c; z) = (1 - z)^{c-1}(1 - 2z)^{a-c}\, _2F_1\Big(\frac{c - a}{2}, \frac{c - a + 1}{2}; c; \frac{4z(z - 1)}{(1 - 2z)^2} \Big),$$

see Abramowitz and Stegun [1], Eq. 15.3.32, p. 561, and setting $a = \frac{3-n}{2}$, $c = 2|\alpha| + 2l - \frac{n+1}{2}$, $z = \frac{1}{2}\big(1 - \sqrt{\lambda_j}\, t/|x|\big)$, we have $\frac{4z(z-1)}{(1-2z)^2} = 1 - \frac{|x|^2}{\lambda_j t^2}$ and the functions $F_j(t, x)$ in (3.2.9) assume the form

$$F_j(t, x) = Y\Big(t - \frac{|x|}{\sqrt{\lambda_j}}\Big) \frac{|x|^{(1-n)/2}}{\sqrt{\lambda_j}} (\sqrt{\lambda_j}\, t - |x|)^{2l+2|\alpha|-(n+3)/2} \times$$

$$\times\, 2^{2|\alpha|+2l-(n+3)/2} B\big(|\alpha| + l - 1, |\alpha| + l - \tfrac{n-1}{2}\big) \times$$

$$_2F_1\Big(\frac{3 - n}{2}, \frac{n - 1}{2}; 2|\alpha| + 2l - \frac{n + 1}{2}; \frac{1}{2}\Big(1 - \frac{\sqrt{\lambda_j}\, t}{|x|}\Big) \Big).$$

(3.2.14)

If we express the hypergeometric function in (3.2.14) by Euler's integral representation and use the doubling formula for the gamma function, we obtain

$$F_j(t,x) = Y\left(t - \frac{|x|}{\sqrt{\lambda_j}}\right) \frac{2^{(n-1)/2}\sqrt{\pi}}{\Gamma(\frac{n-1}{2})\sqrt{\lambda_j}} |x|^{(1-n)/2}(\sqrt{\lambda_j}t - |x|)^{2l+2|\alpha|-(n+3)/2} \times$$

$$\times \frac{(|\alpha|+l-2)!}{\Gamma(|\alpha|+l-\frac{n}{2})} \int_0^1 s^{(n-3)/2}(1-s)^{2|\alpha|+2l-n-1}\left(1 - \frac{s}{2}\left(1 - \frac{\sqrt{\lambda_j}t}{|x|}\right)\right)^{(n-3)/2} ds.$$

Upon employing the linear transformation $s = \frac{v-1}{\sqrt{\lambda_j}t/|x|-1}$ and inserting F_j into (3.2.7), we arrive at Herglotz' representation for the fundamental solution E of the operator $\prod_{j=1}^l (\partial_t^2 - \lambda_j\Delta_n)^{\alpha_j+1}$:

$$E(t,x) = \frac{\Gamma(\frac{n}{2})|x|^{2|\alpha|+2l-n-1}}{2\pi^{n/2}(n-2)!\Gamma(2|\alpha|+2l-n)} \sum_{j=1}^l \frac{1}{\alpha_j!}\left(\frac{\partial}{\partial\lambda_j}\right)^{\alpha_j}\left[\frac{c_j}{\sqrt{\lambda_j}} K\left(\frac{\sqrt{\lambda_j}t}{|x|}\right)\right],$$

$$(3.2.15)$$

where

$$K(u) = Y(u-1)\int_1^u (v^2-1)^{(n-3)/2}(u-v)^{2|\alpha|+2l-n-1}dv,$$

see Herglotz [125], II, Eqs. (171), (175), p. 313; [126], p. 556, for $\alpha = 0$. A completely different derivation of (3.2.15) will be given below employing the "difference device", see Example 3.3.4.

The above derivation of (3.2.15) was performed under the assumptions that $n \geq 2$ and $2|\alpha| + 2l - n > 0$. However, due to the representation

$$\frac{K(u)}{\Gamma(2|\alpha|+2l-n)} = T*F(2|\alpha|+2l-n), \quad T = Y(u-1)(u^2-1)^{(n-3)/2} \in L^1_{loc}(\mathbf{R}),$$

where $\mathbf{C} \to \mathcal{D}'(\mathbf{R}) : z \mapsto F(z)$ is the convolution group given by $F(z) = Y(u)u^{z-1}/\Gamma(z)$ for $\mathrm{Re}\, z > 0$, see Examples 1.4.8 and 1.5.11, we can express E in the form

$$E = \frac{\Gamma(\frac{n}{2})|x|^{2|\alpha|+2l-n-1}}{2\pi^{n/2}(n-2)!} \sum_{j=1}^l \frac{1}{\alpha_j!}\left(\frac{\partial}{\partial\lambda_j}\right)^{\alpha_j}\left[\frac{c_j}{\sqrt{\lambda_j}} (T*F(2|\alpha|+2-n))\left(\frac{\sqrt{\lambda_j}t}{|x|}\right)\right].$$

$$(3.2.16)$$

Formula (3.2.16) is valid for $n \geq 2$ and arbitrary $\alpha \in \mathbf{N}_0^l$. This can be justified by considering

$$E(z) = e^{\sigma t} \mathcal{F}^{-1} \left(\prod_{j=1}^{l} [(i\tau + \sigma)^2 + \lambda_j |\xi|^2]^{\alpha_j + 1 + z} \right), \quad z \in \mathbf{C}, \quad \sigma > 0,$$

similarly as in Example 2.3.6.

Formula (3.2.16), which extends Herglotz' formula (3.2.15) from the case of sufficiently many factors, i.e., $|\alpha| + l > \frac{n}{2}$, to the general case, becomes particularly simple if $|\alpha| + l = \frac{n}{2}$. Then $F(2|\alpha| + 2l - n) = F(0) = \delta$ and hence

$$E = \frac{\Gamma(\frac{n}{2})}{2\pi^{n/2}(n-2)!|x|} \sum_{j=1}^{l} \frac{1}{\alpha_j!} \left(\frac{\partial}{\partial \lambda_j} \right)^{\alpha_j} \left[\frac{c_j}{\sqrt{\lambda_j}} Y\left(t - \frac{|x|}{\sqrt{\lambda_j}} \right) \left(\frac{\lambda_j t^2}{|x|^2} - 1 \right)^{(n-3)/2} \right]$$

$$(3.2.17)$$

is the fundamental solution of $R(\partial) = \prod_{j=1}^{l} (\partial_t^2 - \lambda_j \Delta_n)^{\alpha_j + 1}$ with support in $t \geq 0$ if $|\alpha| + l = \frac{n}{2}$ and $c_j = \prod_{\substack{k=1 \\ k \neq j}}^{l} (\lambda_j - \lambda_k)^{-\alpha_k - 1}$.   □

Let us next state an analogue of Proposition 3.2.1 for the case of elliptic operators with the goal of constructing a fundamental solution of the product $\prod_{j=1}^{l} (\Delta_{n-1} + \lambda_j \partial_n^2)^{\alpha_j + 1}$ of Laplaceans.

Proposition 3.2.6 *Suppose that $P_1(\partial), P_2(\partial)$ are differential operators in $\mathbf{R}^n$ which are homogeneous of degree m and such that each operator*

$$Q_\lambda(\partial) = P_1(\partial) + \lambda P_2(\partial), \quad \lambda \in [c, d] \subset \mathbf{R},$$

is elliptic. Let $\alpha \in \mathbf{N}_0^l$ and $\lambda_0, \ldots, \lambda_l \in [c, d]$ and such that $\lambda_1, \ldots, \lambda_l$ are pairwise different and set $c_j := \prod_{\substack{k=1 \\ k \neq j}}^{l} (\lambda_j - \lambda_k)^{-\alpha_k - 1}$. Furthermore assume that E_λ are fundamental solutions of $Q_\lambda(\partial)^{|\alpha| + l}$ which depend continuously on λ and satisfy the estimate (3.1.13) for $\lambda \in [c, d]$.

Then the parameter integration formula

$$E = (|\alpha| + l - 1) \sum_{j=1}^{l} \frac{1}{\alpha_j!} \left(\frac{\partial}{\partial \lambda_j} \right)^{\alpha_j} \left[c_j \int_{\lambda_0}^{\lambda_j} (\lambda_j - \lambda)^{|\alpha| + l - 2} E_\lambda \, d\lambda \right]$$

yields a fundamental solution E of $R(\partial) = \prod_{j=1}^{l} (P_1(\partial) + \lambda_j P_2(\partial))^{\alpha_j + 1}$, which grows at most as a constant multiple of $|x|^{m(|\alpha| + l) - n} \log |x|$ for $|x| \to \infty$.

Proof As in the proof of Proposition 3.1.7, we conclude that $E_\lambda = \tilde{E}_\lambda + q_\lambda$ where

$$\tilde{E}_\lambda := \mathcal{F}^{-1} \left[\Pf_{\mu = -m(|\alpha| + l)} (Q_\lambda(i\omega)^{-|\alpha| - l} \cdot |\xi|^\mu) \right],$$

$Q_\lambda(\partial)\tilde{E}_\lambda = \delta$, and q_λ is a polynomial of degree at most $m(|\alpha| + l) - n$. As E_λ and $\tilde{E}_\lambda$, also q_λ depends continuously on $\lambda \in [c, d]$, and we therefore obtain, with a polynomial q,

$$\mathcal{F}(E - q) = (|\alpha| + l - 1) \sum_{j=1}^{l} \frac{1}{\alpha_j!} \left(\frac{\partial}{\partial \lambda_j} \right)^{\alpha_j} \left[c_j \int_{\lambda_0}^{\lambda_j} (\lambda_j - \lambda)^{|\alpha| + l - 2} \mathcal{F}\tilde{E}_\lambda \, d\lambda \right]$$

$$= (|\alpha| + l - 1) \sum_{j=1}^{l} \frac{1}{\alpha_j!} \left(\frac{\partial}{\partial \lambda_j} \right)^{\alpha_j} c_j \times$$

$$\times \Pf_{\mu = -m(|\alpha| + l)} \left[\left(\int_{\lambda_0}^{\lambda_j} \frac{(\lambda_j - \lambda)^{|\alpha| + l - 2}}{\big(P_1(i\omega) + \lambda P_2(i\omega)\big)^{|\alpha| + l}} \, d\lambda \right) \cdot |\xi|^\mu \right].$$

If we take into account formula (3.2.1), we obtain

$$\mathcal{F}(E - q) = \Pf_{\mu = -m(|\alpha| + l)} \left[\prod_{j=1}^{l} \big(P_1(i\omega) + \lambda_j P_2(i\omega)\big)^{-\alpha_j - 1} \cdot |\xi|^\mu \right],$$

which implies $R(\partial)(E - q) = \delta$, cf. Proposition 2.4.8.   □

Example 3.2.7 Let us apply now Proposition 3.2.6 to the *product* $R(\partial) = \prod_{j=1}^{l}(\Delta_{n-1} + \lambda_j \partial_n^2)^{\alpha_j + 1}$ *of Laplace operators,* where $\alpha \in \mathbf{N}_0^l$ and $\lambda_1, \ldots, \lambda_l$ are positive and pairwise different.

For $\alpha = 0$, the operator $R(\partial)$ was considered for the first time in Herglotz [125], II, § 10, p. 316; [126], p. 559. Herglotz' expression for a fundamental solution of $R(\partial)$ is not completely explicit insofar as it contains a quadrature, see Herglotz [125], II, (199a), (200a), p. 318. Explicit results were given in Ortner [201], p. 182. For even n, a different formula was derived in Galler [86], Sätze 15.1, 15.2, pp. 59, 60.

From Proposition 3.2.6, we obtain the following representation of a fundamental solution E of $R(\partial)$:

$$E = (|\alpha| + l - 1) \sum_{j=1}^{l} \frac{1}{\alpha_j!} \left(\frac{\partial}{\partial \lambda_j} \right)^{\alpha_j} \left[c_j \int_{\lambda_0}^{\lambda_j} (\lambda_j - \lambda)^{|\alpha| + l - 2} G\left(x', \frac{x_n}{\sqrt{\lambda}} \right) \frac{d\lambda}{\sqrt{\lambda}} \right],$$

$$(3.2.18)$$

where $c_j = \prod_{\substack{k=1 \\ k\neq j}}^{l}(\lambda_j-\lambda_k)^{-\alpha_k-1}$, $\lambda_0 > 0$, $x' = (x_1,\dots,x_{n-1})$ and G is a fundamental solution of $\Delta_n^{|\alpha|+l}$. According to Eqs. (1.6.19) and (1.6.20) in Example 1.6.11, we can set

$$G(x) = \begin{cases} \dfrac{(-1)^{|\alpha|+l}\Gamma(\frac{n}{2} - |\alpha| - l)}{2^{2(|\alpha|+l)}(|\alpha| + l - 1)!\pi^{n/2}}\,|x|^{2(|\alpha|+l)-n}, & \text{if } n \text{ is odd or } |\alpha| + l < \frac{n}{2}, \\[4mm] \dfrac{(-1)^{n/2-1}|x|^{2(|\alpha|+l)-n}\log|x|}{2^{2(|\alpha|+l)-1}(|\alpha| + l - 1)!(|\alpha| + l - \frac{n}{2})!\pi^{n/2}}, & \text{if } n \text{ is even and } |\alpha| + l \geq \frac{n}{2}. \end{cases}$$

$$(3.2.19)$$

Let us observe that E is locally integrable and $\mathcal{C}^\infty$ outside the origin.

(a) Let us now first investigate the case where $|\alpha| + l < \frac{n}{2}$ and $n = 2k$ is even. Then we can set $\lambda_0 = 0$ in (3.2.18) and obtain

$$E = \frac{(-1)^{|\alpha|+l}\Gamma(\frac{n}{2} - |\alpha| - l)}{2^{2(|\alpha|+l)}(|\alpha| + l - 2)!\pi^{n/2}} \sum_{j=1}^{l} \frac{1}{\alpha_j!}\left(\frac{\partial}{\partial\lambda_j}\right)^{\alpha_j}(c_jF_j), \qquad (3.2.20)$$

where, for $x \neq 0$,

$$F_j(x) = \int_0^{\lambda_j}(\lambda_j - \lambda)^{|\alpha|+l-2}\left(|x'|^2 + \frac{x_n^2}{\lambda}\right)^{|\alpha|+l-k}\frac{d\lambda}{\sqrt{\lambda}}.$$

With the abbreviation $\rho = |x'| = \sqrt{x_1^2 + \cdots + x_{n-1}^2}$, we infer

$$F_j(x) = \frac{1}{(k - |\alpha| - l - 1)!}\left(-\frac{\partial}{2\rho\,\partial\rho}\right)^{k-|\alpha|-l-1}\int_0^{\lambda_j}\frac{(\lambda_j - \lambda)^{|\alpha|+l-2}\sqrt{\lambda}\,d\lambda}{x_n^2 + \lambda\rho^2}.$$

If we apply the binomial expansion to

$$(\lambda_j - \lambda)^{|\alpha|+l-2} = \rho^{4-2(|\alpha|+l)}\cdot\left[x_n^2 + \lambda_j\rho^2 - (x_n^2 + \lambda\rho^2)\right]^{|\alpha|+l-2},$$

we obtain

$$\int_0^{\lambda_j}\frac{(\lambda_j - \lambda)^{|\alpha|+l-2}\sqrt{\lambda}\,d\lambda}{x_n^2 + \lambda\rho^2}$$

$$= \left(\frac{x_n^2}{\rho^2} + \lambda_j\right)^{|\alpha|+l-2}\int_0^{\lambda_j}\frac{\sqrt{\lambda}\,d\lambda}{x_n^2 + \lambda\rho^2} + \sum_{i=1}^{|\alpha|+l-2}\binom{|\alpha| + l - 2}{i}\rho^{4-2(|\alpha|+l)}\times$$

$$\times (x_n^2 + \lambda_j\rho^2)^{|\alpha|+l-2-i}(-1)^i\int_0^{\lambda_j}(x_n^2 + \lambda\rho^2)^{i-1}\sqrt{\lambda}\,d\lambda$$

$$= \frac{2x_n}{\rho^3}\Big(\frac{x_n^2}{\rho^2} + \lambda_j\Big)^{|\alpha|+l-2}\Big[\frac{\sqrt{\lambda_j}\rho}{x_n} - \arctan\Big(\frac{\sqrt{\lambda_j}\rho}{x_n}\Big)\Big] + \rho^{-2}\sum_{i=1}^{|\alpha|+l-2}\binom{|\alpha|+l-2}{i}$$

$$(-1)^i\Big(\frac{x_n^2}{\rho^2} + \lambda_j\Big)^{|\alpha|+l-2-i}\sum_{r=0}^{i-1}\binom{i-1}{r}\frac{\lambda_j^{i-r+1/2}}{i-r+\frac{1}{2}}\Big(\frac{x_n}{\rho}\Big)^{2r}.$$

This implies that, in the case $n = 2k$ and $k > |\alpha| + l$, a fundamental solution E of $R(\partial)$ has the form

$$E = \frac{(-1)^{k-1}}{2^{|\alpha|+l+k-1}(|\alpha| + l - 2)!\pi^k}\sum_{j=1}^{l}\frac{1}{\alpha_j!}\Big(\frac{\partial}{\partial\lambda_j}\Big)^{\alpha_j}\Big[c_j\Big(\frac{\partial}{\rho\,\partial\rho}\Big)^{k-|\alpha|-l-1} \times$$

$$\times\rho^{-2}\Big\{\frac{2x_n}{\rho}\Big(\frac{x_n^2}{\rho^2} + \lambda_j\Big)^{|\alpha|+l-2}\Big[\frac{\sqrt{\lambda_j}\rho}{x_n} - \arctan\Big(\frac{\sqrt{\lambda_j}\rho}{x_n}\Big)\Big]$$

$$+ \sum_{i=1}^{|\alpha|+l-2}\binom{|\alpha|+l-2}{i}(-1)^i\Big(\frac{x_n^2}{\rho^2} + \lambda_j\Big)^{|\alpha|+l-2-i} \times$$

$$\times\sum_{r=0}^{i-1}\binom{i-1}{r}\frac{\lambda_j^{i-r+1/2}}{i-r+\frac{1}{2}}\Big(\frac{x_n}{\rho}\Big)^{2r}\Big\}\Big],$$

(3.2.21)

cf. Ortner [201], p. 182, no. 1. We postpone the case of even n with $n \leq 2(|\alpha|+l)$ to Example 5.2.5. Note that E in (3.2.21) is homogeneous of degree $2(|\alpha| + l) - n < 0$, and thus E is the only homogeneous fundamental solution of $R(\partial)$, or, put differently, E is the only fundamental solution which vanishes at infinity.

In particular, for $\alpha = 0$, $l = 2$, $n = 6$, we obtain for the fundamental solution E of $(\Delta_5 + \lambda_1\partial_6^2)(\Delta_5 + \lambda_2\partial_6^2)$, $\lambda_j > 0$, $\lambda_1 \neq \lambda_2$, the formula

$$E = \frac{1}{8\pi^3(\lambda_1 - \lambda_2)}\Big[\frac{\sqrt{\lambda_1} - \sqrt{\lambda_2}}{\rho^2} - \frac{x_6}{\rho^3}\Big(\arctan\Big(\frac{\sqrt{\lambda_1}\rho}{x_6}\Big) - \arctan\Big(\frac{\sqrt{\lambda_2}\rho}{x_6}\Big)\Big)\Big],$$

where $\rho = \sqrt{x_1^2 + \cdots + x_5^2}$.

(b) Let us consider now odd $n = 2k + 1 \geq 3$, but still under the hypothesis of $|\alpha| + l < \frac{n}{2}$, i.e., $|\alpha| + l \leq k$. As before, the fundamental solution becomes unique under the assumption of homogeneity.

Then formula (3.2.20) persists, where now

$$F_j(x) = \frac{\sqrt{\pi}}{\Gamma(k - |\alpha| - l + \frac{1}{2})}\Big(-\frac{\partial}{2\rho\,\partial\rho}\Big)^{k-|\alpha|-l}\int_0^{\lambda_j}\frac{(\lambda_j - \lambda)^{|\alpha|+l-2}}{\sqrt{x_n^2 + \lambda\rho^2}}\,\mathrm{d}\lambda.$$

The last integral originates by $|\alpha| + l - 1$ indefinite integrations of the function $(x_n^2 + \lambda \rho^2)^{-1/2}$, and hence it coincides with

$$B(|\alpha| + l - 1, \tfrac{1}{2})(x_n^2 + \lambda_j \rho^2)^{|\alpha|+l-3/2} \rho^{2-2|\alpha|-2l}$$

up to a polynomial in λ_j of degree at most $|\alpha| + l - 2$. Since

$$\sum_{j=1}^{l} \frac{1}{\alpha_j!} \left(\frac{\partial}{\partial \lambda_j}\right)^{\alpha_j} (c_j \lambda_j^i) = 0 \quad \text{for } i = 0, \dots, |\alpha| + l - 2, \qquad (3.2.22)$$

as a consequence of formula (3.2.2) for $t = 0$, we obtain for the case $n = 2k + 1$, $|\alpha| + l \le k$,

$$E = \frac{(-1)^k \pi^{1/2-k}}{2^{|\alpha|+l+k} \Gamma(|\alpha| + l - \tfrac{1}{2})} \sum_{j=1}^{l} \frac{1}{\alpha_j!} \left(\frac{\partial}{\partial \lambda_j}\right)^{\alpha_j}$$

$$\left[c_j \left(\frac{\partial}{\rho \, \partial \rho}\right)^{k-|\alpha|-l} \frac{(x_n^2 + \lambda_j \rho^2)^{|\alpha|+l-3/2}}{\rho^{2(|\alpha|+l-1)}} \right],$$

cf. Ortner [201], p. 182, no. 2 (where a numerical factor should be corrected).

In particular, for $\alpha = 0$, $l = 2$, $n = 5$ and positive λ_j, we obtain the following formula for the homogeneous fundamental solution E of $(\Delta_4 + \lambda_1 \partial_5^2)(\Delta_4 + \lambda_2 \partial_5^2)$:

$$E = \frac{1}{8\pi^2 \left[\sqrt{x_5^2 + \lambda_1 \rho^2} + \sqrt{x_5^2 + \lambda_2 \rho^2} \right]}, \qquad \rho^2 = x_1^2 + \dots + x_4^2.$$

(c) Let us next stick to the case of odd dimension $n = 2k + 1$, $k \ge 1$, but assume that $|\alpha| + l > k$. Then formula (3.2.20) still holds with

$$F_j(x) = \int_{\lambda_0}^{\lambda_j} (\lambda_j - \lambda)^{|\alpha|+l-2} \left(\rho^2 + \frac{x_n^2}{\lambda}\right)^{|\alpha|+l-k-1/2} \frac{d\lambda}{\sqrt{\lambda}}$$

for some fixed $\lambda_0 > 0$. Note that E is the only fundamental solution of $R(\partial)$ which is homogeneous and even according to Proposition 2.4.8, (2), and hence E is independent of the choice of λ_0.

If we employ the substitution $x_n^2 + \lambda \rho^2 = u^2$, $d\lambda = 2u \, du/\rho^2$, we obtain

$$F_j(x) = \int_{\lambda_0}^{\lambda_j} (\lambda_j - \lambda)^{|\alpha|+l-2} (x_n^2 + \lambda \rho^2)^{|\alpha|+l-k-1/2} \lambda^{k-|\alpha|-l} \, d\lambda$$

$$= 2\rho^{2-2k} \int_{\sqrt{x_n^2 + \lambda_0 \rho^2}}^{\sqrt{x_n^2 + \lambda_j \rho^2}} (x_n^2 + \lambda_j \rho^2 - u^2)^{|\alpha|+l-2} \frac{u^{2(|\alpha|+l-k)} \, du}{(u^2 - x_n^2)^{|\alpha|+l-k}}$$

$$= 2\rho^{2-2k} \sum_{i=0}^{|\alpha|+l-2} \binom{|\alpha|+l-2}{i} (-1)^i (x_n^2 + \lambda_j \rho^2)^{|\alpha|+l-2-i} \times$$

$$\times \int_{\sqrt{x_n^2+\lambda_0\rho^2}}^{\sqrt{x_n^2+\lambda_j\rho^2}} \frac{u^{2(|\alpha|+l-k+i)}\,du}{(u^2 - x_n^2)^{|\alpha|+l-k}}.$$

The last integral over a rational function can easily be evaluated, see Gröbner and Hofreiter [116], Eqs. 15.21 b) and f):

$$F_j(x) = 2\rho^{2-2k} \sum_{i=0}^{|\alpha|+l-2} \binom{|\alpha|+l-2}{i} (-1)^i (x_n^2 + \lambda_j\rho^2)^{|\alpha|+l-2-i} \times$$

$$\left[-u^{2(|\alpha|+l-k+i)+1} \sum_{r=1}^{|\alpha|+l-k-1} \frac{(2i+3)(2i+5)\ldots(2i+2r-1)}{(|\alpha|+l-k-1)\ldots(|\alpha|+l-k-r)} \frac{(2x_n^2)^{-r}}{(u^2-x_n^2)^{|\alpha|+l-k-r}} \right.$$

$$+ \frac{(2i+3)(2i+5)\ldots(2(|\alpha|+l-k+i)-1)}{(|\alpha|+l-k-1)!(2x_n^2)^{|\alpha|+l-k-1}} \times$$

$$\times \sum_{r=0}^{|\alpha|+l-k+i-1} \frac{x_n^{2r} u^{2(|\alpha|+l-k+i)-2r-1}}{2(|\alpha|+l-k+i)-2r-1}$$

$$+ \frac{(2i+3)(2i+5)\ldots(2(|\alpha|+l-k+i)-1)x_n^{2i+1}}{(|\alpha|+l-k-1)!\, 2^{|\alpha|+l-k}}$$

$$\left. \log\left(\frac{u-x_n}{u+x_n}\right) \right]\Bigg|_{u=\sqrt{x_n^2+\lambda_0\rho^2}}^{u=\sqrt{x_n^2+\lambda_j\rho^2}}.$$

As in case (b), we use Eq. (3.2.22) in order to conclude that the sum of the contributions from the lower integration limit $\sqrt{x_n^2 + \lambda_0\rho^2}$ vanishes. Therefore, (3.2.20) yields for the homogeneous fundamental solution E of $R(\partial) = \prod_{j=1}^{l}(\Delta_{n-1} + \lambda_j \partial_n^2)^{\alpha_j+1}$ in the case $n = 2k+1$, $|\alpha| + l > k$, the following:

$$E = \frac{(-1)^{|\alpha|+l}\Gamma(k-|\alpha|-l+\frac{1}{2})}{2^{2(|\alpha|+l)-1}(|\alpha|+l-2)!\pi^{n/2}} \sum_{j=1}^{l} \frac{1}{\alpha_j!}\left(\frac{\partial}{\partial\lambda_j}\right)^{\alpha_j} c_j \sum_{i=0}^{|\alpha|+l-2} \binom{|\alpha|+l-2}{i}(-1)^i$$

$$\times \left\{ -(x_n^2+\lambda_j\rho^2)^{2(|\alpha|+l)-k-3/2} \sum_{r=1}^{|\alpha|+l-k-1} \frac{(2i+3)(2i+5)\ldots(2i+2r-1)}{(|\alpha|+l-k-1)\ldots(|\alpha|+l-k-r)} \right.$$

$$\times \lambda_j^{-|\alpha|-l+k+r}(2x_n^2)^{-r}\rho^{-2(|\alpha|+l-r-1)}$$

$$+ \frac{(2i+3)(2i+5)\ldots(2(|\alpha|+l-k+i)-1)}{(|\alpha|+l-k-1)!\, 2^{|\alpha|+l-k-1}} \times$$

$$\times \sum_{r=0}^{|\alpha|+l-k+i-1} \frac{x_n^{2(r-|\alpha|-l+k+1)} \rho^{2-2k}}{2(|\alpha|+l-k+i)-2r-1}(x_n^2+\lambda_j\rho^2)^{2(|\alpha|+l)-k-r-5/2}$$

$$+\frac{(2i+3)(2i+5)\ldots(2(|\alpha|+l-k+i)-1)}{(|\alpha|+l-k-1)!\,2^{|\alpha|+l-k}}\,x_n^{2i+1}\rho^{2-2k}(x_n^2+\lambda_j\rho^2)^{|\alpha|+l-i-2}$$

$$\times \log\left(\frac{\sqrt{x_n^2+\lambda_j\rho^2}-x_n}{\sqrt{x_n^2+\lambda_j\rho^2}+x_n}\right)\Bigg\}.$$

In particular, for $n = 3$, i.e., $k = 1$, and $\alpha = 0$, $l = 2$, positive $\lambda_1 \neq \lambda_2$, we obtain the uniquely determined homogeneous and even fundamental solution E of $(\Delta_2 + \lambda_1\partial_3^2)(\Delta_2 + \lambda_2\partial_3^2)$:

$$E = -\frac{1}{4\pi}\left[\frac{\rho^2}{\sqrt{x_3^2+\lambda_1\rho^2}+\sqrt{x_3^2+\lambda_2\rho^2}} - \frac{x_3}{\lambda_1-\lambda_2}\log\left(\frac{\sqrt{x_3^2/\lambda_1+\rho^2}+x_3/\sqrt{\lambda_1}}{\sqrt{x_3^2/\lambda_2+\rho^2}+x_3/\sqrt{\lambda_2}}\right)\right],$$

$$(3.2.23)$$

where $\rho^2 = x_1^2 + x_2^2$, cf. Garnir [100], p. 1140; Brillouin [27], § 6; Bureau [35], p. 483; Bureau [36], p. 23; Wagner [285], Ch. III, 4. Bsp, pp. 45–46.

Let us finally observe that the above formulas for E remain valid by analytic continuation as long as $\lambda_1, \ldots, \lambda_l \in \mathbf{C} \setminus (-\infty, 0]$ are pairwise different. $\quad\square$

3.3 The Difference Device

For a polynomial $Q(z) = \prod_{j=1}^{l}(z-\lambda_j)^{\alpha_j+1}$ in one complex variable z (with $\lambda_j \in \mathbf{C}$ pairwise different and $\alpha \in \mathbf{N}_0^l$), we have shown already in Proposition 1.4.4 how to derive fundamental solutions of the operator $Q(P(\partial)) = \prod_{j=1}^{l}(P(\partial)-\lambda_j)^{\alpha_j+1}$ from fundamental solutions E_λ of $P(\partial) - \lambda$. (Proposition 1.4.4 refers to the case of real λ_j, but it holds indeed also for E_λ depending holomorphically on complex λ.) This essentially algebraic formula relies on the partial fraction decomposition of $Q(z)^{-1}$.

In this section, we shall consider instead operators of the form $Q(P_1(\partial), P_2(\partial))$ for homogeneous polynomials $Q(z_1, z_2)$, i.e., $Q(z_1, z_2) = \prod_{j=1}^{l}(z_1 - \lambda_j z_2)^{\alpha_j+1}$. In this case, the algebraic decomposition procedure stops short of producing a fundamental solution of $Q(P_1(\partial), P_2(\partial))$, but it yields instead a solution F of the equation

$$Q(P_1(\partial), P_2(\partial))F = (P_1(\partial) - \lambda_0 P_2(\partial))^{|\alpha|+l-1}\delta,$$

cf. Ortner [202], Prop. 2, p. 86. We emphasize that this so-called "difference device" works independently of any assumptions on the involved operators, as, e.g., quasihyperbolicity or ellipticity.

Proposition 3.3.1 *Let $P_1(\partial), P_2(\partial)$ be linear constant coefficient differential operators in $\mathbf{R}^n$ and $E : U \longrightarrow \mathcal{D}'(\mathbf{R}^n)$ such that $E(\lambda)$ is a fundamental solution of
$P_1(\partial) - \lambda P_2(\partial)$ for $\lambda \in U$. We assume that either $U \subset \mathbf{R}$ is open, $m \in \mathbf{N}$, and
E depends C^m on λ, or that $U \subset \mathbf{C}$ is open and E depends holomorphically on λ.
Then*

$$F = \sum_{j=1}^{l} \frac{1}{\alpha_j!} \left(\frac{\partial}{\partial \lambda_j}\right)^{\alpha_j} (d_j E(\lambda_j)), \qquad d_j := (\lambda_j - \lambda_0)^{|\alpha|+l-1} \prod_{\substack{k=1 \\ k \neq j}}^{l} (\lambda_j - \lambda_k)^{-\alpha_k - 1},$$

satisfies

$$\prod_{j=1}^{l} \big(P_1(\partial) - \lambda_j P_2(\partial)\big)^{\alpha_j + 1} F = \big(P_1(\partial) - \lambda_0 P_2(\partial)\big)^{|\alpha|+l-1} \delta \qquad (3.3.1)$$

*for $\lambda_0 \in \mathbf{C}$ and pairwise different $\lambda_1, \dots, \lambda_l \in U$ and $\alpha \in \mathbf{N}_0^l$ with $\alpha_j \leq m$, $j =
1, \dots, l$ (in case E depends only C^m on λ).*

Proof For $i = 0, \dots, \alpha_j$, let us define

$$d_j^i := \frac{1}{i!(\alpha_j - i)!} \left(\frac{\partial}{\partial \lambda_j}\right)^{\alpha_j - i} d_j.$$

Then the Leibniz formula yields

$$F = \sum_{j=1}^{l} \sum_{i=0}^{\alpha_j} d_j^i \cdot \frac{\partial^i E}{\partial \lambda^i}(\lambda_j).$$

On the other hand, by differentiation of the equation $\big(P_1(\partial) - \lambda P_2(\partial)\big) E(\lambda) = \delta$
with respect to λ, we obtain

$$\big(P_1(\partial) - \lambda P_2(\partial)\big)^{i+1} \frac{\partial^i E(\lambda)}{\partial \lambda^i} = i! P_2(\partial)^i \delta.$$

Therefore,

$$\prod_{j=1}^{l} \big(P_1(\partial) - \lambda_j P_2(\partial)\big)^{\alpha_j + 1} F = \sum_{j=1}^{l} \sum_{i=0}^{\alpha_j} d_j^i \left(\prod_{\substack{k=1 \\ k \neq j}}^{l} \big(P_1(\partial) - \lambda_k P_2(\partial)\big)^{\alpha_k + 1}\right)$$

$$\times \big(P_1(\partial) - \lambda_j P_2(\partial)\big)^{\alpha_j - i} \big(P_1(\partial) - \lambda_j P_2(\partial)\big)^{i+1} \frac{\partial^i E}{\partial \lambda^i}(\lambda_j)$$

$$= \sum_{j=1}^{l} \prod_{\substack{k=1 \\ k \neq j}}^{l} \big(P_1(\partial) - \lambda_k P_2(\partial)\big)^{\alpha_k + 1} \sum_{i=0}^{\alpha_j} d_j^i i! \big(P_1(\partial) - \lambda_j P_2(\partial)\big)^{\alpha_j - i} P_2(\partial)^i \delta.$$

By Fourier transformation in $\mathcal{E}'$ and setting $s = P_1(i\xi)$, $t = -P_2(i\xi)$, $\xi \in \mathbf{R}^n$, (3.3.1) reduces to

$$(s + \lambda_0 t)^{|\alpha|+l-1} = \sum_{j=1}^{l} \left(\prod_{\substack{k=1 \\ k \neq j}}^{l} (s + \lambda_k t)^{\alpha_k+1} \right) \sum_{i=0}^{\alpha_j} d_j^i i! (s + \lambda_j t)^{\alpha_j-i}(-t)^i. \qquad (3.3.2)$$

Upon division by $\prod_{j=1}^{l}(s + \lambda_j t)^{\alpha_j+1}$, (3.3.2) is equivalent to

$$(s + \lambda_0 t)^{|\alpha|+l-1} \prod_{j=1}^{l} (s + \lambda_j t)^{-\alpha_j-1} = \sum_{j=1}^{l} \sum_{i=0}^{\alpha_j} d_j^i i! (-t)^i (s + \lambda_j t)^{-i-1}. \qquad (3.3.3)$$

If we abbreviate $c_j = \prod_{\substack{k=1 \\ k \neq j}}^{l}(\lambda_j - \lambda_k)^{-\alpha_k-1}$ as in Proposition 3.2.1, then

$$d_j^i i! = \frac{1}{(\alpha_j - i)!} \left(\frac{\partial}{\partial \lambda_j} \right)^{\alpha_j-i} \left(c_j(\lambda_j - \lambda_0)^{|\alpha|+l-1} \right).$$

Furthermore,

$$(s + \lambda_j t)^{-i-1}(-t)^i = \frac{1}{i!} \left(\frac{\partial}{\partial \lambda_j} \right)^i (s + \lambda_j t)^{-1}$$

and hence the right-hand side of (3.3.3) yields, by Leibniz' rule,

$$\sum_{j=1}^{l} \frac{1}{\alpha_j!} \left(\frac{\partial}{\partial \lambda_j} \right)^{\alpha_j} \left[\frac{c_j(\lambda_j - \lambda_0)^{|\alpha|+l-1}}{s + \lambda_j t} \right].$$

Therefore, (3.3.3) is a consequence of formula (3.2.2). $\qquad \square$

Let us observe that the limiting cases of $\lambda_0 = 0$ and $\lambda_0 \to \infty$, respectively, lead to the formulas

$$R(\partial) \sum_{j=1}^{l} \frac{1}{\alpha_j!} \partial_{\lambda_j}^{\alpha_j} (\lambda_j^{|\alpha|+l-1} c_j E(\lambda_j)) = P_1(\partial)^{|\alpha|+l-1} \delta$$

and

$$R(\partial) \sum_{j=1}^{l} \frac{1}{\alpha_j!} \partial_{\lambda_j}^{\alpha_j} (c_j E(\lambda_j)) = P_2(\partial)^{|\alpha|+l-1} \delta, \qquad (3.3.4)$$

wherein $R(\partial) = \prod_{j=1}^{l}\big(P_1(\partial) - \lambda_j P_2(\partial)\big)^{\alpha_j+1}$ and $c_j = \prod_{\substack{k=1 \\ k\neq j}}^{l}(\lambda_j - \lambda_k)^{-\alpha_k-1}$. In particular, for $P_2(\partial) = 1$, formula (3.3.4) reduces to the assertion in Proposition 1.4.4.

We shall next apply Proposition 3.3.1 to *homogeneous operators in the plane*, i.e., to $R(\partial) = \prod_{j=1}^{l}(\partial_1 - \lambda_j\partial_2)^{\alpha_j+1}$. For $\alpha = 0$, a fundamental solution of this operator was constructed in Somigliana [258], § 2, (5), p. 147; Herglotz [125], I, (98), p. 111; Bureau [32], 8, (15'), p. 12; Wagner [285], Satz 4, p. 40. For arbitrary α, a formula was presented first in Galler [86], (4.1), p. 15.

Proposition 3.3.2 *Let $l > 1$ and $R(\partial) = \prod_{j=1}^{l}(\partial_1 - \lambda_j\partial_2)^{\alpha_j+1}$, $\alpha \in \mathbf{N}_0^l$, for pairwise different $\lambda_1, \ldots, \lambda_l \in \mathbf{C} \setminus \mathbf{R}$. Let $\log z$ be defined as usually for $z \in \mathbf{C} \setminus (-\infty, 0]$, i.e., $\log(re^{i\varphi}) = \log r + i\varphi$ for $r > 0$ and $\varphi \in (-\pi, \pi)$. For $j = 1, \ldots, l$, set*

$$c_j = \prod_{\substack{k=1 \\ k\neq j}}^{l}(\lambda_j - \lambda_k)^{-\alpha_k-1} \text{ and } P_j(x_1, x_2) = \frac{1}{\alpha_j!}\partial_{\lambda_j}^{\alpha_j}\big[c_j(x_2 + \lambda_j x_1)^{|\alpha|+l-2}\big].$$

Then a fundamental solution E of $R(\partial)$ is given by

$$E = \frac{i}{2\pi(|\alpha| + l - 2)!} \sum_{j=1}^{l} \operatorname{sign}(\operatorname{Im}\lambda_j)P_j(x)\log(x_2 + \lambda_j x_1). \qquad (3.3.5)$$

In particular, if $R(\partial)$ has real coefficients and $x_1 \neq 0$, then

$$E(x) = \frac{1}{2\pi(|\alpha| + l - 2)!} \sum_{\substack{j=1 \\ \operatorname{Im}\lambda_j > 0}}^{l} \bigg[-\operatorname{Im}P_j \cdot \log(x_2^2 + 2x_1 x_2 \operatorname{Re}\lambda_j + |\lambda_j|^2 x_1^2)$$

$$+ 2\operatorname{Re}P_j \cdot \arctan\Big(\frac{x_2 + x_1\operatorname{Re}\lambda_j}{x_1\operatorname{Im}\lambda_j}\Big)\bigg].$$
$$(3.3.6)$$

Proof

(a) In order to apply Proposition 3.3.1 to the operator $R(\partial)$, we set $\lambda_0 = -i$ and consequently $d_j = (\lambda_j + i)^{|\alpha|+l-1}c_j$. Then $F = \sum_{j=1}^{l}\alpha_j!^{-1}\partial_{\lambda_j}^{\alpha_j}\big(d_j E(\lambda_j)\big)$ fulfills $R(\partial)F = (\partial_1 + i\partial_2)^{|\alpha|+l-1}\delta$ if $E(\lambda)$ is a fundamental solution of $\partial_1 - \lambda\partial_2$ depending holomorphically on λ for λ in an open neighborhood of $\{\lambda_1, \ldots, \lambda_l\}$.

By a linear transformation (see Proposition 1.3.19) of the fundamental solution $\frac{1}{2\pi(x_1+ix_2)}$ of the Cauchy–Riemann operator $\partial_1 + i\partial_2$, see Example 1.3.14 (b), we infer that we can set $E(\lambda) = -\operatorname{sign}(\operatorname{Im}\lambda)/\big(2\pi i(x_2 + \lambda x_1)\big)$, $\lambda \in \mathbf{C}\setminus\mathbf{R}$, and hence

$$F = \frac{i}{2\pi}\sum_{j=1}^{l}\frac{\operatorname{sign}(\operatorname{Im}\lambda_j)}{\alpha_j!}\Big(\frac{\partial}{\partial\lambda_j}\Big)^{\alpha_j}\frac{(\lambda_j + i)^{|\alpha|+l-1}c_j}{x_2 + \lambda_j x_1}.$$

(b) If E is defined by Eq. (3.3.5), then $(\partial_1 + i\partial_2)^{|\alpha|+l-1}E = F$. In fact, this is obvious in the open set $U = \{x \in \mathbf{R}^2; x_1 \neq 0 \text{ or } x_2 > 0\}$, since there $\log(x_2 + \lambda_j x_1)$, $j = 1, \ldots, l$, and hence also E are C^∞ and can be classically differentiated. Along the half-ray $x_1 = 0$, $x_2 < 0$, the function $\log(x_2 + \lambda x_1) \in L^1_{\text{loc}}(\mathbf{R}^2)$ is discontinuous with the jump $2\pi i \operatorname{sign}(\operatorname{Im}\lambda)$, and hence, by the jump formula (1.3.9),

$$(\partial_1 + i\partial_2) \log(x_2 + \lambda x_1) = 2\pi i \operatorname{sign}(\operatorname{Im}\lambda)\delta(x_1) \otimes Y(-x_2) + \frac{\lambda + i}{x_2 + \lambda x_1}.$$

Furthermore, for $t = 0$, the identity (3.2.2) yields

$$1 = \sum_{j=1}^{l} \frac{1}{\alpha_j!} \left(\frac{\partial}{\partial \lambda_j}\right)^{\alpha_j} [c_j(\lambda_j - \lambda_0)^{|\alpha|+l-1}]$$

for arbitrary $\lambda_0 \in \mathbf{C}$, and hence

$$\sum_{j=1}^{l} \frac{1}{\alpha_j!} \left(\frac{\partial}{\partial \lambda_j}\right)^{\alpha_j} [\lambda_j^k c_j] = \begin{cases} 0, & \text{if } k = 0, \ldots, |\alpha| + l - 2, \\ 1, & \text{if } k = |\alpha| + l - 1. \end{cases} \tag{3.3.7}$$

This implies that the sum over the jump terms vanishes when the differentiations in $(\partial_1 + i\partial_2)^{|\alpha|+l-1}E$ are performed, i.e., $(\partial_1 + i\partial_2)^{|\alpha|+l-1}E = F$ holds in $\mathcal{D}'(\mathbf{R}^2)$.

(c) The equation $R(\partial)F = (\partial_1 + i\partial_2)^{|\alpha|+l-1}\delta$ then furnishes

$$(\partial_1 + i\partial_2)^{|\alpha|+l-1}[R(\partial)E - \delta] = 0.$$

From Liouville's theorem, we therefore conclude by induction that $R(\partial)E = \delta$ since $R(\partial)E$ and its derivatives vanish at infinity. Hence E is a fundamental solution of $R(\partial)$, which by Proposition 2.4.8, is uniquely determined up to polynomials of degree $|\alpha| + l - 2$ by the growth condition (2.4.4).

(d) Finally, if $R(\partial)$ has real coefficients, and e.g., $\lambda_2 = \overline{\lambda_1}$, $\operatorname{Im}\lambda_1 > 0$, $\alpha_1 = \alpha_2$, then $P_2 = \overline{P_1}$, and we can add the terms for $j = 1, 2$:

$$\operatorname{sign}(\operatorname{Im}\lambda_1)P_1(x)\log(x_2 + \lambda_1 x_1) + \operatorname{sign}(\operatorname{Im}\lambda_2)P_2(x)\log(x_2 + \lambda_2 x_1)$$

$$= 2i \operatorname{Im}(P_1(x) \cdot \log(x_2 + \lambda_1 x_1))$$

$$= i \operatorname{Im} P_1 \cdot \log(x_2^2 + 2x_1 x_2 \operatorname{Re}\lambda_1 + |\lambda_1|^2 x_1^2) + 2i \operatorname{Re} P_1 \cdot \arg(x_2 + \lambda_1 x_1).$$

Here $\log z = \log|z| + i \arg z$ is determined as in the proposition, i.e., $\arg z \in (-\pi, \pi)$ for $z \in \mathbf{C} \setminus (-\infty, 0]$. Note that then

$$\arg z = -\arctan\left(\frac{\operatorname{Re} z}{\operatorname{Im} z}\right) + \frac{\pi}{2}\operatorname{sign}(\operatorname{Im} z)$$

holds for $z \in \mathbf{C} \setminus \mathbf{R}$, and hence, due to $\operatorname{Im} \lambda_1 > 0$,

$$\arg(x_2 + \lambda_1 x_1) = -\arctan\left(\frac{x_2 + x_1 \operatorname{Re} \lambda_1}{x_1 \operatorname{Im} \lambda_1}\right) + \frac{\pi}{2} \operatorname{sign} x_1$$

holds for $x_1 \neq 0$. Therefore, for $x_1 \neq 0$,

$$E(x) = \frac{i}{2\pi(|\alpha| + l - 2)!} \sum_{j=1}^{l} \operatorname{sign}(\operatorname{Im} \lambda_j) P_j(x) \log(x_2 + \lambda_j x_1)$$

$$= \frac{1}{2\pi(|\alpha| + l - 2)!} \sum_{\substack{j=1 \\ \operatorname{Im} \lambda_j > 0}}^{l} \Big[-\operatorname{Im}(P_j) \log(x_2^2 + 2x_1 x_2 \operatorname{Re} \lambda_j + |\lambda_j|^2 x_1^2)$$

$$+ 2\operatorname{Re}(P_j) \arctan\left(\frac{x_2 + x_1 \operatorname{Re} \lambda_j}{x_1 \operatorname{Im} \lambda_j}\right) - \pi \operatorname{Re}(P_j) \operatorname{sign} x_1 \Big].$$

$$(3.3.8)$$

Due to formula (3.3.7), $\sum_{j=1}^{l} \operatorname{Re} P_j = \frac{1}{2} \sum_{j=1}^{l} P_j = 0$, and thus the last term in (3.3.8) can be omitted. This yields formula (3.3.6) and concludes the proof. $\quad\square$

Example 3.3.3 Let us apply Proposition 3.3.2 to the operator

$$R(\partial) = \prod_{j=1}^{m} (\partial_1^2 + b_j^2 \partial_2^2), \qquad b_j > 0, \quad b_j \neq b_k \text{ for } j \neq k.$$

This amounts to setting $l = 2m$, $\alpha = 0$ and $\lambda_j = ib_j$, $\lambda_{j+m} = -ib_j$, $j = 1, \ldots, m$. For $j = 1, \ldots, m$, we then obtain $c_j = (-1)^m i d_j / (2b_j)$, where $d_j = \prod_{\substack{k=1 \\ k \neq j}}^{m} (b_j^2 - b_k^2)^{-1}$. Furthermore, $P_j = c_j (x_2 + ib_j x_1)^{2m-2}$ and hence

$$\operatorname{Re} P_j = \frac{(-1)^{m+1} d_j}{2b_j} \operatorname{Im}(x_2 + ib_j x_1)^{2m-2} = \frac{(-1)^{m+1} d_j}{2b_j} q(b_j x_1, x_2)$$

with

$$q(x_1, x_2) = \sum_{k=0}^{m-2} \binom{2m-2}{2k+1} (-1)^k x_1^{2k+1} x_2^{2m-2k-3},$$

and, similarly,

$$\operatorname{Im} P_j = \frac{(-1)^m d_j}{2b_j} \operatorname{Re}(x_2 + ib_j x_1)^{2m-2} = \frac{(-1)^m d_j}{2b_j} p(b_j x_1, x_2)$$

with

$$p(x_1, x_2) = \sum_{k=0}^{m-1} \binom{2m-2}{2k} (-1)^k x_1^{2k} x_2^{2m-2k-2}.$$

Therefore we obtain as fundamental solution E of $\prod_{j=1}^{m}(\partial_1^2 + b_j^2 \partial_2^2)$ the expression

$$E = \frac{(-1)^{m+1}}{4\pi(2m-2)!} \sum_{j=1}^{m} \frac{d_j}{b_j} \left\{ p(b_j x_1, x_2) \log(x_2^2 + b_j^2 x_1^2) + 2q(b_j x_1, x_2) \arctan\left(\frac{x_2}{b_j x_1}\right) \right\}.$$

(3.3.9)

Formula (3.3.9) appears for the first time in Galler [86], Section 11, p. 48 (with $x_1 = y$, $x_2 = x$).

In particular, if we set $m = 2$, $b_1 = 1$ and $b_2 = \sqrt{a_2}$, $a_2 > 0$, formula (3.3.9) furnishes the expression in (3.1.15) for $a_1 = 1$.   □

Example 3.3.4 Let us next apply the difference device, i.e. formula (3.3.1) in Proposition 3.3.1, to the *product of wave operators* $R(\partial) = \prod_{j=1}^{l}(\partial_t^2 - \lambda_j \Delta_n)^{\alpha_j+1}$, $\alpha \in \mathbb{N}_0^l$, for λ_j positive and pairwise different. This operator has been treated in Examples 3.2.2 to 3.2.5 with the method of parameter integration.

Setting $\lambda_0 = 0$ in Proposition 3.3.1 we obtain $R(\partial)F = \partial_t^{2|\alpha|+2l-2}\delta$ if

$$F = \sum_{j=1}^{l} \frac{1}{\alpha_j!} \partial_{\lambda_j}^{\alpha_j} \left(\lambda_j^{|\alpha|+l-1} c_j E(\lambda_j) \right),$$

$c_j = \prod_{\substack{k=1 \\ k \neq j}}^{l}(\lambda_j - \lambda_k)^{-\alpha_k-1}$ and $E(\lambda)$ is the forward fundamental solution of $\partial_t^2 - \lambda \Delta_n$.

Since F and the fundamental solution $(2|\alpha| + 2l - 3)!^{-1} t_+^{2|\alpha|+2l-3} \otimes \delta(x)$ of $\partial_t^{2|\alpha|+2l-2}$ are convolvable by support, we can represent the forward fundamental solution E of $R(\partial)$ in the form

$$E = \sum_{j=1}^{l} \frac{1}{\alpha_j!} \partial_{\lambda_j}^{\alpha_j} \left[\lambda_j^{|\alpha|+l-1} c_j E(\lambda_j) * \left(\frac{t_+^{2|\alpha|+2l-3}}{(2|\alpha| + 2l - 3)!} \otimes \delta(x) \right) \right].$$

If $2|\alpha| + 2l - 3 \geq n - 2$, i.e., $2|\alpha| + 2l - n > 0$, then

$$E(\lambda) * \left(\frac{t_+^{2|\alpha|+2l-3}}{(2|\alpha| + 2l - 3)!} \otimes \delta(x) \right) = E_1(\lambda) * \left(\frac{t_+^{2|\alpha|+2l-n-1}}{(2|\alpha| + 2l - n - 1)!} \otimes \delta(x) \right),$$

where $E_1(\lambda)$ is the forward fundamental solution of $\partial_t^{n-2}(\partial_t^2 - \lambda\Delta_n)$, $\lambda > 0$. From the subsequent Lemma 3.3.5 (for the case $k = 1$), we infer that

$$E_1(\lambda) = \frac{Y(t - |x|/\sqrt{\lambda})|x|^{2-n}(t^2 - \frac{|x|^2}{\lambda})^{(n-3)/2}}{(n-2)!|S^{n-1}|\lambda},$$

and hence

$$E(t,x) = \frac{\Gamma(\frac{n}{2})}{2\pi^{n/2}(n-2)!\,\Gamma(2|\alpha| + 2l - n)} \sum_{j=1}^{l} \frac{1}{\alpha_j!}\left(\frac{\partial}{\partial\lambda_j}\right)^{\alpha_j}$$

$$\left[c_j\lambda_j^{|\alpha|+l-2}|x|^{2-n}Y\left(t - \frac{|x|}{\sqrt{\lambda_j}}\right)\int_{|x|/\sqrt{\lambda_j}}^{t}(t-\tau)^{2|\alpha|+2l-n-1}\left(\tau^2 - \frac{|x|^2}{\lambda_j}\right)^{(n-3)/2}d\tau\right].$$

$$(3.3.10)$$

Up to the substitution $\tau = \frac{v|x|}{\sqrt{\lambda_j}}$, formula (3.3.10) coincides with Herglotz' representation of E in (3.2.15). $\qquad\square$

Lemma 3.3.5 *Let $n, k \in \mathbf{N}$ with $n \geq 2k$. Then the fundamental solution G of the operator $\partial_t^{n-2k}(\partial_t^2 - \Delta_n)^k$ with support in $t \geq 0$ is given by the locally integrable function*

$$G(t,x) = \frac{\Gamma(\frac{n}{2}+1-k)Y(t-|x|)|x|^{2k-n}(t^2-|x|^2)^{(n-1)/2-k}}{(k-1)!(n-2k)!\,2^{2k-1}\pi^{n/2}}. \qquad (3.3.11)$$

Proof Let us repeat first that, according to Example 2.3.6, the forward fundamental solution of $(\partial_t^2 - \Delta_n)^k$ is the value for $\lambda = -k$ of the entire function

$$\mathbf{C} \longrightarrow S'(\mathbf{R}^{n+1}) : \lambda \longmapsto E(\lambda)$$

which, for $\mathrm{Re}\,\lambda < \frac{1-n}{2}$, is locally integrable and given by

$$E(\lambda) = \frac{2^{2\lambda+1}Y(t-|x|)(t^2-|x|^2)^{-\lambda-(n+1)/2}}{\pi^{(n-1)/2}\Gamma(-\lambda)\Gamma(-\lambda-\frac{n-1}{2})},$$

see (2.3.11).

On the other hand, the distribution-valued function

$$\{\lambda \in \mathbf{C}; \mathrm{Re}\,\lambda > -\tfrac{n}{2}\} \longrightarrow S'(\mathbf{R}^{n+1}) : \lambda \longmapsto F(\lambda) := \frac{t_+^{2\lambda+n-1}}{\Gamma(2\lambda+n)} \otimes \delta(x)$$

can be extended to an entire function, cf. Example 1.5.11, and $E(\lambda)$, $F(\lambda)$ are convolvable by support for each $\lambda \in \mathbf{C}$. Due to

$$\partial_t^{n-2k}(\partial_t^2 - \Delta_n)^k[E(-k) * F(-k)] = \delta,$$

we conclude that the forward fundamental solution G of $\partial_t^{n-2k}(\partial_t^2 - \Delta_n)^k$ is given by $G = E(-k) * F(-k)$.

If $-\frac{n}{2} < \text{Re}\,\lambda < \frac{1-n}{2}$, then $E(\lambda)$ is locally integrable and the convolution $E(\lambda) * F(\lambda)$ can be represented by the absolutely convergent integral

$$E(\lambda) * F(\lambda) = \frac{2^{2\lambda+1}Y(t-|x|)}{\pi^{(n-1)/2}\Gamma(-\lambda)\Gamma(-\lambda-\frac{n-1}{2})\Gamma(2\lambda+n)}$$

$$\times \int_{|x|}^t (t-\tau)^{2\lambda+n-1}(\tau^2-|x|^2)^{-\lambda-(n+1)/2}d\tau.$$

From Eq. 421.4 in Gröbner and Hofreiter [115], we conclude that

$$E(\lambda) * F(\lambda) = \frac{2^{1-n}|x|^{-2\lambda-n}}{\pi^{(n-1)/2}\Gamma(-\lambda)\Gamma(\lambda+\frac{n+1}{2})} \cdot Y(t-|x|)(t^2-|x|^2)^{\lambda+(n-1)/2}.$$

(3.3.12)

The right-hand side in (3.3.12) is locally integrable for $-\frac{n+1}{2} < \text{Re}\,\lambda < 0$, and coincides, for these λ, with $E(\lambda) * F(\lambda)$ since $\lambda \mapsto E(\lambda) * F(\lambda)$ is also entire.

In particular, setting $\lambda = -k$ and using the doubling formula of the gamma function we obtain

$$G = E(-k) * F(-k) = \frac{2^{1-n}|x|^{2k-n}}{\pi^{(n-1)/2}(k-1)!\Gamma(-k+\frac{n+1}{2})} \cdot Y(t-|x|)(t^2-|x|^2)^{-k+(n-1)/2}$$

$$= \frac{\Gamma(\frac{n}{2}+1-k)Y(t-|x|)|x|^{2k-n}(t^2-|x|^2)^{(n-1)/2-k}}{(k-1)!(n-2k)!\,2^{2k-1}\pi^{n/2}}.$$

$\square$

Example 3.3.6 From (3.3.11) for $k = 1$, we obtain the following representation of the solution u of the Cauchy problem

$$(\partial_t^2 - \Delta_n)u = 0 \text{ for } t > 0, \qquad u(0,x) = 0, \ (\partial_t u)(0,x) = u_1(x)$$

by convolution:

$$u(t,x) = \frac{\partial^{n-2}}{\partial t^{n-2}}[G * (\delta(t) \otimes u_1(x))]$$

$$= \frac{1}{(n-2)!}\frac{\partial^{n-2}}{\partial t^{n-2}}\left(\int_0^t (t^2-r^2)^{(n-3)/2}\left(\frac{1}{|\mathbf{S}^{n-1}|}\int_{\mathbf{S}^{n-1}} u_1(x-r\omega)\,d\sigma(\omega)\right)r\,dr\right).$$

(3.3.13)

In this form, G already appears in Herglotz [125], II, (177),(178), p. 314; [126], p. 557; John [151], (2.32), p. 33; Courant and Hilbert [52], Ch. VI, § 12, 1., Eq. (2), p. 682. We note that our deduction of formula (3.3.13) is, in contrast to the references, independent of the parity of the dimension n. A different unified deduction of (3.3.13) for all dimensions n by means of partial Fourier transformation is contained in Shilov [250], Section 4.7.1, pp. 290–292.   □

3.4 Products of Operators from Higher-Dimensional Affine Subspaces

Generalizing the considerations in Sect. 3.2, which refer to operators of the form

$$R(\partial) = \prod_{j=1}^{l} \left(P_1(\partial) + \lambda_j P_2(\partial) \right)^{\alpha_j + 1}, \qquad \alpha \in \mathbf{N}_0^l, \ \lambda \in \mathbf{R}^l,$$

we consider here products of the form

$$R(\partial) = \prod_{j=1}^{l} \left(P_0(\partial) + \sum_{i=1}^{m} \lambda_{ij} P_i(\partial) \right), \qquad \Lambda = (\lambda_{ij}) \in \mathbf{R}^{m \times l},$$

i.e., all the factors of $R(\partial)$ are contained in the affine subspace of $\mathbf{C}[\partial]$ spanned by $P_0(\partial) + \lambda P_i(\partial)$, $\lambda \in \mathbf{R}$, $i = 1, \dots, m$. The corresponding formulas were derived in Ortner and Wagner [211].

Proposition 3.4.1 *Let $l, m \in \mathbf{N}$, $m \le l - 1$, and let $\Lambda = (\lambda_{ij})$ be a real $m \times l$ matrix. We denote by Λ_j the j-th column of Λ, i.e., $\Lambda_j = (\lambda_{1j}, \dots, \lambda_{mj})^T \in \mathbf{R}^m$ and we suppose that $\Lambda_{j_1} - \Lambda_k, \dots, \Lambda_{j_m} - \Lambda_k$ are linearly independent in $\mathbf{R}^m$ for pairwise different indices $j_1, \dots, j_m, k \in \{1, \dots, l\}$. Let $\Lambda_0 \in \mathbf{R}^m$ and define*

$$T_j := \left\{ \rho_0 \Lambda_0 + \sum_{i=1}^{m} \rho_i \Lambda_{j_i}; \ \rho = (\rho_0, \dots, \rho_m) \in \Sigma_m \right\}$$

where Σ_m is the standard simplex in $\mathbf{R}^{m+1}$, see 3.1, and $j := (j_1, \dots, j_m)$, $1 \le j_1 < \cdots < j_m \le l$.

We suppose that the operators $Q_\mu(\partial) = P_0(\partial) + \mu_1 P_1(\partial) + \cdots + \mu_m P_m(\partial)$ are uniformly quasihyperbolic with respect to $N \in \mathbf{R}^n \setminus \{0\}$ for $\mu = (\mu_1, \dots, \mu_m) \in M := \cup_j T_j$, and we denote by E and by E_μ the fundamental solutions of

$$R(\partial) = \prod_{j=1}^{l} \left(P_0(\partial) + \sum_{i=1}^{m} \lambda_{ij} P_i(\partial) \right) \qquad and \ of \ \ Q_\mu(\partial)^l,$$

respectively, which satisfy $e^{-\sigma x N} E, e^{-\sigma x N} E_\mu \in \mathcal{S}'(\mathbf{R}^n)$ *for* $\sigma > \sigma_0$, σ_0 *as in* (3.1.3) *with respect to* $\mu \in M$. *Then* E_μ *continuously depends on* $\mu \in M$ *and*

$$E = \frac{(l-1)!}{(l-m-1)!} \sum_{1 \leq j_1 < \cdots < j_m \leq l} c_j \int_{T_j} \det^{l-m-1}\left(\Lambda_{j_1} - \mu, \ldots, \Lambda_{j_m} - \mu\right) E_\mu \, d\mu$$

(3.4.1)

if the constants c_j *are given by*

$$c_j := \text{sign}\left[\det\left(\Lambda_{j_1} - \Lambda_0, \ldots, \Lambda_{j_m} - \Lambda_0\right)\right] \prod_{\substack{k=1 \\ k \notin \{j_1, \ldots, j_m\}}}^{l} \det^{-1}\left(\Lambda_{j_1} - \Lambda_k, \ldots, \Lambda_{j_m} - \Lambda_k\right).$$

Proof According to Proposition 3.1.2, E can be represented by the formula

$$E = (l-1)! \int_{\Sigma_{l-1}} E_{\Lambda\rho} \, d\sigma(\rho),$$

(3.4.2)

since, for $\rho \in \Sigma_{l-1}$,

$$\sum_{j=1}^{l} \rho_j\left[P_0(\partial) + \sum_{i=1}^{m} \lambda_{ij} P_i(\partial)\right] = P_0(\partial) + \sum_{i=1}^{m} \mu_i P_i(\partial), \qquad \mu = \Lambda\rho.$$

Upon application to a test function $\phi \in \mathcal{D}(\mathbf{R}^n)$, the assertion therefore follows from the representation of the "Dirichlet averages" $\int_{\Sigma_{l-1}} f(\Lambda\rho) \, d\sigma(\rho)$, $f \in \mathcal{C}(M)$, in the following lemma. $\qquad \square$

Lemma 3.4.2 *Let* $l, m, \Lambda, \Lambda_0, T_j, M, c_j$ *be as in* Proposition 3.4.1 *and assume that* $f \in \mathcal{C}(M)$. *Then we have*

$$\int_{\Sigma_{l-1}} f(\Lambda\rho) \, d\sigma(\rho) = \frac{1}{(l-m-1)!}$$

$$\times \sum_{1 \leq j_1 < \cdots < j_m \leq l} c_j \int_{T_j} \det^{l-m-1}\left(\Lambda_{j_1} - \mu, \ldots, \Lambda_{j_m} - \mu\right) f(\mu) \, d\mu. \quad (3.4.3)$$

Proof

(a) First we note that the linear combinations of functions of the form

$$f(\mu) = (t + v \cdot \mu)^{-l}, \qquad v \in \mathbf{R}^m, \; t > \max\{-v \cdot \mu; \; \mu \in M\}, \quad (3.4.4)$$

are dense in the Banach algebra $\mathcal{C}(M)$ due to the Stone–Weierstraß theorem, see Hewitt and Stromberg [130], Thm. 7.30, p. 95. In fact, products of functions of the form in (3.4.4) can be approximated by functions of the same kind by differentiation and integration processes according to the formula

$$\frac{1}{(t+v\cdot\mu)^l(s+w\cdot\mu)^l} = \frac{(2l-1)!}{(l-1)!^2}\int_0^1 \frac{x^{l-1}(1-x)^{l-1}\,dx}{[xt+(1-x)s+(xv+(1-x)w)\cdot\mu]^{2l}},$$

cf. Gröbner and Hofreiter [115], Eq. 421.4. Hence it suffices to verify formula (3.4.3) in the lemma for functions of the form given in (3.4.4).

(b) Let us fix now $v \in \mathbf{R}^m$ and $t > \max\{-v\cdot\mu; \mu \in M\}$ and consider the function $f(\mu) = (t+v\cdot\mu)^{-l}$ in $C(M)$. Due to Feynman's formula (3.1.2), we have

$$\int_{\Sigma_{l-1}} f(\Lambda\rho)\,d\sigma(\rho) = \int_{\Sigma_{l-1}} \frac{d\sigma(\rho)}{(t+v\cdot\Lambda\rho)^l} = \frac{1}{(l-1)!}\prod_{k=1}^l (t+v\cdot\Lambda_k)^{-1}.$$

(3.4.5)

We next reduce the product on the right-hand side of (3.4.5) to a sum over products with $m+1$ factors only. This is accomplished by means of the many-dimensional version of Lagrange's interpolation formula in Lemma 3.4.3 below. This yields

$$\prod_{k=1}^l (t+v\cdot\Lambda_k)^{-1} = (t+v\cdot\Lambda_0)^{m-l} \sum_{1\le j_1<\cdots<j_m\le l} e_j \prod_{i=1}^m (t+v\cdot\Lambda_{j_i})^{-1}, \quad (3.4.6)$$

where

$$e_j := \prod_{\substack{k=1 \\ k\notin\{j_1,\dots,j_m\}}}^l \frac{\det(\Lambda_{j_1}-\Lambda_0,\dots,\Lambda_{j_m}-\Lambda_0)}{\det(\Lambda_{j_1}-\Lambda_k,\dots,\Lambda_{j_m}-\Lambda_k)}.$$

(In Lemma 3.4.3, we use $P = z_0^{l-m}$, $z = v$, $z_0 = t+v\cdot\Lambda_0$, and we replace Λ_j by $\Lambda_0 - \Lambda_j$ for $j = 1,\dots,l$.)

The products on the right-hand side of (3.4.6) can in turn be represented by Feynman integrals by employing (3.1.5):

$$(t+v\cdot\Lambda_0)^{m-l}\prod_{i=1}^m (t+v\cdot\Lambda_{j_i})^{-1} = \frac{(l-1)!}{(l-m-1)!}\int_{\Sigma_m} \frac{\rho_0^{l-m-1}\,d\sigma(\rho)}{(t+\rho_0 v\cdot\Lambda_0+\sum_{i=1}^m \rho_i v\cdot\Lambda_{j_i})^l}.$$

The last integral is transformed into one over T_j, which is the convex hull of $\Lambda_0, \Lambda_{j_1},\dots,\Lambda_{j_m}$, by the linear substitution $\mu = \rho_0\Lambda_0 + \sum_{i=1}^m \rho_i\Lambda_{j_i}$. In fact, this substitution has the Jacobian

$$\det\left(\left(\frac{\partial\rho_i}{\partial\mu_k}\right)_{i,k=1,\dots,m}\right) = \det{}^{-1}(\Lambda_{j_1}-\Lambda_0,\dots,\Lambda_{j_m}-\Lambda_0),$$

and ρ_0 is determined by μ through the system of linear equations

$$\rho_0 + \rho_1 + \cdots + \rho_m = 1, \qquad \rho_0 \Lambda_0 + \sum_{i=1}^{m} \rho_i \Lambda_{j_i} = \mu,$$

and hence Cramer's rule implies

$$\rho_0 = \frac{\det(\Lambda_{j_1} - \mu, \ldots, \Lambda_{j_m} - \mu)}{\det(\Lambda_{j_1} - \Lambda_0, \ldots, \Lambda_{j_m} - \Lambda_0)}.$$

Therefore,

$$\int_{\Sigma_{l-1}} \frac{d\sigma(\rho)}{(t + v \cdot \Lambda\rho)^l} = \frac{1}{(l-m-1)!}$$

$$\times \sum_{\substack{1 \le j_1 < \cdots < j_m \le l}} c_j \int_{T_j} \det^{l-m-1}(\Lambda_{j_1} - \mu, \ldots, \Lambda_{j_m} - \mu)(t + v \cdot \mu)^{-l} \, d\mu,$$

which is formula (3.4.3) in the special case of $f(\mu) = (t + v \cdot \mu)^{-l}$.   □

The following many-dimensional version of Lagrange's interpolation formula was first formulated and proven in Ortner and Wagner [212], Lemma 1, p. 86; it has proven useful in quantum field theory, cf. Wagner [289], (2), p. 2429; Metzner and Neumayr [178], p. 623.

Lemma 3.4.3 *Let* $1 \le m < l$, $\Lambda_1, \ldots, \Lambda_l \in \mathbf{C}^m$ *such that* $\Lambda_{j_1}, \ldots, \Lambda_{j_m}$ *as well as* $\Lambda_{j_1} - \Lambda_k, \ldots, \Lambda_{j_m} - \Lambda_k$ *are linearly independent for pairwise different indices* $j_1, \ldots, j_m, k \in \{1, \ldots, l\}$. *For* $j = (j_1, \ldots, j_m)$, $1 \le j_1 < \cdots < j_m \le l$, *determine* $z(j) \in \mathbf{C}^m$ *by the system of linear equations* $\Lambda_{j_i} \cdot z(j) = 1$, $i = 1, \ldots, m$. *Then, for each complex homogeneous polynomial* $P(z_0, z)$ *in* $m+1$ *variables of the degree* $l - m$, *we have*

$$P(z_0, z) = \sum_{\substack{j=(j_1, \ldots, j_m) \\ 1 \le j_1 < \cdots < j_m \le l}} P(1, z(j)) \prod_{\substack{k=1 \\ k \notin \{j_1, \ldots, j_m\}}}^{l} \frac{(z_0 - \Lambda_k \cdot z) \det(\Lambda_{j_1}, \ldots, \Lambda_{j_m})}{\det(\Lambda_{j_1} - \Lambda_k \ldots \Lambda_{j_m} - \Lambda_k)}.$$

$$(3.4.7)$$

Proof For $j = (j_1, \ldots, j_m)$, $1 \le j_1 < \cdots < j_m \le l$, define the polynomial P_j by

$$P_j(z_0, z) := \prod_{\substack{k=1 \\ k \notin \{j_1, \ldots, j_m\}}}^{l} (z_0 - \Lambda_k \cdot z).$$

By the definition of $z(j)$, we have $P_j(1, z(j')) = 0$ for $j \ne j'$. The vector $(z_0, z) := (1 - \Lambda_k \cdot z(j), z(j))$ is the solution of the system of linear equations $z_0 + \Lambda_k \cdot z = 1$,

$\Lambda_{j_1} \cdot z = 1, \ldots, \Lambda_{j_m} \cdot z = 1$, and hence we deduce from Cramer's rule that

$$1 - \Lambda_k \cdot z(j) = \frac{\det(\Lambda_{j_1} - \Lambda_k, \ldots, \Lambda_{j_m} - \Lambda_k)}{\det(\Lambda_{j_1}, \ldots, \Lambda_{j_m})}.$$

This yields

$$P_j(1, z(j)) = \prod_{\substack{k=1 \\ k \notin \{j_1, \ldots, j_m\}}}^{l} (1 - \Lambda_k \cdot z(j)) = \prod_{\substack{k=1 \\ k \notin \{j_1, \ldots, j_m\}}}^{l} \frac{\det(\Lambda_{j_1} - \Lambda_k, \ldots, \Lambda_{j_m} - \Lambda_k)}{\det(\Lambda_{j_1}, \ldots, \Lambda_{j_m})} \neq 0.$$

Therefore, the set

$$M := \{P_j : j = (j_1, \ldots, j_m), \ 1 \leq j_1 < \cdots < j_m \leq l\}$$

is linearly independent in the complex vector space H which consists of all homogeneous polynomials in $(z_0, z) \in \mathbf{C}^{m+1}$ of the degree $l - m$. Since H has the dimension $\binom{l}{m}$, this implies that M constitutes a basis of H. Now (3.4.7) is nothing else than the co-ordinate representation of a polynomial $P \in H$ with respect to M.
□

The number m of integrations in the parameter integral in (3.4.1) representing the fundamental solution of the operator $R(\partial) = \prod_{j=1}^{l}\big(P_0(\partial) + \sum_{i=1}^{m} \lambda_{ij} P_i(\partial)\big)$ can further be reduced by one if one of the operators $P_1(\partial), \ldots, P_l(\partial)$ is a constant, a case which covers our successive examples, i.e., products of transport, heat or Klein–Gordon operators. The assumption that one of the operators $P_1(\partial), \ldots, P_l(\partial)$ is a constant also allows to comprise the case of iterated factors, i.e., of higher multiplicities.

Proposition 3.4.4 *Let $1 \leq m < l$, $\alpha \in \mathbf{N}_0^l$ and let $\Lambda = (\lambda_{ij})$ be a real $m \times l$ matrix with the columns $\Lambda_j = (\lambda_{1j}, \ldots, \lambda_{mj})^T \in \mathbf{R}^m, j = 1, \ldots, l$. Let $d_1, \ldots, d_l \in \mathbf{R}$ such that the vectors*

$$\begin{pmatrix} \Lambda_{j_1} - \Lambda_k \\ d_{j_1} - d_k \end{pmatrix}, \ldots, \begin{pmatrix} \Lambda_{j_{m+1}} - \Lambda_k \\ d_{j_{m+1}} - d_k \end{pmatrix} \in \mathbf{R}^{m+1}$$

are linearly independent for pairwise different indices $j_1, \ldots, j_{m+1}, k \in \{1, \ldots, l\}$. We suppose that the operators $P_0(\partial) + \sum_{i=1}^{m} \mu_i P_i(\partial) - \nu$ are uniformly quasihyperbolic with respect to $N \in \mathbf{R}^n \setminus \{0\}$ when $(\mu, \nu) = (\mu_1, \ldots, \mu_m, \nu)$ varies in the convex hull of $\binom{\Lambda_1}{d_1}, \ldots, \binom{\Lambda_l}{d_l}$. Denote by $E_{\mu,\nu}$ the fundamental solution of $(P_0(\partial) + \sum_{i=1}^{m} \mu_i P_i(\partial) - \nu)^{m+1}$ satisfying $e^{-\sigma x N} E_{\mu,\nu} \in \mathcal{S}'(\mathbf{R}^n), \sigma > \sigma_0$. Then $E_{\mu,\nu}$ continuously depends on (μ, ν), and the fundamental solution E of the operator

$$R(\partial) = \prod_{j=1}^{l}\left(P_0(\partial) + \sum_{i=1}^{m} \lambda_{ij} P_i(\partial) - d_j\right)^{\alpha_j+1}$$

is given by

$$E = m! \sum_{1 \leq j_1 < \cdots < j_{m+1} \leq l} \left(\prod_{i=1}^{m+1} \frac{1}{\alpha_{j_i}!} \partial_{d_{j_i}}^{\alpha_{j_i}} \right) \left[c_j \int_{\Sigma_m} E_{\mu_j(\rho), \nu_j(\rho)} \, d\sigma(\rho) \right], \qquad (3.4.8)$$

where, for $j = (j_1, \ldots, j_{m+1})$,

$$c_j := \prod_{\substack{k=1 \\ k \notin \{j_1, \ldots, j_{m+1}\}}}^{l} \frac{\det^{\alpha_k+1} \left(\Lambda_{j_2} - \Lambda_{j_1}, \ldots, \Lambda_{j_{m+1}} - \Lambda_{j_1} \right)}{\det^{\alpha_k+1} \begin{pmatrix} d_{j_1} - d_k & \cdots & d_{j_{m+1}} - d_k \\ \Lambda_{j_1} - \Lambda_k & \cdots & \Lambda_{j_{m+1}} - \Lambda_k \end{pmatrix}}, \qquad (3.4.9)$$

and $\mu_j(\rho) := \sum_{i=1}^{m+1} \rho_i \Lambda_{j_i} \in \mathbf{R}^m$, $\nu_j(\rho) := \sum_{i=1}^{m+1} \rho_i d_{j_i} \in \mathbf{R}$.

Proof If $E(\lambda)$ is the fundamental solution given by (2.4.13) of an operator of the form $\prod_{j=1}^{l} (P_j(\partial) - \lambda_j)$ which is uniformly quasihyperbolic for $\lambda \in U \subset \mathbf{R}^l$ open, then $E(\lambda)$ depends $\mathcal{C}^\infty$ on λ and the operator $\prod_{j=1}^{l} (P_j(\partial) - \lambda_j)^{\alpha_j+1}$ has the fundamental solution $\alpha!^{-1} \partial_\lambda^\alpha E(\lambda)$, see Proposition 1.4.2. The use of this fact reduces the proof of the general case of higher multiplicities to the one where $\alpha = 0$. We also note that the case $l = m + 1$, where $j = (1, \ldots, l), c_j = 1, \mu_j(\rho) = \Lambda\rho$ and $\nu_j(\rho) = \rho \cdot d$, is an immediate consequence of Proposition 3.1.2. We therefore assume in the following that $l > m + 1$.

Furthermore, we observe that the operator $R(\partial)$ as well as the expressions in (3.4.8) and (3.4.9) remain unchanged if P_0 and $\binom{\Lambda_k}{d_k}$ are replaced by $P_0 + \sum_{i=1}^{m} b_i P_i - b_{m+1}$ and by $\binom{\Lambda_k}{d_k} - b$, respectively, for a fixed vector b in the convex hull of $\binom{\Lambda_1}{d_1}, \ldots, \binom{\Lambda_l}{d_l}$. After these replacements, 0 is contained in the convex hull of $\binom{\Lambda_1}{d_1}, \ldots, \binom{\Lambda_l}{d_l}$, and we can thus apply Proposition 3.4.1 with $\Lambda_0 = 0$. This implies

$$E = \frac{(l-1)!}{(l-m-2)!} \sum_{1 \leq j_1 < \cdots < j_{m+1} \leq l} \text{sign} \left[\det \begin{pmatrix} \Lambda_{j_1} & \cdots & \Lambda_{j_{m+1}} \\ d_{j_1} & \cdots & d_{j_{m+1}} \end{pmatrix} \right] \times$$

$$\times \prod_{\substack{k=1 \\ k \notin \{j_1, \ldots, j_{m+1}\}}}^{l} \det^{-1} \begin{pmatrix} \Lambda_{j_1} - \Lambda_k & \cdots & \Lambda_{j_{m+1}} - \Lambda_k \\ d_{j_1} - d_k & \cdots & d_{j_{m+1}} - d_k \end{pmatrix} \times$$

$$\times \int_{T_j} \det^{l-m-2} \begin{pmatrix} \Lambda_{j_1} - \mu & \cdots & \Lambda_{j_{m+1}} - \mu \\ d_{j_1} - \nu & \cdots & d_{j_{m+1}} - \nu \end{pmatrix} \frac{m!}{(l-1)!} \partial_\nu^{l-m-1} E_{\mu,\nu} \, d\mu d\nu,$$

where T_j is the convex hull of the vectors $\binom{0}{0}, \binom{\Lambda_{j_1}}{d_{j_1}}, \ldots, \binom{\Lambda_{j_{m+1}}}{d_{j_{m+1}}}$ and $\frac{m!}{(l-1)!} \partial_\nu^{l-m-1} E_{\mu,\nu}$ is the fundamental solution of the operator $\left(P_0(\partial) + \sum_{i=1}^{m} \mu_i P_i(\partial) - \nu \right)^l$.

We evaluate the ν − integral by partial integrations. (Let us remark, in parentheses, that integrals over continuous, distribution-valued functions like the one above can be handled in exactly the same way as ordinary integrals, since we can think of the distributions as being evaluated on a test function.) Notice that $\det\begin{pmatrix} \Lambda_{j_1} - \mu & \ldots & \Lambda_{j_{m+1}} - \mu \\ d_{j_1} - \nu & \ldots & d_{j_{m+1}} - \nu \end{pmatrix}$ vanishes along that part of the boundary ∂T_j which lies on the affine hyperplane through the points $\binom{\Lambda_{j_1}}{d_{j_1}}, \ldots, \binom{\Lambda_{j_{m+1}}}{d_{j_{m+1}}}$ in $\mathbf{R}^{m+1}$. Therefore, we obtain

$$
\text{sign}\left[\det\begin{pmatrix} \Lambda_{j_1} & \ldots & \Lambda_{j_{m+1}} \\ d_{j_1} & \ldots & d_{j_{m+1}} \end{pmatrix} \right] \times
$$

$$
\times \int_{T_j} \det^{l-m-2}\begin{pmatrix} \Lambda_{j_1} - \mu & \ldots & \Lambda_{j_{m+1}} - \mu \\ d_{j_1} - \nu & \ldots & d_{j_{m+1}} - \nu \end{pmatrix} \partial_\nu^{l-m-1} E_{\mu,\nu}\, d\mu d\nu =
$$

$$
= (l-m-2)!\, \det^{l-m-1}(\Lambda_{j_1} - \Lambda_{j_2}, \ldots, \Lambda_{j_1} - \Lambda_{j_{m+1}}) \int_{\Sigma_m} E_{\mu_j(\rho),\nu_j(\rho)}\, d\sigma(\rho) +
$$

$$
+ \sum_{i=0}^{l-m-2} \frac{(l-m-2)!}{(l-m-2-i)!} \det^i(\Lambda_{j_1} - \Lambda_{j_2}, \ldots, \Lambda_{j_1} - \Lambda_{j_{m+1}}) \times
$$

$$
\times \sum_{k=1}^{m+1} (-1)^{m+k}\, \text{sign}\left[\det\left(\Lambda_{j_1}, \ldots, \Lambda_{j_{k-1}}, \Lambda_{j_{k+1}}, \ldots, \Lambda_{j_{m+1}} \right) \right] \times
$$

$$
\times \int_{T_{j,k}} \det^{l-m-2-i}\begin{pmatrix} \Lambda_{j_1} - \mu & \ldots & \Lambda_{j_{m+1}} - \mu \\ d_{j_1} - \nu_{j,k}(\mu) & \ldots & d_{j_{m+1}} - \nu_{j,k}(\mu) \end{pmatrix} \left(\partial_\nu^{l-m-2-i} E_{\mu,\nu} \right)\big|_{\nu=\nu_{j,k}(\mu)}\, d\mu
$$

$$
(3.4.10)
$$

where $T_{j,k}$ is the part of the boundary of T_j which lies in the subspace of $\mathbf{R}^{m+1}$ spanned by $\binom{\Lambda_{j_1}}{d_{j_1}}, \ldots, \binom{\Lambda_{j_{k-1}}}{d_{j_{k-1}}}, \binom{\Lambda_{j_{k+1}}}{d_{j_{k+1}}}, \ldots, \binom{\Lambda_{j_{m+1}}}{d_{j_{m+1}}}$ and $\nu_{j,k}(\mu)$ is determined by the requirement $\binom{\mu}{\nu_{j,k}(\mu)} \in T_{j,k}$. (Here we have used that $T_{j,k}$ is part of the lower integration border with respect to ν iff the bases $\binom{\Lambda_{j_1}}{d_{j_1}}, \ldots, \binom{\Lambda_{j_{m+1}}}{d_{j_{m+1}}}$ and $\binom{\Lambda_{j_1}}{d_{j_1}}, \ldots, \binom{\Lambda_{j_{k-1}}}{d_{j_{k-1}}}, \binom{0}{1}, \binom{\Lambda_{j_{k+1}}}{d_{j_{k+1}}}, \ldots, \binom{\Lambda_{j_{m+1}}}{d_{j_{m+1}}}$ induce the same orientation in $\mathbf{R}^{m+1}$.)

The first term on the right-hand side of Eq. (3.4.10) already yields the desired expression for E. Thus we have to show that the remaining terms cancel each other if summed up for all $j = (j_1, \ldots, j_{m+1})$, $1 \le j_1 < \cdots < j_{m+1} \le l$. Without loss of generality, it is sufficient to consider that part of the boundary which lies in the subspace spanned by $\binom{\Lambda_1}{d_1}, \ldots, \binom{\Lambda_m}{d_m}$. That means, we gather all terms pertaining to

$j = (1, \ldots, m, k)$ with $k \in \{m+1, \ldots, l\}$, to a fixed $i \in \{0, \ldots, l - m - 2\}$, and to a fixed $\binom{\mu}{\nu(\mu)}$, where $\nu(\mu) = \nu_{j,m+1}(\mu)$. Hence it remains to show that

$$\sum_{k=m+1}^{l} \det{}^i(\Lambda_1 - \Lambda_2, \ldots, \Lambda_1 - \Lambda_m, \Lambda_1 - \Lambda_k) \times$$

$$\times \det{}^{l-m-2-i} \begin{pmatrix} \Lambda_1 - \mu & \cdots & \Lambda_m - \mu & \Lambda_k - \mu \\ d_1 - \nu(\mu) & \cdots & d_m - \nu(\mu) & d_k - \nu(\mu) \end{pmatrix} \times$$

$$\times \prod_{\substack{s=m+1 \\ s \neq k}}^{l} \det{}^{-1} \begin{pmatrix} \Lambda_1 - \Lambda_s & \cdots & \Lambda_m - \Lambda_s & \Lambda_k - \Lambda_s \\ d_1 - d_s & \cdots & d_m - d_s & d_k - d_s \end{pmatrix} = 0. \qquad (3.4.11)$$

Since $\mu = \sum_{r=1}^{m} \rho_r \Lambda_r$, $\nu(\mu) = \sum_{r=1}^{m} \rho_r d_r$, for some positive $\rho_1, \ldots, \rho_m$, it follows that

$$\det \begin{pmatrix} \Lambda_1 - \mu & \cdots & \Lambda_m - \mu & \Lambda_k - \mu \\ d_1 - \nu(\mu) & \cdots & d_m - \nu(\mu) & d_k - \nu(\mu) \end{pmatrix} = c \det \begin{pmatrix} \Lambda_1 & \cdots & \Lambda_m & \Lambda_k \\ d_1 & \cdots & d_m & d_k \end{pmatrix},$$

for some real number c which depends on μ but not on k. Furthermore, if we replace $\binom{\Lambda_k}{d_k}$ by $\binom{\Lambda_k + \Lambda_1}{d_k + d_1}$, $k = 2, \ldots, l$, we see that (3.4.11) is equivalent to the following equation:

$$\sum_{k=m+1}^{l} \det{}^i(\Lambda_2, \ldots, \Lambda_m, \Lambda_k) \det{}^{l-m-2-i} \begin{pmatrix} \Lambda_1 & \cdots & \Lambda_m & \Lambda_k \\ d_1 & \cdots & d_m & d_k \end{pmatrix} \times$$

$$\times \prod_{\substack{s=m+1 \\ s \neq k}}^{l} \det{}^{-1} \begin{pmatrix} \Lambda_s & \Lambda_2 & \cdots & \Lambda_m & \Lambda_k \\ d_s & d_2 & \cdots & d_m & d_k \end{pmatrix} = 0.$$

If we set $\Lambda_k = \sum_{r=1}^{m} \alpha_{kr} \Lambda_r$, then the identity to be proven reduces to

$$\sum_{k=m+1}^{l} \alpha_{k1}^i \left(d_k - \sum_{r=1}^{m} \alpha_{kr} d_r \right)^{l-m-2-i} \times$$

$$\times \prod_{\substack{s=m+1 \\ s \neq k}}^{l} \left[\alpha_{s1} \left(d_k - \sum_{r=2}^{m} \alpha_{kr} d_r \right) - \alpha_{k1} \left(d_s - \sum_{r=2}^{m} \alpha_{sr} d_r \right) \right]^{-1} = 0. \qquad (3.4.12)$$

Identity (3.4.12) in turn is implied by the classical equation,

$$\sum_{k=1}^{p} z_k^i \prod_{\substack{s=1 \\ s \neq k}}^{p} (z_k - z_s)^{-1} = 0, \qquad 0 \leq i \leq p-2,$$

(for pairwise different complex numbers $z_1, \ldots, z_p$), which can be proven either by applying the residue theorem to the meromorphic function $z^i \prod_{k=1}^{p}(z - z_k)^{-1}$ on $\overline{\mathbf{C}}$ or by invoking Lagrange's interpolation theorem, cf. also (3.3.7). Thus the proof is complete. $\square$

Example 3.4.5 Let us consider now *products of transport operators*, i.e.,

$$R(\partial) = \prod_{j=1}^{l} \left(d_j + \sum_{i=1}^{n} a_{ij} \partial_i \right)^{\alpha_j + 1}, \qquad A = (a_{ij}) \in \mathbf{R}^{n \times l}, \, d \in \mathbf{C}^l, \, \alpha \in \mathbf{N}_0^l,$$

(3.4.13)

in the case $l > n$, which we have postponed in Example 3.1.3.

As in (3.1.7), we may suppose, without loss of generality, that $N \in \mathbf{R}^n$ can be chosen such that $\sum_{i=1}^{n} a_{ij} N_i > 0$ for $j = 1, \ldots, l$. We shall furthermore assume that

$$A_{j_1}, \ldots, A_{j_n} \in \mathbf{R}^n \text{ and } \begin{pmatrix} A_{j_1} \\ d_{j_1} \end{pmatrix}, \ldots, \begin{pmatrix} A_{j_n} \\ d_{j_n} \end{pmatrix}, \begin{pmatrix} A_k \\ d_k \end{pmatrix} \in \mathbf{C}^{n+1}, \text{ respectively,}$$

are linearly independent for pairwise different indices $j_1, \ldots, j_n, k \in \{1, \ldots, l\}$.

(3.4.14)

Here, similarly as before, A_j, $j = 1, \ldots, l$, denote the columns of the matrix A, i.e., $A_j = (a_{1j}, \ldots, a_{nj})^T$.

Let us assume first that $N = (1, 0, \ldots, 0)^T$, $d \in \mathbf{R}^l$, and $a_{1j} = 1$, $j = 1, \ldots, l$. If we set $\lambda_{ij} = a_{i+1,j}$, then Proposition 3.4.4 yields the following formula for the fundamental solution E of $R(\partial)$ with support in $x_1 > 0$:

$$\langle \phi, E \rangle = (n-1)! \sum_{1 \leq j_1 < \cdots < j_n \leq l} \left(\prod_{i=1}^{n} \frac{(-1)^{\alpha_{j_i}}}{\alpha_{j_i}!} \partial_{d_{j_i}}^{\alpha_{j_i}} \right) \left[c_j \int_{\Sigma_{n-1}} \langle \phi, E_{\mu_j(\rho), v_j(\rho)} \rangle \, d\sigma(\rho) \right],$$

where $\phi \in \mathcal{D}(\mathbf{R}^n)$ and

$$c_j = \prod_{\substack{k=1 \\ k \notin \{j_1, \ldots, j_n\}}}^{l} \frac{\det^{\alpha_k+1} \left(\Lambda_{j_2} - \Lambda_{j_1}, \ldots, \Lambda_{j_n} - \Lambda_{j_1} \right)}{\det^{\alpha_k+1} \begin{pmatrix} d_k - d_{j_1} & \cdots & d_k - d_{j_n} \\ \Lambda_{j_1} - \Lambda_k & \cdots & \Lambda_{j_n} - \Lambda_k \end{pmatrix}}$$

$$
= \prod_{\substack{k=1 \\ k \notin \{j_1,\dots,j_n\}}}^{l} \frac{\det^{\alpha_k+1}(A_{j_1},\dots,A_{j_n})}{\det^{\alpha_k+1}\begin{pmatrix} d_k \ d_{j_1} \ \cdots \ d_{j_n} \\ A_k \ A_{j_1} \ \cdots \ A_{j_n} \end{pmatrix}}
$$

and $\mu_j(\rho) := \sum_{i=1}^{n} \rho_i \Lambda_{j_i} \in \mathbf{R}^{n-1}$, $v_j(\rho) := \sum_{i=1}^{n} \rho_i d_{j_i} \in \mathbf{R}$.
Furthermore, in our case,

$$
\langle \phi, E_{\mu_j(\rho),v_j(\rho)} \rangle = \frac{1}{(n-1)!} \int_0^\infty t^{n-1} \phi(t, \mu_j(\rho)t) e^{-v_j(\rho)t} \, dt
$$

according to (2.5.2). Employing the substitution $x = (t, \mu_j(\rho)t)^T = \sum_{i=1}^{n} t \rho_i A_{j_i}$ we obtain for the Jacobian

$$
\frac{\partial(x_1,\dots,x_n)}{\partial(t,\rho_2,\dots,\rho_n)} = t^{n-1} \det(A_{j_1},\dots,A_{j_n})
$$

and hence

$$
\int_{\Sigma_{n-1}} \langle \phi, E_{\mu_j(\rho),v_j(\rho)} \rangle \, d\sigma(\rho) = \frac{1}{(n-1)!} \left| \det(A_{j_1},\dots,A_{j_n}) \right|^{-1} \times
$$

$$
\times \int_{C_j} \phi(x) \, e^{-(d_{j_1},\dots,d_{j_n})(A_{j_1},\dots,A_{j_n})^{-1}x} \, dx,
$$

where C_j is the cone spanned by $A_{j_1},\dots,A_{j_n}$.

The case of arbitrary a_{1j} is reduced to the foregoing one by putting $\lambda_{ij} := a_{i+1,j}/a_{1j}$. This is legitimate if $N = (1,0,\dots,0)$, and hence $a_{1j} > 0, j = 1,\dots,l$. By rotational invariance and analytic continuation with respect to d_k, $k = 1,\dots,l$, we then infer that the operator $R(\partial)$ in (3.4.13) has the following uniquely determined fundamental solution E with support in $H_N = \{x \in \mathbf{R}^n; Nx \geq 0\}$, provided $\sum_{i=1}^{n} a_{ij} N_i > 0$ for $j = 1,\dots,l > n$, and the condition in (3.4.14) is satisfied:

$$
E(x) = \sum_{1 \leq j_1 < \cdots < j_n \leq l} \chi_{C_j}(x) \left(\prod_{i=1}^{n} \frac{(-1)^{\alpha_{j_i}}}{\alpha_{j_i}!} \partial_{d_{j_i}}^{\alpha_{j_i}} \right) \left[\tilde{c}_j \, e^{-(d_{j_1},\dots,d_{j_n})(A_{j_1},\dots,A_{j_n})^{-1}x} \right],
$$

$$
(3.4.15)
$$

where $\chi_{C_j}(x)$ denotes the characteristic function of the cone C_j spanned by $A_{j_1},\dots,A_{j_n}$, and

$$
\tilde{c}_j := \left| \det(A_{j_1},\dots,A_{j_n}) \right|^{-1} \prod_{\substack{k=1 \\ k \notin \{j_1,\dots,j_n\}}}^{l} \frac{\det^{\alpha_k+1}(A_{j_1},\dots,A_{j_n})}{\det^{\alpha_k+1}\begin{pmatrix} d_k \ d_{j_1} \ \cdots \ d_{j_n} \\ A_k \ A_{j_1} \ \cdots \ A_{j_n} \end{pmatrix}}.
$$

Formula (3.4.15) goes back to Ortner and Wagner [211], Prop. 5, p. 315. As in this reference, let us remark that the formula in (3.4.15) implies that the (analytic) singular support of E coincides with the union of the cones spanned by subsets of cardinality $n-1$ of the set of vectors $\{A_1, \dots, A_l\}$. Of course, this is also immediate from the theory of singularities and lacunas for hyperbolic operators in Atiyah, Bott and Gårding [5], which we shall expound in Chap. 4 below.   □

Because of the condition (3.4.14), the representation of E in (3.4.15) is not applicable if the constants d_k vanish. In this case, we shall use Proposition 3.4.1:

Example 3.4.6 Let us consider here *products of homogeneous transport operators*, i.e.,

$$R(\partial) = \prod_{j=1}^{l}\left(\sum_{i=1}^{n} a_{ij}\partial_i\right), \qquad A = (a_{ij}) \in \mathbf{R}^{n \times l}, \tag{3.4.16}$$

still in the case $l > n$.

Similarly as before, we assume that

$$\sum_{i=1}^{n} a_{ij}N_i > 0, \quad j = 1, \dots, l, \text{ and } \det(A_{j_1}, \dots, A_{j_n}) \neq 0 \text{ for } 1 \leq j_1 < \dots < j_n \leq l.$$
$$\tag{3.4.17}$$

As in Example 3.5.5, we then reduce the general case to the case where $N = (1, 0, \dots, 0)$ and $a_{1j} = 1$, $j = 1, \dots, l$. We choose an arbitrary vector $A_0 = (a_{10}, \dots, a_{n0})^T \in \mathbf{R}^n$ such that $\sum_{i=1}^{n} a_{i0}N_i > 0$ and define C_j as the cone generated by $A_0, A_{j_1}, \dots, A_{j_{n-1}}$ and χ_{C_j} as its characteristic function for $1 \leq j_1 < \dots < j_{n-1} \leq l$. In this way, we obtain the following formula for the uniquely determined fundamental solution E of the operator $R(\partial)$ in (3.4.16) and with support in $H_N = \{x \in \mathbf{R}^n; Nx \geq 0\}$, provided (3.4.17) is valid:

$$E(x) = \frac{1}{(l-n)!} \sum_{1 \leq j_1 < \dots < j_{n-1} \leq l} c_j \chi_{C_j}(x) \det^{l-n}(x, A_{j_1}, \dots, A_{j_{n-1}}), \tag{3.4.18}$$

where

$$c_j := \text{sign}\big(\det(A_0, A_{j_1}, \dots, A_{j_{n-1}})\big) \prod_{\substack{k=1 \\ k \notin \{j_1, \dots, j_{n-1}\}}}^{l} \det^{-1}\big(A_k, A_{j_1}, \dots, A_{j_{n-1}}\big).$$

Formula (3.4.18) was given in Ortner and Wagner [211], p. 317.

As in this reference, let us also specialize (3.4.18) to the following quartic operator in $\mathbf{R}^3_{t,x,y}$:

$$R(\partial) = \prod_{\epsilon \in \{\pm 1\}^2} (\partial_t + \epsilon_1 \partial_x + \epsilon_2 \partial_y) = \partial_t^4 - 2\partial_t^2(\partial_x^2 + \partial_y^2) + (\partial_x^2 - \partial_y^2)^2.$$

If we choose $N = A_0 = (1,0,0)^T$, then formula (3.4.18) readily yields

$$E = \frac{1}{8}(t - \max\{|x|,|y|\}) \, Y(t - \max\{|x|,|y|\}) \in C(\mathbf{R}^3). \tag{3.4.19}$$

The singular support of E is the union of the wedges spanned by each two of the four vectors $(1, \epsilon_1, \epsilon_2)^T$, $\epsilon \in \{\pm 1\}^2$, in accordance with the Atiyah-Bott-Gårding theory we have hinted at above and which we shall study in Chap. 4. □

Example 3.4.7 Turning now to the case of *products of heat operators*, let us observe that the representation in Example 3.1.4 of the fundamental solution of

$$R(\partial) = \prod_{j=1}^{l} \left(\partial_t - \nabla^T A_j \nabla - d_j + \sum_{k=1}^{n} b_{jk} \partial_k \right)^{\alpha_j + 1},$$

contains $l - 1$ integrations. This is well-suited only if the number l of factors is smaller than the dimension $\binom{n+2}{2}$ of the space of polynomials in x of degree at most 2. A different case occurs in particular if $R(\partial)$ is isotropic, i.e., if

$$R(\partial) = \prod_{j=1}^{l} (\partial_t - a_j \Delta_n - d_j)^{\alpha_j + 1}, \qquad a_j > 0, \ d_j \in \mathbf{C}, \ j = 1, \ldots, l. \tag{3.4.20}$$

In this case, we will apply Proposition 3.4.4, and we shall assume therefore that $a_1, \ldots, a_l$ are pairwise different, and that $\det \begin{pmatrix} a_i - a_k & a_j - a_k \\ d_i - d_k & d_j - d_k \end{pmatrix} \neq 0$, $1 \leq i < j < k \leq l$.

In fact, setting $m = 1$, $P_0(\partial) = \partial_t$, $P_1(\partial) = -\Delta_n$ and renaming $j = (j_1, j_2)$ as (i,j), we only have to insert into formula (3.4.8) the fundamental solution $E_{\mu,\nu}$ of the iterated heat operator $(\partial_t - \mu \Delta_n - \nu)^2$, which, by Example 2.5.5, is given by

$$E_{\mu,\nu} = \frac{Y(t) \, t^{1-n/2}}{(4\pi\mu)^{n/2}} \exp\left(\nu t - \frac{|x|^2}{4\mu t}\right).$$

Finally, with the substitution

$$u = t\mu_{(i,j)}(\rho) = t(a_i \rho_1 + a_j \rho_2), \qquad (\rho_1, \rho_2) \in \Sigma_1,$$

which implies

$$t\nu_{(i,j)}(\rho) = \frac{a_i d_j - a_j d_i}{a_i - a_j} t + \frac{d_i - d_j}{a_i - a_j} u,$$

we obtain the following formula for the uniquely determined fundamental solution E of the operator $R(\partial)$ in (3.4.20) satisfying $e^{-\sigma t}E \in \mathcal{S}'(\mathbf{R}^{n+1})$ for $\sigma > \sigma_0$:

$$E = \frac{Y(t)}{(4\pi)^{n/2}} \sum_{1 \le i < j \le l} \frac{1}{\alpha_i! \alpha_j!} \left(\frac{\partial}{\partial d_i}\right)^{\alpha_i} \left(\frac{\partial}{\partial d_j}\right)^{\alpha_j} \times$$

$$\times \left[(a_i - a_j)^{|\alpha| - \alpha_i - \alpha_j + l - 3} \left(\prod_{\substack{k=1 \\ k \ne i, k \ne j}}^{l} \det^{-\alpha_k - 1} \begin{pmatrix} a_i - a_k & a_j - a_k \\ d_i - d_k & d_j - d_k \end{pmatrix} \right) \times \right.$$

$$\left. \times \exp\left(\frac{t}{a_i - a_j} \det \begin{pmatrix} a_i & a_j \\ d_i & d_j \end{pmatrix}\right) \int_{a_j t}^{a_i t} u^{-n/2} \exp\left(-\frac{|x|^2}{4u} + \frac{d_i - d_j}{a_i - a_j} u\right) du \right].$$

$$(3.4.21)$$

Formula (3.4.21) goes back to Ortner and Wagner [211], Prop. 7, p. 318. As in this reference, let us yet deduce the case of equal constants d_j from (3.4.21) by a limit argument. This special case appears in Galler [86], Satz 17.1, p. 66.

For $\alpha = 0$, (3.4.21) furnishes

$$E = \frac{Y(t)}{(4\pi)^{n/2}} \sum_{i=1}^{l} e^{td_i} \int_0^{a_i t} u^{-n/2} \exp\left(-\frac{|x|^2}{4u}\right) \sum_{\substack{j=1 \\ j \ne i}}^{l} c_{ij} \exp\left(\frac{d_i - d_j}{a_i - a_j}(u - a_i t)\right) du,$$

where

$$c_{ij} = (a_i - a_j)^{l-3} \prod_{\substack{k=1 \\ k \ne i, k \ne j}}^{l} \det^{-1} \begin{pmatrix} a_i - a_k & a_j - a_k \\ d_i - d_k & d_j - d_i \end{pmatrix} = -c_{ji}.$$

Taking into account that $\sum_{j=1, j \ne i}^{l} c_{ij} \left(\frac{d_i - d_j}{a_i - a_j}\right)^s = 0$ for $s = 0, \ldots, l - 3$, we see that the first $l - 2$ terms of the power series expansion of $\exp\left(\frac{d_i - d_j}{a_i - a_j}(u - a_i t)\right)$ can be dispensed with. Substituting d_i by $d + \epsilon d_i$ and letting ϵ tend to 0 we therefore obtain the $(l - 1) - $ st term of this power series, i.e.,

$$E = \frac{Y(t) e^{td}}{(4\pi)^{n/2}(l-2)!} \sum_{i=1}^{l} \int_0^{a_i t} u^{-n/2} (u - a_i t)^{l-2} e^{-|x|^2/(4u)} du \sum_{\substack{j=1 \\ j \ne i}}^{l} c_{ij} \left(\frac{d_i - d_j}{a_i - a_j}\right)^{l-2}.$$

Since the last sum yields $(-1)^{l-2} \prod_{k=1, k \ne i}^{l} (a_i - a_k)^{-1}$, we obtain the following representation for the fundamental solution E of the operator $R(\partial) = \prod_{j=1}^{l} (\partial_t -$

$a_j \Delta_n - d$) satisfying $e^{-\sigma t} E \in \mathcal{S}'(\mathbf{R}^{n+1})$ for $\sigma > \sigma_0$, provided $a_1, \ldots, a_l$ are positive and pairwise different and $d \in \mathbf{C}$:

$$E = \frac{Y(t)\, e^{td}}{(4\pi)^{n/2}(l-2)!} \sum_{i=1}^{l} \left(\prod_{\substack{k=1 \\ k \neq i}}^{l} (a_i - a_k)^{-1} \right) \int_0^{a_i t} u^{-n/2}\, (a_i t - u)^{l-2}\, e^{-|x|^2/(4u)}\, du.$$

$$(3.4.22)$$

Of course, formula (3.4.22) could also easily be deduced from Proposition 3.2.1.

□

Example 3.4.8 As our final example, we give integral representations for the fundamental solutions E of *products of isotropic Klein–Gordon operators*, i.e.,

$$R(\partial) = \prod_{j=1}^{l} (\partial_t^2 - a_j \Delta_n - d_j)^{\alpha_j + 1}, \qquad l \geq 2,\ a_j > 0, j = 1, \ldots, l,\ d \in \mathbf{C}^l,\ \alpha \in \mathbf{N}_0^l.$$

$$(3.4.23)$$

We restrict ourselves to treating space dimensions $n = 1, 2, 3, 4$, since then E is locally integrable.

Similarly as in Example 3.4.7, we invoke Proposition 3.4.4 and rename again $j = (j_1, j_2)$ as (i, j). Here $E_{\mu,\nu}$ is the fundamental solution of $(\partial_t^2 - \mu \Delta_n - \nu)^2$, which, according to Example 2.3.7, is given by

$$E_{\mu,\nu} = \frac{\nu^{(n-3)/4}\, Y(t - |x|/\sqrt{\mu})}{2^{(n+3)/2} \pi^{(n-1)/2}\, \mu^{n/2}} \left(t^2 - \frac{|x|^2}{\mu} \right)^{(3-n)/4} I_{(3-n)/2}\left(\sqrt{\nu \left(t^2 - \frac{|x|^2}{\mu} \right)} \right).$$

In (3.4.8), we substitute $\mu = \mu_{(i,j)}(\rho) = a_i \rho_1 + a_j \rho_2$, $(\rho_1, \rho_2) \in \Sigma_1$, as a new integration variable, and this yields

$$h_{ij}(\mu) := \nu_{(i,j)}(\rho) = \frac{1}{a_i - a_j} \det \begin{pmatrix} a_i\ a_j \\ d_i\ d_j \end{pmatrix} + \frac{d_i - d_j}{a_i - a_j}\, \mu. \qquad (3.4.24)$$

As in Example 3.4.7, we replace the integral from a_j to a_i by two integrals from 0 to a_i and 0 to a_j, respectively. This can be accounted for by changing the summation from $1 \leq i < j \leq m$ to $1 \leq i, j \leq m$, $i \neq j$. Finally, note that the Heaviside function $Y(t - |x|/\sqrt{\mu})$ in the integrand brings about the factor $Y(t - |x|/\sqrt{a_i})$ and changes the lower limit of integration from 0 to $|x|^2/t^2$.

Hence we obtain the following formula for the forward fundamental solution E of the operator $R(\partial)$ in (3.4.23), provided that $n \leq 4$, $a_i, i = 1, \ldots, l$, are pairwise different and $\det \begin{pmatrix} a_i - a_k & a_j - a_k \\ d_i - d_k & d_j - d_k \end{pmatrix} \neq 0$ for $1 \leq i < j < k \leq l$:

$$E = \frac{1}{2^{(n+3)/2}\,\pi^{(n-1)/2}} \sum_{\substack{1 \leq i,j \leq l \\ i \neq j}} \frac{1}{\alpha_i!\,\alpha_j!} \left(\frac{\partial}{\partial d_i}\right)^{\alpha_i} \left(\frac{\partial}{\partial d_j}\right)^{\alpha_j} \times$$

$$\times \left[(a_i - a_j)^{|\alpha|-\alpha_i-\alpha_j+l-3} \left(\prod_{\substack{k=1 \\ k \neq i, k \neq j}}^{l} \det^{-\alpha_k-1} \begin{pmatrix} a_i - a_k & a_j - a_k \\ d_i - d_k & d_j - d_k \end{pmatrix} \right) Y\left(t - \frac{|x|}{\sqrt{a_i}}\right) \times$$

$$\times \int_{|x|^2/t^2}^{a_i} \mu^{-n/2} h_{ij}(\mu)^{(n-3)/4} \left(t^2 - \frac{|x|^2}{\mu}\right)^{(3-n)/4} I_{(3-n)/2}\left(\sqrt{h_{ij}(\mu)\left(t^2 - \frac{|x|^2}{\mu}\right)}\right) d\mu \right],$$

$$(3.4.25)$$

where h_{ij} is defined in (3.4.24).

Formula (3.4.25) goes back to Ortner and Wagner [211], Prop. 8, p. 320. □

3.5 Parameter Integration for Indecomposable Operators

In this section, we deduce a parameter integral which represents the fundamental solution of a quasihyperbolic operator of the form $R(\partial) = P_0(\partial)^2 - P_1(\partial)^2 - \cdots - P_l(\partial)^2$ by means of the fundamental solutions of the operators $Q_\mu(\partial)^2$, $|\mu| < 1$, where $Q_\mu(\partial) = P_0(\partial) + \sum_{j=1}^{l} \mu_j P_j(\partial)$. This method was devised and applied to various examples in Ortner and Wagner [207]. Instead of Feynman's formula, see (3.1.2), we shall employ the integral representation

$$\left(b^2 - \sum_{j=1}^{l} a_j^2\right)^\lambda = \frac{\Gamma(\frac{1}{2} - \lambda)}{\pi^{l/2}\Gamma(-\frac{l-1}{2} - \lambda)} \int_{\{\mu \in \mathbf{R}^l;\, |\mu| < 1\}} \frac{(b + \mu_1 a_1 + \cdots + \mu_l a_l)^{2\lambda}\, d\mu}{(1 - |\mu|^2)^{\lambda + (l+1)/2}},$$

$$(3.5.1)$$

valid for suitable complex $b, a_1, \ldots, a_l, \lambda$ according to the following lemma.

Lemma 3.5.1 *Let us denote by C the open forward light cone $\{(t,x) \in \mathbf{R}^{l+1};\, t > |x|\}$ and let $\lambda \in \mathbf{C}$ with $\operatorname{Re}\lambda < \frac{1-l}{2}$. Then the identity (3.5.1) holds for each (b,a) in the tube domain $C + i\mathbf{R}^{l+1}$.*

Proof

(1) Let us first assume that $(b, a) \in C$. Apparently, the integral on the right-hand side of (3.5.1) is rotationally symmetric with respect to a and hence coincides with

$$\int_{\{\mu \in \mathbf{R}^l; |\mu| < 1\}} \frac{(b + \mu_l |a|)^{2\lambda} \, d\mu}{(1 - |\mu|^2)^{\lambda + (l+1)/2}}$$

$$= \int_{\mu_l = -1}^1 (b + \mu_l |a|)^{2\lambda} \int_{\{\mu' \in \mathbf{R}^{l-1}; |\mu'| < \sqrt{1 - \mu_l^2}\}} \frac{d\mu' d\mu_l}{(1 - \mu_l^2 - |\mu'|^2)^{\lambda + (l+1)/2}}$$

$$= |\mathbf{S}^{l-2}| \int_{\mu_l = -1}^1 (b + \mu_l |a|)^{2\lambda} \frac{d\mu_l}{(1 - \mu_l^2)^{\lambda + 1}} \cdot \int_0^1 r^{l-2} \frac{dr}{(1 - r^2)^{\lambda + (l+1)/2}}$$

$$= |\mathbf{S}^{l-2}| \cdot \frac{\Gamma(-\lambda)^2 (b^2 - |a|^2)^\lambda}{2^{2\lambda + 1} \Gamma(-2\lambda)} \cdot \frac{\Gamma(\frac{l-1}{2}) \Gamma(-\lambda - \frac{l-1}{2})}{2 \Gamma(-\lambda)}$$

$$= \frac{\pi^{l/2} \Gamma(-\lambda - \frac{l-1}{2})}{\Gamma(-\lambda + \frac{1}{2})} \cdot (b^2 - |a|^2)^\lambda,$$

where we have used Gröbner and Hofreiter [115], Eqs. 421.4, 431.1, and the doubling formula for the gamma function. (In this deduction, we have assumed $l \geq 2$, but the result holds also in the case $l = 1$.)

(2) The second step consists in extending the validity of Eq. (3.5.1) by analytic continuation with respect to (b, a). For $(b, a) \in C + i\mathbf{R}^{l+1}$, we note that $(\operatorname{Re} b)^2 - \sum_{j=1}^l (\operatorname{Re} a_j)^2 > 0$ and

$$(\operatorname{Im} b)^2 - \sum_{j=1}^l (\operatorname{Im} a_j)^2 \leq 0 \quad \text{if} \quad \operatorname{Re} b \operatorname{Im} b = \sum_{j=1}^l \operatorname{Re} a_j \operatorname{Im} a_j.$$

Hence

$$b^2 - \sum_{j=1}^l a_j^2 = (\operatorname{Re} b)^2 - \sum_{j=1}^l (\operatorname{Re} a_j)^2 - \left[(\operatorname{Im} b)^2 - \sum_{j=1}^l (\operatorname{Im} a_j)^2 \right]$$

$$+ 2i \left[\operatorname{Re} b \operatorname{Im} b - \sum_{j=1}^l \operatorname{Re} a_j \operatorname{Im} a_j \right]$$

belongs to $\mathbf{C} \setminus (-\infty, 0]$. Therefore

$$\left(b^2 - \sum_{j=1}^l a_j^2 \right)^\lambda = \exp\left(\lambda \log\left(b^2 - \sum_{j=1}^l a_j^2 \right) \right)$$

is well-defined when we take the principal branch of the logarithm on $\mathbf{C} \setminus (-\infty, 0]$.

Similarly, the numerator in the integral in (3.5.1) is well-defined since $\mathrm{Re}\,(b + \sum_{j=1}^{l} \mu_j a_j) > 0$ for $(b, a) \in C + i\mathbf{R}^{l+1}$ and $|\mu| \le 1$. □

In Lemma 3.5.1, we assumed that the power λ fulfills $\mathrm{Re}\,\lambda < \frac{1-l}{2}$, as otherwise the factor $(1-|\mu|^2)^{-\lambda-(l+1)/2}$ in the integral in (3.5.1) ceases to be locally integrable. In the next lemma, we formulate what Eq. (3.5.1) amounts to when $\lambda = -1$.

Lemma 3.5.2 *Let* $(b, a_1, \dots, a_l) \in \mathbf{C}^{l+1}$ *such that* $b + \sum_{j=1}^{l} \mu_j a_j \neq 0$ *for each* $\mu \in \mathbf{R}^l$ *with* $|\mu| \le 1$. *Denote by* χ_+^λ *the holomorphic distribution-valued function*

$$\mathbf{C} \longrightarrow \mathcal{D}'_{[0,\infty)}(\mathbf{R}) : \lambda \longmapsto \chi_+^\lambda := \begin{cases} x_+^{\lambda-1}/\Gamma(\lambda), & \text{if } \lambda \in \mathbf{C} \setminus -\mathbf{N}_0, \\ \delta^{(k)}, & \text{if } \lambda = -k \in -\mathbf{N}_0, \end{cases}$$

see Example 1.5.11 *and Hörmander* [139], (3.2.17), p. 73.

Then $\lambda \mapsto \chi_+^\lambda \circ (1 - |\mu|^2)$, $\mu \in \mathbf{R}^l$, *is a holomorphic function with values in* $\mathcal{E}'(\mathbf{R}^l)$ *and*

$$\left(b^2 - \sum_{j=1}^{l} a_j^2\right)^{-1} = \frac{1}{2} \pi^{(1-l)/2} \langle \left(b + \sum_{j=1}^{l} \mu_j a_j\right)^{-2}, \chi_+^{(3-l)/2} \circ (1 - |\mu|^2)\rangle. \quad (3.5.2)$$

In particular, for $l = 2$,

$$\left(b^2 - a_1^2 - a_2^2\right)^{-1} = \frac{1}{2\pi} \int_{|\mu|<1} (b + \mu_1 a_1 + \mu_2 a_2)^{-2} \frac{d\mu}{\sqrt{1 - |\mu|^2}}, \quad (3.5.3)$$

and, for $l = 3$,

$$\left(b^2 - a_1^2 - a_2^2 - a_3^2\right)^{-1} = \frac{1}{4\pi} \int_{\mathbf{S}^2} (b + \mu_1 a_1 + \mu_2 a_2 + \mu_3 a_3)^{-2} d\sigma(\mu). \quad (3.5.4)$$

Proof For $(b, a) \in C + i\mathbf{R}^{l+1}$ as in Lemma 3.5.1, we can conceive formula (3.5.1) as an evaluation of the distribution $T_\lambda = \chi_+^{-\lambda-(l-1)/2} \circ (1 - |\mu|^2) \in \mathcal{E}'(\mathbf{R}^l_\mu)$ on the test function $(b + \sum_{j=1}^{l} \mu_j a_j)^{2\lambda} \in \mathcal{E}(\mathbf{R}^l_\mu)$, i.e.,

$$\left(b^2 - \sum_{j=1}^{l} a_j^2\right)^\lambda = \frac{\Gamma(\frac{1}{2} - \lambda)}{\pi^{l/2}} \langle \left(b + \sum_{j=1}^{l} \mu_j a_j\right)^{2\lambda}, T_\lambda \rangle, \ \mathrm{Re}\,\lambda < \frac{1-l}{2}. \quad (3.5.5)$$

By analytic continuation with respect to λ, formula (3.5.5) holds for $(b, a) \in C + i\mathbf{R}^{l+1}$ and $\lambda \in \mathbf{C} \setminus (\frac{1}{2} + \mathbf{N}_0)$.

Since the domain

$$\Omega = \left\{ (b, a) \in \mathbf{C}^{l+1}; \ \forall \mu \in \mathbf{R}^l \text{ with } |\mu| \leq 1 : b + \sum_{j=1}^{l} \mu_j a_j \neq 0 \right\}$$

is connected and contains $C + i\mathbf{R}^{l+1}$, formula (3.5.2) then follows from (3.5.5) by setting $\lambda = -1$ and using analytic continuation with respect to (b, a). Finally, (3.5.3) follows directly from (3.5.2), whereas (3.5.4) is a consequence of

$$\langle \phi, \delta \circ (1 - |\mu|^2) \rangle = \frac{1}{2} \int_{\mathbf{S}^2} \phi(\mu) \, d\sigma(\mu), \qquad \phi \in \mathcal{E}(\mathbf{R}^3),$$

cf. formula (1.2.2). □

Let us apply now Eqs. (3.5.3) and (3.5.4) in order to represent the fundamental solutions of quasihyperbolic operators of the form $P_0(\partial)^2 - P_1(\partial)^2 - \cdots - P_l(\partial)^2$, $l = 2, 3$.

Proposition 3.5.3 *Suppose that the operators* $Q_\mu(\partial) = P_0(\partial) + \sum_{j=1}^{3} \mu_j P_j(\partial)$, $\mu \in \Lambda = \mathbf{S}^2$, *are uniformly quasihyperbolic with respect to* $N \in \mathbf{R}^n \setminus \{0\}$. *Then also the operator* $R(\partial) = P_0(\partial)^2 - P_1(\partial)^2 - P_2(\partial)^2 - P_3(\partial)^2$ *is quasihyperbolic with respect to* N. *Let* $\sigma_0 \in \mathbf{R}$ *be as in* (3.1.3) *and denote by* E *and by* E_μ *the fundamental solutions of* $R(\partial)$ *and of* $Q_\mu(\partial)^2$, $\mu \in \mathbf{S}^2$, *respectively, which satisfy* $e^{-\sigma x N} E$, $e^{-\sigma x N} E_\mu \in \mathcal{S}'(\mathbf{R}^n)$ *for* $\sigma > \sigma_0$. *Then* E_μ *continuously depends on* $\mu \in \mathbf{S}^2$ *and* $E = \frac{1}{4\pi} \int_{\mathbf{S}^2} E_\mu \, d\sigma(\mu)$.

Proof If $\sigma > \sigma_0$ and $\xi \in \mathbf{R}^n$ and $b = P_0(i\xi + \sigma N)$, $a_j = P_j(i\xi + \sigma N)$, then $b + \sum_{j=1}^{3} \mu_j a_j$ does not vanish for $\mu \in \mathbf{S}^2$, and hence formula (3.5.4) holds and yields

$$R(i\xi + \sigma N)^{-1} = \frac{1}{4\pi} \int_{\mathbf{S}^2} Q_\mu(i\xi + \sigma N)^{-2} \, d\sigma(\mu).$$

This incidentally implies that $R(i\xi + \sigma N) \neq 0$ and hence that $R(\partial)$ is quasihyperbolic.

Applying the Seidenberg–Tarski lemma as in the proof of Proposition 2.3.5 we conclude that E_μ continuously depends on $\mu \in \mathbf{S}^2$ and that, by Fourier transformation, $E = \frac{1}{4\pi} \int_{\mathbf{S}^2} E_\mu \, d\sigma(\mu)$. This completes the proof. □

In particular, if $P_3 = 0$, then Proposition 3.5.3 reduces, in accordance with formula (3.5.3), to the representation

$$E = \frac{1}{2\pi} \int_{\{(\lambda, \mu) \in \mathbf{R}^2; \, \lambda^2 + \mu^2 < 1\}} E_{\lambda, \mu} \frac{d\lambda d\mu}{\sqrt{1 - \lambda^2 - \mu^2}} \tag{3.5.6}$$

for the fundamental solution E of $P_0(\partial)^2 - P_1(\partial)^2 - P_2(\partial)^2$ in terms of the fundamental solutions $E_{\lambda,\mu}$ of the uniformly quasihyperbolic operators $\left(P_0(\partial) + \lambda P_1(\partial) + \mu P_2(\partial)\right)^2$, $\lambda^2 + \mu^2 \le 1$.

Example 3.5.4 As a first application of Eq. (3.5.6), let us calculate the fundamental solution of *Timoshenko's beam operator*, viz.

$$P(\partial) = \partial_t^2 + \frac{EI}{\rho A}\,\partial_x^4 + \frac{\rho I}{GA\kappa}\,\partial_t^4 - \frac{I}{A}\left(1 + \frac{E}{G\kappa}\right)\partial_t^2\partial_x^2. \qquad (3.5.7)$$

The operator $P(\partial)$ describes the transversal deflection of a vibrating beam taking into account the effects of rotatory inertia and of shearing forces, see Timoshenko and Young [269], Eq. (129), p. 331. The constants have the following meaning: E is Young's modulus, ρ is density, A is the area of cross-section, J is the moment of inertia, G is the shear modulus, and κ is Timoshenko's shear coefficient.

The operator $R(\partial) = GA\kappa I^{-1}\rho^{-1} \cdot P(\partial)$ can be written in the form

$$R(\partial) = (\partial_t^2 - a\partial_x^2 + b)^2 - (c\partial_x^2 - d)^2 - e^2, \qquad (3.5.8)$$

where $a = \dfrac{G\kappa + E}{2\rho}$, $b = \dfrac{GA\kappa}{2\rho I}$, $c = \dfrac{|G\kappa - E|}{2\rho}$, $d = b\,\dfrac{G\kappa + E}{|G\kappa - E|}$, $e = 2ib\,\dfrac{\sqrt{G\kappa E}}{|G\kappa - E|}$.

For the validity of formula (3.5.6), we have to assume that the operators

$$Q_{\lambda,\mu}(\partial) = \partial_t^2 - a\partial_x^2 + b + \lambda(c\partial_x^2 - d) + \mu e, \qquad \lambda^2 + \mu^2 \le 1,$$

are uniformly quasihyperbolic with respect to t. This is satisfied due to $a \ge |c|$.

By formula (2.3.14), the fundamental solution $E_{\lambda,\mu}$ of $Q_{\lambda,\mu}(\partial)^2$ with support in $t \ge 0$ is given by

$$E_{\lambda,\mu} = \frac{Y(\sqrt{a - \lambda c}\,t - |x|)\sqrt{(a - \lambda c)\,t^2 - x^2}}{4(a - \lambda c)\sqrt{b - \lambda d + \mu e}}\,J_1\left(\sqrt{(b - \lambda d + \mu e)\left(t^2 - \frac{x^2}{a - \lambda c}\right)}\right).$$

If we insert $E_{\lambda,\mu}$ into (3.5.6) and substitute $\mu = \sqrt{1 - \lambda^2}\,v$, we obtain an integral of the form

$$\int_{-1}^{1} J_1\left(\sqrt{A + Bv}\right)\frac{dv}{\sqrt{A + Bv}\,\sqrt{1 - v^2}} = \int_0^\pi \frac{J_1\left(\sqrt{A + B\cos\varphi}\right)}{\sqrt{A + B\cos\varphi}}\,d\varphi$$

$$= -2\frac{\partial}{\partial A}\int_0^\pi J_0\left(\sqrt{A + B\cos\varphi}\right)d\varphi$$

$$= -2\pi\frac{\partial}{\partial A}\left[J_0\left(\sqrt{\tfrac{1}{2}\left(A + \sqrt{A^2 - B^2}\right)}\right)J_0\left(\sqrt{\tfrac{1}{2}\left(A - \sqrt{A^2 - B^2}\right)}\right)\right],$$

where we have used Gradshteyn and Ryzhik [113], Eq. 6.684.1. Since $J_0(z)$ and $J_1(z)/z$ are entire functions of z, these equations make sense and are valid for complex values of A and B as well.

With the abbreviations $C_\pm := \sqrt{\frac{1}{2}\left(A \pm \sqrt{A^2 - B^2}\,\right)}$, this yields

$$\int_{-1}^{1} J_1\left(\sqrt{A + Bv}\,\right) \frac{dv}{\sqrt{A + Bv}\,\sqrt{1 - v^2}}$$

$$= \frac{\pi}{\sqrt{A^2 - B^2}}\left[C_+ J_1(C_+)J_0(C_-) - C_- J_1(C_-)J_0(C_+)\right].$$

In our case,

$$A := (b - \lambda d)\left(t^2 - \frac{x^2}{a - \lambda c}\right), \quad B := e\sqrt{1 - \lambda^2}\left(t^2 - \frac{x^2}{a - \lambda c}\right),$$

and, on account of $d^2 + e^2 = b^2$, we obtain

$$\sqrt{A^2 - B^2} = \left(t^2 - \frac{x^2}{a - \lambda c}\right)(d - \lambda b),$$

$$A \pm \sqrt{A^2 - B^2} = \left(t^2 - \frac{x^2}{a - \lambda c}\right)(b \pm d)(1 \mp \lambda).$$

Therefore, the fundamental solution E (with support in $t \geq 0$) of the operator $R(\partial)$ in (3.5.8) is given by the following definite integral:

$$E = \frac{1}{8}\int_{-1}^{1}\left[C_+ J_1(C_+)J_0(C_-) - C_- J_1(C_-)J_0(C_+)\right]\frac{Y\left(\sqrt{a - \lambda c}\,t - |x|\right)d\lambda}{\sqrt{a - \lambda c}\,(d - \lambda b)},$$

$$(3.5.9)$$

with

$$C_\pm = \sqrt{\frac{1}{2}\left(t^2 - \frac{x^2}{a - \lambda c}\right)} \cdot \sqrt{(b \pm d)(1 \mp \lambda)}.$$

The singular support of E consists of the rays $|x|/t = \sqrt{a \pm c}$, or, with respect to the physical constants, $|x|/t = \sqrt{E/\rho}$ and $|x|/t = \sqrt{G\kappa/\rho}$, which rays correspond to the different velocities of pressure and of shear waves. In relation to these two velocities, there arise three different representations of E :

$$
E(t,x) = \begin{cases} 0, & \text{if } \sqrt{a+c} \le \dfrac{|x|}{t}, \\[2mm] \displaystyle\int_{-1}^{\lambda_0} K(\lambda,t,x)\,d\lambda, & \text{if } \sqrt{a-c} \le \dfrac{|x|}{t} \le \sqrt{a+c}, \\[2mm] \displaystyle\int_{-1}^{1} K(\lambda,t,x)\,d\lambda, & \text{if } \dfrac{|x|}{t} \le \sqrt{a-c}, \end{cases} \tag{3.5.10}
$$

where $\lambda_0 = a/c - x^2/(ct^2)$ and

$$
K(\lambda,t,x) = \frac{C_+ J_1(C_+)J_0(C_-) - C_- J_1(C_-)J_0(C_+)}{8\sqrt{a-\lambda c}\,(d-\lambda b)}.
$$

Formula (3.5.10) was deduced for the first time in Ortner [203], Satz 1, p. 551, by a different method. The derivation above stems from Ortner and Wagner [207], Ex. 4, p. 456, and Prop. 7, p. 457, where a typographical error in the definition of the constant e should be corrected. A further derivation of (3.5.10) can be found in Ortner and Wagner [208], Prop. 3, p. 530. A completely different representation of the fundamental solution E was derived by analytic continuation in Ortner and Wagner [214], Prop. 1, p. 219, see Example 4.1.6 below.

Note that the generalization of the classical Euler–Bernoulli beam operator, i.e., $\partial_t^2 + EI/(\rho A)\partial_x^4$, to the operator $P(\partial)$ in (3.5.7) is usually attributed to S. Timoshenko, but was in fact anticipated by Bresse [23], p. 126, cf. Deresiewicz and Mindlin [57], p. 178; Mindlin [181], p. 320.

The analogous generalization of Lagrange's plate operator $\partial_t^2 + D/(\rho h)\Delta_2^2$, where ρ, h, D denote density, thickness and flexural rigidity, respectively, was given independently in Uflyand [278], p. 291, and in Mindlin [180]. This generalized operator has the form

$$
M(\partial) = \left(\Delta_2 - \frac{h}{\kappa^2 G}\,\partial_t^2\right)\left(D\Delta_2 - \frac{\rho h^3}{12}\,\partial_t^2\right) + \rho h \partial_t^2,
$$

see Mindlin [180], Eq. (37), p. 36; Mindlin [181], p. 320. The fundamental solution of $M(\partial)$ with support in $t \ge 0$ can be represented by a simple definite integral over Bessel functions similarly as in (3.5.10), see Ortner and Wagner [208], Prop. 4, p. 533. □

Example 3.5.5

(a) Let us derive next the fundamental solution E of a three-dimensional analogue of Timoshenko's operator, namely, of

$$
R(\partial) = (a_0\partial_t^2 - b_0\Delta_3 + c_0)(a_1\partial_t^2 - b_1\Delta_3 + c_1) - d^2. \tag{3.5.11}
$$

This operator is hyperbolic for positive a_0, a_1, b_0, b_1 and complex c_0, c_1, d. If $R(\partial)$ is written in the form

$$R(\partial) = \tfrac{1}{4}\big((a_0 + a_1)\partial_t^2 - (b_0 + b_1)\Delta_3 + c_0 + c_1\big)^2$$

$$- \tfrac{1}{4}\big((a_0 - a_1)\partial_t^2 - (b_0 - b_1)\Delta_3 + c_0 - c_1\big)^2 - d^2,$$

then the parameter integration formula (3.5.6) yields

$$E = \frac{1}{2\pi} \iint_{\lambda^2 + \mu^2 < 1} E_{\lambda,\mu}\, \frac{\mathrm{d}\lambda\,\mathrm{d}\mu}{\sqrt{1 - \lambda^2 - \mu^2}}, \tag{3.5.12}$$

where $E_{\lambda,\mu}$ is the forward fundamental solution of the operator

$$Q_{\lambda,\mu}(\partial)^2 = (a_\lambda \partial_t^2 - b_\lambda \Delta_3 + c_\lambda + \mu d)^2,$$

and $a_\lambda = (a_0 + a_1)/2 + \lambda(a_0 - a_1)/2$ and analogously for b_λ and c_λ.
    Making use of Example 2.5.6 we obtain

$$E_{\lambda,\mu} = \frac{1}{8\pi}\, \frac{Y\big(\frac{t}{\sqrt{a_\lambda}} - \frac{|x|}{\sqrt{b_\lambda}}\big)}{\sqrt{a_\lambda}\, b_\lambda^{3/2}}\, J_0\left(\sqrt{c_\lambda + \mu d}\,\sqrt{\frac{t^2}{a_\lambda} - \frac{|x|^2}{b_\lambda}}\right).$$

Hence, inserting $E_{\lambda,\mu}$ into (3.5.12) and substituting $\mu = \sqrt{1 - \lambda^2}\,\cos\varphi$, we infer

$$E = \frac{1}{16\pi^2} \int_{-1}^{1} \frac{\mathrm{d}\lambda}{\sqrt{a_\lambda}\, b_\lambda^{3/2}}\, Y\Big(\frac{t}{\sqrt{a_\lambda}} - \frac{|x|}{\sqrt{b_\lambda}}\Big) \int_0^\pi J_0\big(\sqrt{A + B\cos\varphi}\big)\mathrm{d}\varphi$$

$$= \frac{1}{16\pi} \int_{-1}^{1} Y\Big(\frac{t}{\sqrt{a_\lambda}} - \frac{|x|}{\sqrt{b_\lambda}}\Big) J_0(C_+) J_0(C_-) \frac{\mathrm{d}\lambda}{\sqrt{a_\lambda}\, b_\lambda^{3/2}} \tag{3.5.13}$$

where

$$A = c_\lambda \cdot \Big(\frac{t^2}{a_\lambda} - \frac{|x|^2}{b_\lambda}\Big), \qquad B = d\sqrt{1 - \lambda^2} \cdot \Big(\frac{t^2}{a_\lambda} - \frac{|x|^2}{b_\lambda}\Big),$$

$$C_\pm = \sqrt{\tfrac{1}{2}\big(A \pm \sqrt{A^2 - B^2}\,\big)} = \sqrt{\tfrac{1}{2}\big(c_\lambda \pm \sqrt{c_\lambda^2 - d^2(1 - \lambda^2)}\,\big)}\sqrt{\frac{t^2}{a_\lambda} - \frac{|x|^2}{b_\lambda}}.$$

This representation of E was deduced first by a different method in Ortner [203], Lemma 2, p. 550, and rederived with the method of parameter integration as above in Gawinecki, Kirchner and Łazuka [101], Thm. 2, p. 830.

As in Example 3.5.4, let us yet specify formula (3.5.13) for the case $d^2 = c_0 c_1$. Then

$$C_\pm = \frac{1}{\sqrt{2}} \left\{ \begin{array}{c} \sqrt{c_0(1+\lambda)} \\ \sqrt{c_1(1-\lambda)} \end{array} \right\} \cdot \sqrt{\frac{t^2}{a_\lambda} - \frac{|x|^2}{b_\lambda}}.$$

If $v_i = \sqrt{b_i/a_i}$, $i = 0, 1$, are the wave speeds and $v_0 < v_1$, then

$$E(t,x) = \frac{1}{16\pi} \left\{ \begin{array}{ll} 0, & \text{if } v_1 \le \dfrac{|x|}{t}, \\[3mm] \displaystyle\int_{-1}^{\lambda_0} J_0(C_+) J_0(C_-) \dfrac{d\lambda}{\sqrt{a_\lambda}\, b_\lambda^{3/2}}, & \text{if } v_0 \le \dfrac{|x|}{t} \le v_1, \\[5mm] \displaystyle\int_{-1}^{1} J_0(C_+) J_0(C_-) \dfrac{d\lambda}{\sqrt{a_\lambda}\, b_\lambda^{3/2}}, & \text{if } \dfrac{|x|}{t} \le v_0, \end{array} \right.$$

(3.5.14)

where

$$\lambda_0 = \frac{(b_0 + b_1)t^2 - (a_0 + a_1)|x|^2}{(b_1 - b_0)t^2 + (a_0 - a_1)|x|^2} \in (-1, 1) \text{ for } v_0 < \frac{|x|}{t} < v_1.$$

(b) Let us apply the result in (3.5.14) in order to derive a representation of the fundamental solution of the operator

$$R(\partial) = \partial_t^2 - \alpha \partial_t^2 \Delta_3 + \beta \Delta_3^2 - \delta \Delta_3, \qquad \alpha > 0,\ \beta \ge 0,\ \delta \in \mathbf{C}. \tag{3.5.15}$$

For $\beta = 0$, this is the "Boussinesq operator" in dimension three, see Ortner [203], p. 552; in space dimensions one and two, it describes water waves in the Boussinesq approximation for long waves, see Whitham [301], (1.20), p. 9, and (11.7), p. 366, and Example 3.5.6 below. Let us observe that $R(\partial)$ is quasihyperbolic with respect to t.

The operator $R(\partial)$ in (3.5.15) becomes a limit case of the operator in (3.5.11), if we set $a_0 = 1$, $a_1 = 0$, $b_0 = \beta/\alpha$, $b_1 = \alpha$, $c_0 = \delta/\alpha - \beta/\alpha^2$, $c_1 = 1$, $d^2 = c_0 c_1$. Employing the substitution $u = \dfrac{\beta}{\alpha} + \alpha \dfrac{1-\lambda}{1+\lambda}$ in (3.5.14)

we obtain

$$C_+ = \frac{1}{\alpha} \sqrt{\alpha\delta - \beta} \sqrt{t^2 - \frac{|x|^2}{u}} \quad \text{and} \quad C_- = \frac{1}{\alpha} \sqrt{\alpha u - \beta} \sqrt{t^2 - \frac{|x|^2}{u}},$$

and the integration limits $\lambda = -1, \lambda_0, 1$ yield $u = \infty, |x|^2/t^2, \beta/\alpha$, respectively. Furthermore, $\dfrac{d\lambda}{\sqrt{a_\lambda}\, b_\lambda^{3/2}} = -\dfrac{2du}{\alpha u^{3/2}}$. Hence

$$E(t,x) = \frac{Y(t)}{8\pi\alpha} \int_{u_0}^{\infty} J_0\left(\frac{1}{\alpha}\sqrt{\alpha\delta-\beta}\sqrt{t^2-\frac{|x|^2}{u}}\right) J_0\left(\frac{1}{\alpha}\sqrt{\alpha u-\beta}\sqrt{t^2-\frac{|x|^2}{u}}\right)\frac{du}{u^{3/2}},$$
(3.5.16)

where $u_0 = \max\{\frac{\beta}{\alpha}, \frac{|x|^2}{t^2}\}$.

In the above mentioned special case of water waves in the Boussinesq approximation, we have $\beta = 0$, and we obtain the following representation for the fundamental solution E of $R(\partial) = \partial_t^2 - \alpha\partial_t^2\Delta_3 - \delta\Delta_3$, $\alpha > 0$, $\delta \in \mathbf{C}$:

$$E(t,x) = \frac{Y(t)}{8\pi\alpha} \int_{|x|^2/t^2}^{\infty} J_0\left(\sqrt{\frac{\delta}{\alpha}}\sqrt{t^2-\frac{|x|^2}{u}}\right) J_0\left(\sqrt{\frac{u}{\alpha}}\sqrt{t^2-\frac{|x|^2}{u}}\right)\frac{du}{u^{3/2}}$$

$$= \frac{Y(t)}{4\pi\alpha|x|} \int_0^t J_0\left(\sqrt{\frac{\delta}{\alpha}}\tau\right) J_0\left(\frac{|x|\tau}{\sqrt{\alpha}\sqrt{t^2-\tau^2}}\right)\frac{\tau d\tau}{\sqrt{t^2-\tau^2}}.$$
(3.5.17)

In contrast, if we set $\delta = 0$ in (3.5.15), we obtain a three-dimensional analogue of Rayleigh's operator, see Example 2.4.14. Then (3.5.16) yields

$$E(t,x) = \frac{Y(t)}{8\pi\alpha} \int_{u_0}^{\infty} I_0\left(\frac{\sqrt{\beta}}{\alpha}\sqrt{t^2-\frac{|x|^2}{u}}\right) J_0\left(\frac{1}{\alpha}\sqrt{\alpha u-\beta}\sqrt{t^2-\frac{|x|^2}{u}}\right)\frac{du}{u^{3/2}},$$
(3.5.18)

$u_0 = \max\{\frac{\beta}{\alpha}, \frac{|x|^2}{t^2}\}$, for the fundamental solution E of $\partial_t^2 - \alpha\partial_t^2\Delta_3 + \beta\Delta_3^2$.
Formula (3.5.18) can be tested by setting $\beta = 0$, which furnishes

$$E(t,x) = \frac{Y(t)}{8\pi\alpha} \int_{|x|^2/t^2}^{\infty} J_0\left(\frac{1}{\sqrt{\alpha}}\sqrt{ut^2-|x|^2}\right)\frac{du}{u^{3/2}} = \frac{Y(t)te^{-|x|/\sqrt{\alpha}}}{4\pi\alpha|x|}$$

upon using Gradshteyn and Ryzhik [113], Eq. 6.554.4. The result is in accordance with the fundamental solution of the metaharmonic operator $1 - \alpha\Delta_3$ derived in Example 1.4.11. □

Example 3.5.6 Let us eventually derive formulas for the fundamental solutions of the *Boussinesq operators*

$$B_n(\partial) = \partial_t^2 - \alpha\partial_t^2\Delta_n - \delta\Delta_n, \qquad \alpha > 0, \delta \in \mathbf{C},$$

for the physically relevant dimensions $n = 1, 2$. These formulas appear for the first time in Ortner [203], Sätze 2, 3, p. 553.

As in Examples 2.6.3 and 2.6.6, we employ Hadamard's method of descent: If E_n denotes the fundamental solution of $B_n(\partial)$ satisfying $E_n \cdot e^{-\sigma t} \in \mathcal{S}'(\mathbf{R}^{n+1})$ for $\sigma > \sigma_0$ (see Proposition 2.4.13), then

$$E_2 \otimes 1_{x_3} = E_3 * (\delta_{(t,x_1,x_2)} \otimes 1_{x_3}) \quad \text{and} \quad E_1 \otimes 1_{x_2,x_3} = E_3 * (\delta_{(t,x_1)} \otimes 1_{(x_2,x_3)}).$$

(a) Setting $x' = (x_1, x_2)$, we obtain from (3.5.17)

$$E_2(t, x') = \frac{Y(t)}{2\pi\alpha} \int_0^t J_0\left(\sqrt{\frac{\delta}{\alpha}}\tau\right) \int_0^\infty J_0\left(\frac{\tau\sqrt{|x'|^2 + x_3^2}}{\sqrt{\alpha}\sqrt{t^2 - \tau^2}}\right) \frac{dx_3}{\sqrt{|x'|^2 + x_3^2}} \frac{\tau\, d\tau}{\sqrt{t^2 - \tau^2}}$$

$$= \frac{Y(t)}{2\pi\alpha} \int_0^t J_0\left(\sqrt{\frac{\delta}{\alpha}}\tau\right) \int_0^\infty J_0\left(\frac{\tau|x'|\sqrt{1 + u^2}}{\sqrt{\alpha}\sqrt{t^2 - \tau^2}}\right) \frac{du}{\sqrt{1 + u^2}} \frac{\tau\, d\tau}{\sqrt{t^2 - \tau^2}}$$

$$= -\frac{Y(t)}{4\alpha} \int_0^t J_0\left(\sqrt{\frac{\delta}{\alpha}}\tau\right) J_0\left(\frac{\tau|x'|}{2\sqrt{\alpha}\sqrt{t^2 - \tau^2}}\right) N_0\left(\frac{\tau|x'|}{2\sqrt{\alpha}\sqrt{t^2 - \tau^2}}\right) \frac{\tau\, d\tau}{\sqrt{t^2 - \tau^2}}.$$

Here we have used Eq. 6.596.2 in Gradshteyn and Ryzhik [113], and the result coincides with Ortner [203], Satz 2, p. 553.

(b) Similarly, by integration with respect to x_1, x_2, we obtain from (3.5.17)

$$E_1(t, x_1) = \frac{Y(t)}{2\alpha} \int_0^t J_0\left(\sqrt{\frac{\delta}{\alpha}}\tau\right) \int_0^\infty J_0\left(\frac{\tau\sqrt{x_1^2 + \rho^2}}{\sqrt{\alpha}\sqrt{t^2 - \tau^2}}\right) \frac{\rho\, d\rho}{\sqrt{x_1^2 + \rho^2}} \frac{\tau\, d\tau}{\sqrt{t^2 - \tau^2}}$$

$$= \frac{Y(t)}{2\alpha} \int_0^t J_0\left(\sqrt{\frac{\delta}{\alpha}}\tau\right) \int_{|x_1|/\sqrt{t^2 - \tau^2}}^\infty J_0\left(\frac{\tau v}{\sqrt{\alpha}}\right) dv\, \tau\, d\tau$$

$$= \frac{Y(t)}{2\alpha} \int_{|x_1|/t}^\infty \left(\int_0^{\sqrt{t^2 - |x_1|^2 v^{-2}}} J_0\left(\sqrt{\frac{\delta}{\alpha}}\tau\right) J_0\left(\frac{\tau v}{\sqrt{\alpha}}\right) \tau\, d\tau \right) dv$$

$$= \frac{Y(t)}{2\sqrt{\alpha}} \int_{|x_1|/t}^\infty \frac{\sqrt{t^2 - |x_1|^2 v^{-2}}}{\delta - v^2} \left[\sqrt{\delta} J_1\left(\sqrt{\frac{\delta}{\alpha}}\sqrt{t^2 - \frac{|x_1|^2}{v^2}}\right) \right.$$

$$\times J_0\left(\frac{1}{\sqrt{\alpha}}\sqrt{v^2 t^2 - |x_1|^2}\right) - v J_0\left(\sqrt{\frac{\delta}{\alpha}}\sqrt{t^2 - \frac{|x_1|^2}{v^2}}\right)$$

$$\left. \times J_1\left(\frac{1}{\sqrt{\alpha}}\sqrt{v^2 t^2 - |x_1|^2}\right) \right] dv.$$

Here we have used Eq. 5.54.1 in Gradshteyn and Ryzhik [113], and the result coincides with Ortner [203], Satz 3, p. 553.   □

In order to deduce the fundamental solutions of the Euler–Bernoulli beam and plate operators with elastic embedding, i.e., of $\partial_t^2 + \Delta_n^2 + c^2$, let us first simplify formula (3.5.6) in the special case of operators $P_0(\partial)^2 - P_1(\partial)^2 + c^2$. This goes back to Ortner and Wagner [207], Prop. 3, p. 448.

Proposition 3.5.7 *Let $p_0(\partial), p_1(\partial)$ be operators in $\mathbf{R}^n$ such that*

$$\exists \sigma_0 \in \mathbf{R} : \forall \xi \in \mathbf{R}^n : \operatorname{Re} p_0(i\xi) + |\operatorname{Re} p_1(i\xi)| \leq \sigma_0. \qquad (3.5.19)$$

Then the family $Q_\lambda(\partial) = \partial_t - p_0(\partial_x) + \lambda p_1(\partial_x)$, $\lambda \in [-1, 1]$, of operators in the $n + 1$ variables (t, x) is uniformly quasihyperbolic with respect to t. Also

$$R(\partial) = \big(\partial_t - p_0(\partial_x)\big)^2 - p_1(\partial_x)^2 + c^2, \qquad c \in \mathbf{C},$$

is quasihyperbolic in the direction t.

If E and F_λ denote the fundamental solutions of $R(\partial)$ and of $Q_\lambda(\partial)$, respectively, with $e^{-\sigma t}E, e^{-\sigma t}F_\lambda \in \mathcal{S}'(\mathbf{R}^{n+1})$, $\lambda \in [-1, 1]$, $\sigma > \sigma_0$, then

$$E = \frac{t}{2} \int_{-1}^{1} J_0(ct\sqrt{1 - \lambda^2}) F_\lambda \, d\lambda.$$

Proof

(1) According to Definition 3.1.1, the uniform quasihyperbolicity of $Q_\lambda(\partial)$, $\lambda \in [-1, 1]$, is equivalent to the existence of $\sigma_0 > 0$ such that, for each $\sigma > \sigma_0$ and $(\tau, \xi) \in \mathbf{R}^{n+1}$, the complex numbers $i\tau + \sigma - p_0(i\xi) + \lambda p_1(i\xi)$ do not vanish, i.e., such that

$$\{p_0(i\xi) - \lambda p_1(i\xi); \, \xi \in \mathbf{R}^n\} \cap \{z \in \mathbf{C}; \, \operatorname{Re} z > \sigma_0\} = \emptyset.$$

Evidently, this is equivalent to the condition that the real parts of $p_0(i\xi) \pm p_1(i\xi)$ are bounded by σ_0 for $\xi \in \mathbf{R}^n$, and this yields condition (3.5.19).

(2) If we assume the validity of (3.5.19), we can apply Proposition 3.5.3, and in particular formula (3.5.6) holds. In this formula, $E_{\lambda,\mu}$ now denotes the fundamental solution of $\big(\partial_t - p_0(\partial_x) + \lambda p_1(\partial_x) + ic\mu\big)^2$, and $E_{\lambda,\mu}$ continuously depends on $\lambda \in [-1, 1]$, $\mu \in \mathbf{C}$.

By Proposition 2.5.1 and Lemma 2.5.3, we have $E_{\lambda,\mu} = te^{-i\mu ct}F_\lambda$ and thus

$$E = \frac{t}{2\pi} \int_{-1}^{1} F_\lambda \left(\int_{-\sqrt{1-\lambda^2}}^{\sqrt{1-\lambda^2}} \frac{e^{-i\mu ct} \, d\mu}{\sqrt{1 - \lambda^2 - \mu^2}} \right) d\lambda = \frac{t}{2} \int_{-1}^{1} J_0(ct\sqrt{1 - \lambda^2}) F_\lambda \, d\lambda$$

by Poisson's integral representation of J_0. $\qquad \qquad \square$

Example 3.5.8 As an application of Proposition 3.5.7, let us give a representation of the fundamental solution of the operator $R(\partial) = \partial_t^2 + (\Delta_n + a)^2 + c^2$, $a, c \in \mathbf{C}$.

This operator describes, for $n = 1, 2$ and $a, c \in \mathbf{R}$, the transverse vibrations of *prestressed and elastically supported beams or plates*, respectively, cf. Graff [114], (3:3.11) and (3.3.25), pp. 173, 175.

In this case, $Q_\lambda(\partial) = \partial_t + i\lambda(\Delta_n + a)$, $\lambda \in [-1, 1]$, is a Schrödinger operator, and its fundamental solution is, according to Example 2.5.5,

$$F_\lambda = \frac{Y(t)e^{-i\lambda at}e^{i(\text{sign}\,\lambda)n\pi/4}}{(4\pi t|\lambda|)^{n/2}}\,\exp\!\Big(-\frac{i|x|^2}{4\lambda t}\Big) \in \mathcal{C}\big([0, \infty), \mathcal{D}'(\mathbf{R}^n_x)\big). \tag{3.5.20}$$

Hence the fundamental solution E of $R(\partial)$ is given by

$$E = \frac{Y(t)t^{-n/2+1}}{(4\pi)^{n/2}} \int_0^1 \cos\!\Big(\lambda at + \frac{|x|^2}{4\lambda t} - \frac{n\pi}{4}\Big)J_0\!\big(ct\sqrt{1 - \lambda^2}\big)\frac{\mathrm{d}\lambda}{\lambda^{n/2}}. \tag{3.5.21}$$

Note that, for general n, the integral in (3.5.21) must be conceived as the integral of a continuous distribution-valued function $[0, 1] \longrightarrow \mathcal{D}'(\mathbf{R}^{n+1}_{t,x})$.

For $n = 1$, the integral in (3.5.21) is absolutely convergent, for $n = 2, 3$, it is still conditionally convergent, and it yields in these three cases a locally integrable function of (t, x). Substituting $\tau = \lambda t$ we obtain the following for $n = 1, 2, 3$:

$$E = \frac{Y(t)}{(4\pi)^{n/2}} \int_0^t \cos\!\Big(a\tau + \frac{|x|^2}{4\tau} - \frac{n\pi}{4}\Big)J_0\!\big(c\sqrt{t^2 - \tau^2}\big)\frac{\mathrm{d}\tau}{\tau^{n/2}} \in L^1_{\text{loc}}(\mathbf{R}^{n+1}_{t,x}). \tag{3.5.22}$$

More precisely, $E \in \mathcal{C}(\mathbf{R}^2)$ if $n = 1$ and $E \in L^\infty_{\text{loc}}(\mathbf{R}^3)$ if $n = 2$. For formula (3.5.22) in the case $n = 1$, see also Shreves and Stadler [253], (3.11), p. 202.

Similarly, Proposition 3.5.7 can be applied to operators of the form

$$(\partial_t - a\Delta_n - b)^2 - (c\Delta_n + 2\omega^T\nabla + d)^2 - h^2$$

and, still more generally,

$$(\partial_t - a_1\partial_1^2 - \cdots - a_n\partial_n^2 - 2\omega^T\nabla - b)^2 - (c_1\partial_1^2 + \cdots + c_n\partial_n^2 + 2\eta^T\nabla + d)^2 - h^2,$$

see Ortner and Wagner [207], Prop. 4 and Remark 4, pp. 450, 452. $\square$

Let us generalize now Proposition 3.5.7 so as to yield a representation of the *convolution group* $E(\lambda)$ of an operator of the form $R(\partial) = \big(\partial_t - p_0(\partial_x)\big)^2 - p_1(\partial_x)^2 + c^2$.

Proposition 3.5.9 *Let* $p_0(\partial), p_1(\partial)$ *be operators in* $\mathbf{R}^n$ *such that* (3.5.19) *holds. Let* $E(\lambda)$, $\lambda \in \mathbf{C}$, *denote the convolution group of the quasihyperbolic operator* $R(\partial) = \big(\partial_t - p_0(\partial_x)\big)^2 - p_1(\partial_x)^2 + c^2$, $c \in \mathbf{C}$, *i.e.,*

$$E(\lambda) = e^{\sigma t}\mathcal{F}^{-1}_{\tau,\xi}\Big(\big[(i\tau + \sigma - p_0(i\xi))^2 - p_1(i\xi)^2 + c^2\big]^\lambda\Big), \qquad \sigma > \sigma_0.$$

Then $E(\lambda)$ can be represented by the fundamental solution F_μ of $Q_\mu(\partial) = \partial_t - p_0(\partial_x) + \mu p_1(\partial_x)$ in the following way if $\operatorname{Re} \lambda < 0$:

$$E(\lambda) = \frac{t^{-\lambda} 2^\lambda c^{\lambda+1}}{\Gamma(-\lambda)} \int_{-1}^{1} \frac{J_{-\lambda-1}(ct\sqrt{1-\mu^2})}{(1-\mu^2)^{(\lambda+1)/2}} F_\mu \, d\mu. \tag{3.5.23}$$

(Note that $E(-k)$, $k \in \mathbf{N}$, is the uniquely determined fundamental solution of $R(\partial)^k$ satisfying $e^{-\sigma t}E(-k) \in \mathcal{S}'(\mathbf{R}^{n+1})$ *for $\sigma > \sigma_0$.)*

Proof If σ_0 is as in (3.5.19), $\sigma > \sigma_0$ and $\operatorname{Re} \lambda < -\frac{1}{2}$, then Lemma 3.5.1 implies

$$E(\lambda) = e^{\sigma t} \mathcal{F}_{\tau,\xi}^{-1} \left(\left[(i\tau + \sigma - p_0(i\xi))^2 - p_1(i\xi)^2 + c^2 \right]^\lambda \right)$$

$$= -\frac{\lambda + \frac{1}{2}}{\pi} e^{\sigma t} \mathcal{F}_{\tau,\xi}^{-1} \left(\int_{\mu^2+\nu^2<1} \frac{(i\tau + \sigma - p_0(i\xi) + \mu p_1(i\xi) + icv)^{2\lambda}}{(1-\mu^2-\nu^2)^{\lambda+3/2}} \, d\mu d\nu \right).$$

Since

$$e^{\sigma t} \mathcal{F}_\tau^{-1} \left((i\tau + \sigma + A)^{2\lambda} \right) = e^{-At} \chi_+^{-2\lambda}(t), \tag{3.5.24}$$

see Example 1.5.11, Lemma 3.5.2 and (1.6.8), we obtain

$$E(\lambda) = -\frac{\lambda + \frac{1}{2}}{\pi} \chi_+^{-2\lambda}(t) \mathcal{F}_\xi^{-1} \int_{\mu^2+\nu^2<1} \frac{e^{-t[-p_0(i\xi)+\mu p_1(i\xi)]} \cdot e^{-i\nu ct}}{(1-\mu^2-\nu^2)^{\lambda+3/2}} \, d\mu d\nu.$$

Upon substituting $\nu = \sqrt{1-\mu^2}\,u$ and using Poisson's integral representation of the Bessel function we infer that

$$E(\lambda) = -\frac{\lambda + \frac{1}{2}}{\pi} \chi_+^{-2\lambda}(t) \int_{-1}^{1} (1-\mu^2)^{-\lambda-1} \mathcal{F}_\xi^{-1} \left(e^{-t[-p_0(i\xi)+\mu p_1(i\xi)]} \right) \times$$

$$\times \int_{-1}^{1} (1-u^2)^{-\lambda-3/2} \cos(ct\sqrt{1-\mu^2}u) \, du d\mu$$

$$= \frac{\Gamma(-\lambda+\frac{1}{2})}{\Gamma(-2\lambda)\sqrt{\pi}} \left(\frac{c}{2} \right)^{\lambda+1} Y(t)t^{-\lambda} \int_{-1}^{1} \frac{J_{-\lambda-1}(ct\sqrt{1-\mu^2})}{(1-\mu^2)^{(\lambda+1)/2}} \mathcal{F}_\xi^{-1} \left(e^{-t[-p_0(i\xi)+\mu p_1(i\xi)]} \right) d\mu.$$

If we apply (3.5.24) once more, i.e., if we use

$$F_\mu = e^{\sigma t} \mathcal{F}_{\tau,\xi}^{-1} \left((i\tau + \sigma - p_0(i\xi) + \mu p_1(i\xi))^{-1} \right)$$

$$= \mathcal{F}_\xi^{-1} \left(Y(t)e^{-t[-p_0(i\xi)+\mu p_1(i\xi)]} \right),$$

and employ the doubling formula for the gamma function, we finally conclude that

$$E(\lambda) = \frac{t^{-\lambda}2^{\lambda}c^{\lambda+1}}{\Gamma(-\lambda)} \int_{-1}^{1} \frac{J_{-\lambda-1}(ct\sqrt{1-\mu^2})}{(1-\mu^2)^{(\lambda+1)/2}} F_{\mu}\, d\mu.$$

By analytic continuation, the last formula then holds for $\mathrm{Re}\,\lambda < 0$, and this completes the proof. □

Example 3.5.10 Let us apply Proposition 3.5.9 in order to represent the convolution group $E(\lambda)$ of the operator $R(\partial) = \partial_t^2 + (\Delta_n + a)^2 + c^2$, $a, c \in \mathbf{C}$, which was considered already in Example 3.5.8.

Then formulas (3.5.20) and (3.5.23) imply that

$$E(\lambda) = \frac{Y(t)2^{\lambda+1-n}c^{\lambda+1}}{\pi^{n/2}\Gamma(-\lambda)t^{n/2+\lambda}} \int_0^1 \cos\left(\mu at + \frac{|x|^2}{4\mu t} - \frac{n\pi}{4}\right) \frac{J_{-\lambda-1}(ct\sqrt{1-\mu^2})}{(1-\mu^2)^{(\lambda+1)/2}} \frac{d\mu}{\mu^{n/2}}.$$

(3.5.24)

In particular, $E(-k)$, $k \in \mathbf{N}$, is the fundamental solution of $R(\partial)^k$.

Note that $E(\lambda)$ is an entire distribution-valued function, but that (3.5.24) holds only for $\mathrm{Re}\,\lambda < 0$. Also in this case, (3.5.24) must be interpreted for general n as the integral of a continuous distribution-valued function $[0, 1] \to \mathcal{D}'(\mathbf{R}_{(t,x)}^{n+1})$, compare Example 3.5.8 for the case $\lambda = -1$. □

We finally generalize Proposition 3.5.3 in order to represent the convolution groups of quasihyperbolic operators of the form $R(\partial) = P_0(\partial)^2 - P_1(\partial)^2 - \cdots - P_l(\partial)^2$. As an example, we shall then deduce a second time the convolution group of the Klein–Gordon operator $\partial_t^2 - \Delta_n - c^2$, cf. Example 2.3.7.

Proposition 3.5.11 *Suppose that the operators* $Q_{\mu}(\partial) = P_0(\partial) + \sum_{j=1}^{l}\mu_j P_j(\partial)$, $|\mu| \leq 1$, *are uniformly quasihyperbolic with respect to* $N \in \mathbf{R}^n \setminus \{0\}$. *Then also the operator* $R(\partial) = P_0(\partial)^2 - P_1(\partial)^2 - \cdots - P_l(\partial)^2$ *is quasihyperbolic with respect to* N. *Let* $\sigma_0 \in \mathbf{R}$ *be as in* (3.1.3) *and denote by* $E(\lambda)$ *and by* $F_{\mu}(\lambda)$ *the convolution groups of* $R(\partial)$ *and of* $Q_{\mu}(\partial)$, $|\mu| \leq 1$, *respectively, i.e.,* $E(\lambda) = e^{\sigma Nx}\mathcal{F}_{\xi}^{-1}(R(\sigma N + i\xi)^{\lambda})$, $F_{\mu}(\lambda) = e^{\sigma Nx}\mathcal{F}_{\xi}^{-1}(Q_{\mu}(\sigma N + i\xi)^{\lambda})$, $\sigma > \sigma_0$, *see Sect. 2.3. If, furthermore,* χ_+^{λ} *is defined as in* Lemma 3.5.2, *then*

$$E(\lambda) = \frac{\Gamma(\frac{1}{2} - \lambda)}{\pi^{l/2}} \langle F_{\mu}(2\lambda), \chi_+^{-\lambda-(l-1)/2} \circ (1 - |\mu|^2)\rangle$$

(3.5.25)

holds for $\lambda \in \mathbf{C} \setminus (\frac{1}{2} + \mathbf{N}_0)$.

Proof Upon multiplication by $e^{-\sigma Nx}$ and Fourier transformation, (3.5.25) is equivalent to

$$R(\sigma N + i\xi)^{\lambda} = \frac{\Gamma(\frac{1}{2} - \lambda)}{\pi^{l/2}} \langle Q_{\mu}(\sigma N + i\xi)^{2\lambda}, \chi_+^{-\lambda-(l-1)/2} \circ (1 - |\mu|^2)\rangle,$$

which immediately follows from Eq. (3.5.5) in the proof of Lemma 3.5.2. □

Example 3.5.12 We shall apply Proposition 3.5.11 in order to calculate the convolution group of the Klein–Gordon operator $\partial_t^2 - \Delta_n - c^2$, $c \in \mathbf{C}$. The result was derived already once in Example 2.3.7 with the help of the partial Fourier transformation and the Poisson–Bochner formula.

If we set $l = n + 1$, $P_0(\partial) = \partial_t$, $P_j(\partial) = \partial_j$, $j = 1, \ldots, n = l - 1$, $P_l(\partial) = c$, then $R(\partial) = P_0(\partial)^2 - P_1(\partial)^2 - \cdots - P_l(\partial)^2 = \partial_t^2 - \Delta_n - c^2$ and

$$Q_{\mu,\nu}(\partial) = P_0(\partial) + \sum_{j=1}^{l-1} \mu_j P_j(\partial) + \nu P_l(\partial) = \partial_t + \sum_{j=1}^{n} \mu_j \partial_j + \nu c.$$

Therefore, according to (2.3.19), the convolution group $F_{\mu,\nu}(\lambda)$ of $Q_{\mu,\nu}(\partial)$ is given by

$$F_{\mu,\nu}(\lambda) = e^{\sigma t} \mathcal{F}_{\tau,\xi}^{-1}\left(\left(i\tau + \sigma + \nu c + i\sum_{j=1}^{n} \mu_j \xi_j\right)^{\lambda}\right)$$

$$= \mathcal{F}_{\xi}^{-1}\left(\exp\left(-\nu c t - it\sum_{j=1}^{n} \mu_j \xi_j\right)\right) \cdot \chi_+^{-\lambda}(t) = e^{-\nu c t} \delta(x - \mu t)\chi_+^{-\lambda}(t).$$

Hence Eq. (3.5.25) in Proposition 3.5.11 yields for the convolution group of $\partial_t^2 - \Delta_n - c^2$ the representation

$$E(\lambda) = \frac{\Gamma(\frac{1}{2} - \lambda)}{\pi^{(n+1)/2}} \chi_+^{-2\lambda}(t) \langle e^{-\nu c t}\delta(x - \mu t), \chi_+^{-\lambda - n/2} \circ (1 - |\mu|^2 - \nu^2)\rangle \qquad (3.5.26)$$

for $\lambda \in \mathbf{C} \setminus (\frac{1}{2} + \mathbf{N}_0)$.

For fixed $t > 0$ and $\operatorname{Re}\lambda < -\frac{n}{2}$, we can evaluate (3.5.26) by classical integration:

$$E(\lambda) = \frac{\Gamma(\frac{1}{2} - \lambda)Y(t)t^{-2\lambda - 1 - n}}{\pi^{(n+1)/2}\Gamma(-2\lambda)} \langle e^{-\nu c t}, \chi_+^{-\lambda - n/2} \circ \left(1 - \frac{|x|^2}{t^2} - \nu^2\right)\rangle$$

$$= \frac{2^{2\lambda + 1}Y(t - |x|)t^{-2\lambda - 1 - n}}{\pi^{n/2}\Gamma(-\lambda)\Gamma(-\lambda - \frac{n}{2})} \int_{-\sqrt{1 - |x|^2/t^2}}^{\sqrt{1 - |x|^2/t^2}} e^{-\nu c t}\left(1 - \frac{|x|^2}{t^2} - \nu^2\right)^{-\lambda - n/2 - 1} d\nu$$

$$= \frac{2^{2\lambda + 1}Y(t - |x|)(t^2 - |x|^2)^{-\lambda - (n+1)/2}}{\pi^{n/2}\Gamma(-\lambda)\Gamma(-\lambda - \frac{n}{2})} \int_{-1}^{1} e^{-cu\sqrt{t^2 - |x|^2}}(1 - u^2)^{-\lambda - n/2 - 1} du$$

$$= \frac{2^{\lambda - (n-1)/2}c^{\lambda + (n+1)/2}Y(t - |x|)(t^2 - |x|^2)^{-\lambda/2 - (n+1)/4}}{\pi^{(n-1)/2}\Gamma(-\lambda)} I_{-\lambda - (n+1)/2}(c\sqrt{t^2 - |x|^2}).$$

$$(3.5.27)$$

The last equation follows from Poisson's integral representation for the Bessel function, see Gradshteyn and Ryzhik [113], Eq. 8.431.1.

The result for $E(\lambda)$ in formula (3.5.27) is valid for $\operatorname{Re}\lambda < -\frac{n-1}{2}$ since then the right-hand side is a locally integrable function. For $\lambda = -k$, $k \in \mathbf{N}$, $k > \frac{n-1}{2}$, we obtain for $E(-k)$ the fundamental solution of the iterated Klein–Gordon operator $(\partial_t^2 - \Delta_n - c^2)^k$, and the expression in (3.5.27) was given in Schwartz [246], Eq. (VII, 5; 30), p. 179; de Jager [149], (4.3.25), p. 92.

In order to give a representation for the fundamental solution $E(-1)$, we use, as in Example 2.3.7, the recursion formula

$$E(\lambda + 1) = -\frac{2(\lambda + 1)}{t} \frac{\partial E(\lambda)}{\partial t}.$$

This furnishes, for $n = 2m$, $m \in \mathbf{N}$,

$$E(-1) = 2^{m-1}(m-1)!\left(\frac{\partial}{t\partial t}\right)^{m-1} E(-m)$$

$$= \frac{1}{(2\pi)^m}\left(\frac{\partial}{t\partial t}\right)^{m-1}\left[\frac{Y(t-|x|)}{\sqrt{t^2 - |x|^2}}\cosh\left(c\sqrt{t^2 - |x|^2}\right)\right],$$

cf. Léonard [162], p. 36; Ortner and Wagner [207], Ex. 5, p. 457.

Similarly, for odd $n = 2m - 1$, we obtain

$$E(-1) = \frac{1}{2(2\pi)^{m-1}}\left(\frac{\partial}{t\partial t}\right)^{m-1}\left[Y(t-|x|)I_0\left(c\sqrt{t^2 - |x|^2}\right)\right],$$

cf. Léonard [162], p. 36; Bresters [24], (5.15), p. 580.   □

Chapter 4
Quasihyperbolic Systems

Whereas the method of parameter integration is applicable to both elliptic and hyperbolic operators (but in general relies on the product structure of the operator), the method of Laplace transform is applicable only to quasihyperbolic systems. The first systematic treatment of fundamental solutions by means of the (inverse) Laplace transform dates back to Leray [163].

For the representation of fundamental matrices of *hyperbolic* systems $A(\partial)$ the Laplace transform $\mathcal{L} : S'(\Gamma) \rightarrow H(T^C)$ according to V.S. Vladimirov is better suited. Here C is chosen a priori as the hyperbolicity cone of $A(\partial)$ and $\Gamma = C^*$, $T^C = C + i\mathbf{R}^n$. In contrast, for *quasihyperbolic*, but non-hyperbolic systems, we use the more general Laplace transformation $\mathcal{L} : S'_C \rightarrow H_C$ according to L. Schwartz. Here C is not necessarily a cone.

The representations of fundamental matrices as inverse Laplace transforms lead to the determination of the singularities of these fundamental matrices and, in particular, allow for the investigation of the phenomenon of *conical refraction*, see Sects. 4.2, 4.3. For example, conical refraction occurs in cubic elastodynamics, but does, except for one special case, not occur in hexagonal elastodynamics, see Examples 4.3.9, 4.3.10.

Since many physically relevant hyperbolic systems are also *homogeneous*, we derive in Sect. 4.4 representations of fundamental matrices for such systems involving $n-2$ integrations. These so-called Herglotz–Gårding formulas for strictly hyperbolic homogeneous systems (Proposition 4.4.1, Corollary 4.4.2) generalize the Herglotz–Petrovsky–Leray formulas, which apply to operators. In Proposition 4.4.3 the assumption of strict hyperbolicity is relaxed using the method of parameter integration.

© Springer International Publishing Switzerland 2015
N. Ortner, P. Wagner, *Fundamental Solutions of Linear Partial Differential Operators*, DOI 10.1007/978-3-319-20140-5_4

By means of the Herglotz–Gårding formula, we derive a representation of the fundamental matrix of hexagonal elastodynamics in the form of a simple integral over elementary functions, see Example 4.4.5. In the final Example 4.4.8 we deal with Maxwell's system of crystal optics. We derive an explicit expression for the so-called static term, we investigate the singularities of the fundamental matrix and determine the set of conical refraction, and we present an explicit formula for the fundamental matrix in the uniaxial case.

4.1   Representations by Laplace Inversion

Let us first define the Laplace transformation for distributions similarly as in Vladimirov [280], § 9; Vladimirov, Drozzinov and Zavialov [281], § 2.5. In the following, Γ denotes a convex, acute, closed cone in $\mathbf{R}^n$ with vertex in 0, and we write $\Gamma^* = \{\vartheta \in \mathbf{R}^n; \forall x \in \Gamma : \vartheta \cdot x \geq 0\}$ for the dual cone. We denote by $C = (\Gamma^*)^\circ$ the interior of Γ^* and by $T^C = C + i\mathbf{R}^n$ the corresponding tube domain in $\mathbf{C}^n$ with basis C.

Definition 4.1.1 For Γ and C as above, let us define

$$S'(\Gamma) = \{S \in S'(\mathbf{R}^n); \operatorname{supp} S \subset \Gamma\}$$

and

$$H(T^C) = \{f : T^C \longrightarrow \mathbf{C} \text{ holomorphic}; \exists \mu, \nu, M > 0 : \forall p \in T^C :$$

$$|f(p)| \leq M(1 + |p|^2)^\mu d(\operatorname{Re} p, \partial C)^{-\nu}\}.$$

(Here $d(\vartheta, \partial C) = \min\{|\vartheta - \zeta|; \zeta \in \partial C\}$ denotes the distance from $\vartheta \in C$ to the boundary ∂C of C.)

We equip the space $S'(\Gamma)$ with the topology induced by $S'(\mathbf{R}^n)$, and we consider $H(T^C)$ as the locally convex inductive limit of the Banach spaces

$$H^{\mu,\nu}(T^C) = \{f \in H(T^C); f(p)(1 + |p|^2)^{-\mu} d(\operatorname{Re} p, \partial C)^\nu \text{ is bounded}\}$$

with the norms given by

$$\|f\|_{H^{\mu,\nu}(T^C)} = \sup\{|f(p)|(1 + |p|^2)^{-\mu} d(\operatorname{Re} p, \partial C)^\nu; p \in T^C\}.$$

For an integrable function $S(x) \in L^1(\Gamma)$, we define the Laplace transform in the usual way:

$$(\mathcal{L}S)(p) = \int_\Gamma e^{-px} S(x) \, dx, \qquad p \in T^C.$$

Then $\mathcal{L}S \in H(T^C)$ and

$$(\mathcal{L}S)(\vartheta + i\xi) = \int_{\mathbf{R}^n} e^{-i\xi x} \cdot e^{-\vartheta x} S(x) \, dx = \mathcal{F}\big(e^{-\vartheta x} S(x)\big)(\xi)$$

$$= {}_{\mathcal{D}_{L^\infty}}\langle e^{-i\xi x}, e^{-\vartheta x} S(x)\rangle_{\mathcal{D}'_{L^1}} = {}_{\mathcal{D}_{L^\infty}}\langle 1, e^{-px} S(x)\rangle_{\mathcal{D}'_{L^1}}$$

where $p = \vartheta + i\xi \in T^C$. This leads to the following definition.

Definition and Proposition 4.1.2 *For $S \in \mathcal{S}'(\Gamma)$, the Laplace transform $\mathcal{L}S \in H(T^C)$ is defined by*

$$\mathcal{L}S(p) = \mathcal{F}_x\big(e^{-\vartheta x} S(x)\big)(\xi) = {}_{\mathcal{D}_{L^\infty}}\langle 1, e^{-px} S(x)\rangle_{\mathcal{D}'_{L^1}}$$

for $p = \vartheta + i\xi \in T^C$. Then the mapping $\mathcal{L} : \mathcal{S}'(\Gamma) \longrightarrow H(T^C)$ is an isomorphism of locally convex topological vector spaces. The Laplace inverse $\mathcal{L}^{-1}$ is given explicitly by the formula

$$\mathcal{L}^{-1}(f) = e^{\vartheta x} (2\pi)^{-n} \mathcal{F}_\xi\big(f(\vartheta - i\xi)\big) \qquad (4.1.1)$$

for $f \in H(T^C)$ and arbitrary $\vartheta \in C$.

Proof If we fix $K \subset C$ compact, then

$$\exists \epsilon > 0 : \exists M > 0 : \forall \vartheta \in K : \forall x \in \Gamma : \cosh(\epsilon|x|) \cdot e^{-\vartheta x} \le M,$$

cf. Schwartz [246], p. 302. Therefore, also $\cosh(\epsilon|x|) \cdot e^{-px} \cdot S \in \mathcal{S}'(\mathbf{R}^n)$ if $\vartheta = \operatorname{Re} p \in K$ and $S \in \mathcal{S}'(\Gamma)$. Hence

$$e^{-px} S(x) = \frac{1}{\cosh(\epsilon|x|)} \cdot \cosh(\epsilon|x|) e^{-px} S(x) \in \mathcal{S} \cdot \mathcal{S}' \subset \mathcal{O}'_C \subset \mathcal{D}'_{L^1},$$

and thus $(\mathcal{L}S)(p) = {}_{\mathcal{D}_{L^\infty}}\langle 1, e^{-px} S(x)\rangle_{\mathcal{D}'_{L^1}}$ is well defined.

The holomorphy of $\mathcal{L}S$ on T^C follows from that of e^{-px} with respect to p. In order to show that $\mathcal{L}$ is well defined, it remains to verify for $f = \mathcal{L}S$ the inequality in Definition 4.1.1. This is shown in Vladimirov [280], § 12.2, p. 189, using the representation of temperate distributions as derivatives of slowly increasing functions in Schwartz [246], Ch. VI, Thm. VI, p. 239.

These estimates also yield the continuity of the mapping $\mathcal{L}$. The injectivity of the Laplace transform follows from that of the Fourier transformation. For the surjectivity of $\mathcal{L}$, we refer to Vladimirov [280], § 10.5.

Finally note that, for $S \in \mathcal{S}'(\Gamma)$, the function $f(p) = (\mathcal{L}S)(p) \in H(T^C)$, and hence $f(\vartheta - i\xi) \in \mathcal{O}_M(\mathbf{R}^n_\xi) \subset \mathcal{S}'(\mathbf{R}^n_\xi)$ for each fixed $\vartheta \in C$. This implies

$$(\mathcal{L}^{-1}f)(x) = e^{\vartheta x} \cdot \mathcal{F}_\xi^{-1}\big(f(\vartheta + i\xi)\big) = e^{\vartheta x} \cdot (2\pi)^{-n} \mathcal{F}_\xi\big(f(\vartheta - i\xi)\big)$$

by the Fourier inversion theorem, see Proposition 1.6.5. $\qquad \square$

Let us remark that the Laplace transform in Schwartz [246], Ch. VIII, is defined differently and with different notation. As domain of definition for the Laplace transform, L. Schwartz uses a larger topological vector space, namely

$$\mathcal{S}'_C = \{S \in \mathcal{D}'(\mathbf{R}^n); \ \forall \xi \in C : e^{-\xi x} S \in \mathcal{S}'(\mathbf{R}^n_x)\},$$

where $C \subset \mathbf{R}^n$ is open and convex. Evidently, $\mathcal{S}'(\Gamma) \subset \mathcal{S}'_C$ if $C = (\Gamma^*)^\circ$, but the converse is not true as shows the example $e^{-|x|^2} \in \mathcal{S}'_C$. (Also note that L. Schwartz writes $\mathcal{S}'(C)$ instead of $\mathcal{S}'_C$.) We shall develop the Laplace transform in L. Schwartz' sense in Definition 4.1.7 and in Definition and Proposition 4.1.8.

We also mention that the notation in Vladimirov [280] is slightly different from ours insofar as there $T^C = \mathbf{R}^n + iC$, $(\mathcal{F}g)(\eta) = \int_{\mathbf{R}^n} e^{i\eta x} g(x)\, dx$ for $g \in \mathcal{S}(\mathbf{R}^n)$ and $(\mathcal{L}S)(p) = \int_{\mathbf{R}^n} e^{ipx} S(x)\, dx$ if $p \in \mathbf{R}^n + iC$ and $S \in L^1(\Gamma)$.

For hyperbolic systems $A(\partial) \in \mathbf{C}[\partial]^{l \times l}$, we can determine the fundamental matrix as an inverse Laplace transform in view of formula (2.4.13). For the cone C in Definition 4.1.1, we use the *hyperbolicity cone* $\Gamma(P(\partial), N)$ with $P(\partial) = \det A(\partial)$, see Atiyah, Bott, and Gårding [5], Def. 3.21, p. 132; Gårding [90], p. 222.

Definition 4.1.3 Let $N \in \mathbf{R}^n \setminus \{0\}$ and $P(\partial) = \sum_{|\alpha| \le m} a_\alpha \partial^\alpha$ be an operator of degree m which is hyperbolic in the direction N (see Definition 2.4.10), and denote, as always, by $P_m(\partial)$ its principal part $\sum_{|\alpha|=m} a_\alpha \partial^\alpha$.

(1) The *hyperbolicity cone* $\Gamma(P(\partial), N)$ is the connectivity component containing N of the set $\{\vartheta \in \mathbf{R}^n; P_m(\vartheta) \ne 0\}$.
(2) The dual cone $K(P(\partial), N) = \{x \in \mathbf{R}^n; \ \forall \vartheta \in \Gamma(P(\partial), N) : x\vartheta \ge 0\}$ is called the *propagation cone* of $P(\partial)$.

Let us next collect some properties of the cone $C = \Gamma(P(\partial), N)$, which we shall use when employing the Laplace transform.

Proposition 4.1.4 *Let $P(\partial)$ be hyperbolic in direction N, i.e., $P_m(N) \ne 0$ and $\forall \sigma > \sigma_0 : \forall \xi \in \mathbf{R}^n : P(i\xi + \sigma N) \ne 0$, see Definition 2.4.10.*

(1) $\Gamma(P(\partial), N)$ *is an open convex cone;*
(2) $P(\partial)$ *is hyperbolic in each direction $\vartheta \in \Gamma(P(\partial), N)$; more precisely, if $\sigma_0 = 0$, then $P(i\xi + \vartheta) \ne 0$ for all $\xi \in \mathbf{R}^n$ and $\vartheta \in \Gamma(P(\partial), N)$;*
(3) P_m *is real-valued up to a constant factor, i.e., $P_m/P_m(N)$ is real valued;*
(4) *the polynomial $\sigma \mapsto P_m(\xi + \sigma N)$ has only real roots for each $\xi \in \mathbf{R}^n$.*

For the **proof** we refer to Hörmander [136], Section 5.3; [138], Section 12.4.

Let us apply now the Laplace transform in 4.1.2 to hyperbolic systems.

Proposition 4.1.5 *Let $A(\partial) \in \mathbf{C}[\partial]^{l \times l}$ be a hyperbolic system with respect to the direction $N \in \mathbf{R}^n \setminus \{0\}$, i.e., $P(\partial) = \det A(\partial)$ fulfills $P_m(N) \ne 0$ and $P(i\xi + \sigma N) \ne 0$ for $\xi \in \mathbf{R}^n$ and $\sigma > \sigma_0$, see Definition 2.4.10.*

If we set $C = \Gamma(P(\partial), N)$, *then* $A(p + \sigma_0 N)^{-1} \in H(T^C)^{l \times l}$ *and the forward fundamental matrix* E *of* $A(\partial)$ (*i.e., the one with support in the half-space* $H_N = \{x \in \mathbf{R}^n; xN \geq 0\}$) *is given by the formula*

$$E = e^{\sigma_0 N x} \cdot \mathcal{L}^{-1}(A(p + \sigma_0 N)^{-1}). \tag{4.1.2}$$

Furthermore, $\operatorname{supp} E \subset K(P(\partial), N)$.

Proof For $p = \vartheta + i\xi \in T^C$, i.e., $\vartheta \in C$, $\xi \in \mathbf{R}^n$, Proposition 4.1.4 (2) yields $P(p + \sigma_0 N) = P(i\xi + \vartheta + \sigma_0 N) \neq 0$. Therefore, an application of the Seidenberg–Tarski lemma as in the proof of Proposition 2.3.5 implies that $P(p + \sigma_0 N)^{-1} \in H(T^C)$ and hence also $A(p + \sigma_0 N)^{-1} \in H(T^C)^{l \times l}$.

On the other hand, if $\vartheta = \tau N$, $\tau > 0$, then formula (2.4.13) furnishes, with $\sigma = \sigma_0 + \tau$,

$$E = e^{\vartheta x + \sigma_0 N x} \mathcal{F}^{-1}(A(i\xi + \vartheta + \sigma_0 N)^{-1}) = e^{\sigma_0 N x} \mathcal{L}^{-1}(A(p + \sigma_0 N)^{-1}). \quad \square$$

For scalar hyperbolic operators, formula (4.1.2) coincides with Atiyah, Bott, and Gårding [5], Eq. (4.2), p. 142; Gårding [90], p. 224; [94], (6.2), p. 43. A version for systems is contained in Chazarain and Piriou [48], Ch. VI, Prop. 3.1.6, p. 309.

Example 4.1.6 Let us now treat a second time *Timoshenko's beam operator*

$$P(\partial) = \partial_t^2 + \frac{EI}{\rho A} \partial_x^4 + \frac{\rho I}{GA\kappa} \partial_t^4 - \frac{I}{A}\left(1 + \frac{E}{G\kappa}\right)\partial_t^2\partial_x^2,$$

see (3.5.7), which is hyperbolic in the t-direction. Its forward fundamental solution E was already represented in Example 3.5.4 by integrals over products of Bessel functions using the method of parameter integration. We shall now derive a completely different representation of E by calculating the inverse Laplace transform in formula (4.1.2) by means of analytic continuation, comp. Ortner and Wagner [214].

(a) After renaming the constant factors, Timoshenko's beam operator assumes the form

$$T(\partial) = (a^2\partial_t^2 - \partial_x^2)(b^2\partial_t^2 - \partial_x^2) + c^2\partial_t^2. \tag{4.1.3}$$

We suppose that $0 < a < b$ and $c \in \mathbf{C}\backslash\{0\}$. Upon setting $N = (1, 0)$, formula (4.1.2) yields

$$E = e^{\sigma_0 t} \mathcal{L}^{-1}\left(\frac{1}{[a^2(p_1 + \sigma_0)^2 - p_2^2][b^2(p_1 + \sigma_0)^2 - p_2^2] + c^2(p_1 + \sigma_0)^2}\right).$$

Here $C = \Gamma(T(\partial), N) = \{(\vartheta_1, \vartheta_2) \in \mathbf{R}^2; a\vartheta_1 > |\vartheta_2|\}$ and $p \in T^C = C + i\mathbf{R}^2$.

If we put $\vartheta_2 = 0$, we obtain E from formula (4.1.1) as a one-fold Laplace inverse of a Fourier integral:

$$E = \frac{e^{\sigma_0 t}}{2\pi} \mathcal{L}_{p_1 \to t}^{-1} \left(\int_0^\infty \frac{(e^{ix\xi} + e^{-ix\xi})\, d\xi}{[a^2(p_1 + \sigma_0)^2 + \xi^2][b^2(p_1 + \sigma_0)^2 + \xi^2] + c^2(p_1 + \sigma_0)^2} \right)$$

$$= \frac{1}{2\pi} \mathcal{L}_{p \to t}^{-1} (F(p, ix) + F(p, -ix)),$$

where in the last formula p stands for a single complex variable $p \in (\sigma_0, \infty) + i\mathbf{R}$ and

$$F(p, z) = \int_0^\infty \frac{e^{-z\xi}\, d\xi}{(a^2 p^2 + \xi^2)(b^2 p^2 + \xi^2) + c^2 p^2}, \qquad \mathrm{Re}\, z \geq 0.$$

For positive z, the inverse Laplace transform $\mathcal{L}_{p \to t}^{-1}(F(p, z))$ can be represented by a simple integral. In fact, if $p > \sigma_0$, then the substitution $\xi = ps$ yields

$$F(p, z) = \frac{1}{p} \int_0^\infty \frac{e^{-zps}\, ds}{p^2(a^2 + s^2)(b^2 + s^2) + c^2}, \qquad (4.1.4)$$

and this equation holds by analytic continuation for each p in the tube $(\sigma_0, \infty) + i\mathbf{R}$. Therefore, (4.1.4) implies, still for positive z and $\vartheta > \sigma_0$,

$$\mathcal{L}_{p \to t}^{-1}(F(p, z)) = \frac{1}{2\pi i c^2} \int_0^\infty ds \int_{\vartheta - i\infty}^{\vartheta + i\infty} e^{p(t-zs)} \left(\frac{1}{p} - \frac{p}{p^2 + \frac{c^2}{(a^2 + s^2)(b^2 + s^2)}} \right) dp$$

$$= \frac{1}{c^2} \int_0^\infty Y(t - zs) \left[1 - \cos\left(\frac{c(t - zs)}{\sqrt{(a^2 + s^2)(b^2 + s^2)}} \right) \right] ds$$

$$= \frac{Y(t)}{c^2} \int_0^{t/z} \left[1 - \cos\left(\frac{c(t - zs)}{\sqrt{(a^2 + s^2)(b^2 + s^2)}} \right) \right] ds,$$

cf. Badii and Oberhettinger [7], II, 2.33, p. 219.

E is obtained from $\mathcal{L}_{p \to t}^{-1}(F(p, z))$ by analytic continuation with respect to z. If two semi-circles are chosen as integration paths from 0 to $it/|x|$ and to $-it/|x|$, respectively, and the reflection $s \mapsto -s$ is used in the second one, this yields

$$E(t, x) = \frac{Y(t)}{2\pi c^2} \int_C \left[1 - \cos\left(\frac{c(t + i|x|s)}{\sqrt{(a^2 + s^2)(b^2 + s^2)}} \right) \right] ds, \qquad (4.1.5)$$

where C is the circle through 0 and $it/|x|$ which is symmetric with respect to the imaginary s-axis and oriented in the counterclockwise direction. We observe that the differential form integrated in (4.1.5) has no branch points, but essential singularities at $s = \pm ia, \pm ib$. Therefore the following representation of E by residues is valid:

$$E(t,x) = \begin{cases} 0 & : t < a|x|, \\ -\dfrac{i}{c^2} R_a & : a|x| < t < b|x|, \\ -\dfrac{i}{c^2} (R_a + R_b) & : b|x| < t, \end{cases}$$

(4.1.6)

where

$$R_z = \operatorname*{Res}_{s=iz}\left[\cos\left(\frac{c(t + i|x|s)}{\sqrt{(a^2 + s^2)(b^2 + s^2)}}\right)\right], \qquad z \in \{a, b\}.$$

(b) Representation of E by a definite integral in the inner cone $t > b|x|$.

In this case, we can, by Cauchy's theorem, deform the closed contour C of the integral in (4.1.5) such that it consists of part of the real axis and of a large semi-circle in the upper half-plane. If we let the radius of this semi-circle tend to infinity, the corresponding contribution to E converges to 0, and we conclude that, for $t > b|x|$,

$$\begin{aligned} E(t,x) &= \frac{1}{2\pi c^2} \int_{-\infty}^{\infty} \left[1 - \cos\left(\frac{c(t + i|x|s)}{\sqrt{(a^2 + s^2)(b^2 + s^2)}}\right)\right] ds \\ &= \frac{1}{\pi c^2} \int_0^{\infty} \left[1 - \cos\left(\frac{ct}{\sqrt{(a^2 + s^2)(b^2 + s^2)}}\right) \cosh\left(\frac{cxs}{\sqrt{(a^2 + s^2)(b^2 + s^2)}}\right)\right] ds. \end{aligned}$$

Finally, the real substitution

$$(a^2 + s^2)(b^2 + s^2) = \frac{1}{4v^2}, \quad 0 < v < \frac{1}{2ab}, \quad \text{i.e., } s = \frac{1}{2v} A(v),$$

where

$$A(v) = \sqrt{2v}\sqrt{-(a^2 + b^2)v + \sqrt{1 + (b^2 - a^2)^2 v^2}}, \quad ds = \frac{-dv}{2vA(v)\sqrt{1 + (b^2 - a^2)^2 v^2}},$$

furnishes, in the cone $t > b|x|$, the integral representation $E = E^{(1)}$ with

$$E^{(1)}(t,x) = \frac{1}{2\pi c^2} \int_0^{1/(2ab)} \frac{1 - \cos(2ctv)\cosh(cxA(v))}{vA(v)\sqrt{1 + (b^2 - a^2)^2 v^2}} \, dv.$$

(4.1.7)

(c) Representation of E in the region $b|x| > t > a|x|$.

In this case, the contour C in (4.1.5) is homotopic to the curve consisting of the real axis and of the branch Γ of a hyperbola defined by

$$\Gamma = \left\{ s \in \mathbf{C};\ -\mathrm{Re}\,(s^2) = (\mathrm{Im}\,s)^2 - (\mathrm{Re}\,s)^2 = \frac{a^2 + b^2}{2} \text{ and } \mathrm{Im}\,s > 0 \right\}.$$

If we divide Γ into the two parts $\Gamma_\pm = \{s \in \Gamma;\ \pm\mathrm{Re}\,s > 0\}$ and orient Γ and Γ_+ starting from $(1+\mathrm{i})\infty$, then we obtain, for $b|x| > t > a|x|$ and with $E^{(1)}$ as defined in (4.1.7),

$$E(t,x) = E^{(1)}(t,x) + \frac{1}{2\pi c^2} \int_\Gamma \left[1 - \cos\left(\frac{c(t + \mathrm{i}|x|s)}{\sqrt{(a^2 + s^2)(b^2 + s^2)}} \right) \right] ds$$

$$= E^{(1)}(t,x) + \frac{1}{\pi c^2} \mathrm{Re}\left\{ \int_{\Gamma_+} \left[1 - \cos\left(\frac{c(t + \mathrm{i}|x|s)}{\sqrt{(a^2 + s^2)(b^2 + s^2)}} \right) \right] ds \right\}.$$

The parametrization

$$(a^2 + s^2)(b^2 + s^2) = -\frac{1}{4v^2}, \quad 0 < v < \frac{1}{b^2 - a^2}, \quad \text{i.e., } s = \frac{1}{2v}(B(v) + \mathrm{i}C(v)),$$

where

$$\left.\begin{array}{r} B(v) \\ C(v) \end{array}\right\} = \sqrt{v}\sqrt{\mp(a^2 + b^2)v + \sqrt{1 + 4a^2b^2v^2}}, \quad ds = \frac{-(C(v) + \mathrm{i}B(v))dv}{4v^2\sqrt{1 - (b^2 - a^2)^2v^2}\sqrt{1 + 4a^2b^2v^2}},$$

finally yields, in the region $b|x| > t > a|x|$,

$$E(t,x) = E^{(1)}(t,x) - \frac{1}{4\pi c^2} \int_0^{1/(b^2 - a^2)} \frac{dv}{v^2\sqrt{1 - (b^2 - a^2)^2v^2}\sqrt{1 + 4a^2b^2v^2}} \times$$

$$\times \{C(v)[1 - \cos(cxB(v))\cosh(2ctv - c|x|C(v))] + B(v)\sin(c|x|B(v))\sinh(2ctv - c|x|C(v))\}.$$

$$(4.1.8)$$

$\square$

In order to represent fundamental matrices of quasihyperbolic systems which are not hyperbolic by inverse Laplace transforms, let us introduce now the distributional Laplace transform as it was formulated in Schwartz [246], Ch. VIII.

Definition 4.1.7 For a convex open set $C \subset \mathbf{R}^n$, we set

$$\mathcal{S}'_C = \{S \in \mathcal{D}'(\mathbf{R}^n);\ \forall \vartheta \in C : e^{-\vartheta x}S(x) \in \mathcal{S}'(\mathbf{R}^n_x)\}$$

and

$$H_C = \{f : T^C = C + i\mathbf{R}^n \longrightarrow \mathbf{C} \text{ holomorphic;}$$

$$\forall K \subset C \text{ compact} : \exists \mu, M > 0 : \forall p \in T^K : |f(p)| \le M(1 + |p|^2)^\mu\}.$$

We equip $\mathcal{S}'_C$ with the coarsest topology such that all the mappings

$$\mathcal{S}'_C \longrightarrow \mathcal{S}' : S \longmapsto e^{-\vartheta x}S, \qquad \vartheta \in C,$$

are continuous. On the other hand, H_C is the projective limit of the inductive limits of the Banach spaces

$$H_C^{\mu,K} = \{f \in H^C; f(p) \cdot (1 + |p|^2)^{-\mu} \text{ is bounded on } T^K\}$$

for $\mu > 0$, $K \subset C$ compact, i.e., $H_C = \lim_{\overleftarrow{K}} \lim_{\overrightarrow{\mu}} H_C^{\mu,K}$.

Definition and Proposition 4.1.8 *For $S \in \mathcal{S}'_C$, the Laplace transform $\mathcal{L}S \in H_C$ is defined by*

$$\mathcal{L}S(p) = \mathcal{F}_x\big(e^{-\vartheta x}S(x)\big)(\xi) = {}_{\mathcal{D}_{L^\infty}}\langle 1, e^{-px}S(x)\rangle_{\mathcal{D}'_{L^1}}$$

for $p = \vartheta + i\xi \in T^C$. Then the mapping $\mathcal{L} : \mathcal{S}'_C \longrightarrow H_C$ is an isomorphism of locally convex topological vector spaces. As in 4.1.2, the Laplace inverse $\mathcal{L}^{-1}$ is given explicitly by the formula $\mathcal{L}^{-1}(f) = e^{\vartheta x}(2\pi)^{-n}\mathcal{F}_\xi\big(f(\vartheta - i\xi)\big)$ for $f \in H_C$ and arbitrary fixed $\vartheta \in C$.

Proof For $K \subset C$ compact, the reasoning in the proof of 4.1.2 shows that if $S \in \mathcal{S}'_C$ then $e^{-px}S(x)$ belongs to $\mathcal{O}'_C$ and is bounded therein for $\vartheta = \operatorname{Re} p \in K$. Hence $(\mathcal{L}S)(p) = {}_{\mathcal{D}_{L^\infty}}\langle 1, e^{-px}S(x)\rangle_{\mathcal{D}'_{L^1}}$ is well defined and $(\mathcal{L}S)(\vartheta + i\xi)$ is bounded in $\mathcal{O}_M(\mathbf{R}^n_\xi)$ for $\vartheta \in K$. Furthermore, $\mathcal{L}S$ is holomorphic in p and hence $\mathcal{L}S \in H_C$.

Conversely, for $f \in H_C$ and $\vartheta \in C$,

$$S_\vartheta := \mathcal{F}_\xi^{-1}\big(f(\vartheta + i\xi)\big) = (2\pi)^{-n}\mathcal{F}_\xi\big(f(\vartheta - i\xi)\big) \in \mathcal{O}'_C(\mathbf{R}^n)$$

and $e^{\vartheta x}S_\vartheta \in \mathcal{D}'(\mathbf{R}^n)$ is independent of $\vartheta \in C$, see Schwartz [246], Prop. 5, p. 305. This shows that $\mathcal{L}$ is surjective and hence an isomorphism of linear spaces.

In order to show that $\mathcal{L}$ is also a topological isomorphism, one first verifies that the topology on $\mathcal{S}'_C$ coincides with the projective limit topology (see Robertson and Robertson [236], Ch. V, § 4, p. 84) with respect to the mappings

$$\mathcal{S}'_C \longrightarrow \mathcal{O}'_C : S \longmapsto e^{-\vartheta x}S, \qquad \vartheta \in C.$$

Hence $\mathcal{L}$ is an isomorphism if H_C is equipped with the projective limit topology with respect to the mappings

$$H_C \longrightarrow \mathcal{O}_M : f \longmapsto f(\vartheta + i\xi), \qquad \vartheta \in C.$$

Cauchy's inequalities (cf. Schwartz [246], p. 306) then imply that H_C is also the projective limit (with respect to $\vartheta \in C$) of the inductive limits (with respect to $\mu > 0$) of $H_C^{\mu,\{\vartheta\}}$. Finally, if $K \subset C$ is the convex envelope of the points $\vartheta_1, \ldots, \vartheta_m \in C$, then the set

$$\left\{ a(\vartheta, x) = e^{-\vartheta x} \Big/ \Big(\sum_{j=1}^m e^{-\vartheta_j x} \Big); \ \vartheta \in K \right\}$$

is bounded in $\mathcal{D}_{L^\infty}(\mathbf{R}_x^n)$ (cf. Schwartz [246], p. 301), and hence

$$\left\{ f(\vartheta + i\xi) = \sum_{j=1}^m \mathcal{F}\big(a(\vartheta, x) \cdot e^{-\vartheta_j x} S \big)(\xi); \ \vartheta \in K \right\}$$

is a bounded subset of $\mathcal{O}_M(\mathbf{R}_\xi^n)$. Therefore, the above topology on H_C coincides with the one given after Definition 4.1.7. □

Analogously to Proposition 4.1.5, we can apply Proposition 4.1.8 to systems $A(\partial) \in \mathbf{C}[\partial]^{l \times l}$ such that $P(\partial) = \det A(\partial)$ is quasihyperbolic with respect to $N \in \mathbf{R}^n \setminus \{0\}$ and fulfills $P(\vartheta + i\xi) \neq 0$ if $\xi \in \mathbf{R}^n$ and $\vartheta \in C$ for a convex open set in $\mathbf{R}^n$ containing $\{\sigma N; \sigma > \sigma_0\}$. If E denotes the uniquely determined fundamental matrix of $A(\partial)$ satisfying $e^{-\sigma Nx} E \in \mathcal{S}'(\mathbf{R}^n)^{l \times l}$ for $\sigma > \sigma_0$ (see Proposition 2.4.13), then $E = \mathcal{L}^{-1}(A(p)^{-1})$.

Example 4.1.9 Let us resume *Rayleigh's system,* which was introduced already in Example 2.4.14.

(a) According to Example 2.4.14, Rayleigh's system is given by the quasihyperbolic matrix

$$B(\partial) = \begin{pmatrix} \alpha^2 \partial_t^2 - \partial_x^2 & \partial_x \\ \beta^2 \partial_x & \partial_x^2 - \beta^2 \end{pmatrix}, \qquad \alpha, \beta > 0, \qquad (4.1.9)$$

with the determinant

$$P(\partial) = \det B(\partial) = -\partial_x^4 + \alpha^2 \partial_t^2 \partial_x^2 - \gamma^2 \partial_t^2, \qquad \gamma = \alpha\beta.$$

In order to apply the inverse Laplace transform as in Proposition 4.1.8, let us show first that $P(p)^{-1} \in H_C$ for a suitable non-empty open convex set $C \subset \mathbf{R}^2$. Since the subsets

$$A_1 = \{p_1^2; \ p_1 \in \mathbf{C}, \operatorname{Re} p_1 \geq N\}, \qquad A_2 = \left\{ \frac{p_2^4}{\alpha^2 p_2^2 - \gamma^2}; \ p_2 \in \mathbf{C}, |\operatorname{Re} p_2| \leq \epsilon \right\}$$

of the complex plane are disjoint if $0 < \epsilon < \frac{\gamma}{\alpha}$ and $N = N(\epsilon, \alpha, \gamma)$ is large enough, we can use a set C of the form $C = \{\vartheta \in \mathbf{R}^2;\ \vartheta_1 > N, |\vartheta_2| < \epsilon\}$. This reasoning also shows that $P(\vartheta + i\xi)$ does not vanish for $\xi \in \mathbf{R}^2$ and $\vartheta = \sigma\binom{1}{0}$, $\sigma > 0$, and hence $B(\partial)$ and $P(\partial)$ are quasihyperbolic in the direction $\binom{1}{0}$, cf. also Example 2.4.14.

(b) Let us next derive by inverse Laplace transformation a representation of the uniquely determined fundamental solution F of $P(\partial)$ fulfilling $F \in \mathcal{S}'(\mathbf{R}^2)$ and $F = 0$ for $t < 0$.

By Proposition 4.1.8,

$$
F = \mathcal{L}^{-1}\big(P(p)^{-1}\big)
$$

$$
= \frac{e^{\vartheta x}}{(2\pi)^2} \mathcal{F}_\xi\left(\frac{1}{-(\vartheta_2 - i\xi_2)^4 + \alpha^2(\vartheta_1 - i\xi_1)^2(\vartheta_2 - i\xi_2)^2 - \gamma^2(\vartheta_1 - i\xi_1)^2}\right) \in \mathcal{S}'_C,
$$
$$\tag{4.1.10}$$

where C is as above. Note that $P(p)^{-1} = -[p_2^4 - \alpha^2 p_1^2 p_2^2 + \gamma^2 p_1^2]^{-1}$ is the limit of

$$
-T(p)^{-1} = -[(a^2 p_1^2 - p_2^2)(b^2 p_1^2 - p_2^2) + c^2 p_1^2]^{-1}
$$

in $H(T^C)$ for $a \searrow 0, b = \alpha, c = \gamma$ if $T(\partial)$ denotes the Timoshenko operator in (4.1.3). Therefore, we obtain from formulae (4.1.7) and (4.1.8) the following integral representation of F by performing the limit in $\mathcal{S}'_C$:

$$
F = -\frac{Y(t)}{2\pi\gamma^2} \int_0^\infty \frac{1 - \cos(2\gamma t v)\cosh(\gamma x A(v))}{v A(v)\sqrt{1 + \alpha^4 v^2}}\, dv \tag{4.1.11}
$$

$$
+ \frac{Y(t) - Y(t - \alpha|x|)}{4\pi\gamma^2} \int_0^{1/\alpha^2} \left\{ \frac{1 - \cos(\gamma x\sqrt{v - \alpha^2 v^2})\cosh(2\gamma t v - \gamma|x|\sqrt{v + \alpha^2 v^2})}{\sqrt{v - \alpha^2 v^2}} \right.
$$

$$
\left. + \frac{\sin(\gamma|x|\sqrt{v - \alpha^2 v^2})\sinh(2\gamma t v - \gamma|x|\sqrt{v + \alpha^2 v^2})}{\sqrt{v + \alpha^2 v^2}} \right\} \frac{dv}{v}, \tag{4.1.12}
$$

where $A(v) = \sqrt{-2\alpha^2 v^2 + 2v\sqrt{1 + \alpha^4 v^2}}$.

In contrast to the general Timoshenko operator, we are able to express the integral (4.1.11) in terms which can be combined with the integral in (4.1.12). In fact, the function

$$
f(v) = \frac{1 - e^{2i\gamma t v}\cosh(\gamma x A(v))}{v\sqrt{1 + \alpha^4 v^2} A(v)}
$$

can analytically be continued from the positive real axis (where each square root involved is positive) to the slit plane $U = \mathbf{C} \setminus \{iw;\ -\alpha^{-2} \le w \le \alpha^{-2}\}$. Note that $\sqrt{1 + \alpha^4 v^2}$ then assumes negative values on the negative real axis and thus

$f(-v) = \overline{f(\bar{v})}$ for $v \in U$. Since $f(v)v^{3/2}$ is bounded in the half-plane Im $v \geq 0$ for $t > 0$ and $x \in \mathbf{R}$ fixed, Cauchy's theorem allows us to express $\int_{-\infty}^{\infty} f(v)\, dv$ by an integral along the upper half of the branch cut, i.e., along $\Gamma_{\pm} = \pm 0 + i[0, \alpha^{-2}]$, which paths we orient in the direction towards $+i\infty$. Hence, for $t > 0$,

$$\int_{-\infty}^{\infty} f(v)\, dv = \int_{\Gamma_+} f(v)\, dv - \int_{\Gamma_-} f(v)\, dv = 2\,\mathrm{Re}\left(\int_{\Gamma_+} f(v)\, dv\right).$$

Upon parameterizing Γ_+ by $v = iw$, $0 < w < \alpha^{-2}$, and using the identity

$$A(v)\big|_{\Gamma_+} = \sqrt{-2\alpha^2 v^2 + 2v\sqrt{1 + \alpha^4 v^2}}\;\bigg|_{\Gamma_+} = \sqrt{w + \alpha^2 w^2} + i\sqrt{w - \alpha^2 w^2}$$

we obtain

$$F^{(1)} = -\frac{1}{2\pi\gamma^2}\int_0^{\infty} \frac{1 - \cos(2\gamma tv)\cosh(\gamma x A(v))}{v A(v)\sqrt{1 + \alpha^4 v^2}}\, dv = -\frac{1}{4\pi\gamma^2}\int_{-\infty}^{\infty} f(v)\, dv$$

$$= \frac{1}{4\pi\gamma^2}\int_0^{1/\alpha^2}\left\{\frac{-1 + e^{-2\gamma tw}\cos(\gamma x\sqrt{w - \alpha^2 w^2})\cosh(\gamma x\sqrt{w + \alpha^2 w^2})}{\sqrt{w - \alpha^2 w^2}}\right.$$

$$\text{(4.1.13)}$$

$$\left. + \frac{e^{-2\gamma tw}\sin(\gamma x\sqrt{w - \alpha^2 w^2})\sinh(\gamma x\sqrt{w + \alpha^2 w^2})}{\sqrt{w + \alpha^2 w^2}}\right\}\frac{dw}{w}.$$

Hence, inside the cone $t > \alpha|x|$, $F = F^{(1)}$ coincides with the integral in (4.1.13).

On the other hand, for $0 < t < \alpha|x|$, we have to add to $F^{(1)}$ the integral in (4.1.12). Therefore, the identities

$$\cosh(a - b) - e^{-a}\cosh b = \sinh(a - b) + e^{-a}\sinh b = e^{-b}\sinh a, \qquad a, b \in \mathbf{C},$$

furnish

$$F = \frac{1}{4\pi\gamma^2}\int_0^{1/\alpha^2} \sinh(2\gamma tv)e^{-\gamma|x|\sqrt{v + \alpha^2 v^2}}\left[\frac{\sin(\gamma|x|\sqrt{v - \alpha^2 v^2})}{\sqrt{v + \alpha^2 v^2}} - \frac{\cos(\gamma x\sqrt{v - \alpha^2 v^2})}{\sqrt{v - \alpha^2 v^2}}\right]\frac{dv}{v},$$

valid for $0 < t < \alpha|x|$. Altogether this yields

$$F = \frac{Y(t - \alpha|x|)}{4\pi\gamma^2}\int_0^{1/\alpha^2}\left\{\frac{-1 + e^{-2\gamma tv}\cos(\gamma x\sqrt{v - \alpha^2 v^2})\cosh(\gamma x\sqrt{v + \alpha^2 v^2})}{\sqrt{v - \alpha^2 v^2}}\right.$$

$$\left. + \frac{e^{-2\gamma tv}\sin(\gamma x\sqrt{v - \alpha^2 v^2})\sinh(\gamma x\sqrt{v + \alpha^2 v^2})}{\sqrt{v + \alpha^2 v^2}}\right\}\frac{dv}{v}.$$

$$+\frac{Y(t)-Y(t-\alpha|x|)}{4\pi\gamma^2}\int_0^{1/\alpha^2}\sinh(2\gamma tv)e^{-\gamma|x|\sqrt{v+\alpha^2v^2}}\times \qquad (4.1.14)$$

$$\times\left[\frac{\sin(\gamma|x|\sqrt{v-\alpha^2v^2})}{\sqrt{v+\alpha^2v^2}}-\frac{\cos(\gamma x\sqrt{v-\alpha^2v^2})}{\sqrt{v-\alpha^2v^2}}\right]\frac{dv}{v},$$

comp. Ortner and Wagner [214], Prop. 2, p. 226.

(c) Let us finally derive a formula for the first column of the fundamental matrix E of Rayleigh's system $B(\partial)$ in (4.1.9). This column represents the vector $\binom{u}{\psi}$ in (2.4.14/2.4.15) caused by an instantaneous point force $q=\delta(t,x)$.

By formula (2.1.1),

$$E=B^{\mathrm{ad}}(\partial)F=\begin{pmatrix}\partial_x^2-\frac{\gamma^2}{\alpha^2} & -\partial_x \\ -\frac{\gamma^2}{\alpha^2}\partial_x & \alpha^2\partial_t^2-\partial_x^2\end{pmatrix}F.$$

Let us first show that F and $\partial_x F$ are continuous functions, i.e., $F,\partial_x F\in\mathcal{C}(\mathbf{R}^2)$. This is a consequence of formula (4.1.10) upon showing that

$$P(\vartheta_1-i\xi_1,-i\xi_2)^{-1}=-[\xi_2^4+\alpha^2\xi_2^2(\vartheta_1-i\xi_1)^2+\gamma^2(\vartheta_1-i\xi_1)^2]^{-1}$$

belongs to $L^1(\mathbf{R}_\xi^2)$, and that the same holds for $\xi_2 P(\vartheta_1-i\xi_1,-i\xi_2)^{-1}$. This can be verified by successive integration of the function

$$r|P(\vartheta_1-i\xi_1,-i\xi_2)|^{-1}=r\big((\xi_2^2-r\xi_1)^2+r^2\vartheta_1^2\big)^{-1/2}\big((\xi_2^2+r\xi_1)^2+r^2\vartheta_1^2\big)^{-1/2},$$

where $r=\sqrt{\alpha^2\xi_2^2+\gamma^2}$, first with respect to ξ_1, then with respect to ξ_2.

Starting from the representation of F in (4.1.14) we obtain

$$\partial_x F(t,x)=\frac{Y(t-\alpha|x|)}{2\pi\gamma}\int_0^{\alpha^{-2}}\cos(\gamma x\sqrt{v-\alpha^2v^2})\sinh(\gamma x\sqrt{v+\alpha^2v^2})\frac{e^{-2\gamma tv}\,dv}{v\sqrt{1-\alpha^4v^2}}$$

$$+\frac{[Y(t)-Y(t-\alpha|x|)]\operatorname{sign}x}{2\pi\gamma}\int_0^{\alpha^{-2}}\cos(\gamma x\sqrt{v-\alpha^2v^2})e^{-\gamma|x|\sqrt{v+\alpha^2v^2}}\frac{\sinh(2\gamma tv)\,dv}{v\sqrt{1-\alpha^4v^2}}.$$

$$(4.1.15)$$

(Note that delta terms along the cone $t=\alpha|x|$ do not occur due to the continuity of the fundamental solution F.)

Similarly,

$$\partial_x^2 F(t,x) = \frac{Y(t - \alpha|x|)}{2\pi} \int_0^{\alpha^{-2}} \left[\frac{\cos(\gamma x \sqrt{v - \alpha^2 v^2}) \cosh(\gamma x \sqrt{v + \alpha^2 v^2})}{\sqrt{v - \alpha^2 v^2}} - \right.$$

$$\left. - \frac{\sin(\gamma x \sqrt{v - \alpha^2 v^2}) \sinh(\gamma x \sqrt{v + \alpha^2 v^2})}{\sqrt{v + \alpha^2 v^2}} \right] e^{-2\gamma t v} \, dv$$

$$- \frac{Y(t) - Y(t - \alpha|x|)}{2\pi} \int_0^{\alpha^{-2}} \left[\frac{\cos(\gamma x \sqrt{v - \alpha^2 v^2})}{\sqrt{v - \alpha^2 v^2}} + \right. \qquad (4.1.16)$$

$$\left. + \frac{\sin(\gamma |x| \sqrt{v - \alpha^2 v^2})}{\sqrt{v + \alpha^2 v^2}} \right] e^{-\gamma |x| \sqrt{v + \alpha^2 v^2}} \sinh(2\gamma t v) \, dv.$$

Combining the terms in (4.1.14) to (4.1.16) we obtain the following formula for the first column of the fundamental matrix:

$$\begin{pmatrix} E_{11} \\ E_{21} \end{pmatrix} = \begin{pmatrix} (\partial_x^2 - \frac{\gamma^2}{\alpha^2})F \\ -\frac{\gamma^2}{\alpha^2} \partial_x F \end{pmatrix} = Y(t - \alpha|x|) \int_0^{\alpha^{-2}} \begin{pmatrix} g_1(t,x,v) \\ g_2(t,x,v) \end{pmatrix} \frac{dv}{v} +$$

$$+ [Y(t) - Y(t - \alpha|x|)] \int_0^{\alpha^{-2}} \begin{pmatrix} h_1(t,x,v) \\ h_2(t,x,v) \end{pmatrix} \frac{dv}{v},$$

where

$$g_1 = \frac{1}{4\pi\alpha^2} \left[\frac{1 + (2\alpha^2 v - 1)e^{-2\gamma t v} \cos(\gamma x \sqrt{v - \alpha^2 v^2}) \cosh(\gamma x \sqrt{v + \alpha^2 v^2})}{\sqrt{v - \alpha^2 v^2}} - \right.$$

$$\left. - \frac{(2\alpha^2 v + 1)e^{-2\gamma t v} \sin(\gamma x \sqrt{v - \alpha^2 v^2}) \sinh(\gamma x \sqrt{v + \alpha^2 v^2})}{\sqrt{v + \alpha^2 v^2}} \right],$$

$$h_1 = -\frac{e^{-\gamma |x| \sqrt{v + \alpha^2 v^2}} \sinh(2\gamma t v)}{4\pi\alpha^2} \left[\frac{(2\alpha^2 v - 1)\cos(\gamma x \sqrt{v - \alpha^2 v^2})}{\sqrt{v - \alpha^2 v^2}} + \right.$$

$$\left. + \frac{(2\alpha^2 v + 1)\sin(\gamma |x| \sqrt{v - \alpha^2 v^2})}{\sqrt{v + \alpha^2 v^2}} \right],$$

$$g_2 = -\frac{\gamma}{2\pi\alpha^2} \cos(\gamma x \sqrt{v - \alpha^2 v^2}) \sinh(\gamma x \sqrt{v + \alpha^2 v^2}) \frac{e^{-2\gamma t v}}{\sqrt{1 - \alpha^4 v^2}},$$

$$h_2 = -\frac{\gamma \operatorname{sign} x}{2\pi\alpha^2} \cos(\gamma x \sqrt{v - \alpha^2 v^2}) e^{-\gamma |x| \sqrt{v + \alpha^2 v^2}} \frac{\sinh(2\gamma t v)}{\sqrt{1 - \alpha^4 v^2}}.$$

$$\square$$

Example 4.1.10 In a similar way as Rayleigh's system, let us treat the quasihyperbolic, but not hyperbolic operator $P(\partial) = \partial_t^2 - \partial_t \partial_x^2 - \partial_x^2$, which is also known as *Stokes' operator*, see Gel'fand and Shilov [105], Ch. III, 4.1, Ex. 3, p. 134; Duff [63], p. 473; Morrison [183], p. 154, (7); Nardini [188]; Dautray and Lions [54], (3.81), p. 49, p. 285.

Similarly as in the last example, the two sets

$$A_1 = \left\{ \frac{p_1^2}{1 + p_1}; \, p_1 \in \mathbf{C}, \, \operatorname{Re} p_1 \geq N \right\}, \qquad A_2 = \{ p_2^2; \, p_2 \in \mathbf{C}, \, |\operatorname{Re} p_2| \leq 1 \}$$

are disjoint if N is large enough and hence $P(p)^{-1} \in H_C$ for $C = \{ \vartheta \in \mathbf{R}^2; \, \vartheta_1 > N, \, |\vartheta_2| \leq 1 \}$. We therefore obtain a fundamental solution E fulfilling $E \in \mathcal{S}'(\mathbf{R}^2)$ and $E = 0$ for $t < 0$ by inverse Laplace transform:

$$E = \frac{1}{2\pi} \mathcal{L}_{p \to t}^{-1} \left(\int_0^\infty \frac{e^{ix\xi} + e^{-ix\xi}}{p^2 + p\xi^2 + \xi^2} \, d\xi \right) = \frac{1}{2\pi} \mathcal{L}^{-1} \big(F(p, ix) + F(p, -ix) \big)$$

where p here stands for a single complex variable in $(0, \infty) + i\mathbf{R}$.

For positive p and z, we substitute $\xi = ps$ in the integral for $F(p, z)$ and obtain

$$F(p, z) = \int_0^\infty \frac{e^{-z\xi}}{p^2 + p\xi^2 + \xi^2} \, d\xi = \int_0^\infty \frac{e^{-zps}}{p(1 + s^2 + ps^2)} \, ds$$

$$= \int_0^\infty e^{-zps} \left[\frac{1}{p} - \frac{1}{p + 1 + s^{-2}} \right] \frac{ds}{1 + s^2}.$$

By analytic continuation, this holds for p in the complex right half-plane (and positive z), and hence

$$\mathcal{L}^{-1}(F(p, z)) = \frac{1}{2\pi i} \int_0^\infty \frac{ds}{1 + s^2} \int_{\vartheta - i\infty}^{\vartheta + i\infty} e^{p(t - zs)} \left[\frac{1}{p} - \frac{1}{p + 1 + s^{-2}} \right] dp$$

$$= \int_0^\infty Y(t - zs) \big[1 - e^{-(1 + s^{-2})(t - zs)} \big] \frac{ds}{1 + s^2}$$

$$= Y(t) \int_0^{t/z} \big[1 - e^{-(1 + s^{-2})(t - zs)} \big] \frac{ds}{1 + s^2}.$$

E is obtained from $\mathcal{L}^{-1}(F(p, z))$ by analytic continuation with respect to z. As in Example 4.1.6, we choose two semi-circles as integration paths from 0 to $\pm it/|x|$ and use the reflection $s \mapsto -s$ in the second integral. This yields

$$E = \frac{Y(t)}{2\pi} \int_C \big[1 - e^{-(1 + s^{-2})(t + i|x|s)} \big] \frac{ds}{1 + s^2}$$

where C is the circle through 0 and $\mathrm{i}t/|x|$ which is symmetric with respect to the imaginary s-axis and oriented in the counterclockwise direction. Note that we cannot replace C by the real axis as we did in Example 4.1.6 since the integrand tends to infinity for $x \neq 0$ if s tends to infinity in the upper half-plane. Therefore we substitute $u = -1/s$ and let u run over the straight line $\operatorname{Im} u = 2$ parallel to the real axis. This implies

$$
E = -\frac{Y(t)}{2\pi} \int_{2\mathrm{i}-\infty}^{2\mathrm{i}+\infty} \exp\!\left(-(1+u^2)(t - \tfrac{\mathrm{i}|x|}{u})\right) \frac{\mathrm{d}u}{1+u^2}.
$$

In particular, for $x = 0$, we can shift the integration to the real axis and obtain

$$
E(t,0) = Y(t)\left[\frac{1}{2} - \frac{\mathrm{e}^{-t}}{\pi}\int_0^\infty \mathrm{e}^{-u^2 t}\,\frac{\mathrm{d}u}{1+u^2}\right] = \frac{Y(t)}{2}\,\operatorname{Erf}(\sqrt{t}) = \frac{Y(t)}{2\sqrt{\pi}}\int_0^t \mathrm{e}^{-s}\,\frac{\mathrm{d}s}{\sqrt{s}}
$$

by Gröbner and Hofreiter [115], Eq. 314.8b.

In Ortner and Wagner [207], Prop. 4, p. 450, a fundamental solution of the more general operator

$$
(\partial_t - a\Delta_n - b)^2 - (c\Delta_n + 2\langle\omega, \nabla_n\rangle + d)^2 - h^2
$$

was represented by a simple integral over Bessel and exponential functions. In the special case of the operator $P(\partial)$ above, this yields the following representation of E:

$$
E(t,x) = \frac{Y(t)\sqrt{t}}{2\sqrt{2\pi}}\int_{-1}^1 J_0(t\sqrt{1-\lambda^2})\mathrm{e}^{-\lambda t - x^2/(2t(1-\lambda))}\,\frac{\mathrm{d}\lambda}{\sqrt{1-\lambda}},
$$

see Ortner and Wagner [207], p. 450, Rem. 1.   □

Example 4.1.11 As our final example, let us investigate the system of *dynamic linear thermoelasticity*, cf. Ortner and Wagner [209].

(a) We consider a homogeneous, isotropic elastic medium in $\mathbf{R}^3$. As in Example 2.1.3, we denote by ρ, f the densities of mass and of exterior force, respectively, and by $u = (u_1, u_2, u_3)^T$ the displacements. Whereas, in the absence of heat sources, u satisfies Lamé's system, i.e.,

$$
\rho\partial_t^2 u - \mu\Delta_3 u - (\lambda + \mu)\nabla \cdot \nabla^T u = \rho f,
$$

see (2.1.2), variations of the temperature $T(t,x)$ lead to the so-called *Duhamel–Neumann law*

$$
\rho\partial_t^2 u - \mu\Delta_3 u - (\lambda + \mu)\nabla \cdot \nabla^T u + \beta\nabla T = \rho f, \tag{4.1.17}
$$

where $\beta = (3\lambda + 2\mu)\alpha$ and α is the coefficient of thermal expansion, see Sokolnikoff [257], Eq. (99.5), p. 359; Sneddon [254], Eq. (1.5.13), p. 23; Nowacki [191], Eq. (18), p. 41; [192], Eq. (2.21), p. 267; Boley and Weiner [20], p. 31.

An equation describing the effects of strain on the diffusion of heat was derived in Biot [14] on the basis of the theory of irreversible thermodynamics:

$$\partial_t T - \kappa\Delta_3 T + \eta\partial_t\nabla^T \cdot u = Q. \tag{4.1.18}$$

Here Q measures the supply of heat (more precisely : $Q = (\kappa/\lambda) \times$ amountof heat generated per unit time per unit volume), κ is the temperature conductivity coefficient, $\eta = \beta T_0/(c\rho)$, and c, T_0 denote, respectively, the specific heat per unit mass and the temperature at rest (cf. Sneddon [254], Eq. (1.5.4), p. 21; Nowacki [191], Eq. (17), p. 41; [192], Eq. (2.20), p. 267; Carlson [43], Eq. (7.24), pp. 310, 328.

The four equations in (4.1.17) and (4.1.18) can be written in matrix form as

$$A(\partial)\begin{pmatrix} u \\ T \end{pmatrix} = \begin{pmatrix} \rho f \\ Q \end{pmatrix} \text{ where } A(\partial) = \begin{pmatrix} (\rho\partial_t^2 - \mu\Delta_3)I_3 - (\lambda + \mu)\nabla\cdot\nabla^T & \beta\nabla \\ \eta\partial_t\nabla^T & \partial_t - \kappa\Delta_3 \end{pmatrix}. \tag{4.1.19}$$

As we shall see below, $A(\partial)$ and, equivalently, $D(\partial) = \det A(\partial)$ are quasihyperbolic in the t-direction, see (2.2.3) and Definition 2.4.13, and we denote the corresponding fundamental matrix of $A(\partial)$ by E_A and the corresponding fundamental solution of $D(\partial)$ by E_D. By formula (2.1.1), we have $E_A = A^{\text{ad}}(\partial)E_D$.

A straight-forward calculation yields $D(\partial) = (\rho\partial_t^2 - \mu\Delta_3)^2 P(\partial)$ where

$$P(\partial) = (\rho\partial_t^2 - (\lambda + 2\mu)\Delta_3)(\partial_t - \kappa\Delta_3) - \beta\eta\partial_t\Delta_3.$$

On the other hand,

$$A^{\text{ad}}(\partial) = W(\partial)\begin{pmatrix} P(\partial)I_3 + H(\partial)\nabla\cdot\nabla^T & -\beta W(\partial)\nabla \\ -\eta W(\partial)\partial_t\nabla^T & W(\partial)(\rho\partial_t^2 - (\lambda + 2\mu)\Delta_3) \end{pmatrix},$$

where $W(\partial) = \rho\partial_t^2 - \mu\Delta_3$ and $H(\partial) = (\lambda + \mu + \eta\beta)\partial_t - (\lambda + \mu)\kappa\Delta_3$. By Example 1.4.12 (b) (see also Example 1.6.17), the forward fundamental solution of the wave operator $W(\partial)$ is given by

$$E_W = \frac{1}{4\pi\mu|x|} \cdot \delta\left(t - \sqrt{\frac{\rho}{\mu}}|x|\right).$$

By convolution, we therefore obtain $E_D = E_W * E_W * E_P$ if E_P is the fundamental solution of $P(\partial)$. This furnishes

$$E_A = \begin{pmatrix} I_3 E_W + H(\partial)\nabla\cdot\nabla^T E_W * E_P & -\beta\nabla E_P \\ -\eta\partial_t\nabla^T E_P & (\rho\partial_t^2 - (\lambda + 2\mu)\Delta_3)E_P \end{pmatrix}, \tag{4.1.20}$$

cf. Ortner and Wagner [209], Eq. (4), p. 527.

In order to simplify the further procedure, we introduce dimensionless variables, comp. Sneddon [254], Ch. 2. We abbreviate by E_ϵ the fundamental solution of the operator

$$P_\epsilon(\partial) = (\partial_t - \Delta_3)(\partial_t^2 - \Delta_3) - \epsilon \partial_t \Delta_3 = \partial_t^3 - \partial_t^2 \Delta_3 - (1 + \epsilon)\partial_t \Delta_3 + \Delta_3^2, \quad \epsilon \in \mathbf{C},$$
(4.1.21)

which will be called *thermoelastic operator* in the sequel. Linear transformations of the coordinates yield $a^{-1}b^{-3}d^{-1}E_\epsilon(t/a, x/b)$ as fundamental solution of the operator

$$a^3 d \partial_t^3 - a^2 b^2 d \partial_t^2 \Delta_3 - ab^2 d(1 + \epsilon)\partial_t \Delta_3 + b^4 d \Delta_3^2$$

for $a, b, d > 0$. Therefore, putting

$$a = \frac{\kappa \rho}{\lambda + 2\mu}, \qquad b = \kappa \sqrt{\frac{\rho}{\lambda + 2\mu}}, \qquad d = \frac{(\lambda + 2\mu)^3}{\kappa^3 \rho^2}, \qquad \epsilon = \frac{\eta \beta}{\lambda + 2\mu}$$

yields a representation of E_P in terms of E_ϵ :

$$E_P(t, x) = \frac{1}{\kappa \sqrt{\rho}\sqrt{\lambda + 2\mu}} E_\epsilon\left(\frac{\lambda + 2\mu}{\kappa \rho} t, \frac{\sqrt{\lambda + 2\mu}}{\kappa \sqrt{\rho}} x\right).$$

Note that $t/a, x/b$ and $E_\epsilon = ab^3 dE_P$ are dimensionless quantities. The quantity a^{-1} has been called *characteristic frequency* by Chadwick and Sneddon (see Sneddon [254], p. 41; Nowacki [193], p. 207). The physical values of ϵ for four metals are given in Chadwick [44], p. 279; Chadwick and Sneddon [47], p. 228, and lie in the range from 10^{-4} to 10^{-1}.

(b) In order to apply Proposition 4.1.8 to $P_\epsilon(\partial)$ and represent its fundamental solution E_ϵ by an inverse Laplace transform, let us first determine an open convex set $C \subset \mathbf{R}^4$ such that $P_\epsilon(p)^{-1} \in H_C$. For $p = \vartheta + i\xi \in \mathbf{C}^4$, let us set $\zeta = p_1$ and $z = \sum_{j=2}^4 p_j^2$. Then we have

$$P_\epsilon(p) = \zeta^3 - \zeta^2 z - (1 + \epsilon)\zeta z + z^2 = Q(\zeta, z).$$

For $z \to \infty$, the three roots of $\zeta \mapsto Q(\zeta, z)$ have the following asymptotic expansion:

$$\zeta_1 = z + \epsilon + O(z^{-1}), \qquad \zeta_{2,3} = \pm\sqrt{z} - \frac{\epsilon}{2} + O(z^{-1/2}).$$

Therefore, if $|z|$ is large and $Q(\zeta, z) = 0$, then either $|\zeta - z| \leq 1 + |\epsilon|$ or $|\zeta \pm \sqrt{z}| \leq 1 + \frac{|\epsilon|}{2}$, and hence, setting $\vartheta' = (\vartheta_2, \vartheta_3, \vartheta_4)$, we conclude that $P_\epsilon(p) \neq 0$ if $|z| > a$, $\mathrm{Re}\,\zeta > |\vartheta'|^2 + a$ and a is sufficiently large. Considering that the roots of $\zeta \mapsto Q(\zeta, z)$ are bounded for bounded z we obtain that $P_\epsilon(p)$ does not vanish in the tube domain T^C where $C = \{\vartheta \in \mathbf{R}^4;\ \vartheta_1 > |\vartheta'|^2 + a\}$ for sufficiently large $a > 0$. Hence $P_\epsilon(p)^{-1} \in H_C$ by the Seidenberg–Tarski lemma, cf. the proof of Proposition 2.3.5.

(c) According to Proposition 4.1.8, we can represent the fundamental solution E_ϵ of $P_\epsilon(\partial)$ as inverse Laplace transform in the sense of L. Schwartz: $E_\epsilon = \mathcal{L}^{-1}(P_\epsilon(p)^{-1})$. If we choose $\vartheta_2 = \vartheta_3 = \vartheta_4 = 0$ and $\vartheta_1 > a$, then

$$
E_\epsilon = \mathcal{L}_{p \to t}^{-1}\left(\frac{1}{(2\pi)^3}\,\mathcal{F}_{\xi \to x}\left(\frac{1}{(p + |\xi|^2)(p^2 + |\xi|^2) + \epsilon p|\xi|^2}\right)\right)
$$

$$
= \frac{1}{8\pi^3}\mathcal{L}_{p \to t}^{-1}\big(F(p, i|x|) + F(p, -i|x|)\big),
$$

where now p stands for a simple complex variable $p \in (a, \infty) + i\mathbf{R}$ and

$$
F(p, z) = -\frac{2\pi}{z}\int_0^\infty \frac{e^{-zr}r\,dr}{(p + r^2)(p^2 + r^2) + \epsilon pr^2}, \qquad \mathrm{Re}\,z \geq 0.
$$

For positive z, the inverse Laplace transform $\mathcal{L}_{p \to t}^{-1}(F(p, z))$ can be represented by a simple integral. In fact, if $p > a$, then the substitution $r = ps$ yields

$$
F(p, z) = -\frac{2\pi}{pz}\int_0^\infty \frac{e^{-zps}s\,ds}{(1 + ps^2)(1 + s^2) + \epsilon s^2} \tag{4.1.22}
$$

and this representation holds by analytic continuation for each $p \in (a, \infty) + i\mathbf{R}$. Therefore, (4.1.22) implies, still for positive z, that

$$
\mathcal{L}_{p \to t}^{-1}(F(p, z)) = \frac{i}{z}\int_0^\infty s\,ds \int_{\vartheta - i\infty}^{\vartheta + i\infty} \frac{e^{p(t - zs)}}{1 + s^2(1 + \epsilon)}\left(\frac{1}{p} - \frac{s^2(1 + s^2)}{(1 + ps^2)(1 + s^2) + \epsilon s^2}\right)dp
$$

$$
= -\frac{2\pi}{z}\int_0^\infty \frac{sY(t - zs)}{1 + s^2(1 + \epsilon)}\left(1 - e^{-(1 + s^2(1 + \epsilon))(t - zs)/(s^2(1 + s^2))}\right)ds
$$

$$
= -\frac{2\pi Y(t)}{z}\int_0^{t/z}\left(1 - e^{-(1 + s^2(1 + \epsilon))(t - zs)/(s^2(1 + s^2))}\right)\frac{s\,ds}{1 + s^2(1 + \epsilon)}.
$$

If, as in Example 4.1.6, we choose semi-circles from 0 to $it/|x|$ and from 0 to $-it/|x|$, respectively, as integration paths, and use the reflection $s \mapsto -s$ in the second one, we arrive at the representation

$$
E_\epsilon = -\frac{Y(t)i}{4\pi^2|x|}\int_C \left(1 - e^{-(1 + s^2(1 + \epsilon))(t + i|x|s)/(s^2(1 + s^2))}\right)\frac{s\,ds}{1 + s^2(1 + \epsilon)}, \tag{4.1.23}
$$

where C is the circle through 0 and $it/|x|$ which is symmetric with respect to the imaginary s-axis and oriented in the counterclockwise direction. We observe that the differential form integrated in (4.1.23) has essential singularities in 0 and in $\pm i$.

As in Example 4.1.6, we have to distinguish two cases: If $t > |x|$, then the circle C is homotopic, in $\overline{\mathbf{C}}\setminus\{0, i, -i\}$, to the real axis, and hence $E_\epsilon = E_{\epsilon,1}$ for $t > |x|$, where

$$E_{\epsilon,1} = \frac{1}{2\pi^2|x|} \int_0^\infty \sin\left(\frac{|x|(1 + (1 + \epsilon)s^2)}{s(1 + s^2)}\right) \exp\left(-\frac{t(1 + (1 + \epsilon)s^2)}{s^2(1 + s^2)}\right) \frac{s\,ds}{1 + (1 + \epsilon)s^2}.$$

(4.1.24)

On the other hand, if $0 < t < |x|$, then C is homotopic to the real axis and a loop around the essential singularity $s = i$ oriented in the clockwise sense. Hence $E_\epsilon = E_{\epsilon,1} + E_{\epsilon,2}$ for $0 < t < |x|$ and $\epsilon \neq 0$, where

$$E_{\epsilon,2} = \frac{1}{2\pi|x|} \operatorname*{Res}_{s=i} f(s), \qquad f(s) = \frac{s}{1 + (1 + \epsilon)s^2} \cdot \exp\left(-\frac{(1 + (1 + \epsilon)s^2)(t + i|x|s)}{s^2(1 + s^2)}\right).$$

(4.1.25)

The representation of E_ϵ in (4.1.24/4.1.25) was derived by a different method in Ortner and Wagner [209], pp. 538–542, see in particular formulae (12), (13).

Let us specialize formulae (4.1.24) and (4.1.25) to the uncoupled case $\epsilon = 0$. Hence E_0 is the fundamental solution of $(\partial_t - \Delta_3)(\partial_t^2 - \Delta_3)$. It was derived first by W. Nowacki, see Nowacki [191], Eqs. (10), (11), p. 267; [192], Eqs. (2.29), (2.30), p. 269; [193], Eq. (4.25), p. 198. For $\epsilon = 0$, we have

$$E_{0,1} = \frac{1}{2\pi^2|x|} \int_0^\infty \sin\left(\frac{|x|}{s}\right) \exp\left(-\frac{t}{s^2}\right) \frac{s\,ds}{1 + s^2} = \frac{1}{2\pi^2|x|} \int_0^\infty \sin(|x|v) e^{-tv^2} \frac{dv}{v(1 + v^2)}$$

$$= \frac{1}{8\pi|x|}\left[2\operatorname{Erf}\left(\frac{|x|}{2\sqrt{t}}\right) + e^{t+|x|}\operatorname{Erfc}\left(\sqrt{t} + \frac{|x|}{2\sqrt{t}}\right) - e^{t-|x|}\operatorname{Erfc}\left(\sqrt{t} - \frac{|x|}{2\sqrt{t}}\right)\right]$$

by using a well-known Fourier sine transform, see Oberhettinger [196], (3.29), p. 126.

Furthermore,

$$E_{0,2} = \frac{1}{2\pi|x|} \operatorname*{Res}_{s=i}\left\{\frac{s}{1 + s^2} \cdot \left(\exp\left(-\frac{(t + i|x|s)}{s^2}\right) - 1\right)\right\} = \frac{1}{4\pi|x|}\left(e^{t-|x|} - 1\right),$$

and hence

$$E_0(t, x) = \frac{Y(t)}{8\pi|x|}\left[2\operatorname{Erf}\left(\frac{|x|}{2\sqrt{t}}\right) + e^{t+|x|}\operatorname{Erfc}\left(\sqrt{t} + \frac{|x|}{2\sqrt{t}}\right)\right.$$

$$\left. - e^{t-|x|}\operatorname{Erfc}\left(\sqrt{t} - \frac{|x|}{2\sqrt{t}}\right) + 2Y(|x| - t)\left(e^{t-|x|} - 1\right)\right].$$

(4.1.26)

A representation of E_0 as a definite integral appears already in Bureau [38], Eq. (50.8), p. 197.

(d) Let us yet derive a representation for the temperature caused by a hot spot in a thermoelastic medium. This is expressed by the element $T(t,x) = (E_A)_{44} = (\rho\partial_t^2 - (\lambda + 2\mu)\Delta_3)E_P$ in the lower right corner of the fundamental matrix E_A in (4.1.20). Scaling to dimensionless variables as in (a) yields

$$T(t,x) = \frac{(\lambda + 2\mu)^{3/2}}{\kappa^3 \rho^{3/2}} T_\epsilon\left(\frac{t}{a}, \frac{x}{b}\right) \text{ where } T_\epsilon = (\partial_t^2 - \Delta_3)E_\epsilon.$$

When differentiating E_ϵ, we employ the fact that $E_{\epsilon,2}$ vanishes for $t = |x| = r$ and that the same holds for $(\partial_t - \partial_r)E_{\epsilon,2}$. Therefore, the formulae (4.1.24/4.1.25) imply

$$T_\epsilon(t,x) = \frac{Y(t)}{2\pi^2|x|} \int_0^\infty \sin\left(\frac{|x|(1 + (1+\epsilon)s^2)}{s(1+s^2)}\right) \exp\left(-\frac{t(1 + (1+\epsilon)s^2)}{s^2(1+s^2)}\right) \frac{1 + (1+\epsilon)s^2}{s^3(1+s^2)} \, ds$$

$$+ \frac{Y(t)Y(|x| - t)}{2\pi|x|} \operatorname*{Res}_{s=i}\left\{\frac{1 + (1+\epsilon)s^2}{s^3(1+s^2)} \exp\left(-\frac{(1 + (1+\epsilon)s^2)(t + \mathrm{i}|x|s)}{s^2(1+s^2)}\right)\right\}.$$

The constant and the linear term of the Taylor expansion of T_ϵ with respect to ϵ can be represented by error functions as in (4.1.26). This linear expansion was derived for the first time in Hetnarski [128], Eq. (4.28), p. 935; cf. also Ortner and Wagner [209], p. 548. For the remaining entries of the fundamental matrix E_A, we refer to Wagner [288].

4.2 Singularities of Fundamental Solutions of Quasihyperbolic Operators

In the following, we present part of the singularity theory which was developed for hyperbolic operators in Atiyah, Bott, and Gårding [5] and Hörmander [138]. We restrict ourselves to estimates for the singular support, leaving aside the analogous estimates for the wave front set developed by L. Hörmander. At some instances, we give generalizations to quasihyperbolic operators.

Definition 4.2.1 For a polynomial $P(\eta)$ and $\xi \in \mathbf{R}^n$, the polynomial $P_\xi(\eta)$ denotes the *localization at infinity of P in the direction ξ*, i.e., $P_\xi(\eta)$ is the lowest non-vanishing coefficient with respect to t in the MacLaurin series of $t^m P(\eta + \frac{\xi}{t})$, $m = \deg P \geq 0$. Thus $t^m P(\eta + \frac{\xi}{t}) = t^p P_\xi(\eta) + O(t^{p+1})$ for $t \to 0$. Herein, $p = m_\xi(P)$ is called the *multiplicity of ξ with respect to P*. For $P = 0$, we set $P_\xi = 0$.

For Definition 4.2.1, cf. Atiyah, Bott, and Gårding [5], Def. 3.36, p. 135; Gårding [90], p. 223.

Example 4.2.2

(a) If $\xi = 0$, then $P_\xi = P$ and $m_\xi(P) = \deg P$.
(b) If $P = \sum_{|\alpha|\le m} a_\alpha \eta^\alpha$ and $P_k = \sum_{|\alpha|=k} a_\alpha \eta^\alpha$, $0 \le k \le m$, are the homogeneous components of P, then

$$t^m P\left(\eta + \frac{\xi}{t}\right) = P_m(\xi) + t\big[\eta^T \cdot \nabla P_m(\xi) + P_{m-1}(\xi)\big]$$

$$+ t^2\big[\tfrac{1}{2}\eta^T \cdot \nabla\nabla^T P_m(\xi) \cdot \eta + \eta^T \cdot \nabla P_{m-1}(\xi) + P_{m-2}(\xi)\big] + O(t^3), \quad t \to 0,$$

where ∇P_m is the gradient of P_m and $\nabla\nabla^T P_m$ is the Hesse matrix of P_m. Hence, if $P_m(\xi) \ne 0$, then $m_\xi(P) = 0$ and $P_\xi(\eta)$ is the constant $P_m(\xi)$. If $P_m(\xi) = 0$, but $\nabla P_m(\xi) \ne 0$ or $P_{m-1}(\xi) \ne 0$, then $m_\xi(P) = 1$ and $P_\xi(\eta) = \eta^T \cdot \nabla P_m(\xi) + P_{m-1}(\xi)$, a linear polynomial. If $P_m(\xi) = P_{m-1}(\xi) = 0$ and $\nabla P_m(\xi) = 0$, but $\nabla\nabla^T P_m(\xi) \ne 0$, then $m_\xi(P) = 2$ and $P_\xi(\eta)$ is a second-order polynomial with principal part $\tfrac{1}{2}\eta^T \cdot \nabla\nabla^T P_m(\xi) \cdot \eta$.   □

For a quasihyperbolic operator $P(-i\partial)$, we can estimate the singular support of the fundamental solution E, which was defined in 2.4.13, from below by means of the localizations P_ξ. We point out that—in contrast to the hyperbolic case—the operators $P_\xi(-i\partial)$ are not necessarily quasihyperbolic. (For example, if $P(\xi_1, \xi_2) = 1 + \xi_1\xi_2$, then $P(-i\partial)$ is quasihyperbolic in the direction $N = (1, 0)$, whereas $P_{(1,0)}(-i\partial) = -i\partial_2$ is not.) The next proposition generalizes the "Localization Theorem" in Atiyah, Bott, and Gårding [5], 4.10, p. 144. It was formulated in Ortner and Wagner [209], Prop. 3, p. 534.

Proposition 4.2.3 *Suppose that $P(-i\partial)$ and the localization $P_\xi(-i\partial)$ are quasihyperbolic operators in the direction N for some $\xi \in \mathbf{R}^n \setminus \{0\}$. Denote its fundamental solutions according to* Proposition 2.4.13 *by E and E_ξ, respectively. Then* $\operatorname{supp} E_\xi \subset \operatorname{sing\,supp} E$.

Proof Assume that σ_0 is chosen such that $P(\eta - i\sigma N) \ne 0$ and $P_\xi(\eta - i\sigma N) \ne 0$ for all $\xi \in \mathbf{R}^n$ and $\sigma > \sigma_0$. The representation of E in (2.4.13) implies, for $\phi \in \mathcal{D}$ and $t \in \mathbf{R} \setminus \{0\}$, that

$$\langle \phi, e^{-i\xi x/t} E \rangle = \int_{\mathbf{R}^n} \frac{\mathcal{F}^{-1}(e^{\sigma Nx}\phi)(\eta)}{P(\eta + \xi/t - i\sigma N)}\, d\eta. \qquad (4.2.1)$$

For $p = m_\xi(P)$ and $m = \deg P$, the polynomial $t^{m-p} P(\eta + \xi/t - i\sigma N)$ converges to the localization $P_\xi(\eta - i\sigma N)$ if t tends to 0. As in (2.3.9), we use the Seidenberg–Tarski lemma to obtain the estimate

$$t^{m-p}|P(\eta + \xi/t - i\sigma N)| \ge k^{-1}(1 + |\eta| + |t|)^{-k} \qquad (4.2.2)$$

for some positive constant k and all $\eta \in \mathbf{R}^n$ and $t \in \mathbf{R} \setminus \{0\}$. (Here we employ the quasihyperbolicity of P and P_ξ, which implies that the polynomial $Q(t, \eta) = t^{m-p} P(\eta + \xi/t - i\sigma N)$ has no real zeros.) The inequality (4.2.2) enables us to use

Lebesgue's dominated convergence theorem in Eq. (4.2.1) in order to conclude that

$$\lim_{t \to 0} \langle \phi, t^{p-m} e^{-i\xi x/t} E \rangle = \int_{\mathbf{R}^n} \frac{\mathcal{F}^{-1}(e^{\sigma Nx}\phi)(\eta)}{P_\xi(\eta - i\sigma N)} \, d\eta = \langle \phi, E_\xi \rangle.$$

Outside of sing supp E, the distribution $t^{p-m} e^{-i\xi x/t} E$ converges to 0 if $t \to 0$ in virtue of the Riemann–Lebesgue lemma, and hence E_ξ vanishes on $\mathbf{R}^n \setminus$ sing supp E. This implies the assertion of Proposition 4.2.3 and completes the proof.   □

The content of Proposition 4.2.3 can be expressed by the following bound for the singular support of E from below:

$$\bigcup_{\substack{\xi \in \mathbf{R}^n \setminus \{0\} \\ P_\xi(-i\partial) \text{ quasihyperbolic}}} \text{supp} \, E_\xi \subset \text{sing supp} \, E. \qquad (4.2.3)$$

Note that if the operator $P(-i\partial)$ is *hyperbolic* in the direction N, then the localizations $P_\xi(-i\partial)$ are necessarily also hyperbolic in this direction (see Atiyah, Bott, and Gårding [5], Lemma 3.42, (3.45), p. 136), and hence

$$P(-i\partial) \text{ hyperbolic} \implies \bigcup_{\xi \in \mathbf{R}^n \setminus \{0\}} \text{supp} \, E_\xi \subset \text{sing supp} \, E. \qquad (4.2.4)$$

In order to give a bound for sing supp E from above, let us define the so-called *wavefront surface*, a notion introduced first in Atiyah, Bott, and Gårding [5], Def. 5.15, p. 155.

Definition 4.2.4 For an operator $P(-i\partial)$ which is hyperbolic with respect to $N \in \mathbf{R}^n \setminus \{0\}$, the union of the propagation cones of its localizations is called the *wavefront surface* $W(P(-i\partial), N)$, i.e.,

$$W(P(-i\partial), N) = \bigcup_{\xi \in \mathbf{R}^n \setminus \{0\}} K(P_\xi(-i\partial), N).$$

Proposition 4.2.5 *Let $P(-i\partial)$ be hyperbolic with respect to $N \in \mathbf{R}^n \setminus \{0\}$, and let E denote the forward fundamental solution of $P(-i\partial)$. Then E is real-analytic outside the wavefront surface $W(P(-i\partial), N)$.*

For a **proof** we refer to Atiyah, Bott, and Gårding [5], Thm. 7.24, p. 177, or Hörmander [138], Thm. 12.6.6, p. 132.

Let us summarize the different bounds for the singular support of the forward fundamental solution E of a hyperbolic differential operator $P(-i\partial)$. If sing supp$_A$ E denotes the *analytic singular support* of E, i.e., the complement of the open set where E is real-analytic, and $E(Q(\partial), N)$ denotes the forward fundamental solution of the operator $Q(\partial)$ with respect to N, then Propositions 4.2.3 and 4.2.5 imply the following sequence of inclusions:

$$\bigcup_{\xi \in \mathbf{R}^n \setminus \{0\}} \operatorname{supp} E(P_\xi(-i\partial), N) \subset \operatorname{sing} \operatorname{supp} E \subset \operatorname{sing} \operatorname{supp}_A E \subset$$

$$\subset \bigcup_{\xi \in \mathbf{R}^n \setminus \{0\}} K(P_\xi(-i\partial), N) = W(P(-i\partial), N) \qquad (4.2.5)$$

cf. Gårding [90], Eq. 4, p. 225.

In most of the physically relevant cases, all the inclusions in (4.2.5) are identities. In particular, sing supp $E(P(-i\partial), N) = W(P(-i\partial), N)$ holds if the dimension n does not exceed four (since then either the principal part of P is complete, or the reduced dimension of it is at most three), see Atiyah, Bott, and Gårding [6], Thm. 7.7, p. 175.

Let us now give a mathematical definition of the term *conical refraction*, see also Gårding [91], pp. 24, 25; [92], pp. 360–362; Liess [165], Ch. 6; Musgrave [186], 11.3, p. 143. For the terms "slowness surface" and "normal surface," see Duff [62], p. 251; [64], p. 50.

Definition 4.2.6 Let $P(-i\partial)$ be hyperbolic with respect to $N \in \mathbf{R}^n \setminus \{0\}$ and with principal part $P_m(-i\partial)$, $m = \deg P$. We suppose that $P_m(\xi)$ does not contain multiple factors. As always $H_N = \{x \in \mathbf{R}^n; xN \geq 0\}$.

(1) The set $\Xi = \{\xi \in \mathbf{R}^n; P_m(\xi) = 0\}$ is called the *slowness surface* or *normal surface* of $P(-i\partial)$.
(2) The operator $P(-i\partial)$ is called *strictly hyperbolic* iff $\forall \xi \in \Xi \setminus \{0\} : \nabla P_m(\xi) \neq 0$.
(3) The hypersurface Ξ^* *dual* to Ξ is defined by

$$\Xi^* = \{t \cdot \nabla P_m(\xi); \ t \in \mathbf{R}, \xi \in \mathbf{R}^n, P_m(\xi) = 0\},$$

   and it is called the *characteristic surface* or *wave surface* of $P(-i\partial)$.
(4) We say that *conical refraction* occurs if and only if $\Xi^* \cap H_N$ is a proper subset of sing supp E where E is the forward fundamental solution of $P(-i\partial)$.

Note that $\Xi^* \cap H_N$ is contained in sing supp E due to the inclusion in (4.2.3). Indeed, $P_\xi(-i\partial)$ coincides with the first-order operator $-i\nabla P_m(\xi)^T \cdot \nabla + P_{m-1}(\xi)$ if $P_m(\xi) = 0$ and $\nabla P_m(\xi) \neq 0$ and hence $\nabla P_m(\xi) \in \Xi^*$, see Example 4.2.2.

Thus if sing supp $E = W(P(-i\partial), N)$ (which holds in most physically relevant examples), then

$$\operatorname{sing} \operatorname{supp} E = (\Xi^* \cap H_N) \cup C \quad \text{where} \quad C = \bigcup_{\substack{\xi \in \mathbf{R}^n \setminus \{0\} \\ P_m(\xi) = 0, \nabla P_m(\xi) = 0}} K(P_\xi(-i\partial), N).$$

In other words, C is the union of the propagation cones of the localizations $P_\xi(-i\partial)$ in the directions of the *singular* or *conical points* ξ on the slowness surface, and conical refraction occurs if C is not contained in Ξ^*.

Evidently, for a strictly hyperbolic operator, Ξ is non-singular, the set C is empty, and hence sing supp $E = \Xi^* \cap H_N$. We also mention that the strict hyperbolicity of $P(-i\partial)$ is equivalent to the condition that $(P+Q)(-i\partial)$ is hyperbolic for each lower order operator Q, i.e., for arbitrary Q with $\deg Q < \deg P$, see Hörmander [138], Cor. 12.4.10, p. 118.

Let us also remark that some authors call the projective hypersurfaces Ξ and Ξ^* *slowness cone* and *characteristic cone*, respectively, in order to distinguish them from their affine representations $X = \{\xi \in \Xi; \xi \cdot N = 1\}, X^* = \{x \in \Xi^*; x \cdot N = 1\}$, which they call "surfaces," cf. Duff [62], pp. 251, 252. We shall indifferently call both Ξ and X *slowness surface.*

Example 4.2.7 Let us illustrate the notions of Definition 4.2.6 in the simple example of $P(\partial)$ being the product of two anisotropic wave operators in $\mathbf{R}^3$, i.e.,

$$P(\partial) = P(-i\partial) = (\partial_t^2 - \Delta_3)(\partial_t^2 - a\Delta_2 - b\partial_3^2), \qquad a, b > 0.$$

We set $N = (1, 0, 0, 0)$ and denote by E the forward fundamental solution of $P(\partial)$. This operator was also considered in Bureau [34].

More generally, the fundamental solution of the operator

$$(\partial_t^2 - \Delta_3)(\partial_t^2 - a_1\partial_1^2 - a_2\partial_2^2 - a_3\partial_3^2), \qquad a_j > 0,$$

was represented by elliptic integrals in Herglotz [125], III, § 10, (184), p. 104. This representation immediately results by the method of parameter integration from Proposition 3.2.1. We restrict ourselves here to the simpler case $a_1 = a_2 = a, a_3 = b$ since our main goal here is to illustrate the propagation of singularities and the occurrence of conical refraction. In this simpler case, the elliptic integrals reduce to elementary transcendental functions, see the formulas below.

(a) Let us first consider E from the qualitative viewpoint and determine the singular support of E.

If either $0 < a, b < 1$ or $1 < a, b$, then the slowness surface

$$\Xi = \{(\tau, \xi) \in \mathbf{R}^4; \tau^2 = |\xi|^2 \text{ or } \tau^2 = a(\xi_1^2 + \xi_2^2) + b\xi_3^2\}$$

is a non-singular projective variety and hence, by (4.2.5),

$$\text{sing supp} E = W(P(\partial), N) = \Xi^* \cap H_N$$

$$= \left\{(t, x) \in \mathbf{R}^4; t = |x| \text{ or } t = \sqrt{\tfrac{1}{a}(x_1^2 + x_2^2) + \tfrac{1}{b}x_3^2}\right\}.$$

On the other hand, if e.g., $0 < a < 1$ and $b > 1$, then Ξ is singular in the intersection points of the two hypersurfaces $\tau^2 = |\xi|^2$ and $\tau^2 = a(\xi_1^2 + \xi_2^2) + b\xi_3^2$, i.e., on the set

$$M = \left\{(\tau, \xi) \in \mathbf{R}^4; \tau^2 = |\xi|^2 \text{ and } (\xi_1^2 + \xi_2^2)\frac{1-a}{b-1} = \xi_3^2\right\}.$$

In order to determine the localizations $P_{(\tau,\xi)}$ in the directions $(\tau,\xi) \in M$, let us observe first, that, generally,

$$(P_1 \cdot P_2)_\xi = (P_1)_\xi \cdot (P_2)_\xi \quad \text{and} \quad m_\xi(P_1 \cdot P_2) = m_\xi(P_1) + m_\xi(P_2),$$

for arbitrary polynomials P_1, P_2 in n variables and $\xi \in \mathbf{R}^n$, cf. Atiyah, Bott, and Gårding [5], (3.40/41), p. 136. Therefore, setting $P_1(\tau,\xi) = \tau^2 - |\xi|^2$, $P_2(\tau,\xi) = \tau^2 - a(\xi_1^2 + \xi_2^2) - b\xi_3^2$ and taking $(\tau,\xi) \in M$, we obtain

$$P_{(\tau,\xi)}(\eta) = (P_1)_{(\tau,\xi)}(\eta) \cdot (P_2)_{(\tau,\xi)}(\eta) = \left[\eta^T \cdot (\nabla P_1)(\tau,\xi) \right] \cdot \left[\eta^T \cdot (\nabla P_2)(\tau,\xi) \right]$$

$$= 4(\tau\eta_0 - \xi_1\eta_1 - \xi_2\eta_2 - \xi_3\eta_3)(\tau\eta_0 - a\xi_1\eta_1 - a\xi_2\eta_2 - b\xi_3\eta_3)$$

and hence

$$P_{(\tau,\xi)}(-i\partial) = -4(\tau\partial_t - \xi_1\partial_1 - \xi_2\partial_2 - \xi_3\partial_3)(\tau\partial_t - a\xi_1\partial_1 - a\xi_2\partial_2 - b\xi_3\partial_3). \tag{4.2.6}$$

The set C of conical refraction is therefore given by

$$C = \bigcup_{(\tau,\xi)\in M\setminus\{0\}} K(P_{(\tau,\xi)}(-i\partial), N)$$

$$= \left\{ \lambda(\tau,-\xi) + (1-\lambda)(\tau,-a\xi_1,-a\xi_2-b\xi_3); \ \lambda \in [0,1], (\tau,\xi) \in M, \tau > 0 \right\}.$$

If we consider the surface $\{x \in \mathbf{R}^3; \ (1,x) \in \text{sing supp}\, E\}$, it consists of the characteristic surface

$$X^* = \left\{ x \in \mathbf{R}^3; \ |x| = 1 \text{ or } \frac{x_1^2 + x_2^2}{a} + \frac{x_3^2}{b} = 1 \right\},$$

which is the union of the unit sphere and an intersecting ellipsoid, as well as the two frusta $\{x \in \mathbf{R}^3; \ (1,x) \in C\}$, which lie on the convex envelope of X^*, see the broken lines in Fig. 4.1, right part. These frusta are contained in circular cones around the x_3-axis with vertices in the points $(0,0,\pm\sqrt{(b-a)/(1-a)})$.

Hence (4.2.5) implies that the singular support of E is given by

$$\text{sing supp}\, E = W(P(\partial), N) = (\Xi^* \cap H_N) \cup C$$

$$= \left\{ (t,x) \in \mathbf{R}^4; \ t = |x| \text{ or } t = \sqrt{\tfrac{1}{a}(x_1^2 + x_2^2) + \tfrac{1}{b}x_3^2} \right.$$

$$\text{or } (t,x) = \left(|\xi|, \lambda\xi_1 + (1-\lambda)a\xi_1, \lambda\xi_2 + (1-\lambda)a\xi_2, \lambda\xi_3 + (1-\lambda)b\xi_3 \right),$$

$$\left. 0 \le \lambda \le 1, \xi \in \mathbf{R}^3, \xi_3^2(b-1) = (\xi_1^2 + \xi_2^2)(1-a) \right\}.$$

This equation also holds in the case where $a > 1$ and $0 < b < 1$.

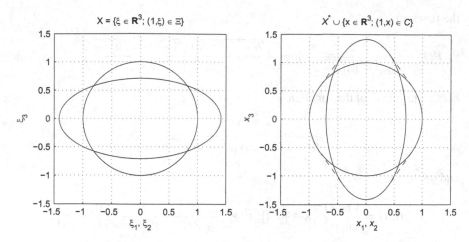

Fig. 4.1 Slowness surface and wavefront surface for $(\partial_t^2 - \Delta_3)(\partial_t^2 - \frac{1}{2}\Delta_2 - 2\partial_3^2)$

(b) We eventually derive an explicit representation of E. By applying the method of parameter integration according to Proposition 3.2.1, we obtain $E = \int_0^1 E_\lambda \, d\lambda$, where E_λ is the forward fundamental solution of

$$\left[\partial_t^2 - (\lambda + a(1 - \lambda))\Delta_2 - (\lambda + b(1 - \lambda))\partial_3^2\right]^2 = \left(P_1(\partial) + \lambda P_2(\partial)\right)^2,$$

wherein $P_1(\partial) = \partial_t^2 - a\Delta_2 - b\partial_3^2$ and $P_2(\partial) = (a - 1)\Delta_2 + (b - 1)\partial_3^2$.

By formula (2.3.12) and a linear substitution, we have

$$E_\lambda = \frac{1}{8\pi(\lambda + a(1 - \lambda))\sqrt{\lambda + b(1 - \lambda)}} Y\left(t - \sqrt{\frac{|x'|^2}{\lambda + a(1 - \lambda)} + \frac{x_3^2}{\lambda + b(1 - \lambda)}}\right),$$

where $x' = (x_1, x_2)$, i.e.,

$$E = \frac{1}{8\pi} \int_0^1 \frac{Y\left(t - \sqrt{\frac{|x'|^2}{a + \lambda(1 - a)} + \frac{x_3^2}{b + \lambda(1 - b)}}\right)}{(a + \lambda(1 - a))\sqrt{b + \lambda(1 - b)}} \, d\lambda.$$

Let us first investigate the case when $0 < a < b < 1$. Then conical refraction does not appear, see (a). From Gradshteyn and Ryzhik [113], Eq. 2.246, we infer

$$E = \frac{Y(t - |x|)}{8\pi}\left[F(\lambda_2(t, x)) - F(\lambda_1(t, x))\right], \tag{4.2.7}$$

the function

$$F(\lambda) = \frac{2}{\sqrt{(1-a)(b-a)}} \log\left(\frac{\sqrt{b+\lambda(1-b)}\sqrt{1-a} - \sqrt{b-a}}{\sqrt{a+\lambda(1-a)}} \right) \qquad (4.2.8)$$

being a primitive of the function

$$\frac{1}{(a+\lambda(1-a))\sqrt{b+\lambda(1-b)}}$$

and

$$\lambda_{1,2}(t,x) = \left\{ \begin{matrix} \min \\ \max \end{matrix} \right\} \left\{ \lambda \in [0,1]; \ t^2 \geq \frac{|x'|^2}{a+\lambda(1-a)} + \frac{x_3^2}{b+\lambda(1-b)} \right\}.$$

In this case, the set $\mathrm{supp}\,E\backslash\mathrm{sing}\ \mathrm{supp}\,E$ consists of two connectivity components: In the inner cone

$$M_1 = \left\{ (t,x) \in \mathbf{R}^4; \ t > \sqrt{\tfrac{1}{a}|x'|^2 + \tfrac{1}{b}x_3^2} \right\},$$

we have $\lambda_1 = 0$, $\lambda_2 = 1$, and hence E is constant in M_1 and given by

$$E|_{M_1} = \frac{1}{4\pi\sqrt{(1-a)(b-a)}} \log\left(\frac{(\sqrt{1-a}-\sqrt{b-a})\sqrt{a}}{\sqrt{b-ab}-\sqrt{b-a}} \right). \qquad (4.2.9)$$

On the other hand, in the region $M_2 = \{(t,x) \in \mathbf{R}^4; \ t > |x|\} \setminus \overline{M_1}$, we have $0 < \lambda_1(t,x) < 1$ and $\lambda_2 = 1$. More precisely, $\lambda = \lambda_1(t,x)$ is the unique solution in $[0,1]$ of the equation

$$t^2 = \frac{|x'|^2}{a+\lambda(1-a)} + \frac{x_3^2}{b+\lambda(1-b)}. \qquad (4.2.10)$$

Hence, for $(t,x) \in M_2$,

$$E(t,x) = \frac{1}{4\pi\sqrt{(1-a)(b-a)}} \log\left(\frac{\sqrt{a+\lambda_1(t,x)(1-a)}\cdot(\sqrt{1-a}-\sqrt{b-a})}{\sqrt{b+\lambda_1(t,x)(1-b)}\sqrt{1-a}-\sqrt{b-a}} \right).$$
$$(4.2.11)$$

(c) Eventually, let us investigate a case in which conical refraction occurs. We assume that $0 < a < 1$ and $b > 1$. Since $(1-a)(b-a)$ is again positive, the formulas (4.2.7) and (4.2.8) are still valid. However, the set $\mathrm{supp}\,E\backslash\mathrm{sing}\ \mathrm{supp}\,E$ splits into six connected components. As before, in the innermost cone

$$M_1 = \left\{ (t,x) \in \mathbf{R}^4; \ t > |x| \text{ and } t^2 > \frac{|x'|^2}{a} + \frac{x_3^2}{b} \right\},$$

E is constant and given by the same constant as in (4.2.9). In the region

$$M_2 = \left\{(t,x) \in \mathbf{R}^4; \ t > |x| \text{ and } t^2 < \frac{|x'|^2}{a} + \frac{x_3^2}{b}\right\},$$

we have, as before, $\lambda_2 = 1$ and E is given by formula (4.2.11) as in case (b).

In the two regions

$$M_{3,\pm} = \left\{(t,x) \in \mathbf{R}^4; \ 0 < t < |x| \text{ and } t^2 > \frac{|x'|^2}{a} + \frac{x_3^2}{b} \text{ and } \pm x_3 > 0\right\},$$

we obtain that $\lambda_1 = 0$ and $\lambda_2(t,x)$ is the unique solution in $[0,1]$ of the equation in (4.2.10). This furnishes, for $(t,x) \in M_{3,+} \cup M_{3,-}$, the following:

$$E(t,x) = \frac{1}{4\pi\sqrt{(1-a)(b-a)}} \log\left(\frac{[\sqrt{b+\lambda_2(t,x)(1-b)}\sqrt{1-a} - \sqrt{b-a}]\sqrt{a}}{\sqrt{a+\lambda_2(t,x)(1-a)}\cdot(\sqrt{b-ab}-\sqrt{b-a})}\right).$$

Finally, suppose that (t,x) belongs to the regions

$$M_{4,\pm} = \left\{(t,x) \in \operatorname{supp} E; \ t < |x| \text{ and } t^2 < \frac{|x'|^2}{a} + \frac{x_3^2}{b} \text{ and } \pm x_3 > 0\right\},$$

i.e., (t,x) is in the exterior of both of the cones $t > |x|$ and $t > (|x'|^2/a + x_3^2/b)^{1/2}$, but inside the region bounded by the frustum C of conical refraction, see (a). Then the Eq. (4.2.10) has two zeroes $\lambda_1(t,x) < \lambda_2(t,x)$ inside the interval $[0,1]$ and

$$E(t,x) = \frac{1}{8\pi}\big[F(\lambda_2(t,x)) - F(\lambda_1(t,x))\big] = \frac{1}{4\pi\sqrt{(1-a)(b-a)}} \times$$

$$\times \log\left(\frac{\sqrt{b+\lambda_2(t,x)(1-b)}\sqrt{1-a} - \sqrt{b-a}}{\sqrt{b+\lambda_1(t,x)(1-b)}\sqrt{1-a} - \sqrt{b-a}} \cdot \frac{\sqrt{a+\lambda_1(t,x)(1-a)}}{\sqrt{a+\lambda_2(t,x)(1-a)}}\right)$$

holds in $M_{4,\pm}$. □

Let us next generalize slightly Proposition 4.2.5 from hyperbolic to quasihyperbolic operators. This will allow to determine the singularities of the fundamental solutions of Rayleigh's operator in Eq. (2.4.16) as well as of the thermoelastic operator in Eq. (4.1.21). For this purpose, we shall accommodate Thm. 10.2.11, p. 23, in Hörmander [138] to our needs. We first adopt the notations in Hörmander [138], Section 10.2.

Definition 4.2.8 Let $P(\eta)$ be a polynomial in $\mathbf{R}^n$.

(1) We denote by $L(P)$ the set of all its *localizations* in the following sense:

$$L(P) = \bigcap_{N=1}^{\infty} \overline{B_N} \quad \text{where} \quad B_N = \left\{ \frac{\tau_\xi P}{\|\tau_\xi P\|}; \, \xi \in \mathbf{R}^n, \, |\xi| \geq N \right\},$$

$$(\tau_\xi P)(\eta) = P(\eta - \xi), \quad \|Q\|^2 = \sum_{\alpha \in \mathbf{N}_0^n} |(\partial^\alpha Q)(0)|^2$$

and the closure $\overline{B_N}$ of B_N is taken in the finite dimensional vector space of all polynomials of degree smaller or equal to that of P.

(2) The *lineality* $\Lambda(P)$ is the set $\{\xi \in \mathbf{R}^n; \, \tau_\xi P = P\}$. We denote by $\Lambda(P)^\perp$ its orthogonal complement, i.e.,

$$\Lambda(P)^\perp = \{\xi \in \mathbf{R}^n; \, \forall \eta \in \Lambda(P) : \xi^T \eta = 0\}.$$

Note that $L(P)$ contains the set $L_1(P) = \{P_\xi / \|P_\xi\|; \, \xi \in \mathbf{R}^n \setminus \{0\}\}$ of normalized localizations at infinity of P in specific directions (see Definition 4.2.1). In general, $L(P)$ is strictly larger than $L_1(P)$. For example, if $P(\eta) = \eta_1 - \eta_2^2 - \cdots - \eta_n^2$ corresponds to the Schrödinger operator, then $L_1(P) = \{1, -1\}$, but $L(P) = \{\pm 1, \omega^T \cdot \eta; \, \omega \in \mathbf{S}^{n-1}, \omega_1 = 0\}$. In fact, if $\omega \in \mathbf{S}^{n-1}$ with $\omega_1 = 0$ and $\xi_t = (t^2, -t\omega_2, \ldots, -t\omega_n)$, then

$$\lim_{t \to \infty} \frac{P(\eta + \xi_t)}{\|P(\eta + \xi_t)\|} = \lim_{t \to \infty} \frac{P(\eta) + 2t(\eta_2 \omega_2 + \cdots + \eta_n \omega_n)}{\sqrt{4n - 3 + 4t^2}} = \omega^T \cdot \eta.$$

Also note that $\Lambda(\omega^T \cdot \eta)^\perp = \mathbf{R} \cdot \omega$ for $\omega \in \mathbf{R}^n \setminus \{0\}$.

In the following proposition, which goes back to Ortner and Wagner [209], Prop. 4, p. 535, we show that, under an additional condition, the union of the linear subspaces $\Lambda(Q)^\perp$, where Q runs through the set $L(P)$ of localizations of P, yields an upper bound for the singular support of the fundamental solution of a quasihyperbolic operator $P(-i\partial)$.

Proposition 4.2.9 *Suppose that $P(-i\partial)$ is a quasihyperbolic operator in the direction N, i.e., $P(\eta - i\sigma N) \neq 0$ for all $\eta \in \mathbf{R}^n$ and $\sigma > \sigma_0$. Assume further that $P(\eta + \zeta - i\sigma N) \neq 0$ for some $\sigma > \sigma_0$, all real η and all ζ in a ball $Z \subset \mathbf{C}^n$ with center in 0, and let E denote the fundamental solution of $P(-i\partial)$ according to Proposition 2.4.13. Then*

$$\operatorname{sing\,supp} E \subset \bigcup_{Q \in L(P)} \Lambda(Q)^\perp.$$

Proof Let us consider the shifted polynomial $P_1(\eta) = P(\eta - i\sigma N)$ and take $Q \in L(P_1)$. If ξ_k is a sequence with $|\xi_k| \to \infty$ such that

$$Q(\eta) = \lim_{k \to \infty} \frac{P_1(\eta - \xi_k)}{\|\tau_{\xi_k} P_1\|},$$

then all the coefficients of the polynomial $\eta \mapsto P_1(\eta - \xi_k)$ converge when divided by the factor $\|\tau_{\xi_k} P_1\|$. Hence it follows that

$$Q(\eta + i\sigma N) = \lim_{k \to \infty} \frac{P(\eta - \xi_k)}{\|\tau_{\xi_k} P_1\|}$$

holds in the Banach space of polynomials of degree at most $\deg P$ considered in Definition 4.2.8 (1). Moreover, the sequences of positive numbers $\|\tau_{\xi_k} P_1\| / \|\tau_{\xi_k} P\|$ and $\|\tau_{\xi_k} P\| / \|\tau_{\xi_k} P_1\|$ are bounded due to Taylor's formula (see Hörmander [138], Eq. (10.1.8), p. 5), and hence we conclude, passing to an appropriate subsequence of ξ_k, that a positive multiple of $Q(\eta + i\sigma N)$ belongs to $L(P)$. Since $\Lambda(Q(\eta + i\sigma N)) = \Lambda(Q)$, we therefore obtain

$$\bigcup_{Q \in L(P)} \Lambda(Q)^\perp = \bigcup_{Q \in L(P_1)} \Lambda(Q)^\perp.$$

Furthermore, the fundamental solution of $P_1(-i\partial)$ according to Proposition 2.4.13 is given by $e^{-\sigma x N} E$, which has the same singular support as E. Therefore, instead of P, we may consider the shifted polynomial P_1, and hence we can assume from the outset that $\sigma_0 = 0$. But in this case, E coincides with the fundamental solution constructed in Hörmander [139], Eq. (7.3.22), p. 190, since, with the notations adopted there, we have

$$\int_Z \frac{\hat{\phi}(-\xi - \zeta)}{P(\xi + \zeta)} \, \Phi(P(\xi + \zeta), \zeta) \, d\lambda(\zeta) = \frac{\hat{\phi}(-\xi)}{P(\xi)}$$

in virtue of Hörmander [139], Eq. (7.3.19), p. 189, and of the assumption $P(\xi + \zeta) \neq 0$ for $\xi \in \mathbf{R}^n$, $\zeta \in Z$. Note that Φ can be chosen so as to have its support contained in the ball Z, cf. Hörmander [139], Lemma 7.3.12, (ii), p. 190. Finally, the proof is completed by invoking Hörmander [138], Thm. 10.2.11, p. 23. $\square$

Let us note that the condition

$$\exists \sigma > \sigma_0 : \forall \zeta \in Z : \forall \eta \in \mathbf{R}^n : P(\eta + \zeta - i\sigma N) \neq 0, \qquad (4.2.12)$$

which is assumed in Proposition 4.2.9, is always valid if $P(-i\partial)$ is a *hyperbolic* operator (in the direction N), cf. Hörmander [138], Thm. 12.4.4, p. 114, but fails to be true for every quasihyperbolic operator. An example for this is provided by the Schrödinger operator $P(-i\partial) = \partial_t - i\Delta_n, N = (1, 0, \dots, 0)$.

Example 4.2.10   Let us apply now Propositions 4.2.3 and 4.2.9 in order to determine the singular support of the fundamental solution F of *Rayleigh's operator*

$$R(-i\partial) = P(\partial) = (\alpha^2\partial_t^2 - \partial_x^2)\partial_x^2 - \gamma^2\partial_t^2, \qquad \alpha, \gamma > 0,$$

see Examples 2.4.14, 4.1.9. Of course, sing supp F could also be read off from the explicit formula in (4.1.14).

(a) If we set $N = (1, 0)$ and $R(\tau, \eta) = (\alpha^2\tau^2 - \eta^2)\eta^2 + \gamma^2\tau^2$, then $R_{(1,0)}(\tau, \eta) = \alpha^2\eta^2 + \gamma^2$ and $R_{(1,\pm\alpha)}(\tau, \eta) = 2\alpha^3(\alpha\tau \mp \eta)$. Hence, with notations as in (4.2.5),

$$\text{sing supp } F \supset \text{supp } E(-\alpha^2\partial_x^2 + \gamma^2, N) \cup \text{supp } E(\alpha\partial_t \mp \partial_x, N)$$

$$= \{(t, x) \in \mathbf{R}^2; t = 0 \text{ or } t = \alpha|x|\}$$

by Proposition 4.2.3.

(b) Let us first note that the condition (4.2.12) is satisfied, i.e.,

$$R(\tau + \zeta_1 - i\sigma, \eta + \zeta_2) \neq 0 \quad \text{for} \quad (\tau, \eta) \in \mathbf{R}^2, \ \zeta \in \mathbf{C}^2, \ |\zeta| < C, \ \sigma > \sigma_0$$

and appropriate positive constants C, σ_0. This is a consequence of the reasoning in Example 4.1.9 (a) if we set $p_1 = i(\tau + \zeta_1 - i\sigma)$, $p_2 = i(\eta + \zeta_2)$.

Let us next determine $L(R)$. If $Q \in L(R)$, i.e., if $Q = \lim \tau_\xi R / \|\tau_\xi R\|$ for $\xi \in \mathbf{R}^2$ with $|\xi| \to \infty$, then we can assume that $\xi/|\xi|$ converges to $\omega = (a, b) \in S^1$. If $\alpha^2 a^2 \neq b^2$ and $b \neq 0$, i.e., if the principal part of R does not vanish in ω, then Q is one of the constants ± 1 and $\Lambda(Q)^\perp = \{0\}$.

Furthermore, if $\alpha a = \pm b$, then $\|\tau_\xi R\|$ grows at least as $|\xi|^3$ and Q belongs to the vector space of polynomials spanned by 1 and R_ω. Hence

$$\Lambda(Q)^\perp \subset \mathbf{R} \cdot (\pm\alpha, 1) = \{(t, x) \in \mathbf{R}^2; t = \pm\alpha x\}.$$

Finally, if $b = 0$ and ξ_2 remains bounded, then $\|\tau_\xi R\|$ grows as a multiple of ξ_1^2 and

$$Q = C_1(\alpha^2(\eta - C_2)^2 + \gamma^2), \qquad C_1 > 0, \ C_2 \in \mathbf{R}.$$

On the other hand, for $|\xi_2| \to \infty$ and $\xi_2/\xi_1 \to 0$, we have $Q = 1$.

Therefore, by Proposition 4.2.9,

$$\text{sing supp } F \subset \{(t, x) \in \mathbf{R}^2; t = 0 \text{ or } t = \pm\alpha x\}.$$

Combining this estimate with the one in (a) we obtain the following precise description of the singular support of F :

$$\text{sing supp } F = \{(t, x) \in \mathbf{R}^2; \ t = 0 \text{ or } t = \alpha |x|\}.$$

Hence the singular support of F does not depend on the lower order term $-\gamma^2 \partial_t^2$.

□

Example 4.2.11 Similarly, let us yet determine from Propositions 4.2.3 and 4.2.9 the singular support of the fundamental solution E_ϵ of the *thermoelastic operator* $P_\epsilon(\partial)$ in (4.1.21).

First we note that condition (4.2.12) is satisfied, i.e., if $R(-i\partial) = P_\epsilon(\partial)$ with

$$R(\tau, \eta) = -i\tau^3 - \tau^2 |\eta|^2 + (1 + \epsilon)i\tau |\eta|^2 + |\eta|^4, \qquad (\tau, \eta) \in \mathbf{R}^4,$$

then

$$R(\tau + u_0 - i\sigma, \eta + u) \neq 0 \quad \text{for} \quad (\tau, \eta) \in \mathbf{R}^4, \ (u_0, u) \in \mathbf{C}^4, \ |(u_0, u)| < C, \ \sigma > \sigma_0.$$

This follows from the reasoning in Example 4.1.11 (b) if we set $\zeta = \sigma + i(\tau + u_0)$, $z = -\sum_{j=1}^3 (\eta_j + u_j)^2$, $\vartheta' = -\text{Im } u$.

Similarly as in Example 4.2.10, the set $L(R)$ is determined by the localizations R_ω, where $\omega = (\omega_0, \omega') \in \mathbf{S}^3$ fulfills $|\omega_0| = |\omega'|$. In fact, if $Q = \lim \tau_\xi R / \|\tau_\xi R\|$ for $\xi \in \mathbf{R}^4$ with $|\xi| \to \infty$ and $\xi/|\xi| \to \omega$, then Q is constant if $|\omega_0| \neq |\omega'|$ and else contained in the vector space of polynomials spanned by 1 and R_ω. Hence Propositions 4.2.3 and 4.2.9 imply that

$$\text{sing supp } E_\epsilon = \{(t, x) \in \mathbf{R}^4; \ t = |x|\}.$$

□

4.3 Singularities of Fundamental Matrices of Hyperbolic Systems

If one tries to transfer the singularity theory for hyperbolic scalar operators in Sect. 4.2 to systems, one encounters the fact that some singularities of the forward fundamental solution $E(P(-i\partial), N)$ of the determinant operator $P(-i\partial) = \det A(-i\partial)$ can disappear in the fundamental matrix $E(A(-i\partial), N)$ of $A(-i\partial)$.

Let us illustrate this phenomenon by the example of the simple diagonal system

$$A(-i\partial) = \begin{pmatrix} \partial_t^2 - \Delta_3 & 0 \\ 0 & \partial_t^2 - a\Delta_2 - b\partial_3^2 \end{pmatrix}, \quad 0 < a < 1, b > 1. \qquad (4.3.1)$$

For $N = (1, 0, 0, 0)$, we have

$$E_A = E(A(-i\partial), N) = \frac{1}{4\pi t} \begin{pmatrix} \delta(t - |x|) & 0 \\ 0 & \frac{1}{a\sqrt{b}} \delta\left(t - \sqrt{\frac{1}{a}(x_1^2 + x_2^2) + \frac{1}{b}x_3^2}\right) \end{pmatrix}.$$

Hence

$$\text{sing supp}\, E_A = \left\{ (t, x) \in \mathbf{R}^4;\ t = |x| \text{ or } t = \sqrt{\tfrac{1}{a}(x_1^2 + x_2^2) + \tfrac{1}{b}x_3^2} \right\},$$

and this set coincides with the part $\Xi^* \cap H_N$ of the characteristic surface Ξ^* already considered in Example 4.2.7.

On the other hand, we have constructed the forward fundamental solution $E_P = E(P(-i\partial), N)$ of

$$P(-i\partial) = \det A(-i\partial) = (\partial_t^2 - \Delta_3)(\partial_t^2 - a\Delta_2 - b\partial_3^2)$$

in Example 4.2.7, and we have noted that its singular support consists of $\text{sing supp}\, E_A$ and, additionally, of the two frusta C of conical refraction, i.e., $\text{sing supp}\, E_P = (\Xi^* \cap H_N) \cup C$, where

$$C = \{(t, x) = (|\xi|, \lambda\xi_1 + (1 - \lambda)a\xi_1, \lambda\xi_2 + (1 - \lambda)a\xi_2, \lambda\xi_3 + (1 - \lambda)b\xi_3);$$

$$\xi \in \mathbf{R}^3, 0 \leq \lambda \leq 1, \xi_3^2(b - 1) = (\xi_1^2 + \xi_2^2)(1 - a)\}.$$

A physically more relevant example of this phenomenon of extinction of singularities will be observed in Example 4.3.10 below for the system of elastic waves in hexagonal media.

Similarly to Definition 4.2.1, let us now define *localizations* of $l \times l$ matrices $A(\eta)$ of polynomials.

Definition 4.3.1 For an $l \times l$ matrix $A(\eta)$ of polynomials in $\mathbf{R}^n$ and $\xi \in \mathbf{R}^n$, the *localization* $A_\xi(\eta)$ is the lowest non-vanishing coefficient with respect to t in the MacLaurin series of $t^m A(\eta + \frac{\xi}{t})$, $m = \deg A \geq 0$. Thus $t^m A(\eta + \frac{\xi}{t}) = t^p A_\xi(\eta) + O(t^{p+1})$ for $t \to 0$, and $p = m_\xi(A)$ is called the *multiplicity* of ξ with respect to A.

In the next lemma, we establish a connection between the localizations of A and of its determinant P.

Lemma 4.3.2 *Let A be a square matrix of polynomials on $\mathbf{R}^n$, $P = \det A$, and $\xi \in \mathbf{R}^n$. Then the following holds:*

(i) *If $\det A_\xi$ does not vanish identically, then $P_\xi = \det A_\xi$;*
(ii) *if $(A_\xi)^{\text{ad}}$ does not vanish identically, then $(A^{\text{ad}})_\xi = (A_\xi)^{\text{ad}}$.*

Proof Let $p = m_\xi(A)$ and A be of size $l \times l$ and of degree m. Then

$$t^{m-p}A(\eta + \tfrac{\xi}{t}) = A_\xi(\eta) + O(t) \quad \text{for } t \to 0,$$

and hence

$$t^{l(m-p)}P(\eta + \tfrac{\xi}{t}) = \det(t^{m-p}A(\eta + \tfrac{\xi}{t})) = \det A_\xi(\eta) + O(t) \quad \text{for } t \to 0.$$

This shows that $P_\xi = \det A_\xi$ if $\det A_\xi$ does not vanish identically.

The second assertion follows analogously from

$$t^{(l-1)(m-p)}A^{\mathrm{ad}}(\eta + \tfrac{\xi}{t}) = \left(t^{m-p}A(\eta + \tfrac{\xi}{t})\right)^{\mathrm{ad}} = A_\xi(\eta)^{\mathrm{ad}} + O(t) \quad \text{for } t \to 0.$$

$\square$

Example 4.3.3 Let us investigate the localizations of the matrix

$$A(\eta) = \begin{pmatrix} \eta_0 & \eta_1 \\ \eta_1 & \eta_0 \end{pmatrix} \quad \text{with } \det A = P = \eta_0^2 - \eta_1^2.$$

If $\xi_0^2 \neq \xi_1^2$, then $A_\xi = A(\xi) \in \mathbf{R}^{2\times 2}$ and $P_\xi = P(\xi) = \det A_\xi$ in accordance with Lemma 4.3.2. Similarly, for $\xi = 0$, $A_\xi = A$ and $P_\xi = P = \det A_\xi$.

If, however, $\xi \neq 0$ and $\xi_1 = \pm\xi_0$, then $A_\xi = A(\xi) \in \mathbf{R}^{2\times 2}$ fulfills $\det A_\xi = 0$ whereas $P_\xi = 2(\xi_0\eta_0 - \xi_1\eta_1)$ is a first-order polynomial. Note that this case is excluded in Lemma 4.3.2. $\square$

In the following proposition, we shall give an estimate of $\operatorname{sing\,supp} E(A(-i\partial), N)$ from below for hyperbolic systems $A(-i\partial)$. This "Localization Theorem" generalizes Atiyah, Bott, and Gårding [5], 4.10, p. 144 (see Proposition 4.2.3) from the scalar case to the matrix case. It was stated first without proof in Esser [70], p. 191, last line, and formulated and proved in Ortner and Wagner [221], Prop. 1, p. 1243.

Proposition 4.3.4 *Let A be an $l \times l$ matrix of polynomials on $\mathbf{R}^n$ such that $A(-i\partial)$ is hyperbolic in the direction $N \in \mathbf{R}^n \setminus \{0\}$ and set $P = \det A$. Then, for $1 \leq j, k \leq l$,*

$$\bigcup_{\xi \in \mathbf{R}^n \setminus \{0\}} \operatorname{supp}\left[(A_{jk}^{\mathrm{ad}})_\xi(-i\partial)E(P_\xi(-i\partial), N)\right] \subset \operatorname{sing\,supp} E(A(-i\partial), N)_{jk}.$$

Proof Let $1 \leq j, k \leq l$, and $q, r \in \mathbf{N}_0$ such that

$$t^q A_{jk}^{\mathrm{ad}}(\eta + \tfrac{\xi}{t}) = (A_{jk}^{\mathrm{ad}})_\xi(\eta) + O(t) \quad \text{and} \quad t^r P(\eta + \tfrac{\xi}{t}) = P_\xi(\eta) + O(t) \quad \text{for } t \to 0.$$

According to Proposition 2.4.13, $E(A(-i\partial),N)_{jk} = A_{jk}^{\mathrm{ad}}(-i\partial)E(P(-i\partial),N)$. Furthermore, the operators $P(-i\partial + \frac{\xi}{t})$ are hyperbolic, and, due to

$$P(-i\partial + \tfrac{\xi}{t})e^{-i\xi x/t}E(P(-i\partial),N) = e^{-i\xi x/t}P(-i\partial)E(P(-i\partial),N) = \delta,$$

the uniqueness of the fundamental solution of $P(-i\partial + \frac{\xi}{t})$ with support in H_N implies

$$E(P(-i\partial + \tfrac{\xi}{t}),N) = e^{-i\xi x/t}E(P(-i\partial),N)$$

$$\text{and } E(t^r P(-i\partial + \tfrac{\xi}{t}),N) = t^{-r}e^{-i\xi x/t}E(P(-i\partial),N).$$

This furnishes the following limit relation in $\mathcal{D}'(\mathbf{R}^n)$:

$$\lim_{t\to 0} t^{q-r}e^{-i\xi x/t}E(A(-i\partial),N)_{jk} = \lim_{t\to 0} t^{q-r}e^{-i\xi x/t}A_{jk}^{\mathrm{ad}}(-i\partial)E(P(-i\partial),N)$$

$$= \lim_{t\to 0} t^q A_{jk}^{\mathrm{ad}}(-i\partial + \tfrac{\xi}{t})\big[t^{-r}e^{-i\xi x/t}E(P(-i\partial),N)\big]$$

$$= \lim_{t\to 0} t^q A_{jk}^{\mathrm{ad}}(-i\partial + \tfrac{\xi}{t})E(t^r P(-i\partial + \tfrac{\xi}{t}),N)$$

$$= (A_{jk}^{\mathrm{ad}})_\xi(-i\partial)E(P_\xi(-i\partial),N). \qquad (4.3.2)$$

(For Eq. (4.3.2), we used that

$$\lim_{t\to 0} E(t^r P(-i\partial + \tfrac{\xi}{t}),N) = \lim_{t\to 0} e^{\sigma Nx}\mathcal{F}_\eta^{-1}\big(t^{-r}P(\eta + \tfrac{\xi}{t} - i\sigma N)^{-1}\big)$$

$$= e^{\sigma Nx}\mathcal{F}_\eta^{-1}\big(P_\xi(\eta - i\sigma N)^{-1}\big) = E(P_\xi(-i\partial),N), \quad \sigma > \sigma_0,$$

holds in $\mathcal{D}'(\mathbf{R}^n)$ as a consequence of (2.4.13), the uniform estimate (with respect to t) of $|t^{-r}P(\eta + \frac{\xi}{t} - i\sigma N)^{-1}|$ by a polynomial in η (see Atiyah, Bott, and Gårding [5], Lemma 3.51, p. 137), and Lebesgue's theorem on dominated convergence.)

Finally, we observe that

$$\lim_{t\to 0} t^{q-r}e^{-i\xi x/t}E(A(-i\partial),N)_{jk} = 0$$

holds in $\mathcal{D}'(U)$ if $U = \mathbf{R}^n \setminus \operatorname{sing\ supp} E(A(-i\partial),N)_{jk}$. (In fact, $E(A(-i\partial),N)_{jk}\big|_U$ is represented by a function $f \in \mathcal{C}^\infty(U)$, and $\phi \cdot f \in \mathcal{D}(\mathbf{R}^n)$ for $\phi \in \mathcal{D}(U)$ implies

$$\lim_{t\to 0} \langle \phi, t^{q-r}e^{-i\xi x/t}f(x)\rangle = \lim_{t\to 0} t^{q-r}\mathcal{F}(\phi \cdot f)\big(\tfrac{\xi}{t}\big) = 0$$

due to $\mathcal{F}(\phi \cdot f) \in \mathcal{S}$.) Hence, by (4.3.2), the distribution $(A_{jk}^{\mathrm{ad}})_\xi(-i\partial)E(P_\xi(-i\partial),N)$ vanishes on U. This completes the proof. $\square$

Clearly, the singular support of the fundamental matrix $E(A(-i\partial),N)$ is bounded by that of the fundamental solution $E(P(-i\partial),N)$ of its determinant $P = \det A$. We formulate this fact in the following proposition.

Proposition 4.3.5 *Let A be an l × l matrix of polynomials on* $\mathbf{R}^n$ *such that* $A(-i\partial)$
is hyperbolic in the direction $N \in \mathbf{R}^n \setminus \{0\}$ *and set* $P = \det A$. *Then*

$$\operatorname{sing\,supp} E(A(-i\partial), N) = \bigcup_{j,k=1}^{l} \operatorname{sing\,supp} E(A(-i\partial), N)_{jk} \subset \operatorname{sing\,supp} E(P(-i\partial), N).$$

Proof Due to Proposition 2.4.13, we have

$$E(A(-i\partial), N)_{jk} = A_{jk}^{\mathrm{ad}}(-i\partial)E(P(-i\partial), N),$$

and hence, obviously,

$$\operatorname{sing\,supp} E(A(-i\partial), N)_{jk} \subset \operatorname{sing\,supp} E(P(-i\partial), N)$$

for $1 \le j, k \le l$.   □

For *strictly hyperbolic polynomials* P, the following corollary to Propositions 4.3.4, 4.3.5 shows that the singular supports of $E(A(-i\partial), N)$ and of
$E(P(-i\partial), N)$ coincide, at least if P and A^{ad} are homogeneous. Note the contrast to
the introductory example in Sect. 4.3 where P was *not strictly* hyperbolic.

Corollary 4.3.6 *Let* $A(-i\partial)$ *be an* $l \times l$ *matrix of differential operators such that
each entry of* A^{ad} *is homogeneous and that* $P(-i\partial) = \det A(-i\partial)$ *is homogeneous
and strictly hyperbolic in the direction* $N \in \mathbf{R}^n \setminus \{0\}$, *see Definition 4.2.6. Then*

$$\operatorname{sing\,supp} E(A(-i\partial), N) = \operatorname{sing\,supp} E(P(-i\partial), N).$$

Proof

(a) By Proposition 4.3.5, $\operatorname{sing\,supp} E(A(-i\partial), N)$ is a subset of sing supp
$E(P(-i\partial), N)$. On the other hand, by the strict hyperbolicity of $P(-i\partial)$, we
have

$$\operatorname{sing\,supp} E(P(-i\partial), N) = \bigcup_{\substack{\xi \in \mathbf{R}^n \setminus \{0\} \\ P(\xi)=0}} \operatorname{supp} E(P_\xi(-i\partial), N) = \Xi^* \cap H_N,$$

see (4.2.5). Proposition 4.3.4 implies that

$$M = \bigcup_{\substack{1 \le j,k \le l \\ \xi \in \mathbf{R}^n \setminus \{0\} P(\xi)=0}} \operatorname{supp}\big[(A_{jk}^{\mathrm{ad}})_\xi(-i\partial)E(P_\xi(-i\partial), N)\big]$$

is contained in $\operatorname{sing\,supp} E(A(-i\partial), N)$. Therefore, the assertion in Corollary 4.3.6
is a consequence of $M = \Xi^* \cap H_N$, and this follows if we show that, for each
$\xi \in \mathbf{R}^n \setminus \{0\}$ with $P(\xi) = 0$, there exist $1 \le j, k \le l$ such that $(A_{jk}^{\mathrm{ad}})_\xi$ is a non-
vanishing constant.

(b) Let us assume to the contrary that $\xi \in \mathbf{R}^n \setminus \{0\}$ with $P(\xi) = 0$ and A_{jk}^{ad} either vanishes or $(A_{jk}^{\mathrm{ad}})_\xi$ is of degree ≥ 1 for each $1 \leq j, k \leq l$. Then, in particular, $A^{\mathrm{ad}}(\xi) = 0$. Since P is strictly hyperbolic and homogeneous, there exists $1 \leq j \leq n$ such that $\partial_j P(\xi) \neq 0$. But this yields a contradiction due to

$$I_l \cdot (\partial_j P)(\xi) = \partial_j (A \cdot A^{\mathrm{ad}})(\xi) = (\partial_j A)(\xi) \cdot A^{\mathrm{ad}}(\xi) + A(\xi) \cdot (\partial_j A^{\mathrm{ad}})(\xi)$$

$$= A(\xi)(\partial_j A^{\mathrm{ad}})(\xi)$$

and $0 \neq (\partial_j P(\xi))^l = \det A(\xi) \cdot \det((\partial_j A^{\mathrm{ad}})(\xi)) = 0.$ □

Example 4.3.7 Let us illustrate our concepts by considering the 2×2 system leading to *Timoshenko's beam operator,* which was treated in Examples 3.5.4 and 4.1.6.

According to Boley and Chao [19], p. 579, Graff [114], pp. 181–183, in particular Eqs. (3.4.11/3.4.12), and Timoshenko and Young [269], pp. 330, 331, the transverse vibrations of a homogeneous bar can be described by the system

$$\begin{pmatrix} \rho A \partial_t^2 - \kappa A G \partial_x^2 & \kappa A G \partial_x \\ \kappa A G \partial_x & -\rho I \partial_t^2 + E I \partial_x^2 - \kappa A G \end{pmatrix} \begin{pmatrix} u \\ \psi \end{pmatrix} = \begin{pmatrix} q \\ 0 \end{pmatrix} \tag{4.3.3}$$

where $u(t,x)$ denotes the displacement of the bar at the coordinate x and at time t, and $\psi(t,x)$ is the slope of the deflection curve diminished by the angle of shear at the neutral axis. The parameters $A, I, \rho, E, G,$ and κ stand for the cross-section area, the moment of inertia, the mass density, Young's modulus, the shear modulus, and Timoshenko's shear coefficient, respectively.

With the abbreviations

$$a = \sqrt{\frac{\rho}{E}}, \qquad b = \sqrt{\frac{\rho}{\kappa G}}, \qquad c = \sqrt{\frac{\rho A}{EI}},$$

the system (4.3.3) takes the form

$$\begin{pmatrix} b^2 \partial_t^2 - \partial_x^2 & -\frac{c^2}{b^2} \partial_x \\ \partial_x & a^2 \partial_t^2 - \partial_x^2 + \frac{c^2}{b^2} \end{pmatrix} \begin{pmatrix} \kappa A G u \\ -EI\psi \end{pmatrix} = \begin{pmatrix} q \\ 0 \end{pmatrix}.$$

Hence, if we set

$$A(-i\partial_t, -i\partial_x) = \begin{pmatrix} b^2 \partial_t^2 - \partial_x^2 & -\frac{c^2}{b^2} \partial_x \\ \partial_x & a^2 \partial_t^2 - \partial_x^2 + \frac{c^2}{b^2} \end{pmatrix},$$

then $P(-i\partial) = \det A(-i\partial)$ coincides with the Timoshenko beam operator

$$T(\partial) = (a^2 \partial_t^2 - \partial_x^2)(b^2 \partial_t^2 - \partial_x^2) + c^2 \partial_t^2.$$

Its forward fundamental solution was derived in Example 4.1.6, see Eqs. (4.1.7/4.1.8).

For $a \neq b$, i.e., $E \neq \kappa G$, $P(-i\partial) = T(\partial)$ is strictly hyperbolic. Note, however, that P is not homogeneous and thus Corollary 4.3.6 cannot be applied as it stands. But still the method in the proof of Corollary 4.3.6 works almost without change: We just have to show that $(A^{\mathrm{ad}}_{jk})_{(\tau,\xi)}$ is a non-vanishing constant for $(\tau,\xi) \in \mathbf{R}^2 \setminus \{0\}$ satisfying $P_4(\tau,\xi) = 0$ and suitable j, k. This is obvious since $P_4(\tau,\xi) = 0$ implies $\xi = \pm a\tau$ or $\xi = \pm b\tau$ and thus $(A^{\mathrm{ad}}_{21})_{(\tau,\xi)} = -i\xi \neq 0$ due to $a > 0, b > 0$. Hence we obtain from (4.2.5) that

$$\mathrm{sing\ supp}\, E(A(-i\partial), N) = \mathrm{sing\ supp}\, E(P(-i\partial), N)$$

$$= \{(t, x) \in \mathbf{R}^2; t = a|x| \text{ or } t = b|x|\},$$

where $N = (1, 0)$.   □

Example 4.3.8 Let us determine similarly as in Example 4.3.7 the singular support of the response caused by an instantaneous point load exciting *transverse vibrations in isotropic plates* according to the theory of Ya.S. Uflyand and R.D. Mindlin.

The three "displacement components" u, α_x, α_y obey the system

$$A(-i\partial)(u, \alpha_x, \alpha_y)^T = (q, 0, 0)^T \tag{4.3.4}$$

of linear partial differential equations, where

$$A(-i\partial) = \begin{pmatrix} h\rho\partial_t^2 - a\Delta_2 & a\partial_x & a\partial_y \\ -a\partial_x & \rho J\partial_t^2 - D\partial_x^2 - D\frac{1-v}{2}\partial_y^2 + a & -D\frac{1+v}{2}\partial_x\partial_y \\ -a\partial_y & -D\frac{1+v}{2}\partial_x\partial_y & \rho J\partial_t^2 - D\partial_y^2 - D\frac{1-v}{2}\partial_x^2 + a \end{pmatrix}$$

and ρ, h, D, v, q denote mass density, thickness, flexural rigidity, Poisson's ratio, and the transverse load, respectively, see Uflyand [278], (2.5), p. 291, Mindlin [180], Eq. (16), p. 33. Furthermore, we have $J = h^3/12$, $D = Eh^3/(12(1 - v^2))$ and we have set $a = \frac{2}{3}\mu h$, where μ characterizes some elastic property of the plate.

As above, we define $P(-i\partial) = \det A(-i\partial)$ and $N = (1, 0, 0)$ and obtain for the solution u of (4.3.4) with respect to $q = \delta(t, x, y)$ the following:

$$U = \begin{pmatrix} u \\ \alpha_x \\ \alpha_y \end{pmatrix} = E(A(-i\partial), N) * \begin{pmatrix} \delta \\ 0 \\ 0 \end{pmatrix} = A^{\mathrm{ad}}(-i\partial)E(P(-i\partial), N) * \begin{pmatrix} \delta \\ 0 \\ 0 \end{pmatrix}$$

$$= \begin{pmatrix} A^{\mathrm{ad}}(-i\partial)_{11} \\ A^{\mathrm{ad}}(-i\partial)_{21} \\ A^{\mathrm{ad}}(-i\partial)_{31} \end{pmatrix} E(P(-i\partial), N). \tag{4.3.5}$$

An algebraic calculation yields

$$A^{\mathrm{ad}}(-i\partial)_{11} = (\rho J\partial_t^2 - D\Delta_2 + a)W(\partial), \quad W(\partial) = \rho J\partial_t^2 - D\frac{1-\nu}{2}\Delta_2 + a$$

$$A^{\mathrm{ad}}(-i\partial)_{21} = a\partial_x W(\partial), \quad A^{\mathrm{ad}}(-i\partial)_{31} = a\partial_y W(\partial),$$

and hence

$$
\begin{aligned}
P(-i\partial) &= \det A(-i\partial) = (h\rho\partial_t^2 - a\Delta_2)A^{\mathrm{ad}}(-i\partial)_{11} + a\partial_x A^{\mathrm{ad}}(-i\partial)_{21} + a\partial_y A^{\mathrm{ad}}(-i\partial)_{31} \\
&= W(\partial)\big[(h\rho\partial_t^2 - a\Delta_2)(\rho J\partial_t^2 - D\Delta_2 + a) + a^2\Delta_2\big] \\
&= aW(\partial)\big[(\Delta_2 - \tfrac{h\rho}{a}\partial_t^2)(D\Delta_2 - \rho J\partial_t^2) + \rho h\partial_t^2\big].
\end{aligned}
\tag{4.3.6}
$$

Note that the limit case for $J \to 0, a \to \infty$ of the second factor $M(\partial)$ in (4.3.6) coincides with the Lagrange–Germain plate operator $\rho h\partial_t^2 + D\Delta_2^2$. For the operator $M(\partial)$ in the general case, see Uflyand [278] (2.7), p. 291; Mindlin [180], Eq. (37), p. 36; [181], p. 320, and also compare the remark at the end of Example 3.5.4.

If we insert the expression for $P(-i\partial)$ in (4.3.6) into the representation of U in (4.3.5), we conclude that

$$
U = \begin{pmatrix} u \\ \alpha_x \\ \alpha_y \end{pmatrix} = \begin{pmatrix} \frac{\rho J}{a}\partial_t^2 - \frac{D}{a}\Delta_2 + 1 \\ \partial_x \\ \partial_y \end{pmatrix} E(M(\partial), N).
\tag{4.3.7}
$$

Employing (4.2.5), we then infer from (4.3.7) that

$$
\operatorname{sing\,supp} U = \operatorname{sing\,supp} E(M(\partial), N) = \Big\{ (t,x) \in \mathbf{R}^3;\ t = \sqrt{\frac{\rho J}{D}}|x| \text{ or } t = \sqrt{\frac{h\rho}{a}}|x| \Big\}
$$

if $M(\partial)$ is strictly hyperbolic, i.e., if the velocities $v_1 = \sqrt{\dfrac{D}{\rho J}} = \sqrt{\dfrac{E}{\rho(1-\nu^2)}}$ and

$v_2 = \sqrt{\dfrac{a}{h\rho}} = \sqrt{\dfrac{2\mu}{3\rho}}$ are different (cf. Uflyand [278], (2.7), p. 291). $\square$

Example 4.3.9 Let us illustrate now the application of Propositions 4.3.4, 4.3.5 by determining the singular support of the fundamental matrix of the system of *elastic waves in cubic media*. A treatment by this method was given for the first time in Ortner and Wagner [221], Section 4.3, p. 1256.

This system of differential operators reads as $A(-i\partial) = -I_3\partial_t^2 + B(\nabla)$ where

$$
B(\xi) = c|\xi|^2 I_3 + b\xi\cdot\xi^T - (b-a)\begin{pmatrix} \xi_1^2 & 0 & 0 \\ 0 & \xi_2^2 & 0 \\ 0 & 0 & \xi_3^2 \end{pmatrix},
$$

see Example 2.1.4 (c). Hence $A(-i\partial)$ consists of second-order homogeneous operators.

According to Ortner and Wagner [221], Prop. 2, p. 1253, the system $A(-i\partial)$ is hyperbolic in the direction $N = (1,0,0,0)$ if and only if

$$c \geq 0, \quad a + c \geq 0, \text{ and } -\tfrac{1}{2}(a + 3c) \leq b \leq a + 2c. \tag{4.3.8}$$

In the following, we shall assume that the inequalities in (4.3.8) are satisfied.

(a) Let us determine first sing supp $E(P(-i\partial), N)$ where $P(\tau, \xi) = \det A(\tau, \xi)$ and $N = (1,0,0,0)$.

The slowness surface $\Xi = \{(\tau, \xi) \in \mathbf{R}^4; P(\tau, \xi) = 0\}$ is given geometrically by the vectors (τ, ξ) for which $\det(\tau^2 I_3 - B(\xi)) = 0$, i.e., where τ^2 is an eigenvalue of the matrix $B(\xi)$. Clearly, if $B(\xi)$ has three different non-zero eigenvalues, then (τ, ξ) is a non-singular point on Ξ. Therefore, singular points $(\tau, \xi) \in \Xi$ arise if two eigenvalues coincide, i.e., if $A(\tau, \xi)$ is of rank at most one, or if $\tau = 0$. In the following, we suppose that $b \neq 0$, $c > 0$ and $a^2 \neq b^2$.

If $A(\tau, \xi)$ has rank one, then all the rows of the matrix $A(\tau, \xi)$ must be proportional, and this yields the following fourteen cases:

$$(\alpha) \; \xi_i = \xi_j = 0, \; \xi_k = \pm\frac{\tau}{\sqrt{c}}, \qquad \{i, j, k\} = \{1, 2, 3\},$$

$$(\beta) \; \xi_1^2 = \xi_2^2 = \xi_3^2 = \frac{\tau^2}{a + 3c - b}.$$

In order to determine sing supp $E(P(-i\partial), N)$, we use the inclusion relations in (4.2.5), which in this case will be seen to be equalities. In fact, we will show that

$$\operatorname{supp} E(P_{(\tau, \xi)}(-i\partial), N) = K(P_{(\tau, \xi)}(-i\partial), N)$$

holds for all localizations $P_{(\tau, \xi)}(-i\partial)$, and hence

$$\operatorname{sing supp} E(P(-i\partial), N) = \bigcup_{(\tau, \xi) \in \mathbf{R}^4 \setminus \{0\}} \operatorname{supp} E(P_{(\tau, \xi)}(-i\partial), N). \tag{4.3.9}$$

If (τ, ξ) is a non-singular point on Ξ, then, as always, supp $E(P_{(\tau, \xi)}(-i\partial), N)$ is a half-ray in the subspace $\mathbf{R} \cdot (\partial_\tau P, \nabla P)(\tau, \xi)$ of the characteristic surface Ξ^*.

In order to calculate the localizations in the singular points of P (see $(\alpha), (\beta)$ above), we use the explicit formula for P according to (2.1.11):

$$P(\tau, \xi) = \prod_{j=1}^{3}(\tau^2 - c|\xi|^2 + (b - a)\xi_j^2) - b\sum_{j=1}^{3}\xi_j^2\prod_{k \neq j}(\tau^2 - c|\xi|^2 + (b - a)\xi_k^2).$$

Now, if (τ, ξ) is one of the singular points of type (α), i.e., if $(\tau, \xi) = (1, \frac{1}{\sqrt{c}}, 0, 0)$, say, then

$$P_{(\tau,\xi)}(\eta_0, \eta_1, \eta_2, \eta_3) = -\frac{4a}{c}(\eta_0 - \sqrt{c}\,\eta_1)^2.$$

Hence supp $E(P_{(\tau,\xi)}(-i\partial), N) = K(P_{(\tau,\xi)}(-i\partial), N)$ yields a half-ray on Ξ^* as in the non-singular case.

Let us finally investigate the singular points of type (β), i.e., let us assume that $(\tau, \xi) = (1, \alpha, \alpha, \alpha)$, say, where $\alpha = 1/\sqrt{a + 3c - b}$. Due to

$$A(\eta) = A(\eta_0, \eta') = (\eta_0^2 - c|\eta'|^2)I_3 - b\eta' \cdot \eta'^T + (b - a)\begin{pmatrix} \eta_1^2 & 0 & 0 \\ 0 & \eta_2^2 & 0 \\ 0 & 0 & \eta_3^2 \end{pmatrix},$$

we obtain

$$A(\eta + s^{-1}(\tau, \xi)) = s^{-2}A(\tau, \xi) + s^{-1}D(\eta) + A(\eta).$$

Here $A(\tau, \xi) = -b\alpha^2 e \cdot e^T$, $e = (1, 1, 1)^T$, is a rank-one matrix. Furthermore $D = (d_{jk})_{1 \leq j,k \leq 3}$ with

$$d_{jk} = -b\alpha(\eta_j + \eta_k) \text{ for } j \neq k \text{ and } d_{jj} = 2\left(\eta_0 - a\alpha\eta_j - c\alpha\sum_{k=1}^{3}\eta_k\right).$$

Therefore,

$$s^6 P(\eta + s^{-1}(\tau, \xi)) = s^6 \det[A(\eta + s^{-1}(\tau, \xi))]$$
$$= -s^2 b\alpha^2 e^T \cdot D(\eta)^{\text{ad}} \cdot e + O(s^3), \quad s \to 0.$$

Thus

$$P_{(\tau,\xi)}(\eta) = -b\alpha^2 e^T \cdot D(\eta)^{\text{ad}} \cdot e = -b\alpha^2 \sum_{j=1}^{3}\sum_{k=1}^{3} D(\eta)_{jk}^{\text{ad}}.$$

Setting $\sigma = e^T \cdot \eta' = \eta_1 + \eta_2 + \eta_3$ we obtain

$$\sum_{j=1}^{3}\sum_{k=1}^{3} D(\eta)_{jk}^{\text{ad}} = 12(\eta_0 - c\alpha\sigma)^2 - 8(a - b)\alpha\sigma(\eta_0 - c\alpha\sigma) + 4(a - b)^2\alpha^2 \sum_{1 \leq j < k \leq 3} \eta_j\eta_k;$$

since $\sum_{1 \leq j < k \leq 3} \eta_j\eta_k = \frac{1}{2}(\sigma^2 - |\eta'|^2)$ and $\alpha^2 = 1/(a + 3c - b)$, this yields

$$\sum_{j=1}^{3}\sum_{k=1}^{3} D(\eta)_{jk}^{\text{ad}} = 12\left(\eta_0 - \frac{\sigma}{3\alpha}\right)^2 - 2(a - b)^2\alpha^2\left(|\eta'|^2 - \frac{\sigma^2}{3}\right).$$

If y_1, y_2, y_3 is an orthogonal coordinate system with $y_3 = \frac{1}{\sqrt{3}} e^T \cdot x$, then $\frac{\partial}{\partial y_3} = \frac{1}{\sqrt{3}} (\partial_1 + \partial_2 + \partial_3)$ and hence

$$P_{(\tau,\xi)}(-i\partial) = b\alpha^2 \left[12 \left(\partial_t - \frac{1}{\sqrt{3}\alpha} \partial_3' \right)^2 - 2(a-b)^2 \alpha^2 (\partial_1'^2 + \partial_2'^2) \right], \qquad \partial_j' = \frac{\partial}{\partial y_j},$$

which has the fundamental solution (with respect to $N = (1,0,0,0)$)

$$E(P_{(\tau,\xi)}(-i\partial), N) = \frac{1}{4\pi b\alpha^3 |a-b|} \cdot \frac{Y(\alpha|a-b|t - \sqrt{6(y_1^2 + y_2^2)})}{\sqrt{\alpha^2(a-b)^2 t^2 - 6(y_1^2 + y_2^2)}} \cdot \delta\left(y_3 + \frac{t}{\sqrt{3}\alpha} \right).$$

$$(4.3.10)$$

Therefore, the equality (4.3.9) finally yields

$$\operatorname{sing\,supp} E(P(-i\partial), N) = (\Xi^* \cap H_N) \cup C \qquad (4.3.11)$$

where

$$C = \left\{ (t, x) \in \mathbf{R}^4;\ t = \alpha(\pm x_1 \pm x_2 \pm x_3),\ \sqrt{3|x|^2 - (\pm x_1 \pm x_2 \pm x_3)^2} \le \frac{\alpha|a-b|t}{\sqrt{2}} \right\}.$$

(The characteristic surface Ξ^* has been defined in Definition 4.2.6.)

(b) Let us consider now the singular support of the *fundamental matrix* $E(A(-i\partial), N)$. According to Proposition 4.3.5 and (4.3.11),

$$\operatorname{sing\,supp} E(A(-i\partial), N) \subset (\Xi^* \cap H_N) \cup C,$$

and we will show that this inclusion is in fact an equality. For this respect we will employ Proposition 4.3.4. Let us therefore first determine the adjoint matrix $A^{\mathrm{ad}}(\tau, \xi)$. We obtain

$$A_{ii}^{\mathrm{ad}}(\tau, \xi) = (\tau^2 - c|\xi|^2 - a\xi_j^2)(\tau^2 - c|\xi|^2 - a\xi_k^2) - b^2 \xi_j^2 \xi_k^2,$$

$$A_{jk}^{\mathrm{ad}}(\tau, \xi) = b\xi_j \xi_k (\tau^2 - c|\xi|^2 - (a-b)\xi_i^2),$$

where $\{i, j, k\} = \{1, 2, 3\}$.

If now (τ, ξ) is a non-singular point on the slowness surface Ξ, i.e., $P(\tau, \xi) = 0$ and $\zeta = (\partial_\tau P, \nabla P)(\tau, \xi) \ne 0$, then $\zeta \in \Xi^*$, $P_{(\tau,\xi)}(\eta) = \eta^T \cdot \zeta$ and

$$\operatorname{supp} E(P_{(\tau,\xi)}(-i\partial), N) = \{\lambda \operatorname{sign}(\zeta_0)\zeta;\ \lambda \ge 0\}.$$

Since (τ, ξ) is a non-singular point on Ξ, the matrix $A(\tau, \xi)$ has three different eigenvalues, and $A^{\text{ad}}(\tau, \xi)$ does not vanish. Therefore,

$$\text{supp}\, E(P_{(\tau,\xi)}(-i\partial), N) = \bigcup_{j,k \in \{1,2,3\}} \text{supp}\Big((A_{jk}^{\text{ad}})_{(\tau,\xi)}(-i\partial)E(P_{(\tau,\xi)}(-i\partial), N)\Big)$$

is contained in $\text{sing}\,\text{supp}\, E(A(-i\partial), N)$. Hence $\Xi^* \cap H_N \subset \text{sing}\,\text{supp}\, E(A(-i\partial), N)$.

Let us finally show that the set C of conical refraction is also contained in the singular support of the fundamental matrix $E(A(-i\partial), N)$.

If, as above, $(\tau, \xi) = (1, \alpha, \alpha, \alpha)$ is one of the singular points on Ξ of type (β), then $E(P_{(\tau,\xi)}(-i\partial), N)$ is as in (4.3.10); let us calculate the localization $(A_{12}^{\text{ad}})_{(\tau,\xi)}(\eta)$, $\eta = (\eta_0, \eta') \in \mathbf{R}^4$. From

$$A_{12}^{\text{ad}}(\eta) = b\eta_1 \eta_2 (\eta_0^2 - c|\eta'|^2 - (a-b)\eta_3^2)$$

we obtain

$$A_{12}^{\text{ad}}\Big(\eta_0 + \frac{\tau}{s}, \eta' + \frac{\xi}{s}\Big) = \frac{2b\alpha^2}{s^3}\big(\eta_0 - c\alpha\sigma - (a-b)\alpha\eta_3\big) + O(s^{-2}), \quad s \to 0,$$

where $\sigma = \eta_1 + \eta_2 + \eta_3$. This implies

$$(A_{12}^{\text{ad}})_{(\tau,\xi)}(\eta) = 2b\alpha^2\big(\eta_0 - c\alpha\sigma - (a-b)\alpha\eta_3\big).$$

We conclude that the support of $E(P_{(\tau,\xi)}(-i\partial), N)$ coincides with that of

$$(A_{12}^{\text{ad}})_{(\tau,\xi)}(-i\partial)E(P_{(\tau,\xi)}(-i\partial), N).$$

(Note that $(A_{12}^{\text{ad}})_{(\tau,\xi)}(-i\partial)\delta(y_3 + t/(\sqrt{3}\alpha))$ vanishes and therefore the distribution $(A_{12}^{\text{ad}})_{(\tau,\xi)}(-i\partial)E(P_{(\tau,\xi)}(-i\partial), N)$ is, essentially, a first-order derivative of the forward fundamental solution of a wave operator in two space dimensions and thus has the support $K(P_{(\tau,\xi)}(-i\partial), N)$.) Hence, by Proposition 4.3.4,

$$C = \bigcup_{\substack{(\tau,\xi) \text{ singular} \\ \text{point of type } (\beta)}} \text{supp}\, E(P_{(\tau,\xi)}(-i\partial), N).$$

is contained in $\text{sing}\,\text{supp}\, E(A(-i\partial), N)$. $\square$

Example 4.3.10 As we have already hinted at in the beginning of this section, a physically relevant example of the non-occurrence of conical refraction in the fundamental matrix (in contrast to the fundamental solution of its determinant) appears for *elastic waves in hexagonal media*. This non-occurrence of conical refraction was conjectured first in Payton [226], p. 67, in contrast to a "presage" in Musgrave [187], p. 579, and proven in Ortner and Wagner [220], Prop. 4, p. 424, and, differently, in Ortner and Wagner [221], Section 3.3, p. 1252.

According to Example 2.1.4 (d), the propagation of such waves is described by the matrix $A(-i\partial) = -I_3\partial_t^2 + B(\nabla)$, where B is given by

$$B(\xi) = \begin{pmatrix} a_1\xi_1^2 + a_4\xi_2^2 + a_5\xi_3^2 & (a_1 - a_4)\xi_1\xi_2 & a_3\xi_1\xi_3 \\ (a_1 - a_4)\xi_1\xi_2 & a_4\xi_1^2 + a_1\xi_2^2 + a_5\xi_3^2 & a_3\xi_2\xi_3 \\ a_3\xi_1\xi_3 & a_3\xi_2\xi_3 & a_5(\xi_1^2 + \xi_2^2) + a_2\xi_3^2 \end{pmatrix},$$

compare (2.1.13). According to Ortner and Wagner [220], Prop. 2, p. 419, the system $A(-i\partial)$ is hyperbolic if and only if the elastic constants fulfill the conditions

$$a_1 \geq 0, \ a_2 \geq 0, \ a_4 \geq 0, \ a_5 \geq 0, \ \text{and} \ a_5 + \sqrt{a_1 a_2} \geq |a_3|. \qquad (4.3.12)$$

We assume, moreover, that the inequalities in (4.3.12) are strict, which is equivalent to $\det A(0, \xi) \neq 0$ for $\xi \in \mathbf{R}^3 \setminus \{0\}$, and which, physically, amounts to the positivity of the *propagation speeds* (see Payton [226], p. 5).

As in the prior Example 4.3.9, we subdivide the investigation into two parts: In (a) we calculate sing supp $E(P(-i\partial), N)$ for $P = \det A$, and in (b) we deduce therefrom the shape of sing supp $E(A(-i\partial), N)$. As before, we set $N = (1, 0, 0, 0)$.

(a) The determinant $P(\tau, \xi) = \det A(\tau, \xi)$ splits into two factors: $P = W_1 \cdot R$, where

$$W_1(\tau, \xi) = \tau^2 - a_4\rho^2 - a_5\xi_3^2, \quad \rho^2 = \xi_1^2 + \xi_2^2,$$

corresponds to a wave operator, and

$$R(\tau, \xi) = \tau^4 - \tau^2(a_1\rho^2 + a_2\xi_3^2 + a_5|\xi|^2) + a_1 a_5\rho^4 + (a_1 a_2 - a_3^2 + a_5^2)\rho^2\xi_3^2 + a_2 a_5\xi_3^4$$

corresponds to a homogeneous hyperbolic fourth-order operator, which, in general is irreducible, cf. Example 2.1.4 (d).

The slowness surface $\Xi = P^{-1}(0)$ becomes singular in the points (τ, ξ) where the surfaces $W_1 = 0$ and $R = 0$ intersect. This occurs on the ξ_3-axis, i.e., if (τ, ξ) is a multiple of $(1, 0, 0, \pm 1/\sqrt{a_5})$. Furthermore, depending on the values of $a_1, \ldots, a_5$, Ξ can become singular along "ridge points" on circular cones, compare Fig. 4.2, left side, for titanium boride (a hexagonal medium for which the values of $a_1, \ldots, a_5$ can be found in Ortner and Wagner [220], p. 415) and Figs. 4.3, 4.4 below.

According to Atiyah, Bott, and Gårding [6], Thm. 7.7, p. 175, we have

$$\text{sing supp}\, E(P(-i\partial), N) = W(P(-i\partial), N) = \bigcup_{(\tau, \xi) \in \mathbf{R}^4 \setminus \{0\}} K(P_{(\tau, \xi)}(-i\partial), N)$$

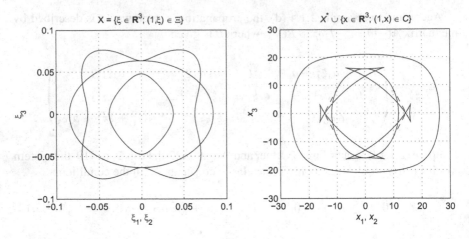

Fig. 4.2 Slowness surface and wavefront surface for titanium boride

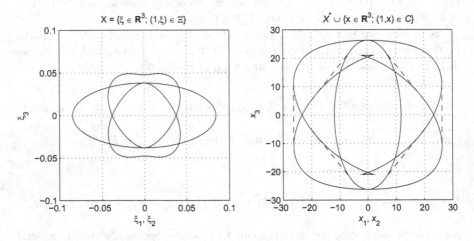

Fig. 4.3 Slowness and wave-front surface for $a_1 = a_5$

i.e., all the inclusions in (4.2.5) are identities. Let us discuss now the propagation cones $K(P_{(\tau,\xi)}(-i\partial), N)$ more in detail. We distinguish four cases:

(α) If $(\tau, \xi) \in \Xi \setminus \{0\}$ is a non-singular point, then $P_{(\tau,\xi)}$ is a first-order operator and $K(P_{(\tau,\xi)}(-i\partial), N)$ is the corresponding half-ray which supports the forward fundamental solution $E(P_{(\tau,\xi)}(-i\partial), N)$. The union of these half-rays yields the wave surface, i.e., the surface Ξ^*, which is dual to Ξ and depicted on the right side of Fig. 4.2.

(β) If (τ, ξ) is a "ridge point" on Ξ, i.e., $W_1(\tau, \xi) = R(\tau, \xi) = 0$, but $(\xi_1, \xi_2) \neq 0$, then $P_{(\tau,\xi)} = (W_1)_{(\tau,\xi)} \cdot R_{(\tau,\xi)}$ is a product of two linearly independent first-order operators, and $K(P_{(\tau,\xi)}(-i\partial), N)$ is the convex hull of the two half-rays $K((W_1)_{(\tau,\xi)}(-i\partial), N)$ and $K(R_{(\tau,\xi)}(-i\partial), N)$. This yields the set C of conical

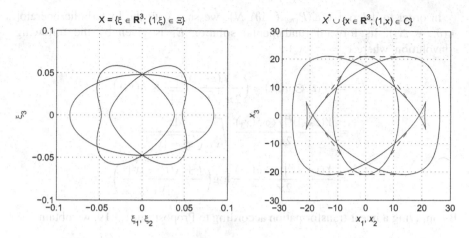

Fig. 4.4 Slowness and wave-front surface for $a_2 = a_5$

refraction (see Definition 4.2.6), which, in Fig. 4.2, right part, is represented by the broken lines corresponding to two frusta on the wavefront surface.

Furthermore, additional ridge points occur in the case $a_1 = a_5$. In this case, the quartic surface $R(\tau, \xi) = 0$ becomes singular along the circular cone $\xi_3 = 0$, $\tau = \pm a_1 \rho$, see Fig. 4.3, left part, $R_{(\tau,\xi)}$ is a product of two linearly independent first-order operators, and $K(R_{(\tau,\xi)}(-i\partial), N)$ yields the two broken vertical lines in Fig. 4.3, right part, which represent a cylindrical lid in the singular support of $E(P(-i\partial), N)$.

(γ) If $a_2 \neq a_5$ and $(\tau, \xi) = (1, 0, 0, \pm\frac{1}{\sqrt{a_5}})$, which lies on the x_3-axis, then $W_1(\tau, \xi) = R(\tau, \xi) = 0$ and $(W_1)_{(\tau,\xi)}$ and $R_{(\tau,\xi)}$ are proportional. Hence $P_{(\tau,\xi)}$ is a square of a first-order operator, similarly as in Example 4.3.9, case (α). Therefore $K(P_{(\tau,\xi)}(-i\partial), N) = \{t \cdot (1, 0, 0, \pm\sqrt{a_5}); t \geq 0\}$ is already contained in the dual surface Ξ^* of Ξ.

(δ) If $a_2 = a_5$ and $(\tau, \xi) = (1, 0, 0, \pm\frac{1}{\sqrt{a_5}})$, then all three sheets of Ξ intersect in (τ, ξ), see Fig. 4.4, left side.

In this case, we can calculate $P_{(\tau,\xi)}$ by means of Lemma 4.3.2. In fact,

$$A_{(\tau,\xi)}(\eta) = \begin{pmatrix} 2\eta_0 \mp 2\sqrt{a_5}\,\eta_3 & 0 & \mp\frac{a_3}{\sqrt{a_5}}\eta_1 \\ 0 & 2\eta_0 \mp 2\sqrt{a_5}\,\eta_3 & \mp\frac{a_3}{\sqrt{a_5}}\eta_2 \\ \mp\frac{a_3}{\sqrt{a_5}}\eta_1 & \mp\frac{a_3}{\sqrt{a_5}}\eta_2 & 2\eta_0 \mp 2\sqrt{a_5}\,\eta_3 \end{pmatrix} \qquad (4.3.13)$$

and hence

$$P_{(\tau,\xi)}(\eta) = \det A_{(\tau,\xi)}(\eta) = 2(\eta_0 \mp \sqrt{a_5}\,\eta_3)\left[4(\eta_0 \mp \sqrt{a_5}\,\eta_3)^2 - \frac{a_3^2}{a_5}\eta'^2\right], \qquad (4.3.14)$$

where $\eta' = (\eta_1, \eta_2)$.

In order to determine $K(P_{(\tau,\xi)}(-i\partial), N)$, we start with the hyperbolic operator $\partial_t(\partial_t^2 - \Delta_2)$. Its forward fundamental solution E_1 is given by the following convolution, where $x' = (x_1, x_2)$:

$$E_1 = \left(Y(t) \otimes \delta(x) \right) * \left(\frac{Y(t - |x'|)}{2\pi \sqrt{t^2 - |x'|^2}} \otimes \delta(x_3) \right)$$

$$= \frac{Y(t - |x'|) \otimes \delta(x_3)}{2\pi} \int_{|x'|}^t \frac{ds}{\sqrt{s^2 - |x'|^2}}$$

$$= \frac{Y(t - |x'|) \otimes \delta(x_3)}{2\pi} \log \left(\frac{t + \sqrt{t^2 - |x'|^2}}{|x'|} \right).$$

By applying a linear transformation according to Proposition 1.3.19, we obtain

$$E(P_{(\tau,\xi)}(-i\partial), N) = \frac{a_5}{4\pi i\, a_3^2} Y(|a_3|t - 2\sqrt{a_5}|x'|)\, \delta(x_3 \pm \sqrt{a_5}t) \times$$

$$\times \log \left(\frac{|a_3|t + \sqrt{a_3^2 t^2 - 4a_5|x'|^2}}{2\sqrt{a_5}|x'|} \right) \tag{4.3.15}$$

as fundamental solution of the operator corresponding to (4.3.14). This implies

$$K(P_{(\tau,\xi)}(-i\partial), N) = \left\{ (t, x) \in \mathbf{R}^4;\ t = \mp \frac{x_3}{\sqrt{a_5}},\ |x'| \le \frac{|a_3|t}{2\sqrt{a_5}} \right\}. \tag{4.3.16}$$

Hence in this case,

$$\text{sing supp}\, E(P(-i\partial), N) = W(P(-i\partial), N) = (\Xi^* \cap H_N) \cup C$$

where Ξ^* is the wave surface (see Definition 4.2.6) and C, the set of conical refraction, consists of the union of the two circular lids given in (4.3.16) and of the two frusta described in (β), see Fig. 4.4, right part, where, as before, C is represented by broken lines.

(b) Finally, we aim at determining the singular support of the fundamental matrix $E(A(-i\partial), N)$. As in Example 4.3.9 (b), we have

$$\text{sing supp}\, E(A(-i\partial), N) \subset (\Xi^* \cap H_N) \cup C,$$

where Ξ^* is the wave surface, and the set C of conical refraction in the case $a_2 \ne a_5$ is either empty, or consists of two frusta (and for $a_1 = a_5$ contains an additional cylindrical lid), see (β) above, and in the case $a_2 = a_5$ contains additionally two circular lids, see (δ) and Fig. 4.4, right part. As in (a), let us discuss now the various types of points $(\tau, \xi) \in \Xi$.

(α) Similarly as in Example 4.3.9 (b), we verify that $\Xi^* \cap H_N$ always pertains to the singular support of $E(A(-i\partial), N)$.

(β) In contrast to what happens in (a), we shall see that the frusta originating from the ridge points on Ξ (see Figs. 4.2 and 4.3, broken lines in the right part) do not appear in sing supp $E(A(-i\partial), N)$. This will be shown in Sect. 4.4 below by means of the Herglotz–Gårding formula for the fundamental matrix, see Example 4.4.5 (b). Hence if $a_2 \neq a_5$, then sing supp $E(A(-i\partial), N)$ coincides with the part $\Xi^* \cap H_N$ of the wave surface.

(γ) As we have seen already before, if $a_2 \neq a_5$, then the singular points of the velocity surface Ξ which lie on the ξ_3-axis do not contribute to the set of conical refraction for $P(-i\partial)$ and neither, consequently, for $A(-i\partial)$.

(δ) If $a_2 = a_5$ and $(\tau, \xi) = (1, 0, 0, \pm\frac{1}{\sqrt{a_5}})$, then we shall employ Proposition 4.3.4 to show that the corresponding circular lids in (4.3.16), see Fig. 4.4, right part, are also present in the singular support of $E(A(-i\partial), N)$.

By Lemma 4.3.2 (ii) and (4.3.13), we have

$$(A_{12}^{\text{ad}})_{(\tau,\xi)}(\eta) = A_{(\tau,\xi)}(\eta)_{12}^{\text{ad}} = \frac{a_3^2}{a_5} \eta_1 \eta_2,$$

and the explicit formula for $E(P_{(\tau,\xi)}(-i\partial), N)$ in (4.3.15) implies that

$$\text{supp}\big((A_{12}^{\text{ad}})_{(\tau,\xi)}(-i\partial)E(P_{(\tau,\xi)}(-i\partial), N)\big) = K(P_{(\tau,\xi)}(-i\partial), N).$$

Hence, by Proposition 4.3.4, the two circular lids in (4.3.16) belong to sing supp $E(A(-i\partial), N)$. In the special case of $a_1 = a_2 = a_5$, this fact was established already in Payton [227].   □

4.4 General Formulas for Fundamental Matrices of Homogeneous Hyperbolic Systems

For a hyperbolic $l \times l$ matrix $A(\partial_t, \nabla)$ of differential operators in $\mathbf{R}_{t,x}^n$, formula (4.1.2) for its forward fundamental matrix, i.e., $E_A = e^{\sigma_0 t} \cdot \mathcal{L}^{-1}\big(A(p + \sigma_0 N)^{-1}\big)$, $N = (1, 0, \ldots, 0)$, yields essentially a representation by n integrations. In the following, we shall reduce the number of integrations to $n - 2$ under the hypothesis that all the polynomials in the elements of the matrix A are homogeneous of the same degree. The primary result we derive is called the *Herglotz–Gårding formula* since it is a modification based on Gårding [92], Thm. 2, p. 375, of the *Herglotz–Petrovsky–Leray formula*.

In concrete examples, the Herglotz–Gårding formula proves advantageous over the Herglotz–Petrovsky–Leray formula, which will be stated in Corollary 4.4.2. The Herglotz–Gårding formula was presented first for scalar operators in Wagner [295], Prop. 1, p. 309, and, for systems, in Ortner and Wagner [220], Prop. 1, p. 415. Note that the integral over $\mathbf{R}^{n-1}$ in this formula, viz. (4.4.2), amounts to an integration over the $(n-2)$-dimensional slowness surface $X = \{\xi \in \mathbf{R}^{n-1}; P(1,\xi) = 0\}$ due to the delta-factor in T. We consider first only systems with strictly hyperbolic determinant, cf. Definition 4.2.6 (2) above, and postpone the general case to Proposition 4.4.3.

Proposition 4.4.1 *Let $A(\tau,\xi) = A(\tau,\xi_1,\ldots,\xi_{n-1})$ be a real $l \times l$ matrix of polynomials which are homogeneous of degree m and suppose that $P(\partial) = \det A(\partial)$ is strictly hyperbolic with respect to t and that $P(\tau,\xi)$ does not contain τ as a factor. Define the measure $T \in \mathcal{D}'(\mathbf{R}^n \setminus \{0\})^{l \times l}$ by*

$$T = A^{\mathrm{ad}}(\tau,\xi)\,\delta\big(P(\tau,\xi)\big)\,\mathrm{sign}\big((\partial_\tau P)(\tau,\xi)\big). \tag{4.4.1}$$

Furthermore, set $s_+^\lambda := Y(s)s^\lambda \in L^1_{\mathrm{loc}}(\mathbf{R}^1_s)$ for $\mathrm{Re}\,\lambda > -1$ and let s_+^{n-m-1} be the finite part evaluated at $n-m-1$ of the meromorphic extension to the whole complex plane of the holomorphic function

$$\{\lambda \in \mathbf{C}; \mathrm{Re}\,\lambda > -1\} \longrightarrow \mathcal{S}'(\mathbf{R}) : \lambda \longmapsto s_+^\lambda,$$

cf. Example 1.4.8.

Then the forward fundamental matrix E of $A(\partial)$ fulfills

$$E(t,x) = -2(2\pi)^{1-n}Y(t)\int_{\mathbf{R}^{n-1}} T(1,\xi)\mathrm{Re}\big[i^{m+1}\mathcal{F}s_+^{n-m-1}\big](t+x\xi)\,d\xi + Y(t)Q(t,x), \tag{4.4.2}$$

where Q vanishes if n is even or $m < n$, and is otherwise an $l \times l$ matrix of homogeneous polynomials of degree $m - n$.

In particular, if $n = 4$ and $m = 2$, then

$$E(t,x) = -\frac{Y(t)}{4\pi^2}\frac{\partial}{\partial t}\int_{\mathbf{R}^3} A^{\mathrm{ad}}(1,\xi)\,\delta\big(P(1,\xi)\big)\,\mathrm{sign}\big((\partial_\tau P)(1,\xi)\big)\delta(t+x\xi)\,d\xi. \tag{4.4.3}$$

Proof

(a) Let us first explain why the formula in (4.4.2) is meaningful.

Due to the homogeneity of T, the restriction of T to the hyperplane $\tau = 1$ is well defined (see Ch. V below); furthermore, the integral $\int_{\mathbf{R}^{n-1}} \cdots d\xi$ in (4.4.2) has to be understood in the distributional sense, i.e., for $\phi \in \mathcal{D}(\mathbf{R}^n_{t,x})$ with $\phi = 0$ for $t < 0$, we have

$$\langle \phi, E \rangle = -2(2\pi)^{1-n}\int_{\mathbf{R}^{n-1}} T(1,\xi)\psi(\xi)\,d\xi + \langle \phi, Q \rangle \in \mathbf{C}^{l \times l}$$

where

$$\psi(\xi) = \langle \phi(t,x), \mathrm{Re}\big[i^{m+1}\mathcal{F}s_+^{n-m-1}\big](t+x\xi) \rangle \in C^\infty(\mathbf{R}^{n-1}_\xi)$$

and $T(1,\xi)\psi(\xi)$ is an $l \times l$ matrix of integrable measures. (In fact the support of $T(1,\xi)$ is contained in the zero set of $P(1,\xi)$ and this set is compact due to the hyperbolicity of $P(\partial)$.) Note also that the multiplication with $Y(t)$ in formula (4.4.2) is well defined since the support of E intersects the hyperplane $t = 0$ in the origin $x = 0$ only, and since a homogeneous distribution of degree $m - n$ can uniquely be continued from $\mathbf{R}^n \setminus \{0\}$ to $\mathbf{R}^n$ (for $m \geq 1$).

(b) In order to deduce the representation of E in (4.4.2), we start from the formula for E in Proposition 2.4.13. By the homogeneity of $P = \det A$, we can set $\sigma_0 = 0$ in (2.4.11). Furthermore, the entries of the inverse matrices $A(i\tau \pm \sigma, i\xi)^{-1}$ grow at most polynomially when $\sigma \searrow 0$. Hence the two limits $\lim_{\sigma \searrow 0} A(i\tau \pm \sigma, i\xi)^{-1}$ exist in $\mathcal{D}'(\mathbf{R}^n)^{l \times l}$ (cf. Atiyah, Bott, and Gårding [5], p. 121) and yield homogeneous distributions of degree $-m$.

Since

$$\mathcal{F}^{-1}\left(\lim_{\sigma \searrow 0} A(i\tau \pm \sigma, i\xi)^{-1}\right)$$

are the two fundamental matrices of $A(\partial)$ with support in $\pm t \geq 0$ respectively, we obtain

$$E = Y(t)\, i^{1-m} 2\pi \mathcal{F}^{-1} T \quad \text{where}$$

$$T = \frac{1}{2\pi i} \lim_{\epsilon \searrow 0}\left(A(\tau - i\epsilon, \xi)^{-1} - A(\tau + i\epsilon, \xi)^{-1}\right) \in \mathcal{S}'(\mathbf{R}^n)^{l \times l}. \qquad (4.4.4)$$

Next we apply Sokhotski's formula (1.1.2) in Example 1.1.12, i.e.,

$$\lim_{\epsilon \searrow 0} \frac{1}{s \pm i\epsilon} = \mathrm{vp}\,\frac{1}{s} \mp i\pi\delta \quad \text{in } \mathcal{S}'(\mathbf{R}^1_s).$$

If we use the pullback by $h : \Omega \longrightarrow \mathbf{R}\ C^\infty$, $\Omega \subset \mathbf{R}^n$ open (see Definition 1.2.12), and $v : \Omega \longrightarrow \mathbf{R}^n$ is a continuous vector field satisfying $v(x)^T \cdot \nabla h(x) \neq 0$ for each $x \in \Omega$ with $h(x) = 0$, then we conclude, by employing a suitable partition of unity, that

$$\lim_{\epsilon \searrow 0} \frac{1}{h(x + i\epsilon v(x))} = \mathrm{vp}\left(\frac{1}{h}\right) - i\pi\,\mathrm{sign}(v^T \cdot \nabla h) \cdot (\delta \circ h) \quad \text{in } \mathcal{D}'(\Omega).$$

In particular,

$$\lim_{\epsilon \searrow 0}\left(\frac{1}{h(x - i\epsilon v(x))} - \frac{1}{h(x + i\epsilon v(x))}\right) = 2\pi i\,\mathrm{sign}(v^T \cdot \nabla h) \cdot (\delta \circ h)$$

and

$$T = \frac{1}{2\pi i} A^{\mathrm{ad}}(\tau, \xi) \lim_{\epsilon \searrow 0} \big(P(\tau - i\epsilon, \xi)^{-1} - P(\tau + i\epsilon, \xi)^{-1}\big)$$

$$= A^{\mathrm{ad}}(\tau, \xi)\, \mathrm{sign}(\partial_\tau P) \cdot \delta(P(\tau, \xi)). \tag{4.4.5}$$

Note that here $\Omega = \mathbf{R}^n \setminus \{0\}$ and that the last expression is defined in $\mathcal{D}'(\Omega)$ as in Definition 1.2.12, i.e.,

$$\langle \phi, \mathrm{sign}(\partial_\tau P) \cdot \delta(P(\tau, \xi)) \rangle = \frac{\mathrm{d}}{\mathrm{d}s} \left(\int_{P(\tau, \xi) < 0} \phi(\tau, \xi)\, \mathrm{sign}(\partial_\tau P(\tau, \xi))\, \mathrm{d}\tau \mathrm{d}\xi \right) \Bigg|_{s=0}$$

for $\phi \in \mathcal{D}(\Omega)$.

(c) In the following, we abbreviate $\eta = (\tau, \xi)$.

Since T is homogeneous in $\mathbf{R}^n$ of degree $-m$, its "characteristic," i.e., the restriction $T|_{\mathbf{S}^{n-1}} \in \mathcal{D}'(\mathbf{S}^{n-1})$ is well defined, and T coincides in $\mathbf{R}^n \setminus \{0\}$ with $T|_{\mathbf{S}^{n-1}} \cdot |\eta|^{-m}$, see Example 1.4.9 and Gårding [89], Lemmes 1.5, 4.1, pp. 393, 400; Ortner and Wagner [219], Thm. 2.5.1, p. 58. Hence the difference

$$U = T - T|_{\mathbf{S}^{n-1}} \cdot |\eta|^{-m}$$

is also homogeneous of degree $-m$ and has support restricted to $\{0\}$. In particular, U vanishes if $m < n$. Due to $\check{T} = (-1)^{m+1} T$, we have also $\check{U} = (-1)^{m+1} U$, and, by the homogeneity of U and since $\mathrm{supp}\, U \subset \{0\}$, $\check{U} = (-1)^{m-n} U$. Therefore U vanishes generally if n is even.

Thus formula (4.4.4) yields

$$E(t, x) = 2\pi Y(t)\, \mathrm{Re}(i^{1-m} \mathcal{F}^{-1} T)$$

$$= 2\pi Y(t)\, \mathrm{Re}\big(i^{1-m} \mathcal{F}^{-1}(T|_{\mathbf{S}^{n-1}} \cdot |\eta|^{-m})\big) + Y(t) Q(t, x), \tag{4.4.6}$$

where $Q = 2\pi\, \mathrm{Re}(i^{1-m} \mathcal{F}^{-1} U)$ is an $l \times l$ matrix of homogeneous polynomials of degree $m - n$, and taking the real part is justified by the reality of A and of U.

Hence it remains to express more explicitly the inverse Fourier transform of the homogeneous distribution $F \cdot |\eta|^\lambda$ for $\lambda = -m$ and $F = T|_{\mathbf{S}^{n-1}} \in \mathcal{D}'(\mathbf{S}^{n-1})$, see Example 1.4.9.

(d) Let us recall that, for $F \in \mathcal{D}'(\mathbf{S}^{n-1})$, the distribution-valued function $\lambda \mapsto F \cdot |\eta|^\lambda$ is analytic in $\mathbf{C} \setminus (-n - \mathbf{N}_0)$ and that we set

$$F \cdot |\eta|^{-n-k} = \Pf_{\lambda = -n-k} (F \cdot |\eta|^\lambda), \qquad k \in \mathbf{N}_0,$$

in the possible poles, see Example 1.4.9. For its inverse Fourier transform, the following formula holds, see Gårding [89], Lemme 6.2, p. 406; Hörmander

[139], Thm. 7.1.24 and formula (7.1.24), p. 172; Ortner and Wagner [219], Cor. 2.6.3, p. 64:

$$\langle \phi, \mathcal{F}^{-1}(F \cdot |\eta|^{\lambda}) \rangle = (2\pi)^{-n}{}_{\mathcal{D}(S^{n-1})}\langle \langle \phi(y), (\mathcal{F}s_{+}^{\lambda+n-1})(y\omega) \rangle, F(-\omega) \rangle_{\mathcal{D}'(S^{n-1})}$$
(4.4.7)

for $\phi \in \mathcal{D}(\mathbf{R}^n)$.

Here $s_{+}^{\lambda+n-1}$ is as defined in Example 1.3.9. Recall that $\lambda \mapsto s_{+}^{\lambda+n-1}$ is analytic except for simple poles in $\lambda = 1 - n - k$, $k \in \mathbf{N}$, in which $s_{+}^{\lambda+n-1}$ is defined as the finite part. Its Fourier transform $\mathcal{F}s_{+}^{\lambda+n-1}$ was calculated in Example 1.6.7. Furthermore, $\langle \phi(y), (\mathcal{F}s_{+}^{\lambda+n-1})(y\omega) \rangle$ depends C^{∞} on $\omega \in \mathbf{S}^{n-1}$.

If we insert formula (4.4.7) with $F = T|_{S^{n-1}}$ into the representation of the fundamental matrix E in (4.4.6), we obtain, for $\phi(t, x) \in \mathcal{D}(\mathbf{R}^n)$ with $\phi = 0$ for $t < 0$,

$$\langle \phi, E \rangle = (2\pi)^{1-n}\text{Re}\big[i^{1-m}\langle \langle \phi(t, x), (\mathcal{F}s_{+}^{n-m-1})(t\tau + x\xi) \rangle, (T|_{S^{n-1}})(-\tau, -\xi) \rangle \big] + \langle \phi, Q \rangle$$
$$= -(2\pi)^{1-n}\langle \langle \phi(t, x), \text{Re}\big[i^{m+1}\mathcal{F}s_{+}^{n-m-1}\big](t\tau + x\xi) \rangle, (T|_{S^{n-1}})(\tau, \xi) \rangle + \langle \phi, Q \rangle.$$

According to Example 1.6.7, we obtain four different expressions for the distribution $\text{Re}\big[i^{m+1}\mathcal{F}s_{+}^{n-m-1}\big] \in \mathcal{S}'(\mathbf{R}_t^1)$ in dependence on the parity and the size of n:

$$\text{Re}\big[i^{m+1}\mathcal{F}s_{+}^{n-m-1}\big] = \begin{cases} \dfrac{(-1)^{(n-1)/2}}{(m-n)!}t^{m-n}\Big(\log|t| + \gamma - \displaystyle\sum_{k=1}^{m-n}\frac{1}{k}\Big) & : m \geq n, n \text{ odd}, \\[2ex] \dfrac{(-1)^{n/2}\pi}{2(m-n)!}t^{m-n}\text{sign}\, t & : m \geq n, n \text{ even}, \\[2ex] (-1)^{(n-1)/2}\Big(\dfrac{\mathrm{d}}{\mathrm{d}t}\Big)^{n-m-1}\text{vp}\,\dfrac{1}{t} & : m < n, n \text{ odd}, \\[2ex] (-1)^{n/2}\pi\,\delta^{(n-m-1)}(t) & : m < n, n \text{ even}, \end{cases}$$

where γ denotes Euler's constant. In particular, formula (4.4.3), i.e., the case $m = 2, n = 4$, follows from (4.4.2) due to $\text{Re}\,[i^3\mathcal{F}s_{+}^1] = \pi\delta'$.

(e) Finally, we make use of the "gnomonian projection," i.e. the diffeomorphism

$$\{-1, 1\} \times \mathbf{R}^{n-1} \longrightarrow \{\eta \in \mathbf{S}^{n-1}; \eta_1 \neq 0\} : (\tau, \xi) \longmapsto \frac{(\tau, \xi)}{|(\tau, \xi)|}.$$

For a function $f \in L^1(\mathbf{S}^{n-1})$, we have

$$\int_{S^{n-1}} f(\eta)\, \mathrm{d}\sigma(\eta) = \int_{\mathbf{R}^{n-1}}\Big[f\Big(\frac{(1, \xi)}{\sqrt{1 + |\xi|^2}}\Big) + f\Big(\frac{(-1, \xi)}{\sqrt{1 + |\xi|^2}}\Big)\Big]\frac{\mathrm{d}\xi}{(1 + |\xi|^2)^{n/2}},$$

$d\sigma$ denoting the surface measure on $\mathbf{S}^{n-1}$. In particular, if $f : \mathbf{R}^n \setminus \{0\} \longrightarrow \mathbf{C}$ is locally integrable, homogeneous of degree $-n$ and even, then

$$\int_{\mathbf{S}^{n-1}} f(\eta)\, d\sigma(\eta) = 2 \int_{\mathbf{R}^{n-1}} f(1, \xi)\, d\xi.$$

Similarly, if μ is a Radon measure on $\mathbf{R}^n \setminus \{0\}$ which fulfills

$$\mu(\{\eta \in \mathbf{R}^n;\ \eta_1 = 0\}) = 0,$$

and which, as an element of $\mathcal{D}'(\mathbf{R}^n \setminus \{0\})$, is even and homogeneous of degree $-n$, then

$$\int_{\mathbf{S}^{n-1}} \mu|_{\mathbf{S}^{n-1}} = 2 \int_{\mathbf{R}^{n-1}} \mu|_{\{1\} \times \mathbf{R}^{n-1}}.$$

By the homogeneity of μ, the restrictions $\mu|_{\mathbf{S}^{n-1}}$ and $\mu|_{\{1\} \times \mathbf{R}^{n-1}}$ are well-defined Radon measures.

In our case,

$$\mu(\eta) = \mu(\tau, \xi) = \langle \phi(t, x), \mathrm{Re}[i^{m+1}\mathcal{F}s_+^{n-m-1}](t\tau + x\xi) \rangle \cdot T(\tau, \xi)$$

is even and the hyperplane $\{\eta \in \mathbf{R}^n;\ \eta_1 = 0\}$ is a null-set with respect to μ, since $P(\tau, \xi)$ does not contain τ as a factor. Furthermore, μ is homogeneous in $\mathbf{R}^n \setminus \{0\}$ except for $m \geq n$ and odd n. In this last case, $\mu(\eta)$ is associated homogeneous (see Gel'fand and Shilov [104], p. 83) and the gnomonian projection changes just the polynomial term $Q(t, x)$.

Hence we conclude that, still for $\phi(t, x) \in \mathcal{D}(\mathbf{R}^n)$ with $\phi = 0$ for $t < 0$,

$$\langle \phi, E \rangle = -2(2\pi)^{1-n} \int_{\mathbf{R}^{n-1}} T(1, \xi)\psi(\xi)\, d\xi + \langle \phi, Q \rangle,$$

where

$$\psi(\xi) = \langle \phi(t, x), \mathrm{Re}[i^{m+1}\mathcal{F}s_+^{n-m-1}](t + x\xi) \rangle$$

is a $\mathcal{C}^\infty$ function,

$$T(1, \xi) = A^{\mathrm{ad}}(1, \xi)\, \delta\big(P(1, \xi)\big) \, \mathrm{sign}\big((\partial_\tau P)(1, \xi)\big)$$

is a matrix of Radon measures and $T(1, \xi)\psi(\xi)$ is a matrix of integrable measures.

$\square$

Let us next derive the Herglotz–Petrovsky–Leray formula from the Herglotz–Gårding formula in Proposition 4.4.1. For its classical form, we refer to Herglotz [126], pp. 609, 610, (4)–(13); [127] (7.58), p. 192; for the distributional generalization, see Leray [163], Ch. IV; Atiyah, Bott, and Gårding [5], pp. 176, 177; Gel'fand and Shilov [104], Ch. I, 6.3, (24), p. 139; for systems, see Ortner and Wagner [217], Thm., p. 324, and formulas (HP), (HPS), p. 325.

Corollary 4.4.2 *Let* $m, n, l, A, P = \det A, s_+^\lambda, Q$ *be as in Proposition 4.4.1. Then the forward fundamental matrix* E *of* $A(\partial)$ *fulfills*

$$E(t, x) = -\frac{2Y(t)}{(2\pi)^{n-1}} \int_X \frac{A(1, \xi)^{\mathrm{ad}}}{(\partial_\tau P)(1, \xi)} \operatorname{Re}\left[i^{m+1} \mathcal{F} s_+^{n-m-1} \right] (t + x\xi) |\gamma|(\xi) + Y(t) Q(t, x)$$

(4.4.8)

where $X = \{ \xi \in \mathbf{R}^{n-1}; P(1, \xi) = 0 \}$ *is the slowness surface,*

$$\gamma = \sum_{j=1}^{n-1} (-1)^{j-1} \xi_j \, d\xi_1 \wedge \cdots \wedge d\xi_{j-1} \wedge d\xi_{j+1} \wedge \cdots \wedge d\xi_{n-1}$$

denotes the Kronecker–Leray form and $|\gamma|$ *the corresponding positive measure on* X. *In particular, if* $m = 2$ *and* $n = 4$, *then*

$$E(t, x) = -\frac{Y(t)}{4\pi^2} \int_X \frac{A(1, \xi)^{\mathrm{ad}}}{(\partial_\tau P)(1, \xi)} \delta'(t + x\xi) |\gamma|(\xi).$$

(4.4.9)

Proof We employ formula (4.4.2) in order to represent E and we apply to it formula (1.2.2) in Example 1.2.14, which expresses the evaluation of $\delta(P(1, \xi))$ on a test function. This yields

$$E = -2(2\pi)^{1-n} Y(t) \int_X A(1, \xi)^{\mathrm{ad}} \operatorname{Re}\left[i^{m+1} \mathcal{F} s_+^{n-m-1} \right] (t + x\xi) \operatorname{sign}(\partial_\tau P(1, \xi)) \frac{d\sigma(\xi)}{|\nabla P(1, \xi)|},$$

where $\nabla P = \left(\frac{\partial P}{\partial \xi_1}, \ldots, \frac{\partial P}{\partial \xi_{n-1}} \right)^T$.

Clearly, the surface measure $d\sigma$ on the slowness surface X is given by $d\sigma = \frac{|\nabla P|}{|\partial P/\partial \xi_1|} d\xi_2 \ldots d\xi_{n-1}$. On the other hand, since the equation $\sum_{j=1}^{n-1} \frac{\partial P(1, \xi)}{\partial \xi_j} d\xi_j = 0$ holds on X, we conclude that

$$\gamma = \frac{1}{\partial P/\partial \xi_1} \sum_{j=1}^{n-1} \xi_j \frac{\partial P(1, \xi)}{\partial \xi_j} d\xi_2 \wedge \cdots \wedge d\xi_{n-1}$$

$$= -\frac{(\partial_\tau P)(1, \xi)}{\partial P/\partial \xi_1} d\xi_2 \wedge \cdots \wedge d\xi_{n-1}$$

holds on X. (Here we have used Euler's equation for the homogeneous function $P(\tau, \xi)$ in the second equality.) Hence

$$\frac{\mathrm{d}\sigma}{|\nabla P(1, \xi)|} = \frac{\mathrm{d}\xi_2 \ldots \mathrm{d}\xi_{n-1}}{|\partial P/\partial \xi_1|} = \frac{|\gamma|}{|(\partial_\tau P)(1, \xi)|}.$$

This implies formula (4.4.8), and the special case (4.4.9) follows again from $\mathrm{Re}\,[i^3 \mathcal{F}s_+^1] = \pi \delta'$.   □

Let us note that in many physically relevant systems as in crystal optics or in elastodynamics, the assumption of strict hyperbolicity in Proposition 4.4.1 and in Corollary 4.4.2 is not satisfied. In order to encompass such cases of *not strictly* hyperbolic systems, the distribution T in formula (4.4.1) has to be defined differently. In fact, $\delta(P(\tau, \xi)) = P^*\delta$ is a priori not meaningful in $\mathcal{D}'(\mathbf{R}^n \setminus \{0\})$ if P is not submersive outside the origin. Already L. Gårding has hinted at the necessary modification of the definition of T in this case, see Gårding [92], p. 375, third note. In the next proposition, we give a representation of T by a parameter integral for general hyperbolic systems. We emphasize that, in general, T is not a measure but a (possibly higher) derivative of a measure for not strictly hyperbolic systems.

Proposition 4.4.3 *Let* $A(\partial_t, \nabla)$ *be an* $l \times l$ *system of linear differential operators with constant coefficients in* $\mathbf{R}^n_{t,x}$. *We assume that the elements of the matrix* $A(\tau, \xi)$ *are real-valued and homogeneous of degree* m *and that* $P(\partial) = \det A(\partial)$ *is even in* τ *and hyperbolic in the direction* t *"without infinite propagation speed," i.e.,* $\forall \xi \in \mathbf{R}^{n-1} \setminus \{0\} : P(0, \xi) \neq 0$.

Then formula (4.4.2) holds for the forward fundamental matrix E *of* $A(\partial)$ *if* $T \in \mathcal{D}'(\mathbf{R}^n \setminus \{0\})^{l \times l}$ *is defined by*

$$T = (-1)^{k-1} A^{\mathrm{ad}}(\tau, \xi) \int_{\Sigma_{k-1}} \delta^{(k-1)}(\tau^2 - g_\lambda(\xi)) \, \mathrm{d}\sigma(\lambda) \qquad (4.4.10)$$

where

$$k = \frac{lm}{2}, \; g_\lambda(\xi) = \sum_{j=1}^k \lambda_j f_j(\xi), \; P(\tau, \xi) = \det A(\tau, \xi) = \prod_{j=1}^k (\tau^2 - f_j(\xi))$$

$$0 \le f_1(\xi) \le f_2(\xi) \le \cdots \le f_k(\xi) \text{ for } \xi \in \mathbf{R}^{n-1},$$

$$\Sigma_{k-1} = \{\lambda \in \mathbf{R}^k; \, \lambda_1 \ge 0, \ldots, \lambda_k \ge 0, \lambda_1 + \cdots + \lambda_k = 1\}, \; \mathrm{d}\sigma(\lambda) = \mathrm{d}\lambda_1 \ldots \mathrm{d}\lambda_{k-1}.$$

Furthermore, $\delta^{(k-1)}(\tau^2 - g_\lambda(\xi))$ *is defined explicitly in (4.4.11/4.4.12) in the proof below.*

Proof First we remark that we cannot compose Sokhotski's formula with P unless P is submersive. Therefore formula (4.4.5) in the proof of Proposition 4.4.1 does not hold any longer if P is not strictly hyperbolic. In order to express the limit

$$S = \frac{1}{2\pi i} \lim_{\epsilon \searrow 0} \left(P(\tau - i\epsilon, \xi)^{-1} - P(\tau + i\epsilon, \xi)^{-1} \right) \in \mathcal{D}'(\mathbf{R}^n_{\tau, \xi}),$$

we use the method of parameter integration, in particular Feynman's first formula (3.1.2). This yields

$$P(\tau \pm i\epsilon, \xi)^{-1} = \prod_{j=1}^{k} \left((\tau \pm i\epsilon)^2 - f_j(\xi) \right)^{-1} = (k-1)! \int_{\Sigma_{k-1}} \left((\tau \pm i\epsilon)^2 - g_\lambda(\xi) \right)^{-k} d\sigma(\lambda)$$

and hence, for $\tau > 0$,

$$\begin{aligned}
S &= \frac{(k-1)!}{2\pi i} \lim_{\epsilon \searrow 0} \int_{\Sigma_{k-1}} \left[\left((\tau - i\epsilon)^2 - g_\lambda(\xi) \right)^{-k} - \left((\tau + i\epsilon)^2 - g_\lambda(\xi) \right)^{-k} \right] d\sigma(\lambda) \\
&= \frac{(-1)^{k-1}}{2^k \pi i} \lim_{\epsilon \searrow 0} \int_{\Sigma_{k-1}} \left(\frac{1}{\tau} \partial_\tau \right)^{k-1} \left[\left((\tau - i\epsilon)^2 - g_\lambda(\xi) \right)^{-1} - \left((\tau + i\epsilon)^2 - g_\lambda(\xi) \right)^{-1} \right] d\sigma(\lambda) \\
&= \frac{(-1)^{k-1}}{2^k \pi i} \int_{\Sigma_{k-1}} \left(\frac{1}{\tau} \partial_\tau \right)^{k-1} 2\pi i \, \delta \left(\tau^2 - g_\lambda(\xi) \right) d\sigma(\lambda) \\
&= (-1)^{k-1} \int_{\Sigma_{k-1}} \delta^{(k-1)} \left(\tau^2 - g_\lambda(\xi) \right) d\sigma(\lambda);
\end{aligned}$$

herein the next to last equation is justified by Sokhotski's formula composed with $\tau^2 - g_\lambda(\xi)$.

Let us mention that the measure $\delta \left(\tau^2 - g_\lambda(\xi) \right) \in \mathcal{D}'(\mathbf{R}^n_{\tau, \xi} \setminus \{0\})$ is well defined for $\phi \in \mathcal{D}(\mathbf{R}^n \setminus \{0\})$ by

$$\langle \phi, \delta(\tau^2 - g_\lambda(\xi)) \rangle = \frac{1}{2} \int_{\mathbf{R}^{n-1}} \left[\phi(\sqrt{g_\lambda(\xi)}, \xi) + \phi(-\sqrt{g_\lambda(\xi)}, \xi) \right] \frac{d\xi}{\sqrt{g_\lambda(\xi)}} \tag{4.4.11}$$

since $g_\lambda(\xi) > 0$ for $\xi \neq 0$ and $\lambda \in \Sigma_{k-1}$. Furthermore,

$$\langle \phi, \delta^{(k-1)}(\tau^2 - g_\lambda(\xi)) \rangle = \langle \psi, \delta(\tau^2 - g_\lambda(\xi)) \rangle \tag{4.4.12}$$

where $\psi = 2^{1-k}(-\partial_\tau \frac{1}{\tau})^{k-1} \phi \in \mathcal{D}((\mathbf{R} \setminus \{0\}) \times \mathbf{R}^{n-1})$, and this function depends continuously on $\lambda \in \Sigma_{k-1}$. $\square$

Example 4.4.4 In order to illustrate the formula (4.4.10) in Proposition 4.4.3 in the case of a scalar operator which misses by far the condition of *strict* hyperbolicity, let us apply it to calculate the forward fundamental solution $E(-k)$ of the *iterated wave operator* $(\partial_t^2 - \Delta_3)^k$, $k \geq 2$, in three space dimensions. According to formula (2.3.12) we should obtain

$$E(-k) = \frac{2^{1-2k}\, Y(t - |x|)(t^2 - |x|^2)^{k-2}}{(k-1)!(k-2)!\pi}.$$

In the notation of Proposition 4.4.3, we have here $P(\tau, \xi) = (\tau^2 - |\xi|^2)^k$,

$$f_j(\xi) = |\xi|^2, \ j = 1, \ldots, k, \qquad g_\lambda(\xi) = |\xi|^2, \ \lambda \in \Sigma_{k-1},$$

$$\text{and } T = \frac{(-1)^{k-1}}{(k-1)!}\, \delta^{(k-1)}(\tau^2 - |\xi|^2) \in \mathcal{D}'(\mathbf{R}^4 \setminus \{0\}).$$

Therefore, by (4.4.10) and (4.4.2),

$$E(-k) = -\frac{2Y(t)}{(2\pi)^3} \int_{\mathbf{R}^3} T(1, \xi) \mathrm{Re}\big[i^{2k+1}\mathcal{F}s_+^{3-2k}\big](t + x\xi)\, d\xi$$

$$= \frac{(-1)^k Y(t)}{8\pi^2(k-1)!(2k-4)!} \int_{\mathbf{R}^3} \delta^{(k-1)}(1 - |\xi|^2)(t + x\xi)^{2k-4}\, \mathrm{sign}(t + x\xi)\, d\xi.$$

$$(4.4.13)$$

The precise meaning of the integral in (4.4.13) is the following: The evaluation of $E(-k)$ on a test function $\phi \in \mathcal{D}((0, \infty) \times \mathbf{R}_x^3)$ is given as

$$\langle \phi, E(-k) \rangle = \frac{(-1)^k}{8\pi^2(k-1)!(2k-4)!} \langle \psi(\xi), \delta^{(k-1)}(1 - |\xi|^2) \rangle$$

where

$$\psi(\xi) = \int_{\mathbf{R}^4} \phi(t, x)(t + x\xi)^{2k-4}\, \mathrm{sign}(t + x\xi)\, dt dx \in C^\infty(\mathbf{R}_\xi^3).$$

Due to $(t + x\xi)^{2k-4}\, \mathrm{sign}(t + x\xi) \in C^{k-1}(\mathbf{R}_\xi^3, L_{\mathrm{loc}}^1(\mathbf{R}_{t,x}^4))$ for $k \geq 3$, we conclude that $E(-k)$ is locally integrable for $k \geq 3$. Therefore, it suffices to determine $E(-k)$ in the two open sets $t > |x|$ and $0 < t < |x|$ in the case $k \geq 3$. The case $k = 2$ can be treated likewise by integrating once with respect to t.

(a) Let us first determine $E(-k)$ inside the forward light cone $C = \{(t, x) \in \mathbf{R}^4;\ t > |x|\}$. Therein, $t + x\xi > 0$ for $|\xi| = 1$ and hence

$$E(-k) = \frac{(-1)^k}{8\pi^2(k-1)!(2k-4)!} {}_{\mathcal{E}(\mathbf{R}_\xi^3)}\langle (t + x\xi)^{2k-4}, \delta^{(k-1)} \circ (1 - |\xi|^2) \rangle_{\mathcal{E}'(\mathbf{R}_\xi^3)}.$$

If we consider the holomorphic function

$$S : \mathbf{C} \longrightarrow \mathcal{E}'(\mathbf{R}^3) : \lambda \longmapsto \chi_+^\lambda \circ (1 - |\xi|^2) = S_\lambda$$

as in Lemma 3.5.2 and use analytic continuation in formula (3.5.1), then we obtain

$$(t^2 - |x|^2)^\lambda = \frac{\Gamma(\frac{1}{2} - \lambda)}{\pi^{3/2}} \langle (t + x\xi)^{2\lambda}, S_{-\lambda-1} \rangle.$$

Setting $\lambda = k - 2$ and employing $\chi_+^{1-k} = \delta^{(k-1)}$, we conclude that

$$E(-k) = \frac{(-1)^k(t^2 - |x|^2)^{k-2}}{8\sqrt{\pi}(k-1)!(2k-4)!\Gamma(\frac{5}{2} - k)} = \frac{2^{1-2k}(t^2 - |x|^2)^{k-2}}{(k-1)!(k-2)!\pi}$$

holds in C.

(b) Let us now verify that the distribution

$$U = \langle (t + x\xi)^{2k-4} \operatorname{sign}(t + x\xi), \delta^{(k-1)} \circ (1 - |\xi|^2) \rangle$$

vanishes in the open set $0 < t < |x|$. Upon introducing spherical coordinates for ξ we obtain

$$U = 2\pi \int_0^\infty \delta^{(k-1)}(1 - \rho^2) \int_0^\pi (t + |x|\rho \cos\theta)^{2k-4} \operatorname{sign}(t + |x|\rho \cos\theta) \sin\theta \, d\theta \, \rho^2 d\rho$$

$$= -\frac{2\pi}{(2k-3)|x|} \int_0^\infty \delta^{(k-1)}(1 - \rho^2) \, (t + |x|\rho \cos\theta)^{2k-3} \operatorname{sign}(t + |x|\rho \cos\theta) \Big|_{\theta=0}^\pi \rho \, d\rho$$

$$= \frac{2\pi}{(2k-3)|x|} \int_0^\infty \rho \, \delta^{(k-1)}(1 - \rho^2) \times$$

$$\times \left[(t + |x|\rho)^{2k-3} \operatorname{sign}(t + |x|\rho) - (t - |x|\rho)^{2k-3} \operatorname{sign}(t - |x|\rho) \right] d\rho$$

$$= \frac{2\pi}{(2k-3)|x|} {}_{\mathcal{E}((0,\infty))} \langle (t + |x|\rho)^{2k-3} + (t - |x|\rho)^{2k-3}, \rho \, \delta^{(k-1)}(1 - \rho^2) \rangle_{\mathcal{E}'((0,\infty))}.$$

$$(4.4.14)$$

In formula (4.4.14), we apply the diffeomorphism

$$h : (0, \infty) \longrightarrow (-\infty, 1) : \rho \longmapsto 1 - \rho^2 = \sigma$$

as in Definition 1.2.7. Then $\delta^{(k-1)}(1 - \rho^2) = \delta^{(k-1)} \circ h$. Furthermore, if

$$\phi(\rho) = (t + |x|\rho)^{2k-3} + (t - |x|\rho)^{2k-3} \in \mathcal{E}((0, \infty)),$$

then $\phi \circ h^{-1}$ is a polynomial in σ of degree $k - 2$. Hence

$$
\begin{aligned}
U &= \frac{2\pi}{(2k-3)|x|} \, \langle \phi, \rho \cdot \delta^{(k-1)}(1-\rho^2) \rangle \\
&= \frac{2\pi}{(2k-3)|x|} \, \langle \left(\frac{\phi \cdot \rho}{|\det h'|} \circ h^{-1} \right)(\sigma), \delta^{(k-1)}(\sigma) \rangle \\
&= \frac{\pi}{(2k-3)|x|} \, \langle \phi \circ h^{-1}, \delta^{(k-1)} \rangle = \frac{\pi(-1)^{k-1}}{(2k-3)|x|} \, (\phi \circ h^{-1})^{(k-1)}(0) = 0.
\end{aligned}
$$

Therefore U vanishes for $0 < t < |x|$. $\square$

Example 4.4.5 Next we illustrate the Herglotz–Gårding formula (4.4.3) in the case of the *system of linear elastodynamics in hexagonal media*, cf. Example 2.1.4 (d) and Example 4.3.10.

Let us first repeat that the displacements $u = (u_1, u_2, u_3)^T$ in such a medium obey the system $A(\partial)u = f$, where f is the density of force and $A(\partial) = I_3 \partial_t^2 - B(\nabla)$, B as in formula (2.1.13). As we have observed already in Example 4.3.10, $A(\partial)$ is *not strictly* hyperbolic due to the various singularities on the slowness surface $\Xi = P^{-1}(0)$, $P = \det A$. Concerning the Herglotz–Gårding representation for the forward fundamental matrix E of $A(\partial)$, i.e.,

$$
E = -\frac{Y(t)}{4\pi^2} \int_{\mathbf{R}^3} T(1, \xi) \delta'(t + x\xi) \, d\xi, \tag{4.4.15}
$$

we first aim in (a) at describing $T \in \mathcal{D}'(\mathbf{R}^4 \setminus \{0\})^{3 \times 3}$ more explicitly than by the parameter integral in (4.4.10). We shall then deduce therefrom in (b) that in general hexagonal media (i.e., for $a_2 \neq a_5$) no conical refraction occurs, compare Example 4.3.10 (b), (β). Finally, in (c), we reduce the three-fold integral in (4.4.15) to a one-fold integral with respect to ξ_3. This remaining integral is a complete Abelian integral pertaining to a Riemannian surface of genus 3 (for general $t, x, a_1, \ldots, a_5$).

In order to ensure that $A(\partial)$ is hyperbolic and that the propagation speeds are positive, we shall generally assume in the sequel that the inequalities

$$
a_1 > 0, \; a_2 > 0, \; a_4 > 0, \; a_5 > 0 \text{ and } |a_3| < a_5 + \sqrt{a_1 a_2}
$$

hold true.

(a) As has been mentioned in Example 2.1.4 (d), the factorization of $P(\tau, \xi) = \det A(\tau, \xi)$ yields

$$
P(\tau, \xi) = W_1(\tau, \xi) \cdot R(\tau, \xi) = \prod_{j=1}^{3} (\tau^2 - f_j(\xi)),
$$

where

$$W_1(\tau, \xi) = \tau^2 - f_1(\xi) = \tau^2 - a_4\rho^2 - a_5\xi_3^2, \quad \rho^2 = \xi_1^2 + \xi_2^2,$$

and

$$R(\tau, \xi) = \prod_{j=2}^{3}(\tau^2 - f_j(\xi)) = \tau^4 - \tau^2(a_1\rho^2 + a_2\xi_3^2 + a_5|\xi|^2)$$

$$+ a_1a_5\rho^4 + (a_1a_2 - a_3^2 + a_5^2)\rho^2\xi_3^2 + a_2a_5\xi_3^4,$$

see (2.1.14).

The general formula for T in Proposition 4.4.3 furnishes $T = A^{\text{ad}}S$ with

$$S = \int_{\Sigma_2} \delta''(\tau^2 - g_\lambda(\xi)) \, d\sigma(\lambda) \in \mathcal{D}'(\mathbf{R}^4 \setminus \{0\}), \qquad g_\lambda(\xi) = \sum_{j=1}^{3} \lambda_j f_j(\xi),$$

$$(4.4.16)$$

i.e.,

$$\langle \phi, S \rangle = \frac{1}{8} \int_{\mathbf{R}^3} \int_{\Sigma_2} \frac{(\partial_\tau \frac{1}{\tau})^2 \phi(\sqrt{g_\lambda(\xi)}, \xi) + (\partial_\tau \frac{1}{\tau})^2 \phi(-\sqrt{g_\lambda(\xi)}, \xi)}{\sqrt{g_\lambda(\xi)}} \, d\sigma(\lambda) d\xi,$$

see (4.4.10–4.4.12).

We emphasize that formula (4.4.16), which determines S as a distribution, is not explicit enough for calculations. First note that the original definition of S as a distributional limit, i.e.,

$$S = \frac{1}{2\pi i} \lim_{\epsilon \searrow 0}[P(\tau - i\epsilon, \xi)^{-1} - P(\tau + i\epsilon, \xi)^{-1}],$$

shows that S coincides with the measure $\delta(P) \operatorname{sign}(\partial_\tau P)$ whenever we can pull back Sokhotski's formula, i.e., outside the set

$$M = \{(\tau, \xi) \in \mathbf{R}^4; \, P(\tau, \xi) = 0, \, \partial_\tau P(\tau, \xi) = 0\}.$$

Furthermore, due to

$$\delta(P) = \delta(W_1 \cdot R) = \frac{\delta(R)}{|W_1|} + \frac{\delta(W_1)}{|R|}$$

(cf. Gel'fand and Shilov [104], p. 236, up to the modulus), we conclude that

$$S = \frac{\delta(R)}{W_1} \operatorname{sign}(\partial_\tau R) + \frac{\delta(W_1)}{R} \operatorname{sign}(\tau)$$

holds in $\mathbf{R}^4 \setminus M$.

Hence S coincides with a Radon measure in $\mathbf{R}^4 \setminus M$. As we shall see below, $T = A^{\mathrm{ad}} \cdot S$ is given in all of $\mathbf{R}^4 \setminus \{0\}$ by a matrix of Radon measures, in spite of the fact that, in general, S is the second derivative of a measure according to formula (4.4.16). Indeed, an easy calculation yields

$$A^{\mathrm{ad}} = \begin{pmatrix} R + \xi_1^2 W_2 & \xi_1\xi_2 W_2 & a_3\xi_1\xi_3 W_1 \\ \xi_1\xi_2 W_2 & R + \xi_2^2 W_2 & a_3\xi_2\xi_3 W_1 \\ a_3\xi_1\xi_3 W_1 & a_3\xi_2\xi_3 W_1 & W_1 W_3 \end{pmatrix}, \tag{4.4.17}$$

with $W_2(\tau, \xi) = (a_1 - a_4)(\tau^2 - a_5\rho^2 - a_2\xi_3^2) + a_3^2\xi_3^2$ and $W_3(\tau, \xi) = \tau^2 - a_1\rho^2 - a_5\xi_3^2$. Moreover, $R = W_1 W_4 - \rho^2 W_2$ for $W_4 = \tau^2 - a_5\rho^2 - a_2\xi_3^2$. From (4.4.17) we infer that A^{ad} vanishes on M since $W_1 = R = 0$ implies that $\rho = 0$ or $W_2 = 0$. Similarly, A^{ad} vanishes to the second order if P has a three-fold zero, which occurs, e.g., if $a_2 = a_5$ and $\rho = 0$, see Example 4.3.10 (b) (δ). Hence $T = A^{\mathrm{ad}} \cdot S$ is a matrix of Radon measures on $\mathbf{R}^4 \setminus \{0\}$.

More explicitly, with $\tilde{\xi} = (\xi_2, -\xi_1, 0)^T$, we obtain

$$T = A^{\mathrm{ad}} \cdot S$$

$$= \begin{pmatrix} R + \xi_1^2 W_2 & \xi_1\xi_2 W_2 & a_3\xi_1\xi_3 W_1 \\ \xi_1\xi_2 W_2 & R + \xi_2^2 W_2 & a_3\xi_2\xi_3 W_1 \\ a_3\xi_1\xi_3 W_1 & a_3\xi_2\xi_3 W_1 & W_1 W_3 \end{pmatrix} \left(\frac{\delta(R)}{W_1} \operatorname{sign}(\partial_\tau R) + \frac{\delta(W_1)}{R} \operatorname{sign}(\tau) \right)$$

$$= \frac{\tilde{\xi}\tilde{\xi}^T}{\rho^2} \delta(W_1) \operatorname{sign}\tau + \begin{pmatrix} \frac{\xi_1^2}{\rho^2} W_4 & \frac{\xi_1\xi_2}{\rho^2} W_4 & a_3\xi_1\xi_3 \\ \frac{\xi_1\xi_2}{\rho^2} W_4 & \frac{\xi_2^2}{\rho^2} W_4 & a_3\xi_2\xi_3 \\ a_3\xi_1\xi_3 & a_3\xi_2\xi_3 & W_3 \end{pmatrix} \delta(R) \operatorname{sign}(\partial_\tau R) \tag{4.4.18}$$

due to $R = -\rho^2 W_2$ for $W_1 = 0$ and

$$\frac{W_2}{W_1} = \frac{W_2 W_4}{W_1 W_4} = \frac{W_2 W_4}{\rho^2 W_2} = \frac{W_4}{\rho^2}$$

for $R = 0, W_1 \neq 0$.

(b) Next we determine the singular support of E by inserting the expression for T in (4.4.18) into the Herglotz–Gårding formula (4.4.15).

The explicit formula for $T = A^{\mathrm{ad}} \cdot S$ in (4.4.15) shows that E is smooth as long as the plane

$$\Pi_{(t,x)} = \{\xi \in \mathbf{R}^3; t + x\xi = 0\}$$

does not contain points on the slowness surface X where $\rho = 0$ and, furthermore, intersects the zero sets of $W_1(1, \xi)$ and of $R(1, \xi)$ both transversely. The last condition means that neither $(t, x) \in \Xi^*$ nor $\Pi_{(t,x)}$ contains a singular point of $R(1, \xi) = 0$.

If $\Pi_{(t,x)}$ contains a point $(0, 0, \xi_3) \in X$ where $\rho = 0$, then $t = \pm x_3 / \sqrt{a_5}$ or $t = \pm x_3 / \sqrt{a_2}$ and the description of the wavefront surface $W(P(-i\partial), N)$, which coincides with the singular support of the forward fundamental solution E_P of $P(-i\partial)$, in Example 4.3.10 (a) shows that E_P and a fortiori E are smooth at such a point (t, x) unless it belongs to $\Xi^* \cap H_N$. This implies that sing supp $E = \Xi^* \cap H_N$, or, in other words, that no conical refraction occurs if R is non-singular, i.e., if $a_1 \neq a_5$, $a_2 \neq a_5$ and $a_3 \neq 0$.

Let us yet show that the same holds generally as long as $a_2 \neq a_5$. For $a_3 = 0$, we have $R = W_3 W_4$ and hence

$$T = \left[\frac{\tilde{\xi}\tilde{\xi}^T}{\rho^2} \delta(W_1) + \frac{\xi'\xi'^T}{\rho^2} \delta(W_3) + \begin{pmatrix} 0 & 0 & 0 \\ 0 & 0 & 0 \\ 0 & 0 & 1 \end{pmatrix} \delta(W_4) \right] \operatorname{sign} \tau$$

where $\xi' = (\xi_1, \xi_2, 0)^T$ and $\tilde{\xi} = (\xi_2, -\xi_1, 0)^T$. This shows that no conical refraction appears in the case $a_3 = 0$.

On the other hand, for $a_1 = a_5$,

$$R = W_5^2 - a_3^2 \rho^2 \xi_3^2 - \left(\frac{a_1 - a_2}{2} \right)^2 \xi_3^4$$

with $W_5 = \tau^2 - a_1 \rho^2 - \frac{a_1 + a_2}{2} \xi_3^2$ and thus $R(1, \xi) = 0$ becomes singular along the circle $\xi_3 = 0$, $\rho = 1/\sqrt{a_1}$. Setting $f(\xi) = a_3^2 \rho^2 + \left(\frac{a_1 - a_2}{2} \right)^2 \xi_3^2$, we can decompose R near this circle, i.e., $R = (W_5 + \xi_3 \sqrt{f})(W_5 - \xi_3 \sqrt{f})$ and obtain

$$\xi_3 \delta(R) \operatorname{sign}(\partial_\tau R) = \frac{\operatorname{sign} \tau}{2\sqrt{f}} \left[\delta(W_5 - \xi_3 \sqrt{f}) - \delta(W_5 + \xi_3 \sqrt{f}) \right].$$

Similarly,

$$W_3 \delta(R) \operatorname{sign}(\partial_\tau R) = \frac{\operatorname{sign} \tau}{2\sqrt{f}} \left[\left(\sqrt{f} + \frac{a_2 - a_1}{2} \xi_3 \right) \delta(W_5 - \xi_3 \sqrt{f}) \right.$$
$$\left. + \left(\sqrt{f} + \frac{a_1 - a_2}{2} \xi_3 \right) \delta(W_5 + \xi_3 \sqrt{f}) \right],$$

and

$$W_4 \delta(R) \operatorname{sign}(\partial_\tau R) = \frac{\operatorname{sign} \tau}{2\sqrt{f}} \left[\left(\sqrt{f} + \frac{a_1 - a_2}{2} \xi_3 \right) \delta(W_5 - \xi_3 \sqrt{f}) \right.$$
$$\left. + \left(\sqrt{f} + \frac{a_2 - a_1}{2} \xi_3 \right) \delta(W_5 + \xi_3 \sqrt{f}) \right].$$

Altogether this yields

$$
T = \operatorname{sign} \tau \left[\frac{\tilde{\xi}\tilde{\xi}^T}{\rho^2} \, \delta(W_1) + \right.
$$

$$
+ \begin{pmatrix} \frac{\xi_1^2}{\rho^2}(\sqrt{f} + \frac{a_1-a_2}{2}\xi_3) & \frac{\xi_1\xi_2}{\rho^2}(\sqrt{f} + \frac{a_1-a_2}{2}\xi_3) & a_3\xi_1 \\ \frac{\xi_1\xi_2}{\rho^2}(\sqrt{f} + \frac{a_1-a_2}{2}\xi_3) & \frac{\xi_2^2}{\rho^2}(\sqrt{f} + \frac{a_1-a_2}{2}\xi_3) & a_3\xi_2 \\ a_3\xi_1 & a_3\xi_2 & \sqrt{f} + \frac{a_2-a_1}{2}\xi_3 \end{pmatrix} \frac{\delta(W_5 - \xi_3\sqrt{f})}{2\sqrt{f}}
$$

$$
\left. + \begin{pmatrix} \frac{\xi_1^2}{\rho^2}(\sqrt{f} + \frac{a_2-a_1}{2}\xi_3) & \frac{\xi_1\xi_2}{\rho^2}(\sqrt{f} + \frac{a_2-a_1}{2}\xi_3) & -a_3\xi_1 \\ \frac{\xi_1\xi_2}{\rho^2}(\sqrt{f} + \frac{a_2-a_1}{2}\xi_3) & \frac{\xi_2^2}{\rho^2}(\sqrt{f} + \frac{a_2-a_1}{2}\xi_3) & -a_3\xi_2 \\ -a_3\xi_1 & -a_3\xi_2 & \sqrt{f} + \frac{a_1-a_2}{2}\xi_3 \end{pmatrix} \frac{\delta(W_5 + \xi_3\sqrt{f})}{2\sqrt{f}} \right].
$$

$$(4.4.19)$$

From (4.4.19) we conclude that $E(t,x)$ is C^∞ unless $\Pi_{(t,x)}$ is tangential to one of the zero sets of the factors $W_5 \pm \xi_3\sqrt{f}$ of R, i.e., that conical refraction does not occur in the case $a_1 = a_5 \neq a_2$.

Note that the key point in the reasoning above is that the adjoint matrix $A^{\mathrm{ad}}(1,\xi)$ cancels factors in the denominators occurring in the additive decomposition of $\delta(\det A)$ near singular points of the slowness surface $\det A = 0$.

(c) Let us next reduce the three-dimensional integral in the Herglotz–Gårding representation (4.4.15) for the fundamental matrix E of the hexagonal system $A(\partial)$ to a one-fold integral with respect to ξ_3.

When we insert the decomposition for $T = A^{\mathrm{ad}}S$ in (4.4.18) as a sum of two surface measures on the ellipsoid $W_1 = 0$ and on the quartic $R = 0$, respectively, into (4.4.15), we obtain $E = E_1 + F$ where

$$
E_1 = -\frac{Y(t)}{4\pi^2} \, \partial_t \int_{\mathbf{R}^3} \frac{\tilde{\xi}\tilde{\xi}^T}{\rho^2} \, \delta(W_1(1,\xi))\delta(t + x\xi) \, d\xi
$$

and

$$
F = -\frac{Y(t)}{4\pi^2} \, \partial_t \int_{\mathbf{R}^3} \phi(\xi) \, \delta(R(1,\xi))\delta(t + x\xi) \, d\xi \qquad (4.4.20)
$$

with $\tilde{\xi} = (\xi_2, -\xi_1, 0)^T$, $\rho = \sqrt{\xi_1^2 + \xi_2^2}$ and

$$
\phi(\xi) = \begin{pmatrix} \frac{\xi_1^2}{\rho^2}W_4(1,\xi) & \frac{\xi_1\xi_2}{\rho^2}W_4(1,\xi) & a_3\xi_1\xi_3 \\ \frac{\xi_1\xi_2}{\rho^2}W_4(1,\xi) & \frac{\xi_2^2}{\rho^2}W_4(1,\xi) & a_3\xi_2\xi_3 \\ a_3\xi_1\xi_3 & a_3\xi_2\xi_3 & W_3(1,\xi) \end{pmatrix} \operatorname{sign}(\partial_\tau R(1,\xi)). \qquad (4.4.21)
$$

Let us start by evaluating the simpler part E_1. Putting $\tilde{\nabla} = (\partial_2, -\partial_1, 0)^T$, we infer

$$E_1 = -\frac{Y(t)}{8\pi^2} \tilde{\nabla}_x \tilde{\nabla}_x^T \int_{\mathbf{R}^3} \frac{1}{\rho^2} \delta(W_1(1,\xi)) \operatorname{sign}(t + x\xi) \, d\xi.$$

Because the integrand in the last integral is rotationally symmetric with respect to ξ_1, ξ_2, we can set therein $x_1 = 0$ and $x_2 = |x'|$ and obtain

$$E_1 = -\frac{Y(t)}{8\pi^2} \tilde{\nabla} \tilde{\nabla}^T \int_{\mathbf{R}^3} \frac{1}{|\xi'|^2} \delta(W_1(1,\xi)) \operatorname{sign}(t + |x'|\xi_2 + x_3\xi_3) \, d\xi$$

$$= -\frac{Y(t)}{4\pi^2} \tilde{\nabla} \frac{\tilde{x}^T}{|x'|} \int_{\mathbf{R}^3} \frac{\xi_2}{|\xi'|^2} \delta(1 - a_4|\xi'|^2 - a_5\xi_3^2) \delta(t + |x'|\xi_2 + x_3\xi_3) \, d\xi.$$

Upon using the homothecy $\eta' = \sqrt{a_4}\xi'/t$, $\eta_3 = \sqrt{a_5}\xi_3/t$, the integral representing E_1 takes the form treated in Example 1.2.15 (b):

$$E_1 = -\frac{Y(t)}{4\pi^2\sqrt{a_4a_5}t} \tilde{\nabla} \frac{\tilde{x}^T}{|x'|} \int_{\mathbf{R}^3} \frac{\eta_2}{|\eta'|^2} \delta(t^{-2} - |\eta|^2) \delta\left(1 + \frac{|x'|\eta_2}{\sqrt{a_4}} + \frac{x_3\eta_3}{\sqrt{a_5}}\right) d\eta$$

$$= -\frac{Y(t)}{4\pi^2\sqrt{a_4a_5}t} \tilde{\nabla} \frac{\tilde{x}^T}{|x'|} \frac{Y\left(\frac{|x'|^2}{a_4} + \frac{x_3^2}{a_5} - t^2\right)}{2\sqrt{\frac{|x'|^2}{a_4} + \frac{x_3^2}{a_5}}} \int_{S^1} \frac{A + B\omega_1}{(A + B\omega_1)^2 + C^2\omega_2^2} \, d\sigma(\omega),$$

where, in the notation of Example 1.2.15, $R = t^{-1}$, $b = \left(0, -\frac{|x'|}{\sqrt{a_4}}, -\frac{x_3}{\sqrt{a_5}}\right)^T$, $|b|^2 = \frac{|x'|^2}{a_4} + \frac{x_3^2}{a_5}$, and hence

$$A = -\frac{|x'|}{\sqrt{a_4}\,|b|^2}, \qquad B = \frac{\sqrt{|b|^2 - t^2}\,x_3}{\sqrt{a_5}\,|b|^2 t}, \qquad C = \frac{\sqrt{|b|^2 - t^2}}{|b|t}.$$

Therefore, $A^2 - B^2 + C^2 = |x'|^2/(a_4|b|^2t^2) > 0$ for $|x'| > 0$. For $A, B, C \in \mathbf{R}$ with $A^2 - B^2 + C^2 > 0$, an application of the residue theorem yields

$$\int_{S^1} \frac{A + B\omega_1}{(A + B\omega_1)^2 + C^2\omega_2^2} \, d\sigma(\omega) = \operatorname{Re} \int_{S^1} \frac{d\sigma(\omega)}{A + B\omega_1 + iC\omega_2}$$

$$= \frac{2\pi Y(|A| - |B|)\operatorname{sign} A}{\sqrt{A^2 - B^2 + C^2}}.$$

Hence

$$E_1 = \frac{Y(t)}{4\pi\sqrt{a_5}}\,\tilde{\nabla}\left[Y\!\left(t-\frac{|x_3|}{\sqrt{a_5}}\right)Y\!\left(\frac{|x'|^2}{a_4}+\frac{x_3^2}{a_5}-t^2\right)\frac{\tilde{x}^T}{|x'|^2}\right]$$

$$= \frac{Y\!\left(t-\frac{|x_3|}{\sqrt{a_5}}\right)-Y\!\left(t-\sqrt{\frac{|x'|^2}{a_4}+\frac{x_3^2}{a_5}}\right)}{4\pi\sqrt{a_5}\,|x'|^4}\cdot(x'x'^T-\tilde{x}\tilde{x}^T)$$

$$+ \frac{\delta\!\left(t-\sqrt{\frac{|x'|^2}{a_4}+\frac{x_3^2}{a_5}}\right)}{4\pi a_4\sqrt{a_5}\,t|x'|^2}\cdot\tilde{x}\tilde{x}^T,$$

and this expression coincides with the formula for E_1 in Ortner and Wagner [220], Prop. 5, p. 429.

Let us finally derive a representation by a one-fold integral for F. Here we use the general formula (1.2.3) in Example 1.2.15 referring to $\langle\phi,\delta\circ h\rangle$ for the function

$$h: \mathbf{R}^3\longrightarrow\mathbf{R}^2:\xi\longmapsto\begin{pmatrix}t+x\xi\\R(1,\xi)\end{pmatrix}.$$

Thereby, $\langle\phi,\delta\circ h\rangle$ is expressed by an integral over the space curve

$$C_{t,x}=h^{-1}(0)=\{\xi\in\mathbf{R}^3;\,t+x\xi=R(1,\xi)=0\}$$

depending on $(t,x)\in\mathbf{R}^4$. From $h'^T=(x,\nabla R(1,\xi))$ we obtain

$$\det(h'\cdot h'^T)=\det\begin{pmatrix}|x|^2 & x^T\nabla R(1,\xi)\\x^T\nabla R(1,\xi) & |\nabla R(1,\xi)|^2\end{pmatrix}=|x\times\nabla R(1,\xi)|^2$$

and hence

$$F=-\frac{Y(t)}{4\pi^2}\,\partial_t\int_{C_{t,x}}\frac{\phi(\xi)\,d\sigma(\xi)}{|x\times\nabla R(1,\xi)|},\tag{4.4.22}$$

where $\phi(\xi)$ is as in (4.4.21) and $d\sigma$ denotes the measure of arc-length on $C_{t,x}$.

If $C_{t,x}$ is locally parameterized by $s=\xi_3$, then $\dot{\xi}(s)$ is perpendicular to x and to $\nabla R(1,\xi)$ and this implies

$$\dot{\xi}(s)=\frac{x\times\nabla R(1,\xi(s))}{(x\times\nabla R(1,\xi(s)))_3}=\frac{x\times\nabla R(1,\xi)}{(x_1\partial_2 R-x_2\partial_1 R)(1,\xi(s))}.$$

By the rotational symmetry of F with respect to x_1, x_2, we can set $x_2 = 0$, $x_1 = |x'|$ and conclude that

$$\frac{d\sigma(\xi)}{|x \times \nabla R(1,\xi)|} = \frac{|\dot{\xi}(s)|\,ds}{|x \times \nabla R(1,\xi)|} = \frac{ds}{|x'| \cdot |\partial_2 R(1,\xi(s))|}$$

and hence

$$F = -\frac{Y(t)}{4\pi^2|x'|}\,\partial_t \int_{J_{t,x}} \frac{\phi(\xi(s))\,ds}{|\partial_2 R(1,\xi(s))|}$$

where the integration is performed over the union $J_{t,x}$ of the intervals in $s = \xi_3$ which parameterize the space curve $C_{t,x}$.

In order to compare this expression with the more explicit formula in Ortner and Wagner [220], Prop. 5, p. 429, where F here corresponds to $E_2 + E_3 + E_4$, let us deduce, by way of example, the element F_{23} in the second row, third column of the matrix F.

From formula (4.4.20) we infer that

$$F_{23} = -\frac{Y(t)a_3}{4\pi^2} \int_{\mathbf{R}^3} \xi_2\xi_3\,\delta(R(1,\xi))\delta'(t+x\xi)\,\mathrm{sign}(\partial_\tau R(1,\xi))\,d\xi$$

$$= -\frac{Y(t)a_3}{4\pi^2}\,\partial_2 \int_{\mathbf{R}^3} \xi_3\,\delta(R(1,\xi))\delta(t+x\xi)\,\mathrm{sign}(\partial_\tau R(1,\xi))\,d\xi$$

$$= -\frac{Y(t)a_3}{4\pi^2}\,\partial_2\frac{1}{|x'|} \int_{J_{t,x}} \frac{s\,\mathrm{sign}(\partial_\tau R(1,\xi(s)))\,ds}{|\partial_2 R(1,\xi(s))|},$$

where the space curve

$$C_{t,x} = \{\xi \in \mathbf{R}^3;\ t + |x'|\xi_1 + x_3\xi_3 = R(1,\xi) = 0\}$$

is parameterized by $s = \xi_3 \in J_{t,x}$.

The equation $R(1,\xi_1,\xi_2,s) = 0$ has the form $\alpha\rho^4 + \beta\rho^2 + \gamma = 0$, where $\rho^2 = \xi_1^2 + \xi_2^2$ and $\alpha = a_1 a_5$, $\beta = 2a_6 s^2 - a_1 - a_5$, $a_6 = \frac{1}{2}(a_1 a_2 + a_5^2 - a_3^2)$, $\gamma = (1 - a_2 s^2)(1 - a_5 s^2)$. This yields the discriminant

$$D(s) = \frac{\beta^2}{4} - \alpha\gamma = \frac{1}{4}(a_1 + a_5 - 2a_6 s^2)^2 - a_1 a_5(1 - a_2 s^2)(1 - a_5 s^2).$$

Hence, the points $\xi \in C_{t,x}$ fulfill $\xi_3 = s$, $\xi_1 = -(t + x_3 s)/|x'|$ and $\xi_2^2 = \mu_{1,2}/(a_1 a_5|x'|^2)$ where

$$\mu_{1,2} = -a_1 a_5(t + x_3 s)^2 + \frac{a_1 + a_5}{2}|x'|^2 - a_6|x'|^2 s^2 \pm |x'|^2\sqrt{D(s)}.$$

The intervals $J_{t,x}$ are determined by the inequalities $D(s) \geq 0$, $\mu_{1,2}(s) \geq 0$.

Finally,

$$\partial_2 R = 4\alpha\rho^2\xi_2 + 2\beta\xi_2 = 2\xi_2(2\alpha\rho^2 + \beta) = \pm 4\xi_2\sqrt{D(s)}.$$

Therefore

$$\frac{1}{|\partial_2 R(1,\xi(s))|} = \frac{\sqrt{a_1 a_5}|x'|}{4\sqrt{\mu_{1,2}}\sqrt{D(s)}}$$

and

$$F_{23} = -\frac{Y(t)a_3\sqrt{a_1 a_5}}{8\pi^2}\partial_2\sum_{j=1}^{2}\epsilon_j\int_{\mathbf{R}}\frac{sY(\mu_j(s))\,Y(D(s))}{\sqrt{\mu_j(s)}\sqrt{D(s)}}\,ds, \qquad \epsilon_j = \pm 1,$$

in agreement with the corresponding entry of the matrix E_3 in Ortner and Wagner [220], Prop. 5, p. 429. (Note that the sign factors $\epsilon_j = \mathrm{sign}(\partial_\tau R(1,\xi))$ are constant along connected components of the curve $C_{t,x}$. They are described in more detail in Ortner and Wagner [220], p. 433. Furthermore, the two possible signs we can choose for ξ_2 account for an additional factor 2.)

Let us yet observe that the integrals representing E are, from the viewpoint of algebraic geometry, complete Abelian integrals of genus $g = 3$ for generic a_j, t, x, see Remark 1 in Ortner and Wagner [220], p. 434. In contrast, the genus g of the curves $C_{t,x}$ reduces to 0 if $R(\tau,\xi)$ is a product of two wave operators. This occurs in the two cases $a_3 = 0$ and $a_3^2 = (a_1 - a_5)(a_2 - a_5)$. In these cases, E is given by algebraic functions, see Ortner and Wagner [220], Section 6. Furthermore, the genus of $C_{t,x}$ is also 0 if (t,x) belongs to the x_3-axis, i.e., if $\rho = 0$, a fact first observed in Payton [226], Ch. 3, 11, p. 105–111, cf. Ortner and Wagner [220], Prop. 6, p. 442. Moreover, $E(t,x)$ is given by complete *elliptic* integrals, i.e., $g = 1$, if $a_1 = a_5$ or if (t,x) belongs to the hyperplane $x_3 = 0$, see Ortner and Wagner [220], Sections 6.3 and 7.1. $\Box$

Due to the two δ-factors occurring in the integral in (4.4.3) in Proposition 4.4.1, this representation of E in the case $m = 2$, $n = 4$ can always be reduced to a one-fold definite integral similarly as this was done in Example 4.4.5 above for the hexagonal system. The resulting integral is an Abelian integral over the algebraic curve $C_{t,x} = \{\xi \in \mathbf{R}^3;\ t + x\xi = P(1,\xi) = 0\}$. In formula (A), p. 327, in Ortner and Wagner [217], this integral was stated as parameterized with respect to ξ_1. In the next proposition, we parameterize it with respect to arc-length. In the case of crystal optics, the respective representation was stated in Burridge and Quian [40], pp. 76, 77, and in Wagner [298], (7.2), p. 2679.

Proposition 4.4.6 *Let $A(\partial_t, \nabla)$ be a strictly hyperbolic (with respect to t) $l \times l$ matrix of differential operators in $\mathbf{R}^4$ which are homogeneous of degree $m = 2$ and set $P(\partial) = \det A(\partial)$. We assume that $P(\tau,\xi)$ does not contain τ as a factor. Then the* *forward fundamental matrix E of A(∂) is given by*

$$E = -\frac{Y(t)}{4\pi^2} \, \partial_t \int_{C_{t,x}} A^{\mathrm{ad}}(1,\xi) \, \mathrm{sign}(\partial_\tau P(1,\xi)) \, \frac{\mathrm{d}\sigma(\xi)}{|x \times \nabla P(1,\xi)|} \qquad (4.4.23)$$

where dσ *denotes the measure of arc-length on the curve* $C_{t,x} = \{\xi \in \mathbf{R}^3;\, t + x\xi = P(1,\xi) = 0\}$.

Proof From (4.4.3) we obtain

$$E = -\frac{Y(t)}{4\pi^2} \, \partial_t \int_{\mathbf{R}^3} A^{\mathrm{ad}}(1,\xi) \, \mathrm{sign}(\partial_\tau P(1,\xi)) \, \delta \circ h \, \mathrm{d}\xi$$

if we set, as in Example 4.4.5 above,

$$h : \mathbf{R}^3 \longrightarrow \mathbf{R}^2 : \xi \longmapsto \begin{pmatrix} t + x\xi \\ R(1,\xi) \end{pmatrix}.$$

From $\det(h' \cdot h'^T) = |x \times \nabla P(1,\xi)|^2$ and Example 1.2.15, we then conclude that

$$\int_{\mathbf{R}^3} \phi(\xi) \, \delta \circ h \, \mathrm{d}\xi = \int_{C_{t,x}} \frac{\phi(\xi) \, \mathrm{d}\sigma(\xi)}{|x \times \nabla P(1,\xi)|}, \qquad \phi \in \mathcal{E}(\mathbf{R}^3).$$

This implies the result. □

Before applying the formulas (4.4.3), (4.4.9), (4.4.23), which all refer to homogeneous hyperbolic systems $A(\partial)$ of degree two in dimension four, let us yet deduce an alternative form in which the adjoint matrix $A^{\mathrm{ad}}(1,\xi)$ on the slowness surface X is expressed by normalized eigenvectors of $A(1,\xi)$. Such a representation was first used in Grünwald [120] for the system of crystal optics and later in Burridge [39] for the elastodynamic system. In the systematic account given in Ortner and Wagner [217], the respective formulas are presented in Section 2.2.2 ("Grünwald's formula") under (G) and (G'), p. 326.

Proposition 4.4.7 *Let* $B(\nabla) = B(\partial_1, \partial_2, \partial_3)$ *be a symmetric* $l \times l$ *matrix of homogeneous differential operators of second degree in* $\mathbf{R}^3$ *and set* $A(\partial) = I_l \partial_t^2 - B(\nabla)$. *We assume that* $P(\partial) = \det A(\partial)$ *is strictly hyperbolic in the direction* $N = (1,0,0,0)$ *and does not contain* ∂_t *as a factor. As before,* $X = \{\xi \in \mathbf{R}^3;\, P(1,\xi) = 0\}$ *denotes the slowness surface,* $C_{t,x} = \{\xi \in X;\, t + x\xi = 0\}$ *for* $(t,x) \in \mathbf{R}^4$ *and* $|\gamma|$ *is the Leray measure, see* Corollary 4.4.2. *For* $\xi \in X$, *let* $e(\xi)$ *be a normalized eigenvector of* $B(\xi)$ *for the eigenvalue one, i.e.,* $|e(\xi)| = 1$ *and* $B(\xi)e(\xi) = e(\xi)$.

Then the forward fundamental matrix E of A(∂) is given by

$$E = -\frac{Y(t)}{8\pi^2} \int_X \delta'(t + x\xi) \, e(\xi) \cdot e(\xi)^T \, |\gamma|(\xi) \qquad (4.4.24)$$

$$= -\frac{Y(t)}{8\pi^2} \, \partial_t \int_{C_{t,x}} e(\xi) \cdot e(\xi)^T \, \frac{|\partial_\tau P(1,\xi)|}{|x \times \nabla P(1,\xi)|} \, \mathrm{d}\sigma(\xi). \qquad (4.4.25)$$

Proof The strict hyperbolicity of $P(\partial)$ implies that $(\partial_\tau P)(1,\xi) \neq 0$ for $\xi \in X$ and hence the symmetric matrix $A(1,\xi)$ has rank $l-1$ for $\xi \in X$. Therefore the eigenvectors $e(\xi)$ are uniquely determined up to a sign. Note, furthermore, that $A(1,\xi) \cdot A^{\mathrm{ad}}(1,\xi) = 0$ and hence the rows of $A^{\mathrm{ad}}(1,\xi)$ are all proportional to $e(\xi)$. This furnishes $A^{\mathrm{ad}}(1,\xi) = ce(\xi) \cdot e(\xi)^T$ where

$$\mathrm{tr}\big(A^{\mathrm{ad}}(1,\xi)\big) = c\,\mathrm{tr}\big(e(\xi) \cdot e(\xi)^T\big) = c|e(\xi)|^2 = c.$$

From

$$\frac{\partial P(\tau,\xi)}{\partial \tau} = \frac{\partial \det A(\tau,\xi)}{\partial \tau} = \sum_{i,j=1}^{l} A_{ji}^{\mathrm{ad}}(\tau,\xi) \cdot \frac{\partial a_{ij}}{\partial \tau} = 2\tau\,\mathrm{tr}\big(A^{\mathrm{ad}}(\tau,\xi)\big),$$

we obtain $\mathrm{tr}\big(A^{\mathrm{ad}}(1,\xi)\big) = \frac{1}{2}(\partial_\tau P)(1,\xi)$ and hence

$$A^{\mathrm{ad}}(1,\xi) = \frac{1}{2}(\partial_\tau P)(1,\xi) \cdot e(\xi) \cdot e(\xi)^T. \tag{4.4.26}$$

Inserting the expression for the adjoint matrix in (4.4.26) into (4.4.9) and (4.4.23), respectively, then furnishes (4.4.24–4.4.25). $\square$

Example 4.4.8 Let us apply now the above formulas to *Maxwell's system of crystal optics*, which describes the propagation of light in homogeneous dielectric media.

(a) As usual, we denote by $\mathcal{E}$ and $\mathcal{H}$ the electric and magnetic field, and by $\mathcal{D}$ and $\mathcal{B}$ the dielectric and magnetic induction, respectively. If, furthermore, ρ, J denote the charge and current densities, respectively, and c is the speed of light, then Maxwell's equations read in the Gauß unit system as follows:

$$\mathrm{curl}\,\mathcal{E} = -\frac{1}{c}\,\partial_t \mathcal{B}, \quad \mathrm{div}\,\mathcal{D} = 4\pi\rho, \quad \mathrm{curl}\,\mathcal{H} = \frac{1}{c}\,\partial_t \mathcal{D} + \frac{4\pi}{c}\,J, \quad \mathrm{div}\,\mathcal{B} = 0 \tag{4.4.27}$$

For anisotropic materials, these equations are supplemented by

$$\mathcal{D} = \epsilon\mathcal{E}, \qquad \mathcal{B} = \mu\mathcal{H} \tag{4.4.28}$$

with $\mu > 0$, $\epsilon = \begin{pmatrix} \epsilon_1 & 0 & 0 \\ 0 & \epsilon_2 & 0 \\ 0 & 0 & \epsilon_3 \end{pmatrix}$ and ϵ_j positive. We consider homogeneous media only, i.e., media wherein c, μ, ϵ_j are constants.

By inserting Eqs. (4.4.28) into (4.4.27) we obtain two uncoupled 3×3-systems of equations for the field vectors $\mathcal{E}$ and $\mathcal{H}$:

$$\epsilon\partial_t^2 \mathcal{E} + \frac{c^2}{\mu}\,\mathrm{curl}(\mathrm{curl}\,\mathcal{E}) = -4\pi\partial_t J, \tag{4.4.29}$$

$$\partial_t^2 \mathcal{H} + \frac{c^2}{\mu}\, \mathrm{curl}(\epsilon^{-1}\,\mathrm{curl}\,\mathcal{H}) = \frac{4\pi c}{\mu}\, \mathrm{curl}(\epsilon^{-1}J) \qquad (4.4.30)$$

For (4.4.30) cf. Gårding [92], (10), p. 363, and p. 377. The system in (4.4.30) has the form

$$A(\partial) = \left(I_3 \partial_t^2 - B_1(\nabla)\right)\mathcal{H} = \frac{4\pi c}{\mu}\, \mathrm{curl}(\epsilon^{-1}J) \qquad (4.4.31)$$

with the symmetric matrix

$$B_1(\xi) = \begin{pmatrix} d_3\xi_2^2 + d_2\xi_3^2 & -d_3\xi_1\xi_2 & -d_2\xi_1\xi_3 \\ -d_3\xi_1\xi_2 & d_3\xi_1^2 + d_1\xi_3^2 & -d_1\xi_2\xi_3 \\ -d_2\xi_1\xi_3 & -d_1\xi_2\xi_3 & d_1\xi_2^2 + d_2\xi_1^2 \end{pmatrix}$$

where we have set $d_j = \dfrac{c^2}{\mu\epsilon_j}$, $j = 1, 2, 3$. In the following, we suppose that $0 < d_1 < d_2 < d_3$.

Similarly, the system in (4.4.29) has the form

$$(\epsilon\partial_t^2 - B_2(\nabla))\mathcal{E} = -4\pi\partial_t J, \qquad B_2(\xi) = \frac{c^2}{\mu}(|\xi|^2 I_3 - \xi \cdot \xi^T).$$

This system for $\mathcal{E}$ could be treated by a modification of Grünwald's formula (see Ortner and Wagner [217], p. 325), but for simplicity, we shall restrict our analysis to the system in (4.4.31) referring to the magnetic field $\mathcal{H}$.

Let us next calculate $P(\tau, \xi) = \det A(\tau, \xi)$. Obviously, we have $B_1(\xi)\xi = 0$ for $\xi \in \mathbf{R}^3$ and hence $\det B_1(\xi) = 0$. On the other hand,

$$\mathrm{tr}\, B_1(\xi) = \sum_{j=1}^{3} \xi_j^2(d_{j+1} + d_{j+2}), \qquad \mathrm{tr}\big(B_1(\xi)^{\mathrm{ad}}\big) = |\xi|^2 \sum_{j=1}^{3} \xi_j^2 d_{j+1}d_{j+2}$$

where $d_{j+3} = d_j$, $j = 1, 2, 3$. This yields $P(\tau, \xi) = \det(I_3\tau^2 - B_1(\xi)) = \tau^2 \cdot R(\tau, \xi)$ where

$$R(\tau, \xi) = \tau^4 - \tau^2 \sum_{j=1}^{3} \xi_j^2(d_{j+1} + d_{j+2}) + |\xi|^2 \sum_{j=1}^{3} \xi_j^2 d_{j+1}d_{j+2}, \qquad (4.4.32)$$

cf. Courant and Hilbert [52], (13), p. 606; Gårding [92], (11), p. 363; Liess [165], (6.7.6), p. 272.

Let us observe that Maxwell's system in (4.4.31) does not fulfill all the requirements for the application of Grünwald's formula in Proposition 4.4.7 due to two circumstances: First, the determinant $P(\partial)$ contains twice the factor ∂_t and, second, neither $P(\partial)$ nor $R(\partial)$ in (4.4.32) are *strictly* hyperbolic. In the following, we shall show how to circumvent these shortcomings by approximation.

(b) Instead of $A(\partial) = I_3 \partial_t^2 - B_1(\nabla)$, let us consider the perturbation $A_\zeta(\partial)$ given by

$$A_\zeta(\partial) = I_3 \partial_t^2 - \zeta \cdot \nabla \cdot \nabla^T - B_1(\nabla), \qquad \zeta > 0,$$

which was already employed in Grünwald [120], p. 519. Due to

$$A_\zeta(\tau, \xi)\xi = (\tau^2 - \zeta \cdot |\xi|^2)\xi \quad \text{and} \quad A_\zeta(\tau, \xi)e = A(\tau, \xi)e \quad \text{for } e \perp \xi,$$

we conclude that $P_\zeta(\tau, \xi) = \det A_\zeta(\tau, \xi) = (\tau^2 - \zeta|\xi|^2)R(\tau, \xi)$ with R as in (4.4.32). Hence $P_\zeta(\tau, \xi)$ does not contain the factor τ. Note, however, that $P_\zeta(\partial)$ is still not strictly hyperbolic, since the part $X_1 = \{\xi \in \mathbf{R}^3; R(1, \xi) = 0\}$ of the slowness surface of P_ζ, which is called *Fresnel's surface* since Hamilton, has four singularities, which we will discuss later on. Nevertheless, Grünwald's formula (4.4.24) can be applied if $|\gamma|$ is replaced by a suitable measure similarly as in Example 4.4.5.

If $X_{0,\zeta} = \{\xi \in \mathbf{R}^3; |\xi| = 1/\sqrt{\zeta}\}$ denotes the trivial part of the slowness surface, then $e(\xi) = \xi/|\xi|$ for $\xi \in X_{0,\zeta}$ and the fundamental matrix E_ζ of $A_\zeta(\partial)$ splits into two parts: $E_\zeta = E_{0,\zeta} + E_1$ where

$$E_{0,\zeta} = -\frac{Y(t)}{8\pi^2} \int_{X_{0,\zeta}} \delta'(t + x\xi) \frac{\xi \cdot \xi^T}{|\xi|^2} |\gamma|(\xi)$$

and

$$E_1 = -\frac{Y(t)}{8\pi^2} \int_{X_1} \delta'(t + x\xi) \, e(\xi) \cdot e(\xi)^T \, |\gamma|(\xi). \tag{4.4.33}$$

We note that E_1 is independent of ζ. If $\xi \in X_{0,\zeta}$ is parameterized by $\omega \in \mathbf{S}^2$, then

$$E_{0,\zeta} = -\frac{Y(t)}{8\pi^2\zeta^{3/2}} \int_{\mathbf{S}^2} \delta'\left(t + \frac{x\omega}{\sqrt{\zeta}}\right)\omega \cdot \omega^T \, d\sigma(\omega)$$

$$= -\frac{Y(t)}{8\pi^2\sqrt{\zeta}} \nabla \cdot \nabla^T \int_{\mathbf{S}^2} Y\left(t + \frac{x\omega}{\sqrt{\zeta}}\right) d\sigma(\omega).$$

The last surface integral gives the area of a spherical cap of height $\min\{2, 1 + \sqrt{\zeta}t/|x|\}$. Therefore

$$
\begin{aligned}
E_{0,\zeta} &= -\frac{Y(t)}{4\pi\sqrt{\zeta}} \nabla \cdot \nabla^T \left[2 - Y(|x| - \sqrt{\zeta}t) \frac{|x| - \sqrt{\zeta}t}{|x|} \right] \\
&= \frac{Y(t)}{4\pi} \left[Y(|x| - \sqrt{\zeta}t)t \cdot \mathrm{vp}\left(\frac{I_3|x|^2 - 3xx^T}{|x|^5} \right) + \frac{xx^T}{\zeta^2 t^3} \delta(|x| - \sqrt{\zeta}t) \right],
\end{aligned}
$$

where we have used formula (1.3.13) to perform the differentiations.

If ζ tends to zero from above, then we obtain the so-called *static term* for the magnetic field of crystal optics:

$$
E_0 = \lim_{\zeta \searrow 0} E_{0,\zeta} = \mathrm{vp}\,\frac{tY(t)}{4\pi|x|^3}\left(I_3 - \frac{3xx^T}{|x|^2} \right) + \frac{1}{3}(tY(t) \otimes \delta(x))I_3. \qquad (4.4.34)
$$

Formula (4.4.34) for the static term E_0 was derived first in Ortner and Wagner [217], pp. 334, 335, see also Wagner [298], (6.2), p. 2676. The corresponding formula for the static term of the electric field, i.e.,

$$
\tilde{E}_0 = -\frac{Y(t)t\sqrt{d_1 d_2 d_3}}{4\pi} \nabla\nabla^T \left(\sum_{j=1}^{3} d_j x_j^2 \right)^{-1/2},
$$

was given in Ortner and Wagner [222], Thm. 1, p. 314. A more complicated expression for $\tilde{E}_0$ was found earlier in Burridge and Quian [40], (5.11), p. 78, and (B19), p. 93.

Let us yet show that E coincides with E_0 for large t, or, more precisely, in the component L containing $N = (1, 0, 0, 0)$ of the complement of $\mathrm{sing\,supp}\,E = W(P(-i\partial), N)$, see Proposition 4.2.5. In fact, if (t, x) belongs to the region L, then $t + x\xi \neq 0$ for $\xi \in X_1$ and hence there E_1 vanishes and $E = E_0$.

(c) Let us now reduce the "non-trivial" part E_1 of the fundamental matrix E, see (4.4.33), to a one-fold integral over the curve $C_{t,x}$ defined as the intersection of the part $X_1 = \{\xi \in \mathbf{R}^3; R(1, \xi) = 0\}$ of the slowness surface with the plane $t + x\xi = 0$.

Similarly as in the proof of Proposition 4.4.7, we conclude that

$$
E_1 = -\frac{Y(t)}{8\pi^2} \partial_t \int_{X_1} \delta(t + x\xi)\, e(\xi) \cdot e(\xi)^T\, |\gamma|(\xi) \qquad (4.4.35)
$$

$$
= -\frac{Y(t)}{8\pi^2} \partial_t \int_{C_{t,x}} e(\xi) \cdot e(\xi)^T\, \frac{|\partial_\tau R(1, \xi)|}{|x \times \nabla R(1, \xi)|}\, d\sigma(\xi), \qquad (4.4.36)
$$

see also (4.4.24–4.4.25). As before, for $\xi \in X_1$, $e(\xi)$ are normalized eigenvectors of $B_1(\xi)$ to the eigenvalue 1.

Note that the affine part of the slowness surface, i.e.,

$$P(1,\xi) = \det(I_3 - B_1(\xi)) = R(1,\xi) = 0$$

is given by

$$X_1 = \left\{ \xi \in \mathbf{R}^3; \ 1 - \sum_{j=1}^{3} \xi_j^2(d_{j+1} + d_{j+2}) + |\xi|^2 \sum_{j=1}^{3} \xi_j^2 d_{j+1} d_{j+2} = 0 \right\}$$

and is called Fresnel's surface.

We observe that X_1 is the union of the two sheets where either one of the two non-zero eigenvalues $\lambda_{2,3}(\xi)$ of the matrix $B_1(\xi)$ equals 1. Therefore, X_1 is non-singular apart from the points where λ_2 and λ_3 coincide, i.e., where $B_1(\xi) - I_3$ has rank one. For such ξ, all vectors orthogonal to ξ must be eigenvectors of $B_1(\xi)$, and inserting $(\xi_2, -\xi_1, 0)$ yields $\xi_1 \xi_2 \xi_3 (d_1 - d_2) = 0$. Due to the assumption that the constants d_j are pairwise different, we conclude that the singular points of X must lie on one of the three coordinate planes. The rank one condition then readily yields that X_1 has exactly four singular points, which—due to the ordering $0 < d_1 < d_2 < d_3$—are given by

$$\xi = \left(\pm\sqrt{\frac{1}{d_2}\frac{d_2 - d_1}{d_3 - d_1}}, 0, \pm\sqrt{\frac{1}{d_2}\frac{d_3 - d_2}{d_3 - d_1}} \right), \tag{4.4.37}$$

cf. Liess [165], p. 273.

Hence X_1 is homeomorphic to two disjoint spheres glued together at the four singular points of X_1, which two by two are pairwise opposite and span the "optical axes," see Fig. 4.5, taken from Fladt and Baur [75], 6. Abschnitt, p. 346, by courtesy of Springer-Verlag. In this figure, the outer sheet of X_1 and two of the four singular

Fig. 4.5 Fresnel's surfaces X_1 and X_1^*, respectively; taken from Fladt and Baur [75], p. 346, © by Friedr. Vieweg & Sohn Verlagsgesellschaft mbH, Braunschweig 1975

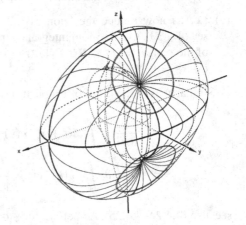

points are plainly visible. (Note that the x-axis in Fig. 4.1 corresponds to the ξ_2-axis in our notation.)

The dual surface of X_1, the wave surface $X_1^* = \{\nabla R(1,\xi)/\partial_\tau R(1,\xi); \ \xi \in X_1\}$, has the same shape as X_1. It is given by

$$X_1^* = \left\{\xi \in \mathbf{R}^3; \ 1 - \sum_{j=1}^{3} \xi_j^2(d_{j+1}^{-1} + d_{j+2}^{-1}) + |\xi|^2 \sum_{j=1}^{3} \xi_j^2 d_{j+1}^{-1} d_{j+2}^{-1} = 0\right\},$$

see Poincaré [230]; Esser [70], (6.a), p. 203; Gårding [92], pp. 359, 360; Wagner [298], (4.5), p. 2671.

Let us next investigate the singular support of the fundamental matrix $E = E_0 + E_1$ of the system $A(\partial) = I_3 \partial_t^2 - B_1(\nabla)$ governing the magnetic field, see (4.4.31). Plainly, sing supp $E \subset$ sing supp $E(P(-i\partial), N)$, where $N = (1, 0, 0, 0)$ and

$$P(\tau, \xi) = \det(I_3 \tau^2 - B_1(\xi)) = \tau^2 \cdot R(\tau, \xi)$$

with $R(\tau, \xi)$ as in (4.4.32). Therefore

$$\text{sing supp } E(P(-i\partial), N) = \{(t, 0); \ t \geq 0\} \cup \text{sing supp } E(R(-i\partial), N).$$

According to Proposition 4.2.5 and formula (4.2.5),

$$\text{sing supp } E(R(-i\partial), N) = W(R(-i\partial), N) = (\Xi_1^* \cap H_N) \cup C,$$

where $\Xi_1^* = \{(t, x); \ t \geq 0, \ \frac{x}{t} \in X_1^*\}$ and the set of conical refraction C is given by the union of the propagation cones $K(R_{(1,\xi)}(-i\partial), N)$ in the directions $(1, \xi)$ of the four singular points on the slowness surface Ξ_1 of R, see (4.4.37).

A straight-forward calculation (see Liess [165], p. 75) yields

$$R_{(1,\xi)}(\tau, \eta) = 4\left[\tau - \tfrac{1}{2}(d_2 + d_3)\xi_1\eta_1 - \tfrac{1}{2}(d_1 + d_2)\xi_3\eta_3\right]^2$$
$$- \left[(d_3 - d_2)\xi_1\eta_1 + (d_1 - d_2)\xi_3\eta_3\right]^2 - \tfrac{1}{d_2}(d_2 - d_1)(d_3 - d_2)\eta_2^2.$$

If y_1, y_2, y_3 is an orthogonal coordinate system with $y_2 = x_2, y_3 = \sqrt{d_2}(\xi_1 x_1 + \xi_3 x_3)$ and y_1 running in the direction of $(d_3 - d_2)\xi_1 x_1 + (d_1 - d_2)\xi_3 x_3$, then $R_{(1,\xi)}(-i\partial)$ becomes a shifted planar wave operator, viz.

$$-R_{(1,\xi)}(-i\partial) = 4\left[\frac{\partial}{\partial t} - \frac{c}{2}\frac{\partial}{\partial y_1} - \sqrt{d_2}\frac{\partial}{\partial y_3}\right]^2 - c^2\left(\frac{\partial^2}{\partial y_1^2} + \frac{\partial^2}{\partial y_2^2}\right)$$

where $c = \sqrt{(d_2 - d_1)(d_3 - d_2)/d_2}$. Its forward fundamental solution is given by

$$E(R_{(1,\xi)}(-i\partial), N) = -\frac{Y(ct - \sqrt{(2y_1 + ct)^2 + 4y_2^2})}{2\pi c\sqrt{c^2t^2 - (2y_1 + ct)^2 - 4y_2^2}} \delta(y_3 + \sqrt{d_2}t),$$

see Example 1.4.12 (b) and Proposition 1.3.19.

Therefore the set C of conical refraction consists of the four lids

$$K(R_{(1,\xi)}(-i\partial), N) = \operatorname{supp} E(R_{(1,\xi)}(-i\partial), N)$$

(with ξ as in (4.4.37)), which are the so-called Hamiltonian circles, cf. Ludwig [172], p. 117. Since $K(R_{(1,\xi)}(-i\partial), N)$ is given by the conditions $y_3 + \sqrt{d_2}t = 0$ and $t \geq \frac{1}{c}\sqrt{(2y_1 + ct)^2 + 4y_2^2}$, their affine representations, i.e., the intersection with $t = 1$, are circles in the planes $\xi_1 x_1 + \xi_3 x_3 + 1 = 0$ with centers $x = -\frac{1}{2}(\xi_1(d_2 + d_3), 0, \xi_3(d_1 + d_2))$ and diameter c, see also Esser [70], p. 206. Two of these circles, which yield the convex hull of the parts of the wave surface X_1^* with negative Gaussian curvature, are visible in Fig. 4.5. Summing up we obtain

$$\operatorname{sing\,supp} E(P(-i\partial), N) = \left\{(t, x) \in \mathbf{R}^4; t \geq 0 \text{ and } \left[x = 0 \text{ or } \frac{x}{t} \in X_1^*\right]\right\} \cup$$

$$\bigcup_{\xi \text{ as in (4.4.37)}} \{(t, x) \in \mathbf{R}^4; t \geq 0, \xi_1 x_1 + \xi_3 x_3 + t = 0,$$

$$|x|^2 + \xi_1(d_2 + d_3)tx_1 + \xi_3(d_1 + d_2)tx_3 + d_2 t^2 \leq 0\}. \qquad (4.4.38)$$

Let us come back to the original goal of determining the singular support of the fundamental matrix E of the system $A(\partial)$ in (4.4.31). Similarly as in Example 4.3.9 (b), one applies Proposition 4.3.4 in order to show that $\operatorname{sing\,supp} E$ coincides with the singular support of $E(P(-i\partial), N)$, which was described in (4.4.38).

Let us yet discuss the relationship of E with the curve integral over $C_{t,x}$ in (4.4.36) in the various connected components of $\operatorname{supp} E \setminus \operatorname{sing\,supp} E$. First note that the propagation cone $K(A(-i\partial), N)$ is the convex hull of the set $\Xi_1^* = \{(t, x) \in \mathbf{R}^4; t > 0, \frac{x}{t} \in X_1^*\}$ and that $E = E_0 + E_1$ vanishes outside this set, i.e., $E_1 = -E_0$ holds in the complement of $\operatorname{supp} E = K(A(-i\partial), N)$. In fact, if $t \geq 0$ and $(t, x) \notin \operatorname{supp} E$, then $C_{t,x}$ consists of two curves and the integral in (4.4.36) has to cancel E_0. In contrast, in the lacunary part L, i.e., in the component of $\mathbf{R}^4 \setminus \operatorname{sing\,supp} E$ containing $N = (1, 0, 0, 0)$, $C_{t,x}$ is empty and hence $E = E_0$, see part (b). If (t, x) lies in between the two sheets of Ξ_1^*, see Fig. 4.5, then $C_{t,x}$ consists only of one curve. If (t, x) belongs to one of the four regions bounded by the outer sheet of Ξ_1^* and by one of the four lids on the convex hull of Ξ_1^*, then again $C_{t,x}$ consists of two curves, which now are both situated on the outer sheet of X_1.

(d) Let us finally calculate the fundamental matrix of the system of crystal optics in the "uniaxial" case where the two optical axes coincide, if, say, $0 < d_1 = d_2 < d_3$, i.e., $0 < \epsilon_3 < \epsilon_1 = \epsilon_2$.

Without restriction of generality, we can set $d_1 = d_2 = 1$, and we denote d_3 by d. Hence the system reads as

$$A(\partial) = \begin{pmatrix} \partial_t^2 - d\partial_2^2 - \partial_3^2 & d\partial_1\partial_2 & \partial_1\partial_3 \\ d\partial_1\partial_2 & \partial_t^2 - d\partial_1^2 - \partial_3^2 & \partial_2\partial_3 \\ \partial_1\partial_3 & \partial_2\partial_3 & \partial_t^2 - \partial_1^2 - \partial_2^2 \end{pmatrix}. \qquad (4.4.39)$$

We shall use formula (4.4.35) in order to calculate E_1. Due to

$$R(1,\xi) = (1 - d|\xi'|^2 - \xi_3^2)(1 - |\xi|^2) \quad \text{where} \quad \xi' = (\xi_1, \xi_2)^T,$$

the part X_1 of the slowness surface now consists of the sphere $|\xi| = 1$ and of the ellipsoid of revolution $d|\xi'|^2 + \xi_3^2 = 1$. Furthermore,

$$e(\xi) = \begin{cases} \left(\dfrac{\xi_1\xi_3}{|\xi'|}, \dfrac{\xi_2\xi_3}{|\xi'|}, -|\xi'| \right)^T : |\xi|^2 = 1, \\[2ex] \left(\dfrac{\xi_2}{|\xi'|}, -\dfrac{\xi_1}{|\xi'|}, 0 \right)^T \quad : d|\xi'|^2 + \xi_3^2 = 1. \end{cases}$$

Consequently, the integral in (4.4.35) splits into two summands: $E_1 = F + G$, where

$$F = -\frac{Y(t)}{8\pi^2} \partial_t \int_{S^2} \begin{pmatrix} \xi_1^2\xi_3^2/|\xi'|^2 & \xi_1\xi_2\xi_3^2/|\xi'|^2 & -\xi_1\xi_3 \\ \xi_1\xi_2\xi_3^2/|\xi'|^2 & \xi_2^2\xi_3^2/|\xi'|^2 & -\xi_2\xi_3 \\ -\xi_1\xi_3 & -\xi_2\xi_3 & |\xi'|^2 \end{pmatrix} \delta(t + x\xi)\, d\sigma(\xi)$$

and

$$G = -\frac{Y(t)}{8\pi^2} \partial_t \int_{\{\xi\in\mathbb{R}^2; d|\xi'|^2+\xi_3^2=1\}} \begin{pmatrix} \xi''\xi''^T/|\xi''|^2 & 0 \\ 0 & 0 \end{pmatrix} \delta(t + x\xi)\, |\gamma|(\xi)$$

$$= -\frac{Y(t)}{8\pi^2 d} \partial_t \int_{S^2} \begin{pmatrix} \eta'\eta'^T/|\eta'|^2 & 0 \\ 0 & 0 \end{pmatrix} \delta(t + \tilde{x}\eta)\, d\sigma(\eta)$$

with $\xi'' = (\xi_2, -\xi_1)^T$ and $\tilde{x} = (x_2/\sqrt{d}, -x_1/\sqrt{d}, x_3)^T$ upon using the substitution $\eta_1 = \xi_2\sqrt{d}, \eta_2 = -\xi_1\sqrt{d}, \eta_3 = \xi_3$.

Let us consider first the matrix F. We split it up into two parts: $F = F_1 + F_2$, with

$$F_1 = \frac{Y(t)}{8\pi^2} \partial_t \int_{S^2} \xi\xi^T \delta(t + x\xi)\, d\sigma(\xi)$$

and $F_2 = -\dfrac{Y(t)}{8\pi^2} \partial_t \displaystyle\int_{S^2} \begin{pmatrix} \xi'\xi'^T/|\xi'|^2 & 0 \\ 0 & 1 \end{pmatrix} \delta(t + x\xi)\, d\sigma(\xi).$

The integral for F_1 is the simplest one:

$$F_1 = \frac{Y(t)}{8\pi^2} \nabla\nabla^T \int_{S^2} Y(t + x\xi)\, d\sigma(\xi)$$

$$= \frac{Y(t)}{8\pi^2} \nabla\nabla^T \left(2\pi \min\left\{2, 1 + \frac{t}{|x|}\right\}\right) = \frac{Y(t)}{4\pi} \nabla\left[Y(|x| - t)\cdot\left(-\frac{tx^T}{|x|^3}\right)\right]$$

$$= -\frac{xx^T}{4\pi t^3} \delta(t - |x|) - \mathrm{vp}\left(\frac{Y(t)Y(|x| - t)t}{4\pi|x|^3}\cdot\left(I_3 - \frac{3xx^T}{|x|^2}\right)\right).$$

Hence

$$E_0 + F_1 = \frac{1}{3}(tY(t)\otimes\delta(x))I_3 - \frac{xx^T}{4\pi t^3}\delta(t - |x|) + \mathrm{vp}\left(\frac{tY(t - |x|)}{4\pi|x|^3}\cdot\left(I_3 - \frac{3xx^T}{|x|^2}\right)\right).$$

$$(4.4.40)$$

The element $(F_2)_{33}$ in the lower right corner of F_2 is also easy to calculate; it yields

$$(F_2)_{33} = -\frac{Y(t)}{8\pi^2} \partial_t^2 \int_{S^2} Y(t + x\xi)\, d\sigma(\xi)$$

$$= -\frac{Y(t)}{4\pi} \partial_t^2 \min\left\{2, 1 + \frac{t}{|x|}\right\} = \frac{Y(t)}{4\pi t}\delta(t - |x|). \qquad (4.4.41)$$

Finally, let us consider the remaining integrals in F_2 and G, which have the same form. By differentiation, we obtain

$$\Delta_2 F_2 = -\frac{Y(t)}{8\pi^2} \int_{S^2} \begin{pmatrix} \xi'\xi'^T & 0 \\ 0 & |\xi'|^2 \end{pmatrix} \delta'''(t + x\xi)\, d\sigma(\xi)$$

$$= -\frac{Y(t)}{8\pi^2} \begin{pmatrix} \nabla'\nabla'^T & 0 \\ 0 & \Delta_2 \end{pmatrix} \int_{S^2} \delta'(t + x\xi)\, d\sigma(\xi)$$

$$= -\frac{Y(t)}{4\pi} \begin{pmatrix} \nabla'\nabla'^T & 0 \\ 0 & \Delta_2 \end{pmatrix} \partial_t^2 \min\left\{2, 1 + \frac{t}{|x|}\right\}$$

$$= \frac{Y(t)}{4\pi t} \begin{pmatrix} \nabla'\nabla'^T & 0 \\ 0 & \Delta_2 \end{pmatrix} \delta(t - |x|) = U.$$

Let us now solve this Poisson equation for F_2 in $\mathbf{R}^2$ (where we consider t, x_3 as parameters) by convolution with the fundamental solution $S = \frac{1}{2\pi} \log |x'|$ of Δ_2 (see Example 1.3.14). This procedure yields F_2 as solution since, on the one hand, S and U are convolvable by support (see Example 1.5.11), and, on the other hand, F_2 converges to 0 for $|x'| \to \infty$, i.e., $\langle F_2, \phi(t, x_3) \rangle \in \dot{\mathcal{B}}'(\mathbf{R}^2)$ for $\phi \in \mathcal{D}(\mathbf{R}^2)$, see Schwartz [246], p. 200, for the definition of $\dot{\mathcal{B}}'$.

Therefore,

$$F_2 = \frac{Y(t)}{4\pi t} \begin{pmatrix} \nabla' \nabla'^T & 0 \\ 0 & \Delta_2 \end{pmatrix} \left[\delta(t - |x|) \underset{x'}{*} \frac{1}{2\pi} \log |x'| \right].$$

In particular, $(F_2)_{33} = \frac{Y(t)}{4\pi t} \delta(t - |x|)$ as we have already obtained above. In contrast,

$$F_2' = \begin{pmatrix} (F_2)_{11} & (F_2)_{12} \\ (F_2)_{21} & (F_2)_{22} \end{pmatrix}$$

is given by

$$F_2' = \frac{Y(t)}{8\pi^2 t} \nabla' \nabla'^T \left[\delta(t - |x|) \underset{x'}{*} \log |x'| \right].$$

Since

$$\frac{\delta(t - |x|)}{2t} = \frac{\delta(t - |x|)}{t + |x|} = \delta(t^2 - |x|^2)$$

$$= \delta(t^2 - x_3^2 - |x'|^2) = \frac{Y(t - |x_3|)}{2\sqrt{t^2 - x_3^2}} \delta(\sqrt{t^2 - x_3^2} - |x'|),$$

we can express F_2' by the logarithmic potential of a uniform charge on a circle in the plane.

If R is the radius of such a uniformly charged circle, then the logarithmic potential is continuous, constant for $|x'| < R$, and equals $2\pi R \cdot \frac{1}{2\pi} \log |x'|$ for $|x'| > R$, i.e.,

$$\frac{1}{2\pi} \log |x'| * \delta(R - |x'|) = R \max\{\log |x'|, \log R\}.$$

Hence

$$F_2' = \frac{Y(t - |x_3|)}{4\pi \sqrt{t^2 - x_3^2}} \nabla' \nabla'^T \left[\delta(\sqrt{t^2 - x_3^2} - |x'|) \underset{x'}{*} \frac{1}{2\pi} \log |x'| \right]$$

$$= \frac{Y(t - |x_3|)}{4\pi} \nabla' \nabla'^T \max \left\{ \log |x'|, \log \sqrt{t^2 - x_3^2} \right\}$$

$$= \frac{Y(t - |x_3|)}{4\pi} \nabla'\left[Y(|x| - t) \cdot \frac{x'^T}{|x'|^2}\right]$$

$$= \frac{\delta(t - |x|)}{4\pi t} \cdot \frac{x'x'^T}{|x'|^2} + \frac{Y(t - |x_3|)Y(|x| - t)}{4\pi} \cdot \mathrm{vp}\left[\frac{1}{|x'|^4}\begin{pmatrix} x_2^2 - x_1^2 & -2x_1x_2 \\ -2x_1x_2 & x_1^2 - x_2^2 \end{pmatrix}\right].$$

$$(4.4.42)$$

Similarly,

$$G = \frac{\delta\left(t - \sqrt{|x'|^2/d + x_3^2}\right)}{4\pi t d} \cdot \frac{1}{|x'|^2}\begin{pmatrix} x_2^2 & -x_1x_2 & 0 \\ -x_1x_2 & x_1^2 & 0 \\ 0 & 0 & 0 \end{pmatrix}$$

$$- \frac{Y(t - |x_3|)Y\left(\sqrt{|x'|^2/d + x_3^2} - t\right)}{4\pi} \cdot \mathrm{vp}\left[\frac{1}{|x'|^4}\begin{pmatrix} x_2^2 - x_1^2 & -2x_1x_2 & 0 \\ -2x_1x_2 & x_1^2 - x_2^2 & 0 \\ 0 & 0 & 0 \end{pmatrix}\right].$$

$$(4.4.43)$$

Collecting all the terms for the fundamental matrix E of the system of uniaxial crystal optics we obtain from (4.4.40–4.4.43)

$$E = E_0 + F_1 + F_2 + G = H_1\delta\left(t - \sqrt{|x'|^2/d + x_3^2}\right) + \frac{1}{3}I_3 t Y(t) \otimes \delta(x)$$

$$+ H_3\delta(t - |x|) + H_4\left[Y\left(t - \sqrt{|x'|^2/d + x_3^2}\right) - Y(t - |x|)\right] + H_5 Y(t - |x|)$$

where $x' = \binom{x_1}{x_2}$ and

$$H_1 = \frac{1}{4\pi d|x'|^2 t}\begin{pmatrix} x_2^2 & -x_1x_2 & 0 \\ -x_1x_2 & x_1^2 & 0 \\ 0 & 0 & 0 \end{pmatrix},$$

$$H_3 = \frac{1}{4\pi t}\left[\frac{1}{|x'|^2}\begin{pmatrix} x_1^2 & x_1x_2 & 0 \\ x_1x_2 & x_2^2 & 0 \\ 0 & 0 & |x'|^2 \end{pmatrix} - \frac{x \cdot x^T}{|x|^2}\right]$$

$$H_4 = \mathrm{vp}\left[\frac{1}{4\pi|x'|^4}\begin{pmatrix} x_2^2 - x_1^2 & -2x_1x_2 & 0 \\ -2x_1x_2 & x_1^2 - x_2^2 & 0 \\ 0 & 0 & 0 \end{pmatrix}\right], \quad H_5 = \mathrm{vp}\left[\frac{t}{4\pi|x|^3}\left(I_3 - \frac{3x \cdot x^T}{|x|^2}\right)\right].$$

This explicit representation of E was already given in Ortner and Wagner [217], Prop. 3, p. 342; there it was derived differently from the fundamental matrix of a particular hexagonal elastodynamic system by a limit process.

Let us mention that the form factors H_1 and H_3 of the two delta terms with support on the two parts of the wave surface $t = |x|$ and $t = \sqrt{|x'|^2/d + x_3^2}$, respectively, also result from the more general formula in Wagner [298], Thm. 10.1, p. 2687, which gives an explicit representation of the so-called sharp waves in the biaxial case of crystal optics.   □

Chapter 5
Fundamental Matrices of Homogeneous Systems

In this last chapter we exploit the homogeneity of a system in order to reduce the number of integrations in the representation of its fundamental matrix by inverse Fourier transform. Let us roughly sketch the idea. If $P(\partial)$ is an elliptic and homogeneous operator of degree m, we obtain for a fundamental solution E the following:

$$ E(x) = \mathcal{F}^{-1}\left(\frac{1}{P(\mathrm{i}\xi)}\right) = \mathrm{i}^{-m}(2\pi)^{-n}\langle 1_\xi, \mathrm{e}^{\mathrm{i}x\xi}P(\xi)^{-1}\rangle. $$

Upon introducing polar coordinates $\xi = r\omega$ this yields

$$ E(x) = \mathrm{i}^{-m}(2\pi)^{-n}\langle r_+^{n-1}|\gamma|(\omega), \mathrm{e}^{\mathrm{i}x\omega r}P(r\omega)^{-1}\rangle $$

$$ = \mathrm{i}^{-m}(2\pi)^{-n}\langle\langle|\gamma|(\omega), P(\omega)^{-1}\langle r_+^{n-m-1}, \mathrm{e}^{\mathrm{i}x\omega r}\rangle\rangle $$

$$ = \mathrm{i}^{-m}(2\pi)^{-n}\langle\langle|\gamma|(\omega), P(\omega)^{-1}\mathcal{F}(t_+^{n-m-1})(-x\omega)\rangle\rangle $$

Since $\mathcal{F}(t_+^{n-m-1})$ is explicitly known from elementary distribution theory, this yields a very symmetrical formula for E, see Proposition 5.2.1 It is, however, of limited practical value for the calculation of E, see Example 5.2.2.

For practical purposes, it is more convenient to single out one (e.g., the last) coordinate and to consider the Fourier transforms $\mathcal{F}T$ of homogeneous distributions T represented in the form $T = F_+(x'/x_n)(x_n)_+^\lambda + F_-(-x'/x_n)(x_n)_-^\lambda$, see Propositions 5.1.6–5.1.8. This leads to a representation of a fundamental matrix of a homogeneous elliptic system $A(\partial)$ as an integral with respect to $\xi' \in \mathbf{R}^{n-1}$ of $A(\xi', 1)^{-1}$ multiplied by a kernel of the type $K(\xi'x' + x_n)$, see Proposition 5.2.3. Among other examples, we employ this formula to calculate the fundamental matrix of hexagonal elastostatics (Example 5.2.6) and to represent the fundamental solution of the operators

$$ \partial_1^4 + \partial_2^4 + \partial_3^4 + 2a\partial_1^2\partial_2^2, \qquad a > -1, $$

© Springer International Publishing Switzerland 2015
N. Ortner, P. Wagner, *Fundamental Solutions of Linear Partial Differential Operators*, DOI 10.1007/978-3-319-20140-5_5

in terms of elliptic integrals (Example 5.2.11). The special case $a = 0$ was treated first in Fredholm [82].

5.1 Homogeneous Distributions and Their Fourier Transforms

The notion of *homogeneity of distributions* was defined in Definition 1.2.9. The representation of a homogeneous $T \in \mathcal{D}'(\mathbf{R}^n \setminus \{0\})$ in the form $T = F \cdot |x|^\lambda$ with a "characteristic" $F \in \mathcal{D}'(\mathbf{S}^{n-1})$ and the formula for its Fourier transform appeared already in several places, see Example 1.4.9, in particular formula (1.4.3), Example 1.4.12, and the proofs of Proposition 2.3.9 and of Proposition 4.4.1. Let us now explain systematically the representation of homogeneous distributions in generalized polar coordinates following Gårding [89], and the formulas for their Fourier transforms first derived in Gel'fand and Shapiro [103], see also Gel'fand and Shilov [104], and Ortner and Wagner [219].

In Example 1.4.9, we embedded $L^1(\mathbf{S}^{n-1})$ into $\mathcal{D}'(\mathbf{S}^{n-1})$ by means of the surface measure $\mathrm{d}\sigma$, i.e.,

$$L^1(\mathbf{S}^{n-1}) \hookrightarrow \mathcal{D}'(\mathbf{S}^{n-1}) : f \longmapsto \left(\phi \mapsto \int_{\mathbf{S}^{n-1}} f\phi \, \mathrm{d}\sigma \right), \qquad \phi \in \mathcal{D}(\mathbf{S}^{n-1}) = \mathcal{C}^\infty(\mathbf{S}^{n-1}),$$

see (1.3.15), and defined $F \cdot |x|^\lambda \in \mathcal{D}'(\mathbf{R}^n)$ for $F \in \mathcal{D}'(\mathbf{S}^{n-1})$ and $\lambda \in \mathbf{C}$ by

$$\langle \phi, F \cdot |x|^\lambda \rangle = \langle \langle \phi(t\omega), F(\omega) \rangle, t_+^{\lambda+n-1} \rangle, \qquad \phi \in \mathcal{D}(\mathbf{R}^n),$$

see (1.4.3).

More generally, let us consider now a $\mathcal{C}^\infty$ function $\rho : \mathbf{R}^n \setminus \{0\} \longrightarrow (0, \infty)$ which is homogeneous of degree one, and let us denote by Γ the compact $\mathcal{C}^\infty$ manifold $\Gamma = \rho^{-1}(1)$. As in Corollary 4.4.2, we use the Kronecker–Leray form

$$\gamma = \sum_{j=1}^n (-1)^{j-1} \xi_j \, \mathrm{d}\xi_1 \wedge \cdots \wedge \mathrm{d}\xi_{j-1} \wedge \mathrm{d}\xi_{j+1} \wedge \cdots \wedge \mathrm{d}\xi_n$$

in order to orient Γ and equip it with the measure $|\gamma|$. Then $L^1(\Gamma)$ is embedded in $\mathcal{D}'(\Gamma) = \mathcal{D}(\Gamma)'$ by means of $|\gamma|$, i.e.,

$$L^1(\Gamma) \hookrightarrow \mathcal{D}'(\Gamma) : f \longmapsto \left(\phi \mapsto \int_\Gamma f(\omega)\phi(\omega) \, |\gamma|(\omega) \right), \qquad \phi \in \mathcal{D}(\Gamma) = \mathcal{C}^\infty(\Gamma).$$

Furthermore, γ is such that the formula

$$\int_{\mathbf{R}^n} f(x)\,\mathrm{d}x = \int_0^\infty t^{n-1}\mathrm{d}t \int_\Gamma f(t\omega)\,|\gamma|(\omega)$$

holds for $f \in L^1(\mathbf{R}^n)$.

Definition 5.1.1 For ρ and Γ as above, $\lambda \in \mathbf{C}$ and $F \in \mathcal{D}'(\Gamma)$, let us define $F \cdot \rho^\lambda \in \mathcal{S}'(\mathbf{R}^n)$ by

$$\langle \phi, F \cdot \rho^\lambda \rangle = {}_{\mathcal{S}(\mathbf{R})}\langle {}_{\mathcal{D}(\Gamma)}\langle \phi(t\omega), F(\omega)\rangle_{\mathcal{D}'(\Gamma)}, t_+^{\lambda+n-1}\rangle_{\mathcal{S}'(\mathbf{R})}, \qquad \phi \in \mathcal{S}(\mathbf{R}^n).$$

Note that, for $F \in L^1(\Gamma)$ and $\operatorname{Re}\lambda > -n$, the distribution $F \cdot \rho^\lambda$ is locally integrable in $\mathbf{R}^n$ and coincides with the function $F(\omega)\rho(x)^\lambda$ in "generalized polar coordinates" $\rho = \rho(x) > 0$, $\omega = \frac{x}{\rho(x)} \in \Gamma$ for $x \in \mathbf{R}^n \setminus \{0\}$.

Proposition 5.1.2 *For ρ, Γ, F as in Definition 5.1.1 and $\lambda \in \mathbf{C} \setminus (-n - \mathbf{N}_0)$, the distribution $F \cdot \rho^\lambda$ is homogeneous of degree λ. Furthermore, the mapping*

$$\mathbf{C} \longrightarrow \mathcal{S}'(\mathbf{R}^n) : \lambda \longmapsto F \cdot \rho^\lambda$$

is meromorphic with at most simple poles in $-n - \mathbf{N}_0$. At $\lambda = -n - k$, $k \in \mathbf{N}_0$, the residues are given by

$$\operatorname*{Res}_{\lambda=-n-k} F \cdot \rho^\lambda = \sum_{\substack{\alpha \in \mathbf{N}_0^n \\ |\alpha|=k}} \frac{(-1)^k}{\alpha!}\,\langle \omega^\alpha, F(\omega)\rangle\,\partial^\alpha\delta,$$

and the finite parts are $\operatorname{Pf}_{\lambda=-n-k} F \cdot \rho^\lambda = F \cdot \rho^{-n-k}$. These finite parts are homogeneous distributions in $\mathbf{R}^n \setminus \{0\}$, but, in case the respective residues do not vanish, only "associated homogeneous" in $\mathbf{R}^n$, i.e.,

$$(F \cdot \rho^{-n-k})(cx) = c^{-n-k} \cdot F \cdot \rho^{-n-k} + c^{-n-k}\log c \cdot \operatorname*{Res}_{\lambda=-n-k}(F \cdot \rho^\lambda), \qquad c > 0.$$

Proof Since the function $\psi(t) = \langle \phi(t\omega), F(\omega)\rangle$ belongs to $\mathcal{S}(\mathbf{R})$ for $\phi \in \mathcal{S}(\mathbf{R}^n)$, the distribution $F \cdot \rho^\lambda$ is well defined, homogeneous of degree λ for $\lambda \in \mathbf{C}\setminus(-n-\mathbf{N}_0)$ and meromorphic in λ due to the properties of the meromorphic function $\lambda \mapsto t_+^{\lambda+n-1}$, cf. Example 1.4.8.

Furthermore, formula (1.4.1) implies that

$$\operatorname*{Res}_{\lambda=-n-k} \langle \phi, F \cdot \rho^\lambda \rangle = \operatorname*{Res}_{\lambda=-n-k} \langle \psi, t_+^{\lambda+n-1} \rangle = \frac{\psi^{(k)}(0)}{k!}$$

$$= \frac{1}{k!}\,\langle ((\omega_1\partial_1 + \cdots + \omega_n\partial_n)^k\phi)(0), F(\omega)\rangle$$

$$= \sum_{\substack{\alpha \in \mathbf{N}_0^n \\ |\alpha|=k}} \frac{(-1)^k}{\alpha!} \langle \omega^\alpha, F(\omega) \rangle \cdot \langle \phi, \partial^\alpha \delta \rangle, \qquad \phi \in \mathcal{S}(\mathbf{R}^n).$$

Evidently,

$$\Pf_{\lambda=-n-k} \langle \phi, F \cdot \rho^\lambda \rangle = \Pf_{\lambda=-n-k} \langle \psi, t_+^{\lambda+n-1} \rangle = \langle \psi, t_+^{-k-1} \rangle = \langle \phi, F \cdot \rho^{-n-k} \rangle.$$

Finally, if $R = \operatorname{Res}_{\lambda=-n-k} F \cdot \rho^\lambda$ for $k \in \mathbf{N}_0$ and $c > 0$, then

$$(F \cdot \rho^{-n-k})(cx) = \lim_{\lambda \to -n-k} \left(F \cdot \rho^\lambda - \frac{R}{\lambda+n+k} \right)(cx)$$

$$= \lim_{\lambda \to -n-k} \left(c^\lambda \cdot F \cdot \rho^\lambda - \frac{c^{-n-k}R}{\lambda+n+k} \right)$$

$$= c^{-n-k} \cdot F \cdot \rho^{-n-k} + \lim_{\lambda \to -n-k} (c^\lambda - c^{-n-k}) F \cdot \rho^\lambda$$

$$= c^{-n-k} \cdot F \cdot \rho^{-n-k} + c^{-n-k} \log c \cdot R. \qquad \square$$

Conversely, the following structure theorem will show that each homogeneous distribution on $\mathbf{R}^n$ has a unique "polar coordinate" representation, i.e., $T = F \cdot \rho^\lambda$, at least if $\lambda \in -n - \mathbf{N}_0$, see also Gel'fand and Shilov [104], pp. 303, 310; Gel'fand and Shapiro [103], pp. 40, 43; Gårding [89], Lemme 1.5, p. 393, and Lemme 4.1, p. 400; Lemoine [161], Thm. 3.1.1, p. 135; Hörmander [139], Thms. 3.2.3/4, pp. 75, 79. A generalization of this structure theorem to quasihomogeneous distributions is contained in Krée [157], (33), (34), pp. 17, 18; von Grudzinski [119], Thm. 4.25′, p. 178; Ortner and Wagner [219], Thm. 2.5.1, p. 58.

Proposition 5.1.3 *Let ρ, Γ be as above and set, for $\lambda \in \mathbf{C}$,*

$$\mathcal{S}'_\lambda(\mathbf{R}^n) = \{ T \in \mathcal{S}'(\mathbf{R}^n); \ T \text{ is homogeneous of degree } \lambda \}.$$

Then the following are isomorphisms of locally convex topological vector spaces:

(1) *For $\lambda \in \mathbf{C} \setminus (-n - \mathbf{N}_0)$:* $\mathcal{D}'(\Gamma) \xrightarrow{\sim} \mathcal{S}'_\lambda(\mathbf{R}^n) : F \longmapsto F \cdot \rho^\lambda$;

(2) *for $\lambda = -n - k$, $k \in \mathbf{N}_0$:* $\mathcal{D}'_k(\Gamma) \times N_k \xrightarrow{\sim} \mathcal{S}'_\lambda(\mathbf{R}^n) : (F, S) \longmapsto F \cdot \rho^\lambda + S$,

where

$$\mathcal{D}'_k(\Gamma) = \{ F \in \mathcal{D}'(\Gamma); \ \langle \omega^\alpha, F \rangle = 0 \text{ for } \alpha \in \mathbf{N}_0^n \text{ with } |\alpha| = k \}$$

and

$$N_k = \{T \in \mathcal{S}'_{-n-k}(\mathbf{R}^n); \, \text{supp}\, T \subset \{0\}\} = \Big\{ \sum_{|\alpha|=k} c_\alpha \partial^\alpha \delta; \, c_\alpha \in \mathbf{C} \Big\}.$$

Proof From the diffeomorphism

$$\chi : \mathbf{R}^n \setminus \{0\} \longrightarrow (0, \infty) \times \Gamma : x \longmapsto \Big(\rho(x), \frac{x}{\rho(x)} \Big), \qquad (5.1.1)$$

we obtain that the mapping

$$\Psi_1 : \mathcal{C}^\infty(\Gamma) \xrightarrow{\sim} \{f \in \mathcal{C}^\infty(\mathbf{R}^n \setminus \{0\}); \, f \text{ homogeneous of degree } \lambda\} : F \mapsto F \cdot \rho^\lambda$$

is an isomorphism of Fréchet spaces. Note that the inverse mapping Ψ_1^{-1} assigns to the homogeneous function f the function

$$(t, \omega) \longmapsto f\big(\chi^{-1}(t, \omega)\big) \cdot t^{-\lambda} = F(t, \omega),$$

which, in effect, is independent of t, i.e., $\Psi_1^{-1}(f) = f|_\Gamma = F \in \mathcal{C}^\infty(\Gamma)$. Applied to a test function $\phi \in \mathcal{D}((0, \infty) \times \Gamma)$, we obtain

$$\langle \phi, F(t, \omega) \rangle = \langle \rho(x)^{-\lambda-n+1} \cdot (\phi \circ \chi)(x), f(x) \rangle.$$

In this way, we can extend the mapping Ψ_1^{-1} in order to yield the isomorphism

$$\Psi_2 : \{T \in \mathcal{D}'(\mathbf{R}^n \setminus \{0\}); \, T \text{ homogeneous of degree } \lambda\} \to \{F \in \mathcal{D}'((0, \infty) \times \Gamma); \, \partial_t F = 0\}$$

$$T \longmapsto (\phi \mapsto \langle \rho(x)^{-\lambda-n+1} \cdot (\phi \circ \chi), T \rangle).$$

(In order to show that Ψ_2 is well defined, one observes that

$$((\partial_t \phi) \circ \chi) \cdot \rho(x)^{-\lambda-n+1} = (x^T \nabla)(\phi \circ \chi) \cdot \rho(x)^{-\lambda-n}$$

and that Euler's equation yields $\nabla^T(x \cdot \rho(x)^{-\lambda-n} T) = 0$. The space $\{F \in \mathcal{D}'((0, \infty) \times \Gamma); \, \partial_t F = 0\}$ can be identified with $\mathcal{D}'(\Gamma)$ in a natural way, and then the inverse mapping of Ψ_2 is given by $\Psi_2^{-1}(F) = F \cdot \rho^\lambda|_{\mathbf{R}^n \setminus \{0\}}$, which was defined in Definition 5.1.1.)

Since $F \cdot \rho^\lambda \in \mathcal{S}'(\mathbf{R}^n)$ for $F \in \mathcal{D}'(\Gamma)$, we note incidentally that each distribution in $\mathcal{D}'(\mathbf{R}^n \setminus \{0\})$ which is homogeneous is the restriction of a temperate distribution to $\mathbf{R}^n \setminus \{0\}$. In particular, if $\lambda \in \mathbf{C} \setminus (-n - \mathbf{N}_0)$, then $F \cdot \rho^\lambda$ is homogeneous of degree λ in $\mathbf{R}^n$ by Proposition 5.1.2, and this shows that actually, for such λ,

$$\mathcal{S}'_\lambda(\mathbf{R}^n) \longrightarrow \{T \in \mathcal{D}'(\mathbf{R}^n \setminus \{0\}); \, T \text{ is homogeneous of degree } \lambda\} : T \mapsto T|_{\mathbf{R}^n \setminus \{0\}}$$

is an isomorphism of topological vector spaces. (Note that $T \in \mathcal{S}'_\lambda(\mathbf{R}^n)$ and $\mathrm{supp}\, T \subset \{0\}$ imply $T = 0$, see Proposition 1.3.15.)

On the other hand, for $\lambda = -n - k$, $k \in \mathbf{N}_0$, Proposition 5.1.2 states that $F \cdot \rho^\lambda$ is homogeneous in $\mathbf{R}^n$ if and only if all the moments $\langle \omega^\alpha, F \rangle$ vanish for $|\alpha| = k$, i.e., if $F \in \mathcal{D}'_k(\Gamma)$, and, furthermore, T with support in 0 is homogeneous of degree λ iff T belongs to N_k. This yields the isomorphism in (2) and concludes the proof.
$\square$

If T is homogeneous of degree $\lambda \in \mathbf{C}$ in $\mathbf{R}^n$, then its Fourier transform $\mathcal{F}T$ is again homogeneous in $\mathbf{R}^n$, the degree of homogeneity being $-\lambda - n$, see Proposition 1.6.6 (1). In contrast, if T is homogeneous in $\mathbf{R}^n \setminus \{0\}$, but not homogeneous in $\mathbf{R}^n$, then $T = F \cdot \rho^{-n-k}$, $k \in \mathbf{N}_0$, and it is associated homogeneous in $\mathbf{R}^n$, see Proposition 5.1.2. Hence its Fourier transform $\mathcal{F}T$ is also just associated homogeneous, and, as we shall see below, $\mathcal{F}T$ has a logarithmic behavior.

In the next proposition, we give a formula for $\mathcal{F}T$ in terms of the one-dimensional Fourier transform $\mathcal{F}x_+^\lambda$, which was investigated in Example 1.6.7. This formula appears in Gel'fand and Shapiro [103], § 5, (16), p. 81; Gårding [89], Lemme 6.2, p. 406; Hörmander [139], Thm. 7.1.24 and equation (7.1.24), p. 172; Ortner and Wagner [219], Cor. 2.6.3, p. 64. An extension to quasihomogeneous distributions was given in Krée [157], (49), p. 24; see also Ortner and Wagner [219], Prop. 2.6.2, p. 62.

Proposition 5.1.4 *As in* Example 1.4.8, *let* $t_+^\lambda = Y(t)t^\lambda \in L^1_{\mathrm{loc}}(\mathbf{R}^1)$ *for* $\mathrm{Re}\,\lambda > -1$ *and denote the holomorphic extension of this distribution-valued function to* $\mathbf{C} \setminus (-\mathbf{N})$ *again by* t_+^λ. *For* $k \in \mathbf{N}$, *we write* $t_+^{-k} = \mathrm{Pf}_{\lambda=-k}\, t_+^\lambda$. *Let* ρ, Γ *be as in* Definition 5.1.1, $\lambda \in \mathbf{C}$ *and* $F \in \mathcal{D}'(\Gamma)$. *Then the formula*

$$\mathcal{F}(F \cdot \rho^\lambda)(x) = \langle (\mathcal{F}t_+^{\lambda+n-1})(\omega \cdot x), F(\omega) \rangle$$

is valid when interpreted in the following sense:

$$\forall \phi \in \mathcal{S}(\mathbf{R}^n) : \langle \phi, \mathcal{F}(F \cdot \rho^\lambda) \rangle = {}_{\mathcal{D}(\Gamma)}\langle \langle \phi(x), (\mathcal{F}t_+^{\lambda+n-1})(\omega \cdot x) \rangle, F(\omega) \rangle_{\mathcal{D}'(\Gamma)}.$$
(5.1.2)

By formulas (1.6.8) and (1.6.5), $\mathcal{F}t_+^{\lambda+n-1} \in \mathcal{S}'(\mathbf{R}^1_\xi)$ *is given explicitly by*

$$\mathcal{F}t_+^{\lambda+n-1} = \begin{cases} \Gamma(\lambda + n) \lim_{\epsilon \searrow 0}(\mathrm{i}\xi + \epsilon)^{-\lambda-n} & : \lambda + n - 1 \in \mathbf{C} \setminus (-\mathbf{N}), \\[2mm] \dfrac{(-\mathrm{i}\xi)^{m-1}}{(m-1)!}\left[\psi(m) - \log|\xi| - \dfrac{\mathrm{i}\pi}{2}\,\mathrm{sign}\,\xi\right] & : \lambda + n - 1 = -m \in -\mathbf{N}. \end{cases}$$
(5.1.3)

Proof Let us first assume that $\phi \in \mathcal{D}(\mathbf{R}^n)$, $F \in L^1(\Gamma)$ and $-n - 1 < \operatorname{Re}\lambda < -n$. Then Fubini's theorem yields due to Definition 5.1.1

$$\langle \phi, \mathcal{F}(F \cdot \rho^\lambda) \rangle = \langle \mathcal{F}\phi, F \cdot \rho^\lambda \rangle$$

$$= \langle \int_\Gamma (\mathcal{F}\phi)(t\omega)F(\omega)|\gamma|(\omega), t_+^{\lambda+n-1} \rangle$$

$$= \int_0^\infty t^{\lambda+n-1}\left[\int_\Gamma F(\omega)\left(\int_{\mathbf{R}^n} \phi(x)(e^{-it\omega x} - 1)dx \right)|\gamma|(\omega)\right]dt$$

$$= \int_\Gamma F(\omega)\left[\int_{\mathbf{R}^n} \phi(x)\left(\int_0^\infty t^{\lambda+n-1}(e^{-it\omega x} - 1)dt \right)dx \right]|\gamma|(\omega)$$

$$= \langle \langle \phi(x), (\mathcal{F}t_+^{\lambda+n-1})(\omega x) \rangle, F(\omega) \rangle.$$

Note that $t_+^{\lambda+n-1} \in \mathcal{D}'_{L^1}(\mathbf{R}^1)$ for $-n - 1 < \operatorname{Re}\lambda < -n$ and hence

$$(\mathcal{F}t_+^{\lambda+n-1})(\xi) = \langle e^{-it\xi}, t_+^{\lambda+n-1} \rangle = \int_0^\infty (e^{-it\xi} - 1)t^{\lambda+n-1}\,dt.$$

Formula (5.1.2) then holds generally for $\phi \in \mathcal{S}$, $\lambda \in \mathbf{C}$ and $F \in \mathcal{D}'(\Gamma)$ by density and analytic continuation. $\qquad\square$

Often in applications, it is advantageous to employ the two hyperplanes $x_n = \pm 1$ instead of the general hypersurface Γ appearing in Proposition 5.1.4. In contrast to (5.1.1), we obtain for $\Gamma = \{x \in \mathbf{R}^n; |x_n| = 1\}$ the diffeomorphism

$$\chi : \mathbf{R}^n \setminus (\mathbf{R}^{n-1} \times \{0\}) \longrightarrow (0, \infty) \times \Gamma : x \longmapsto \left(|x_n|, \frac{x}{|x_n|} \right)$$

where the coordinate plane $x_n = 0$, which in projective geometry represents the points at infinity, is now excluded from the domain of χ. This fact leads to additional requirements on the homogeneous distribution T in order to describe its Fourier transform in terms of the characteristic of T on the hyperplanes $|x_n| = 1$. We mention that we have already made use of the characteristic of homogeneous distributions on the hyperplanes $|x_n| = 1$ in the proof of Proposition 4.4.1 when we composed the "gnomonian projection" with the Fourier transform.

Definition 5.1.5 As in Proposition 5.1.4, we denote by $t_+^\lambda \in \mathcal{D}'(\mathbf{R}^1)$ the analytic extension with respect to λ of $Y(t)t^\lambda$, $\operatorname{Re}\lambda > -1$, outside the poles (i.e., for $\lambda \in \mathbf{C} \setminus (-\mathbf{N})$), and the finite parts otherwise, and we set $t_-^\lambda = (t_+^\lambda)^\vee$. For $n \geq 2$ and $\xi' \in \mathbf{R}^{n-1}$, we then define $\delta(x' \mp \xi' x_n)(x_n)_\pm^\lambda \in \mathcal{S}'(\mathbf{R}^n)$ by

$$\langle \phi, \delta(x' \mp \xi' x_n)(x_n)_\pm^\lambda \rangle = \langle \phi(\pm\xi' t, t), t_\pm^\lambda \rangle, \qquad \phi \in \mathcal{S}(\mathbf{R}^n). \tag{5.1.4}$$

Note that, for $\lambda \in \mathbf{C} \setminus (-n - \mathbf{N}_0)$, $T_{\pm} = \delta(x' \mp \xi' x_n)(x_n)_{\pm}^{\lambda+n-1}$ represent the two homogeneous distributions of degree λ that have the restriction $\delta_{\xi'}$ on the hyperplanes $x_n = \pm 1$, respectively. (In particular, the distributions $T_{\pm}$ are, outside the origin, measures supported by the two half-rays $t(\xi', \pm 1)$, $t \geq 0$, respectively.) Hence a general homogeneous distribution T of degree $\lambda \in \mathbf{C}$ has, at least formally, outside of the hyperplane $x_n = 0$ the representation

$$T = \int_{\mathbf{R}^{n-1}} \left[F_+(\xi')\delta(x' - \xi' x_n)(x_n)_+^{\lambda+n-1} + F_-(\xi')\delta(x' + \xi' x_n)(x_n)_-^{\lambda+n-1} \right] d\xi',$$

$$(5.1.5)$$

where $F_{\pm} = T|_{x_n = \pm 1}$. In order that (5.1.5) is valid, we have to require an extra growth condition on the characteristics $F_{\pm}$, see the next proposition.

Proposition 5.1.6

(1) *Fix $\lambda \in \mathbf{C}$ and set, for $\phi \in \mathcal{S}(\mathbf{R}^n)$,*

$$f_{\pm}^{\phi}(\xi') = \langle \phi, \delta(x' \mp \xi' x_n)(x_n)_{\pm}^{\lambda+n-1} \rangle.$$

Then

$$f_{\pm}^{\phi} \in \begin{cases} (1 + |\xi'|^2)^{-(n+\mathrm{Re}\,\lambda)/2} \mathcal{D}_{L^{\infty}}(\mathbf{R}^{n-1}) & : \lambda \in \mathbf{C} \setminus (-n - \mathbf{N}_0), \\ (1 + |\xi'|^2)^{k/2} \log(2 + |\xi'|^2) \mathcal{D}_{L^{\infty}}(\mathbf{R}^{n-1}) & : \lambda = -n - k,\ k \in \mathbf{N}_0. \end{cases}$$

(2) *Let us suppose that*

$$F_{\pm} \in \begin{cases} (1 + |\xi'|^2)^{(n+\mathrm{Re}\,\lambda)/2} \mathcal{D}'_{L^1}(\mathbf{R}^{n-1}) & : \lambda \in \mathbf{C} \setminus (-n - \mathbf{N}_0), \\ (1 + |\xi'|^2)^{-k/2} \left(\log(2 + |\xi'|^2) \right)^{-1} \mathcal{D}'_{L^1}(\mathbf{R}^{n-1}) & : \lambda = -n - k,\ k \in \mathbf{N}_0, \end{cases}$$

respectively. Then formula (5.1.5), i.e.,

$$\langle \phi, T \rangle = \langle f_+^{\phi}, F_+ \rangle + \langle f_-^{\phi}, F_- \rangle, \qquad \phi \in \mathcal{S}(\mathbf{R}^n), \qquad (5.1.6)$$

yields a distribution $T \in \mathcal{S}'(\mathbf{R}^n)$ which is homogenous of degree λ in $\mathbf{R}^n \setminus \{0\}$ and which coincides with $F_{\pm}$ when restricted to the hyperplanes $x_n = \pm 1$, respectively. In accordance with the corresponding notation for locally integrable restrictions $F_{\pm}$, we shall denote T by

$$T = F_+\left(\frac{x'}{x_n} \right) \cdot (x_n)_+^{\lambda} + F_-\left(-\frac{x'}{x_n} \right) \cdot (x_n)_-^{\lambda}.$$

Proof

(1) Let us first assume that $\mathrm{Re}\,\lambda > -n$. Then the functions $f_\pm^\phi$ are given by the integrals

$$f_\pm^\phi(\xi') = \int_0^\infty \phi(\xi't, \pm t)\, t^{\lambda+n-1}\, dt$$

and hence are clearly $\mathcal{C}^\infty$. Furthermore, for $\xi' \neq 0$, we set $\xi' = |\xi'|\omega$ and substitute $s = |\xi'|t$ which yields

$$f_\pm^\phi(\xi') = |\xi'|^{-\lambda-n} \int_0^\infty \phi\left(\omega s, \pm\frac{s}{|\xi'|}\right) s^{\lambda+n-1}\, ds.$$

This shows that $f_\pm^\phi(\xi')|\xi'|^{\mathrm{Re}\,\lambda+n}$ is bounded. Since

$$\partial^\alpha f_\pm^\phi = \int_0^\infty (\partial^\alpha \phi)(\xi't, \pm t)\, t^{|\alpha|+\lambda+n-1}\, dt, \qquad \alpha \in \mathbf{N}_0^{n-1},$$

we conclude that $f_\pm^\phi \in (1 + |\xi'|^2)^{-(n+\mathrm{Re}\,\lambda)/2}\mathcal{D}_{L^\infty}(\mathbf{R}^{n-1})$.

Next, for $\mathrm{Re}\,\lambda \leq -n$, $\lambda \notin -n - \mathbf{N}_0$, we can use the formula

$$\left(\frac{d}{dt}\right)^k\left(\frac{t_+^{\lambda+k-1}}{\Gamma(\lambda+k)}\right) = \frac{t_+^{\lambda-1}}{\Gamma(\lambda)},$$

see Example 1.3.9, in order to reduce the estimate of $f_\pm^\phi$ to the case above. Similarly, in the case $\lambda = -n - k$, $k \in \mathbf{N}_0$, we employ the formula

$$x_+^{-k-1} = \frac{(-1)^k}{k!}\left[\left(Y(x)\log x\right)^{(k+1)} + \left(\psi(k+1) - \psi(1)\right)\delta^{(k)}\right],$$

see Example 1.4.8, in order to reduce it to the estimation of the integrals

$$g_\pm^\phi(\xi') = \int_0^\infty \phi(\xi't, \pm t)\, \log t\, dt$$

$$= |\xi'|^{-1} \int_0^\infty \phi\left(\omega s, \pm\frac{s}{|\xi'|}\right)[\log s - \log |\xi'|]\, ds, \qquad \xi' \neq 0.$$

From $g_\pm^\phi \in (1 + |\xi'|^2)^{-1/2}\log(2 + |\xi'|^2)\mathcal{D}_{L^\infty}(\mathbf{R}^{n-1})$ we then infer that $f_\pm^\phi \in (1 + |\xi'|^2)^{k/2}\log(2 + |\xi'|^2)\mathcal{D}_{L^\infty}(\mathbf{R}^{n-1})$ for $\lambda = -n - k$, $k \in \mathbf{N}_0$.

(2) By the duality of $\mathcal{D}'_{L^1}$ and $\mathcal{D}_{L^\infty}$, T in (5.1.6) is well defined and fulfills the asserted conditions. $\qquad\square$

The conditions stated in Proposition 5.1.6 can be formulated in terms of *weighted* $\mathcal{D}'_{L^p}$–spaces, which were introduced in Ortner and Wagner [206] and comprehensively investigated in Ortner and Wagner [219], see also Guzmán-Partida, Ortner, and Wagner [122] and Wagner [299].

From the representation of T in formula (5.1.5/6) (under the hypotheses formulated in Proposition 5.1.6), we also obtain a representation of $\mathcal{F}T$ by applying the Fourier transform to the distributions $\delta(x' \mp \xi' x_n)(x_n)_\pm^{\lambda+n-1}$.

Proposition 5.1.7 *Let $\lambda \in \mathbf{C}$ and $F_\pm \in \mathcal{D}'(\mathbf{R}^{n-1})$ such that the hypothesis in Proposition 5.1.6, (2), is satisfied. If $T = F_+\left(\frac{x'}{x_n}\right) \cdot (x_n)_+^\lambda + F_-\left(-\frac{x'}{x_n}\right) \cdot (x_n)_-^\lambda$ is defined as in Proposition 5.1.6, then its Fourier transform has the representation*

$$(\mathcal{F}T)(x) = \langle (\mathcal{F}t_+^{\lambda+n-1})(\xi' x' + x_n), F_+(\xi') \rangle + \langle (\mathcal{F}t_+^{\lambda+n-1})(\xi' x' - x_n), F_-(\xi') \rangle,$$

which has to be understood in the following sense:

$$\langle \phi, \mathcal{F}T \rangle = \sum_\pm \langle \langle \phi(x), (\mathcal{F}t_+^{\lambda+n-1})(\xi' x' \pm x_n) \rangle, F_\pm(\xi') \rangle, \qquad \phi \in \mathcal{S}(\mathbf{R}^n).$$

(Here the outer brackets can be evaluated due to the hypothesis on $F_\pm$, since

$$(\mathcal{F}t_+^{\lambda+n-1})(\xi' x' \pm x_n) = \mathcal{F}\big(\delta(x' \mp \xi' x_n)(x_n)_\pm^{\lambda+n-1}\big) \in \mathcal{S}'(\mathbf{R}_x^n) \qquad (5.1.7)$$

depends on $\xi' \in \mathbf{R}^{n-1}$ as described in Proposition 5.1.6, (1).)

Proof

(a) For $\phi \in \mathcal{S}(\mathbf{R}^n)$, we have

$$\langle \phi, \mathcal{F}T \rangle = \langle \mathcal{F}\phi, T \rangle = \sum_\pm \langle \langle (\mathcal{F}\phi)(x), \delta(x' \mp \xi' x_n)(x_n)_\pm^{\lambda+n-1} \rangle, F_\pm(\xi') \rangle$$

$$= \sum_\pm \langle \langle \phi(x), \mathcal{F}_x\big(\delta(x' \mp \xi' x_n)(x_n)_\pm^{\lambda+n-1}\big) \rangle, F_\pm(\xi') \rangle.$$

Hence it remains to prove the equation in (5.1.7).

(b) For $\xi' \in \mathbf{R}^{n-1}$, let us consider the points $\omega = (\xi', \pm 1)^T / \sqrt{1 + |\xi'|^2}$ on the unit sphere $\mathbf{S}^{n-1}$ and set $\rho(x) = |x|$. Then we obtain directly from the definitions that

$$\delta(x' \mp \xi' x_n)(x_n)_\pm^{\lambda+n-1} = (1 + |\xi'|^2)^{-(\lambda+n)/2} \delta_\omega \cdot \rho^\lambda$$

and hence, by Proposition 5.1.4, that

$$\mathcal{F}\big(\delta(x' \mp \xi' x_n)(x_n)_\pm^{\lambda+n-1}\big) = (1 + |\xi'|^2)^{-(\lambda+n)/2} \mathcal{F}\big(\delta_\omega \cdot \rho^\lambda\big)$$

$$= (1 + |\xi'|^2)^{-(\lambda+n)/2} (\mathcal{F}t_+^{\lambda+n-1})(\omega x)$$

$$= (\mathcal{F}t_+^{\lambda+n-1})(\xi' x' \pm x_n),$$

at least if $t_+^{\lambda+n-1}$ is homogeneous, i.e., for $\lambda \in \mathbf{C} \backslash (-n-\mathbf{N}_0)$. But (5.1.7) remains true also in the case of $\lambda \in -n-\mathbf{N}_0$ by taking the finite part on both sides. □

When applying Proposition 5.1.7 to the calculation of a fundamental solution E of a homogeneous differential operator $P(\partial)$ of degree m, we shall suppose that $E = i^{-m}\mathcal{F}^{-1}T$ where T is homogeneous of degree $\lambda = -m$ in $\mathbf{R}^n \backslash \{0\}$ and solves the division problem $P(\xi)T = 1$ in $\mathbf{R}^n$. Hence $P(\xi', \pm 1)F_{\pm} = 1$, $\xi' \in \mathbf{R}^{n-1}$, if $F_{\pm} = T|_{x_n = \pm 1}$ and therefore we shall assume that $F_- = (-1)^m \check{F}_+$. Let us specialize now Proposition 5.1.7 to this situation.

Proposition 5.1.8 *Let $n \geq 2$, $m \in \mathbf{N}$ and $F_+ \in \mathcal{D}'(\mathbf{R}_{\xi'}^{n-1})$. If $m < n$, we assume that $F_+ \in (1 + |\xi'|^2)^{(n-m)/2}\mathcal{D}'_{L^1}(\mathbf{R}^{n-1})$, and else, if $m \geq n$, we assume that $F_+ \in (1 + |\xi'|^2)^{(n-m)/2}(\log(2 + |\xi'|^2))^{-1}\mathcal{D}'_{L^1}(\mathbf{R}^{n-1})$. We set $F_- = (-1)^m \check{F}_+$ and define $T \in \mathcal{S}'(\mathbf{R}^n)$ by (5.1.6) for $\lambda = -m$. Then*

$$(2\pi)^n i^{-m}(\mathcal{F}^{-1}T)(x) =$$

$$\begin{cases} \dfrac{2(-1)^{n/2}}{(m-n)!}\langle(\xi'x' + x_n)^{m-n}[\psi(m-n+1) - \log|\xi'x' + x_n|], F_+(\xi')\rangle & : m \geq n, \; n \text{ even}, \\[3mm] \dfrac{\pi(-1)^{(n-1)/2}}{(m-n)!}\langle(\xi'x' + x_n)^{m-n}\operatorname{sign}(\xi'x' + x_n), F_+(\xi')\rangle & : m \geq n, \; n \text{ odd}, \\[3mm] 2(-1)^{n/2-1}\langle(\operatorname{vp}\tfrac{1}{t})^{(n-m-1)}(\xi'x' + x_n), F_+(\xi')\rangle & : m < n, \; n \text{ even}, \\[3mm] 2\pi(-1)^{(n-1)/2}\langle\delta^{(n-m-1)}(\xi'x' + x_n), F_+(\xi')\rangle & : m < n, \; n \text{ odd}. \end{cases}$$

Proof According to Proposition 5.1.7, we have

$$(2\pi)^n i^{-m}\mathcal{F}^{-1}T = i^{-m}(\mathcal{F}T)^{\vee}$$

$$= i^{-m}[\langle(\mathcal{F}t_+^{-m+n-1})(-\xi'x' - x_n), F_+(\xi')\rangle + \langle(\mathcal{F}t_+^{-m+n-1})(-\xi'x' + x_n), F_-(\xi')\rangle]$$

$$= i^m\langle(-1)^m(\mathcal{F}t_+^{-m+n-1})^{\vee}(\xi'x' + x_n) + (\mathcal{F}t_+^{-m+n-1})(\xi'x' + x_n), F_+(\xi')\rangle.$$

From Example 1.6.7, in particular formulas (1.6.5) and (1.6.9), we obtain

$$\mathcal{F}t_+^{-m+n-1} = \begin{cases} \dfrac{(-it)^{m-n}}{(m-n)!}\left[\psi(m-n+1) - \log|t| - \dfrac{i\pi}{2}\operatorname{sign}t\right] & : m \geq n, \\[3mm] i^{n-m-2}(\operatorname{vp}\tfrac{1}{t})^{(n-m-1)} + i^{n-m-1}\pi\delta^{(n-m-1)} & : m < n. \end{cases}$$

$$\tag{5.1.8}$$

and hence

$$\mathcal{F}t_+^{-m+n-1} + (-1)^m (\mathcal{F}t_+^{-m+n-1})^{\vee}$$

$$= \begin{cases} \dfrac{2(-it)^{m-n}}{(m-n)!}\,[\psi(m-n+1) - \log|t|] & : m \geq n,\ n \text{ even}, \\[3mm] -\dfrac{i\pi(-it)^{m-n}}{(m-n)!}\,\operatorname{sign} t & : m \geq n,\ n \text{ odd}, \\[3mm] 2i^{n-m-2}\left(\operatorname{vp}\tfrac{1}{t}\right)^{(n-m-1)} & : m < n,\ n \text{ even}, \\[3mm] 2i^{n-m-1}\pi\delta^{(n-m-1)} & : m < n,\ n \text{ odd}. \end{cases}$$

$$(5.1.9)$$

This immediately yields the four expressions for $\mathcal{F}^{-1}T$ in the proposition.   □

Note that the four formulas in Proposition 5.1.8 coincide with the corresponding representation formulas for a fundamental solution of a homogeneous differential operator $P(\partial)$ in Gel'fand and Shilov [104], Ch. I, 6.2, (2)–(6), p. 129, if $F_+ = P(\xi', 1)^{-1}$ and the integration over $\Omega = \mathbf{S}^{n-1}$ is transformed into one over $\xi_n = 1$ by the "gnomonian projection," cf. also the proof of Propositions 4.4.1 and 5.2.1 below.

As the following illustrative example will reveal, it is often advantageous in concrete calculations to use, instead of the kernel in (5.1.9), the one in (5.1.8), which is a boundary value of a holomorphic function in $\operatorname{Im} t < 0$ and thus allows the application of the residue theorem.

Example 5.1.9 In order to illustrate the "projective representation" of homogeneous distributions in Definition 5.1.5 to Proposition 5.1.8, let us calculate a fundamental solution of the *iterated wave operator* $(\partial_t^2 - \Delta_3)^2$.

As a solution of the division problem $(\tau^2 - |\xi|^2)^2 T = 1$, we use

$$T = \lim_{\epsilon \searrow 0}\left((\tau - i\epsilon)^2 - |\xi|^2\right)^{-2},$$

i.e., T has the form in (5.1.6) with $F_\pm = \lim_{\epsilon \searrow 0}\left((1 \mp i\epsilon)^2 - |\xi|^2\right)^{-2}$. More precisely, T coincides in $\mathbf{R}^4 \setminus \{0\}$ with $T_1 = F_+(\tfrac{\xi}{\tau}) \cdot \tau_+^{-4} + F_-(-\tfrac{\xi}{\tau}) \cdot \tau_-^{-4}$ and therefore, by homogeneity, $T = T_1 + c\delta$, $c \in \mathbf{C}$.

Then the fundamental solution $E_1 = \mathcal{F}^{-1}T_1$ of $(\partial_t^2 - \Delta_3)^2$ is, according to Proposition 5.1.7, given by

$$E_1 = (2\pi)^{-4}\left[\langle(\mathcal{F}t_+^{-1})(-\xi x - t), F_+(\xi)\rangle + \langle(\mathcal{F}t_+^{-1})(-\xi x + t), F_-(\xi)\rangle\right]$$

$$= (2\pi)^{-4}\lim_{\epsilon \searrow 0}\Big[\langle(\mathcal{F}t_+^{-1})(\xi x - t), \left((1 - i\epsilon)^2 - |\xi|^2\right)^{-2}\rangle$$

$$+ \langle(\mathcal{F}t_+^{-1})(\xi x + t), \left((1 + i\epsilon)^2 - |\xi|^2\right)^{-2}\rangle\Big].$$

Note that the hypothesis in Proposition 5.1.6, i.e., $F_\pm \in \left(\log(2 + |\xi|^2)\right)^{-1}\mathcal{D}'_{L^1}(\mathbf{R}^3_\xi)$ is satisfied.

Due to Example 1.6.7, $\mathcal{F}t_+^{-1} = \psi(1) - \log(it)$ where $\log z$ denotes the principal branch of the logarithm, i.e., the one with a cut along $(-\infty, 0]$. Hence, by radial symmetry with respect to x,

$$
E_1 = (2\pi)^{-4} \int_{\mathbf{R}^3} \left[\frac{\psi(1) - \log\big(i(\xi x - t)\big)}{\big[(1 - i0)^2 - |\xi|^2\big]^2} + \frac{\psi(1) - \log\big(i(\xi x + t)\big)}{\big[(1 + i0)^2 - |\xi|^2\big]^2} \right] d\xi
$$

$$
= (2\pi)^{-4} \int_{-\infty}^{\infty} \left\{ \big[\psi(1) - \log\big(i(\xi_1|x| - t)\big)\big] \int_0^{\infty} \frac{2\pi\rho\, d\rho}{\big[(1 - i0)^2 - \xi_1^2 - \rho^2\big]^2} \right.
$$

$$
\left. + \big[\psi(1) - \log\big(i(\xi_1|x| + t)\big)\big] \int_0^{\infty} \frac{2\pi\rho\, d\rho}{\big[(1 + i0)^2 - \xi_1^2 - \rho^2\big]^2} \right\} d\xi_1
$$

$$
= \frac{1}{2(2\pi)^3} \int_{-\infty}^{\infty} \left[\frac{\log\big(i(\xi_1|x| - t)\big) - \psi(1)}{(1 - i0)^2 - \xi_1^2} + \frac{\log\big(i(\xi_1|x| + t)\big) - \psi(1)}{(1 + i0)^2 - \xi_1^2} \right] d\xi_1,
$$

where $1 \pm i0$ symbolizes a distributional limit of $1 \pm i\epsilon$ for $\epsilon \searrow 0$.

By changing ξ_1 to $-\xi_1$ in the second integral, we obtain

$$
E_1 = \frac{1}{(2\pi)^3} \operatorname{Re} \int_{-\infty}^{\infty} \frac{\log\big(i(\xi_1|x| - t)\big) - \psi(1)}{(1 - i0)^2 - \xi_1^2} \, d\xi_1
$$

$$
= \frac{1}{2(2\pi)^3} \operatorname{Re} \left[\int_{-\infty}^{\infty} \frac{\log\big(i(\xi_1|x| - t)\big) - \psi(1)}{1 - i0 + \xi_1} \, d\xi_1 \right.
$$

$$
\left. + \int_{-\infty}^{\infty} \frac{\log\big(i(\xi_1|x| - t)\big) - \psi(1)}{1 - i0 - \xi_1} \, d\xi_1 \right],
$$

the last two integrals being understood as conditionally convergent integrals of the form $\lim_{R\to\infty} \operatorname{Re} \int_{-R}^{R}$.

Instead of considering the limit for $\epsilon \searrow 0$, we use in the first integral the complex variable $z = \xi_1 - i0$, $\xi_1 \in \mathbf{R}$, running infinitesimally below the real axis. Since $\log\big(i(z|x| - t)\big)$ is holomorphic for $\operatorname{Im} z < 0$, we can close the contour in the lower half-plane and conclude that the first integral vanishes, i.e.,

$$
\operatorname{Re} \int_{-\infty - i0}^{\infty - i0} \frac{\log\big(i(z|x| - t)\big) - \psi(1)}{1 + z} \, dz = 0.
$$

In fact,

$$\lim_{R\to\infty} \operatorname{Re}\left[i \int_{-\pi}^{0} \frac{\log\big(i(R\,e^{i\varphi}|x|-t)\big)}{1+R\,e^{i\varphi}}\, R\,e^{i\varphi}\, d\varphi\right]$$

$$= \lim_{R\to\infty} \operatorname{Re}\left[i\pi \log(R|x|) - \int_{-\pi}^{0} \left(\tfrac{\pi}{2}+\varphi\right) d\varphi\right] = 0. \qquad (5.1.10)$$

Similarly, closing also the contour for the second integral in the lower half-plane yields

$$\operatorname{Re}\int_{-\infty+i0}^{\infty+i0} \frac{\log\big(i(z|x|-t)\big)-\psi(1)}{1-z}\, dz = \operatorname{Re}\left[-2\pi i \operatorname*{Res}_{z=1} \frac{\log\big(i(z|x|-t)\big)-\psi(1)}{1-z}\right]$$

$$= \operatorname{Re}\left[2\pi i\big[\log\big(i(|x|-t)\big)-\psi(1)\big]\right] = \pi^2 \operatorname{sign}(t-|x|)$$

and, therefore,

$$E_1 = \frac{1}{16\pi}\,\operatorname{sign}(t-|x|).$$

Note that $E = \mathcal{F}^{-1}T = \mathcal{F}^{-1}\big((\tau-i0)^2-|\xi|^2\big)^{-2}$ is the forward fundamental solution of $(\partial_t^2-\Delta_3)^2$, and in fact, $E = (8\pi)^{-1}Y(t-|x|) = E_1 + (16\pi)^{-1}$ in accordance with $T = T_1 + c\delta$. (The forward fundamental solution E is a special case of formula (2.3.12), which refers to the iterated wave operator for arbitrary dimensions.) So we conclude in hindsight that $c = \pi^3$. Of course, the value of the constant c can also be directly inferred from the equation $T = T_1 + c\delta$.

$\square$

5.2   General Formulas for Fundamental Matrices of Elliptic Homogeneous Systems

Let us first apply Proposition 5.1.4 in order to deduce the classical formulas for fundamental matrices of systems of homogeneous partial differential operators according to G. Herglotz, F. John, I.M. Gel'fand, and G.E. Shilov.

Proposition 5.2.1 *Let $A(\xi)$ be an $l \times l$ matrix of polynomials which are homogeneous of degree m. Let ρ, Γ, γ be given as in the beginning of* Sect. 5.1 *and suppose that $\rho = \check{\rho}$, $\det A(\xi)$ does not vanish identically and $F(\omega) \in \mathcal{D}'(\Gamma)^{l\times l}$ fulfills $A(\omega)F(\omega) = I_l$ and $\check{F} = (-1)^m F$. Then a fundamental matrix E of $A(\partial)$*

is given by $E(x) = {}_{\mathcal{D}(\Gamma)}\langle K_{mn}(\omega \cdot x), F(\omega)\rangle_{\mathcal{D}'(\Gamma)}$ *where* $K_{mn} \in \mathcal{S}'(\mathbf{R}_t^1)$ *is defined as*

$$
K_{mn} = \begin{cases}
\dfrac{(-1)^{n/2-1}}{(2\pi)^n(m-n)!}\, t^{m-n} \log|t| & : m \geq n, n \text{ even}, \\[2mm]
\dfrac{(-1)^{(n-1)/2}}{4(2\pi)^{n-1}(m-n)!}\, t^{m-n} \operatorname{sign} t & : m \geq n, n \text{ odd}, \\[2mm]
\dfrac{(-1)^{n/2-1}}{(2\pi)^n} \left(\dfrac{\mathrm{d}}{\mathrm{d}t}\right)^{n-m-1} \operatorname{vp}\dfrac{1}{t} & : m < n, n \text{ even}, \\[2mm]
\dfrac{(-1)^{(n-1)/2}}{2(2\pi)^{n-1}} \delta^{(n-m-1)} & : m < n, n \text{ odd}.
\end{cases}
$$

Proof We apply Proposition 5.1.4 to

$$
E_1 = \mathrm{i}^{-m}\mathcal{F}^{-1}(F \cdot \rho^{-m}) = (2\pi)^{-n}\mathrm{i}^m \mathcal{F}(F \cdot \rho^{-m})
$$
$$
= (2\pi)^{-n}\mathrm{i}^m {}_{\mathcal{D}(\Gamma)}\langle(\mathcal{F}t_+^{n-m-1})(\omega \cdot x), F(\omega)\rangle_{\mathcal{D}'(\Gamma)}.
$$

Since F is even or odd, respectively, in accordance with m, we conclude that

$$
E_1 = \frac{\mathrm{i}^m}{2(2\pi)^n}\langle\left[\mathcal{F}t_+^{n-m-1} + (-1)^m(\mathcal{F}t_+^{n-m-1})^{\vee}\right](\omega x), F(\omega)\rangle.
$$

The above formulas for K_{mn} in the four cases are then implied by (5.1.9) when omitting polynomials of degree $m - n$ in the first case. □

In the special case of $\rho(x) = |x|$, i.e., $\Gamma = \mathbf{S}^{n-1}$ and $\gamma = \mathrm{d}\sigma$, the four formulas for K_{mn} in Proposition 5.2.1 coincide with Gel'fand and Shilov [104], Ch. I, 6.2, (2)–(6), p. 129. In the case of elliptic homogeneous polynomials and $m \geq n$, the first two formulas in Proposition 5.2.1 appear in Herglotz [125], I, (185), p. 125, and Herglotz [126], pp. 528, 610; Gel'fand and Shilov [104], Ch. I, 6.1, (11), (12), pp. 126, 127; John [151], (3.54a), (3.63), pp. 66, 69; Shimakura [251], Ch. III, (1.23), p. 49; Galler [86], (3.1), p. 9.

If P is a real-valued homogeneous polynomial of principal type (i.e., $\forall \xi \in \mathbf{R}^n \setminus \{0\}$: $\nabla P(\xi) \neq 0$, cf. Hörmander [138], Def. 10.4.11, p. 38), then a solution $F \in \mathcal{D}'(\Gamma)$ of $P(\omega)F(\omega) = 1$ can be represented as a principal value distribution:

$$
F = \operatorname{vp} P(\omega)^{-1} = \lim_{\epsilon \searrow 0} Y(|P(\omega)| - \epsilon)P(\omega)^{-1} \in \mathcal{D}'(\Gamma).
$$

In this case, the formulas in Proposition 5.2.1 were deduced first in Borovikov [21], (5a)–(5d), p. 16. Let us also refer to Gårding [89], Note on p. 406, where it is stated that the formulas in Proposition 5.2.1 are valid for general homogeneous operators P due to the solution of the division problem in Hörmander [135]. (In fact, if $P(\xi) = \det A(\xi)$ does not vanish identically, then there exists $G \in \mathcal{D}'(\Gamma)$ with $P \cdot G = 1$ on account of Hörmander [135], Thm. 1; see also the proof of Proposition 2.3.9. Since P is homogeneous of degree lm, we can suppose that G is of parity $(-1)^{lm}$ and that $F = A^{\mathrm{ad}} \cdot G$ fulfills $\check{F} = (-1)^m F$.) By means of "Borovikov's formula," a fundamental solution of the homogeneous operators of degree three in three variables $P(\partial) = \partial_1^3 + \partial_2^3 + \partial_3^3$ was constructed in Wagner [291], and more generally, for $P(\partial) = \partial_1^3 + \partial_2^3 + \partial_3^3 + 3a\partial_1\partial_2\partial_3$, $a \in \mathbf{R}$, in Wagner [290].

Whereas, for general A, the formulas in Propositions 5.1.7 and 5.1.8 are better suited to a further reduction of the number of integrations, the formulas of Proposition 5.2.1 are applied in the literature for rotationally symmetric operators, i.e., for powers of the Laplacean Δ_n^k, and for completeness we shall repeat this calculation in the next example in the case of odd n.

Example 5.2.2 Let $\rho(\xi) = |\xi|$, $\Gamma = \mathbf{S}^{n-1}$, $P(\xi) = |\xi|^{2k}$ for $\xi \in \mathbf{R}^n$ and $k \in \mathbf{N}$. Then $F(\omega) = 1$ for $\omega \in \mathbf{S}^{n-1}$ fulfills $\check{F} = F$ and $\int_{\mathbf{S}^{n-1}} K_{mn}(\omega x)\, d\sigma(\omega)$ is a rotationally invariant fundamental solution of Δ_n^k. For illustration, let us evaluate this integral over the sphere in the case of odd n.

First, if $2k \geq n$, then we obtain, by rotational symmetry,

$$E = \frac{(-1)^{(n-1)/2}}{4(2\pi)^{n-1}(2k-n)!} \int_{\mathbf{S}^{n-1}} |\omega \cdot x|^{2k-n}\, d\sigma(\omega)$$

$$= \frac{(-1)^{(n-1)/2}|\mathbf{S}^{n-2}|}{4(2\pi)^{n-1}(2k-n)!} \int_0^\pi (|x| \cdot |\cos\vartheta|)^{2k-n} \sin^{n-2}\vartheta\, d\vartheta$$

$$= \frac{(-1)^{(n-1)/2} 2\pi^{(n-1)/2}|x|^{2k-n}}{4(2\pi)^{n-1}(2k-n)!\, \Gamma(\frac{n-1}{2})} \int_0^1 v^{k-(n+1)/2}(1-v)^{(n-3)/2}\, dv$$

$$= \frac{(-1)^{(n-1)/2}|x|^{2k-n}}{2^n \pi^{(n-1)/2}\Gamma(2k-n+1)\Gamma(\frac{n-1}{2})} \cdot \frac{\Gamma(k-\frac{n-1}{2})\Gamma(\frac{n-1}{2})}{(k-1)!}$$

$$= \frac{(-1)^{(n-1)/2}|x|^{2k-n}}{2^{2k}\pi^{n/2-1}\Gamma(k-\frac{n}{2}+1)(k-1)!} = \frac{(-1)^k |x|^{2k-n}\Gamma(\frac{n}{2}-k)}{2^{2k}\pi^{n/2}(k-1)!}$$

in accordance with (1.6.19). (In the last two equations, we employed the doubling and the complement formula of Euler's Gamma function, respectively.)

Second, if n is odd and $2k < n$, then

$$E = \frac{(-1)^{(n-1)/2}}{2(2\pi)^{n-1}} \int_{\mathbf{S}^{n-1}} \delta^{(n-2k-1)}(\omega \cdot x)\, d\sigma(\omega)$$

$$= \frac{(-1)^{(n-1)/2}|\mathbf{S}^{n-2}|}{2(2\pi)^{n-1}} \int_0^\pi \delta^{(n-2k-1)}(|x| \cdot \cos\vartheta) \sin^{n-2}\vartheta\, d\vartheta$$

$$= \frac{(-1)^{(n-1)/2} 2\pi^{(n-1)/2}|x|^{2k-n}}{2(2\pi)^{n-1}\Gamma(\frac{n-1}{2})} \langle (1-u^2)^{(n-3)/2}, \delta^{(n-2k-1)}(u) \rangle$$

$$= \frac{(-1)^{(n-1)/2}|x|^{2k-n}}{2^{n-1}\pi^{(n-1)/2}\Gamma(\frac{n-1}{2})} \cdot \frac{(-1)^{(n-1)/2-k}\Gamma(\frac{n-1}{2})\Gamma(n-2k)}{(k-1)!\,\Gamma(\frac{n+1}{2}-k)}$$

$$= \frac{(-1)^k|x|^{2k-n}\Gamma(\frac{n}{2}-k)}{2^{2k}\pi^{n/2}(k-1)!},$$

see (1.6.19). □

For general homogeneous elliptic systems that are not necessarily rotationally invariant, we use Proposition 5.1.8, which distinguishes one variable and hence is not as symmetric as Proposition 5.2.1, but is better suited for calculations, see, e.g., Wagner [292].

Proposition 5.2.3 *Let $A(\xi)$ be an $l \times l$ matrix of polynomials which are homogeneous of degree m and such that $P(\partial) = \det A(\partial)$ is elliptic. Then a fundamental matrix E of $A(\partial)$ is given by the formulas*

$$E(x) = \begin{cases} \dfrac{2(-1)^{n/2-1}}{(2\pi)^n(m-n)!} \displaystyle\int_{\mathbf{R}^{n-1}} (\xi'x' + x_n)^{m-n}\log|\xi'x' + x_n|A(\xi',1)^{-1}\mathrm{d}\xi' & : m \ge n,\ n\,\text{even}, \\[2ex] \dfrac{\pi(-1)^{(n-1)/2}}{(2\pi)^n(m-n)!} \displaystyle\int_{\mathbf{R}^{n-1}} (\xi'x' + x_n)^{m-n}\operatorname{sign}(\xi'x' + x_n)A(\xi',1)^{-1}\mathrm{d}\xi' & : m \ge n,\ n\,\text{odd}, \\[2ex] \dfrac{2(-1)^{n/2-1}}{(2\pi)^n}\big\langle (\operatorname{vp}\tfrac{1}{t})^{(n-m-1)}(\xi'x' + x_n), A(\xi',1)^{-1}\big\rangle & : m < n,\ n\,\text{even}, \\[2ex] \dfrac{(-1)^{(n-1)/2}}{(2\pi)^{n-1}}\big\langle \delta^{(n-m-1)}(\xi'x' + x_n), A(\xi',1)^{-1}\big\rangle & : m < n,\ n\,\text{odd}. \end{cases}$$

If $m < n$, then E is homogeneous of degree $m - n$ and E is uniquely determined by the property of homogeneity. If $m \ge n$ and $n \ge 3$ is odd, then E is the only fundamental matrix which is homogeneous and even.

Proof Due to the ellipticity of $P(\partial)$, the estimate

$$\exists C > 0 : \forall \xi \in \mathbf{R}^n : |P(\xi)| \ge C|\xi|^{ml}$$

holds and therefore

$$F_+(\xi') = A(\xi',1)^{-1} = A^{\mathrm{ad}}(\xi',1)P(\xi',1)^{-1}, \qquad \xi' \in \mathbf{R}^{n-1},$$

fulfills

$$F_+ \in (1 + |\xi'|^2)^{(n-m)/2}(\log(2 + |\xi'|^2))^{-1}L^1(\mathbf{R}^{n-1})$$

and hence the assumptions of Proposition 5.1.8 are satisfied. The uniqueness of E in the case of odd dimensions n follows from Proposition 2.4.8. □

Example 5.2.4 As a first application of Proposition 5.2.3, let us derive a fundamental solution of the operator

$$P(\partial) = (\partial_1^2 - z^2 \partial_2^2)^l (\partial_1^2 - \bar{z}^2 \partial_2^2)^l$$
$$= (\partial_1^2 - 2(\mathrm{Re}\, z)\partial_1 \partial_2 + |z|^2 \partial_2^2)^l (\partial_1^2 + 2(\mathrm{Re}\, z)\partial_1 \partial_2 + |z|^2 \partial_2^2)^l$$

for $z \in \mathbf{C}$ with $\mathrm{Im}\, z > 0$ and $l \in \mathbf{N}$. In particular, if $z = \epsilon + i\sqrt{1 - \epsilon^2}$, $0 < \epsilon < 1$, then

$$P(\partial) = (\partial_1^4 + 2(1 - 2\epsilon^2)\partial_1^2 \partial_2^2 + \partial_2^4)^l$$

is the iterated *operator of the orthotropic plate*, cf. P. Stein [262] for $l = 1$. A fundamental solution of $P(\partial)$ can also be deduced from Proposition 3.3.2, which goes back to Galler [86], (4.1), p. 15.

From the formula in Proposition 5.2.3, we infer, by setting $m = 4l$,

$$E = \frac{1}{2\pi^2(4l - 2)!} \int_{\mathbf{R}} \frac{(\xi x_1 + x_2)^{4l-2} \log |\xi x_1 + x_2|}{(\xi^2 - z^2)^l (\xi^2 - \bar{z}^2)^l}\, d\xi.$$

As usual, we define the complex logarithm in the slit plane $\mathbf{C} \setminus (-\infty, 0]$. For $x_1 \neq 0$, the function $w \mapsto \log(wx_1 + x_2)$ is holomorphic in the upper half-plane and

$$\log |\xi x_1 + x_2| = \lim_{\epsilon \searrow 0} \mathrm{Re}\, \log((\xi + i\epsilon)x_1 + x_2).$$

Therefore, we can apply the residue theorem and conclude with $z_1 = z$, $z_2 = -\bar{z}$ that

$$E = \frac{1}{2\pi^2(4l - 2)!} \mathrm{Re}\left[2\pi i \sum_{j=1}^{2} \underset{w=z_j}{\mathrm{Res}} \frac{(wx_1 + x_2)^{4l-2} \log(wx_1 + x_2)}{(w^2 - z^2)^l (w^2 - \bar{z}^2)^l} \right]$$

$$= -\frac{1}{\pi(4l - 2)!(l - 1)!} \mathrm{Im}\left[\left(\frac{d}{dw}\right)^{l-1} \frac{(wx_1 + x_2)^{4l-2} \log(wx_1 + x_2)}{(w + z)^l (w^2 - \bar{z}^2)^l} \bigg|_{w=z} \right.$$

$$\left. + \left(\frac{d}{dw}\right)^{l-1} \frac{(wx_1 + x_2)^{4l-2} \log(wx_1 + x_2)}{(w - \bar{z})^l (w^2 - z^2)^l} \bigg|_{w=-\bar{z}} \right]. \qquad (5.2.1)$$

A different representation of a fundamental solution of $P(\partial)$ is given in Galler [86], pp. 54–56.

The particular case $l = 1$ in formula (5.2.1) yields

$$E = -\frac{1}{2\pi} \operatorname{Im} \left[\frac{(zx_1 + x_2)^2 \log(zx_1 + x_2)}{2z(z^2 - \bar{z}^2)} + \frac{(-\bar{z}x_1 + x_2)^2 \log(-\bar{z}x_1 + x_2)}{2\bar{z}(z^2 - \bar{z}^2)} \right].$$

For $z = \epsilon + i\sqrt{1 - \epsilon^2}$ with $0 < \epsilon < \frac{1}{\sqrt{2}}$, we obtain

$$E = \frac{1}{32\pi\sqrt{1 - \epsilon^2}} \Big[\big(x_1^2 + \tfrac{2}{\epsilon} x_1 x_2 + x_2^2\big) \log(x_1^2 + 2\epsilon x_1 x_2 + x_2^2) +$$

$$+ \big(x_1^2 - \tfrac{2}{\epsilon} x_1 x_2 + x_2^2\big) \log(x_1^2 - 2\epsilon x_1 x_2 + x_2^2) \Big] + \frac{x_1^2 - x_2^2}{16\pi\epsilon} \arctan\left(\frac{2\epsilon\sqrt{1 - \epsilon^2} x_1^2}{x_1^2(1 - 2\epsilon^2) + x_2^2} \right).$$

In order to deduce the fundamental solution in Wagner [285], p. 44, which is symmetric in x_1, x_2, we have to subtract the polynomial $\frac{1}{16\pi\epsilon}(x_1^2 - x_2^2) \arcsin \epsilon$. This yields

$$\tilde{E} = E - \frac{x_1^2 - x_2^2}{16\pi\epsilon} \arctan \frac{\epsilon}{\sqrt{1 - \epsilon^2}}$$

$$= \frac{1}{32\pi\sqrt{1 - \epsilon^2}} \Big[\big(x_1^2 + \tfrac{2}{\epsilon} x_1 x_2 + x_2^2\big) \log(x_1^2 + 2\epsilon x_1 x_2 + x_2^2) +$$

$$+ \big(x_1^2 - \tfrac{2}{\epsilon} x_1 x_2 + x_2^2\big) \log(x_1^2 - 2\epsilon x_1 x_2 + x_2^2) \Big] + \frac{x_1^2 - x_2^2}{16\pi\epsilon} \arctan\left(\frac{\epsilon}{\sqrt{1 - \epsilon^2}} \frac{x_1^2 - x_2^2}{x_1^2 + x_2^2} \right)$$

as a fundamental solution of the operator

$$\partial_1^4 + 2(1 - 2\epsilon^2)\partial_1^2\partial_2^2 + \partial_2^4, \qquad 0 < \epsilon < 1.$$

By the method of parameter integration, an equivalent result was derived in Example 3.1.8, see in particular formula (3.1.15).   □

Example 5.2.5 Let us apply now the representation formula for a fundamental solution in Proposition 5.2.3 to *products of anisotropic Laplace operators in even space dimension*. Except for the case $n = 2$, which was treated in Example 3.3.3, this was omitted in Example 3.2.7 since the method of parameter integration in this case leads to more complicated expressions, see Ortner [201].

We consider the operator

$$P(\partial) = \prod_{j=1}^{l} (\Delta_{n-1} + b_j^2 \partial_n^2), \qquad n \text{ even}, n \geq 4, \, 2l \geq n,$$

where b_j are positive and pairwise different. The formula in Proposition 5.2.3 yields the following representation of a fundamental solution E of $P(\partial)$:

$$E = \frac{2(-1)^{n/2-1}}{(2\pi)^n(2l-n)!} \int_{\mathbf{R}^{n-1}} \frac{(\xi'x' + x_n)^{2l-n}}{\prod_{j=1}^{l}(|\xi'|^2 + b_j^2)} \log |\xi'x' + x_n| \, d\xi'.$$

If we introduce polar coordinates in $\mathbf{R}^{n-1}$, i.e., if we set $\rho = |\xi'|$, $\xi'x' = \rho|x'|\cos\vartheta$, then we obtain

$$E = \frac{2(-1)^{n/2-1}}{(2\pi)^n(2l-n)!} \cdot \frac{2\pi^{n/2-1}}{\Gamma(\frac{n}{2}-1)} \int_0^\infty \frac{\rho^{n-2}\,d\rho}{\prod_{j=1}^{l}(\rho^2 + b_j^2)} \times$$

$$\times \int_0^\pi (\rho|x'|\cos\vartheta + x_n)^{2l-n} \log |\rho|x'|\cos\vartheta + x_n| \sin^{n-3}\vartheta \, d\vartheta$$

$$= \frac{(-1)^{n/2-1}}{2^{n-1}\pi^{n/2+1}(2l-n)!(\frac{n}{2}-2)!} \int_{-\infty}^\infty \frac{\rho^{n-2}\,d\rho}{\prod_{j=1}^{l}(\rho^2 + b_j^2)} \times$$

$$\times \int_0^\pi (\rho|x'|\cos\vartheta + x_n)^{2l-n} \log |\rho|x'|\cos\vartheta + x_n| \sin^{n-3}\vartheta \, d\vartheta.$$

Upon inverting the order of integration and considering $\log(z|x'|\cos\vartheta + x_n)$ as a holomorphic function of z for $\operatorname{Im} z > 0$, we can apply the residue theorem and conclude that

$$E = \frac{(-1)^{n/2-1}}{2^{n-1}\pi^{n/2+1}(2l-n)!(\frac{n}{2}-2)!} \operatorname{Re}\left[\int_0^\pi 2\pi i \sum_{j=1}^{l} \frac{1}{2}(ib_j)^{n-3} \prod_{k\neq j}(b_k^2 - b_j^2)^{-1}\times\right.$$

$$\left. \times (ib_j|x'|\cos\vartheta + x_n)^{2l-n} \log(ib_j|x'|\cos\vartheta + x_n) \sin^{n-3}\vartheta \, d\vartheta \right]$$

$$= \frac{1}{2^{n-1}\pi^{n/2}(2l-n)!(\frac{n}{2}-2)!} \sum_{j=1}^{l} b_j^{n-3} \prod_{k\neq j}(b_k^2 - b_j^2)^{-1}\times$$

$$\times \int_{-1}^1 (ib_j|x'|u + x_n)^{2l-n} \log(ib_j|x'|u + x_n)(1 - u^2)^{n/2-2}\,du.$$

(Note that the last integral is real valued as the substitution u to $-u$ shows.)

Similarly as in Galler [86], pp. 57, 58, we evaluate the last integral by partial integration and differentiation with respect to $|x'|$. Setting $r = |x'|$ and $b_j = a$, this yields for $n \geq 6$ the following:

$$J = \int_{-1}^1 (iaru + x_n)^{2l-n} \log(iaru + x_n)(1 - u^2)^{n/2-2}\,du$$

$$
= \frac{n-4}{(2l-n+1)iar} \int_{-1}^{1} (iaru + x_n)^{2l-n+1} \left[\log(iaru + x_n) - \frac{1}{2l-n+1} \right] \times
$$
$$
\times u(1 - u^2)^{n/2-3} \, du
$$
$$
= \frac{4-n}{(2l-n+1)(2l-n+2)a^2} \cdot \frac{1}{r} \partial_r \int_{-1}^{1} (iaru + x_n)^{2l-n+2} \times
$$
$$
\times \left[\log(iaru + x_n) - \frac{1}{2l-n+1} - \frac{1}{2l-n+2} \right] (1 - u^2)^{n/2-3} \, du.
$$

Repeating this process $\frac{n}{2} - 2$ times we obtain an explicit formula for J :

$$
J = \frac{(4-n)(6-n)\cdots(-2)}{(2l-n+1)(2l-n+2)\cdots(2l-4)a^{n-4}} \left(\frac{1}{r} \partial_r \right)^{n/2-2}
$$
$$
\int_{-1}^{1} (iaru + x_n)^{2l-4} \left[\log(iaru + x_n) - \sum_{j=2l-n+1}^{2l-4} \frac{1}{j} \right] du
$$
$$
= \frac{(-1)^{n/2} 2^{n/2-2} (\frac{n}{2} - 2)!(2l-n)!}{(2l-3)! \, ia^{n-3}} \left(\frac{1}{r} \partial_r \right)^{n/2-2} \frac{1}{r}
$$
$$
\times \left\{ (x_n + iar)^{2l-3} [\log(x_n + iar) - C_{ln}] - (x_n - iar)^{2l-3} [\log(x_n - iar) - C_{ln}] \right\}
$$
$$
= \frac{(-1)^{n/2} 2^{n/2-1} (\frac{n}{2} - 2)!(2l-n)!}{(2l-3)! \, a^{n-3}} \left(\frac{1}{r} \partial_r \right)^{n/2-2} \frac{1}{r}
$$
$$
\times \operatorname{Im} \left[(x_n + iar)^{2l-3} (\log(x_n + iar) - C_{ln}) \right],
$$

where $C_{ln} = \sum_{j=2l-n+1}^{2l-3} j^{-1}$. This yields

$$
E = \frac{(-1)^{n/2}}{(2\pi)^{n/2}(2l-3)!} \sum_{j=1}^{l} \left(\prod_{k \neq j} (b_k^2 - b_j^2)^{-1} \right) \left(\frac{1}{r} \partial_r \right)^{n/2-2} \frac{1}{r}
$$
$$
\times \operatorname{Im} \left[(x_n + ib_j r)^{2l-3} (\log(x_n + ib_j r) - C_{ln}) \right], \quad (5.2.2)
$$

where $r = |x'| = (x_1^2 + \cdots + x_{n-1}^2)^{1/2}$.

Obviously, the complex logarithm $\log(x_n + ib_j r)$ can be expressed by $\log(x_n^2 + b_j^2 |x'|^2)$ and $\arctan(b_j |x'|/x_n)$; this leads to the formula in Galler [86], (15.1), p. 59.

Let us yet specify the formula in (5.2.2) for the case $n = 4$, $l = 2$. Then we obtain the following fundamental solution E of

$$
P(\partial) = (\Delta_3 + b_1^2 \partial_4^2)(\Delta_3 + b_2^2 \partial_4^2), \qquad b_1 > 0, \ b_2 > 0, \ b_1 \neq b_2.
$$

$$
E = \frac{1}{4\pi^2 r} \cdot \frac{1}{b_2^2 - b_1^2} \sum_{j=1}^{2} (-1)^{j-1} \operatorname{Im} \left[(x_4 + ib_j r)(\log(x_4 + ib_j r) - 1) \right]
$$

$$= \frac{1}{4\pi^2(b_1 + b_2)} + \frac{1}{8\pi^2(b_2^2 - b_1^2)} \left[b_1 \log(b_1^2 r^2 + x_4^2) - b_2 \log(b_2^2 r^2 + x_4^2) \right]$$

$$+ \frac{x_4}{4\pi^2 r(b_2^2 - b_1^2)} \arctan\left(\frac{(b_1 - b_2)rx_4}{b_1 b_2 r^2 + x_4^2} \right)$$

with $r = (x_1^2 + x_2^2 + x_3^2)^{1/2}$. This is in accordance with the example in Galler [86], p. 59.   □

Example 5.2.6 Let us deduce now from Proposition 5.2.3 *Fredholm's formula*, which refers to the case of a system of homogeneous quadratic operators in $\mathbf{R}^3$. We then specify this formula for the *system of hexagonal elastostatics* deriving thereby *Kröner's formula*.

Let $A(\xi)$ be a real-valued $l \times l$ matrix of polynomials in $\mathbf{R}^3$ which are homogeneous of second degree and such that $P(\partial) = \det A(\partial)$ is an elliptic operator. Let $E \in \mathcal{S}'(\mathbf{R}^3)^{l \times l}$ be the unique homogeneous and even fundamental matrix of $A(\partial)$. We aim at deriving the explicit formula for E found first in Fredholm [81], see also Kröner [158], Willis [303], Ortner and Wagner [217], (F), (F'), (F"), pp. 332, 333.

(a) We apply Proposition 5.2.3 and obtain

$$E = -\frac{1}{4\pi^2} \langle \delta(\xi' x' + x_3), A(\xi', 1)^{-1} \rangle$$

$$= -\frac{1}{4\pi^2 |x_2|} \int_{-\infty}^{\infty} A\left(\xi_1, -\frac{\xi_1 x_1 + x_3}{x_2}, 1\right)^{-1} \mathrm{D}\xi_1$$

$$= -\frac{|x_2|}{4\pi^2} \int_{-\infty}^{\infty} A(\xi_1 x_2, -\xi_1 x_1 - x_3, x_2)^{-1} \, \mathrm{d}\xi_1,$$

see Fredholm [81], formula (6), p. 4, for the second-last expression.

Since $P(zx_2, -zx_1 - x_3, x_2)$ is a real-valued polynomial of degree $2l$ in $z \in \mathbf{C}$ which does not vanish for real z and generic x, it generically has, with multiplicity, l complex roots $z_k = z_k(x)$, $k = 1, \ldots, l$, in the upper half-plane. Let us suppose in the following that P has no multiple factors and hence that these l roots are pairwise different for generic $x \in \mathbf{R}^3$. If we set $w_k = w_k(x) = (-z_k(x)x_1 - x_3)/x_2$, then the residue theorem yields

$$E(x) = -\frac{|x_2|i}{2\pi} \sum_{k=1}^{l} \operatorname*{Res}_{z=z_k} A(zx_2, -zx_1 - x_3, x_2)^{-1}$$

$$= -\frac{|x_2|i}{2\pi} \sum_{k=1}^{l} \frac{A(z_k x_2, w_k x_2, x_2)^{\mathrm{ad}}}{\partial_z P(zx_2, -zx_1 - x_3, x_2)|_{z=z_k}}$$

$$= -\frac{i \operatorname{sign} x_2}{2\pi} \sum_{k=1}^{l} \frac{A(z_k, w_k, 1)^{\mathrm{ad}}}{(x_2 \partial_1 P - x_1 \partial_2 P)(z_k, w_k, 1)}, \qquad (5.2.3)$$

see Fredholm [81], (9/10), pp. 6,7; Gårding [95], p. 129; Kröner [158], p. 404; Willis [303], (A12/13/15), p. 433; Ortner and Wagner [217], (F), p. 332.

Formula (5.2.3) has to be interpreted in the following sense: The summands in (5.2.3) are well defined almost everywhere and determine thereby the fundamental matrix E since E is C^∞ outside the origin and homogeneous of degree -1 and hence locally integrable. Note that (z_k, w_k), $k = 1, \ldots, l$, in (5.2.3) are those intersection points of the algebraic curve $P(z, w, 1) = 0$ with the line $zx_1 + wx_2 + x_3 = 0$ which fulfil the condition $\operatorname{Im} z_k > 0$.

(b) Let us yet specify formula (5.2.3) for the system of hexagonal elastostatics. According to formula (2.1.13) in Example 2.1.4 (d), the matrix $B(\xi)$ of the static hexagonal system has the form

$$B(\xi) = \begin{pmatrix} a_1\xi_1^2 + a_4\xi_2^2 + a_5\xi_3^2 & (a_1 - a_4)\xi_1\xi_2 & a_3\xi_1\xi_3 \\ (a_1 - a_4)\xi_1\xi_2 & a_4\xi_1^2 + a_1\xi_2^2 + a_5\xi_3^2 & a_3\xi_2\xi_3 \\ a_3\xi_1\xi_3 & a_3\xi_2\xi_3 & a_5(\xi_1^2 + \xi_2^2) + a_2\xi_3^2 \end{pmatrix}.$$

As has been noted in Examples 2.1.4, 4.3.10, and 4.4.5, the determinant $P(\xi) = \det B(\xi)$ splits, i.e., $P(\xi) = W_1(\xi)R(\xi)$ where $R(\xi) = W_1 W_4 + \rho^2 W_2$ and

$$\rho^2 = \xi_1^2 + \xi_2^2, \quad W_1 = a_4\rho^2 + a_5\xi_3^2, \quad W_2 = (a_1 - a_4)W_4 - a_3^2\xi_3^2, \quad W_4 = a_5\rho^2 + a_2\xi_3^2.$$

As in Example 4.4.5, we shall suppose that

$$a_1 > 0, \ a_2 > 0, \ a_4 > 0, \ a_5 > 0 \text{ and } |a_3| < a_5 + \sqrt{a_1 a_2}.$$

Following Kröner [158], p. 404, we shall make for simplicity the additional assumption that $|a_3| < |a_5 - \sqrt{a_1 a_2}|$. Then the discriminant Λ of the quadratic function

$$R(\xi_1, \xi_2, 1) = a_1 a_5 \rho^4 + (a_1 a_2 - a_3^2 + a_5^2)\rho^2 + a_2 a_5, \quad \rho^2 = \xi_1^2 + \xi_2^2,$$

in ρ^2, i.e.,

$$\Lambda = a_1^2 a_2^2 + a_3^4 + a_5^4 - 2a_3^2(a_1 a_2 + a_5^2) - 2a_1 a_2 a_5^2$$
$$= \left[(\sqrt{a_1 a_2} + a_5)^2 - a_3^2\right] \cdot \left[(\sqrt{a_1 a_2} - a_5)^2 - a_3^2\right]$$

is positive and hence $R(\xi_1, \xi_2, 1)$ is a product of two real-valued positive quadratic functions:

$$R(\xi_1, \xi_2, 1) = a_1 a_5 (\rho^2 + C_2)(\rho^2 + C_3), \quad C_{2,3} = \frac{a_1 a_2 - a_3^2 + a_5^2 \pm \sqrt{\Lambda}}{2a_1 a_5} > 0.$$

The zeros $\xi' = (z_k, w_k)$, $k = 1, 2, 3$, are determined by $P(z_k, w_k, 1) = 0$, $z_k x_1 + w_k x_2 + x_3 = 0$ and $\operatorname{Im} z_k > 0$. In particular, a first zero results from $W_1(\xi', 1) = 0$, i.e., $\rho^2 = -\frac{a_5}{a_4}$, and that yields

$$z_1 = \frac{-x_1 x_3 + \mathrm{i}|x_2|\sqrt{x_3^2 + \frac{a_5}{a_4}|x'|^2}}{|x'|^2}, \qquad w_1 = \frac{-x_2 x_3 - \mathrm{i}x_1 \operatorname{sign}(x_2)\sqrt{x_3^2 + \frac{a_5}{a_4}|x'|^2}}{|x'|^2}.$$

Due to $W_1(z_1, w_1, 1) = 0$, we obtain

$$(x_2 \partial_1 P - x_1 \partial_2 P)(z_1, w_1, 1) = \big(R \cdot (x_2 \partial_1 W_1 - x_1 \partial_2 W_1)\big)(z_1, w_1, 1)$$

$$= (\rho^2 W_2)(z_1, w_1, 1) \cdot 2 a_4 (x_2 z_1 - x_1 w_1).$$

Similarly, the formula

$$B(\xi)^{\mathrm{ad}} = \begin{pmatrix} R - \xi_1^2 W_2 & -\xi_1 \xi_2 W_2 & -a_3 \xi_1 \xi_3 W_1 \\ -\xi_1 \xi_2 W_2 & R - \xi_2^2 W_2 & -a_3 \xi_2 \xi_3 W_1 \\ -a_3 \xi_1 \xi_3 W_1 & -a_3 \xi_2 \xi_3 W_1 & W_1 W_3 \end{pmatrix}, \qquad W_3 = a_1 \rho^2 + a_5 \xi_3^2,$$

(5.2.4)

cf. (4.4.17), yields for $W_1 = 0$, i.e., $\rho^2 = -\frac{a_5}{a_4}$,

$$B(z_1, w_1, 1)^{\mathrm{ad}} = \begin{pmatrix} w_1^2 & -z_1 w_1 & 0 \\ -z_1 w_1 & z_1^2 & 0 \\ 0 & 0 & 0 \end{pmatrix} W_2(z_1, w_1, 1).$$

Since E is necessarily real-valued, we can take the real part of the resulting expressions. When inserting $B(z_1, w_1, 1)^{\mathrm{ad}}$ into formula (5.2.3), this implies the following for the first summand E_1 in the fundamental matrix E of $B(\partial)$:

$$E_1 = \frac{1}{4\pi a_5 \sqrt{x_3^2 + \frac{a_5}{a_4}|x'|^2}} \operatorname{Re} \begin{pmatrix} w_1^2 & -z_1 w_1 & 0 \\ -z_1 w_1 & z_1^2 & 0 \\ 0 & 0 & 0 \end{pmatrix}$$

$$= \frac{1}{4\pi a_5 |x'|^4 r_1} \begin{pmatrix} x_2^2 x_3^2 - x_1^2 r_1^2 & -x_1 x_2(x_3^2 + r_1^2) & 0 \\ -x_1 x_2(x_3^2 + r_1^2) & x_1^2 x_3^2 - x_2^2 r_1^2 & 0 \\ 0 & 0 & 0 \end{pmatrix}$$

with $r_1 = \sqrt{x_3^2 + \frac{a_5}{a_4}|x'|^2}$.

Similarly, for $k = 2, 3$, $\xi'_k = (z_k, w_k)$ are the solutions of $R(z_k, w_k, 1) = 0$, $z_k x_1 + w_k x_2 + x_3 = 0$ with $\operatorname{Im} z_k > 0$. This yields $\rho^2 = -C_{2,3}$ and hence

$$z_k = \frac{-x_1 x_3 + \mathrm{i}|x_2|\sqrt{x_3^2 + C_k|x'|^2}}{|x'|^2}, \qquad w_k = \frac{-x_2 x_3 - \mathrm{i}x_1\,\operatorname{sign}(x_2)\sqrt{x_3^2 + C_k|x'|^2}}{|x'|^2},$$

where here and in the sequel $k = 2, 3$.
Since $R(z_k, w_k, 1) = 0$, we obtain

$$
\begin{aligned}
(x_2\partial_1 P - x_1\partial_2 P)(z_k, w_k, 1) &= \big(W_1 \cdot (x_2\partial_1 R - x_1\partial_2 R)\big)(z_k, w_k, 1) \\
&= (a_5 - a_4 C_k) \cdot (-2a_1 a_5 C_k + a_1 a_2 - a_3^2 + a_5^2) \cdot 2(x_2 z_k - x_1 w_k) \\
&= 2\mathrm{i}\,(a_5 - a_4 C_k)(-1)^{k-1}\sqrt{\Lambda}(\operatorname{sign} x_2)r_k
\end{aligned}
$$

where $r_k = \sqrt{x_3^2 + C_k|x'|^2}$. Also, from (5.2.4) and $R(z_k, w_k, 1) = 0$, we infer

$$
B(z_k, w_k, 1)^{\mathrm{ad}} = \begin{pmatrix} -z_k^2 \tilde{W}_2 & -z_k w_k \tilde{W}_2 & -a_3 z_k \tilde{W}_1 \\ -z_k w_k \tilde{W}_2 & -w_k^2 \tilde{W}_2 & -a_3 w_k \tilde{W}_1 \\ -a_3 z_k \tilde{W}_1 & -a_3 w_k \tilde{W}_1 & \tilde{W}_1 \tilde{W}_3 \end{pmatrix},
$$

where we have set $\tilde{W}_1 = a_5 - a_4 C_k$, $\tilde{W}_3 = a_5 - a_1 C_k$, $\tilde{W}_2 = (a_1 - a_4)(a_2 - a_5 C_k) - a_3^2$.
Therefore, we finally obtain $E = E_1 + E_2 + E_3$ where E_k, $k = 2, 3$, are given by

$$
E_k = \frac{(-1)^{k-1}}{4\pi \tilde{W}_1 \sqrt{\Lambda}\, r_k |x'|^4} \begin{pmatrix} \tilde{W}_2(x_1^2 x_3^2 - x_2^2 r_k^2) & \tilde{W}_2 x_1 x_2 (x_3^2 + r_k^2) & a_3 \tilde{W}_1 x_1 x_3 |x'|^2 \\ \tilde{W}_2 x_1 x_2 (x_3^2 + r_k^2) & \tilde{W}_2(x_2^2 x_3^2 - x_1^2 r_k^2) & a_3 \tilde{W}_1 x_2 x_3 |x'|^2 \\ a_3 \tilde{W}_1 x_1 x_3 |x'|^2 & a_3 \tilde{W}_1 x_2 x_3 |x'|^2 & -\tilde{W}_1 \tilde{W}_3 |x'|^4 \end{pmatrix}.
$$

(Note that the constants $\tilde{W}_i$ also depend on $k = 2, 3$.)
Up to a missing factor $1/(a_1 a_4 a_5)$, this result was deduced first in Kröner [158], (6), p. 405. Further derivations can be found in Willis [303], (36), p. 426; Mura [185], (5.37), p. 29; Chou and Pan [50].   □
We now aim at representing fundamental solutions of homogeneous elliptic operators in three variables by algebraic integrals. For this purpose, we present a variant of the formula in Fredholm [82], (4), p. 3, which was later rederived in Herglotz [125] and in Bureau [33], (40), p. 31.

Proposition 5.2.7 *Let $P(\xi_1, \xi_2, \xi_3)$ be a homogeneous elliptic polynomial of degree $m = 2k$, $k \geq 2$, in three variables. For $(x_1, x_2, \lambda) \in \mathbf{R}^3$, let $\alpha_\nu = \alpha_\nu(x_1, x_2, \lambda)$, $\nu = 1, \ldots, r$, be the solutions of the equation $P(x_2\alpha_\nu, -\lambda - x_1\alpha_\nu, x_2) = 0$ which fulfil $\operatorname{Im}\alpha_\nu > 0$. We suppose that P has no multiple factors and hence α_ν are pairwise*

*different for almost all $x \in \mathbf{R}^3$. Then a homogeneous fundamental solution F of $P(\partial)$
is given by*

$$F = -\frac{i \operatorname{sign} x_2}{2\pi(m-3)!} \int_{-\infty}^{x_3} \sum_{\nu=1}^{r} \frac{(x_3-\lambda)^{m-3}\,d\lambda}{(\partial_1 P)(\alpha_\nu, -\frac{\lambda+x_1\alpha_\nu}{x_2}, 1)\cdot x_2 - (\partial_2 P)(\alpha_\nu, -\frac{\lambda+x_1\alpha_\nu}{x_2}, 1)\cdot x_1}.$$
$$(5.2.5)$$

Formula (5.2.5) has to be interpreted as explained for formula (5.2.3) in Example 5.2.6.

Proof From Proposition 5.2.3, the equation $\operatorname{sign} t = 2Y(t) - 1$ and by omitting a homogeneous polynomial of degree $m - 3$, we obtain the following fundamental solution F of $P(\partial)$:

$$F = -\frac{1}{4\pi^2(m-3)!} \iint_{\{\xi' \in \mathbf{R}^2;\, \xi'x'+x_3 \geq 0\}} (\xi'x' + x_3)^{m-3}\frac{d\xi'}{P(\xi', 1)} \qquad (5.2.6)$$

For $x_2 \neq 0$, the substitution $\alpha = \xi_1$, $\lambda = -\xi'x' = -\xi_1 x_1 - \xi_2 x_2$ and the residue theorem yield

$$F = -\frac{1}{4\pi^2(m-3)!|x_2|} \int_{-\infty}^{x_3} (x_3-\lambda)^{m-3} \int_{-\infty}^{\infty} \frac{d\alpha}{P(\alpha, -\frac{\lambda+x_1\alpha}{x_2}, 1)}\, d\lambda \qquad (5.2.7)$$

$$= -\frac{i}{2\pi(m-3)!|x_2|} \int_{-\infty}^{x_3} (x_3-\lambda)^{m-3} \sum_{\nu=1}^{r} \frac{d\lambda}{(\partial_1 P - \frac{x_1}{x_2}\partial_2 P)(\alpha_\nu, -\frac{\lambda+x_1\alpha_\nu}{x_2}, 1)}.$$

This implies formula (5.2.5), which holds for almost all $x \in \mathbf{R}^3$. Let us yet observe that the integral in (5.2.6) is absolutely convergent and hence the same holds for the integral in (5.2.5). However, in general, it is not legitimate to interchange the integral in (5.2.5) with the sum over ν. $\qquad \Box$

Example 5.2.8 As an introductory example to formula (5.2.5) in Proposition 5.2.7, let us calculate anew the unique homogeneous and even fundamental solution E of a *product of anisotropic Laplaceans in $\mathbf{R}^3$* :

$$P(\partial) = (\Delta_2 + \lambda_1\partial_3^2)(\Delta_2 + \lambda_2\partial_3^2), \qquad \lambda_1 > 0, \lambda_2 > 0, \lambda_1 \neq \lambda_2,$$

see Example 3.2.7 (c).

In this case, the roots $\alpha_\nu = \alpha_\nu(x_1, x_2, \lambda)$ of the polynomial $P(x_2\alpha, -\lambda - x_1\alpha, x_2)$ that lie in the upper half-plane satisfy one of the equations

$$\alpha_\nu^2 + \left(\frac{\lambda+x_1\alpha_\nu}{x_2}\right)^2 + \lambda_\nu = 0, \qquad \nu = 1, 2,$$

and thus are given by

$$\alpha_\nu = \frac{-x_1\lambda + i|x_2|\sqrt{\lambda^2 + \lambda_\nu \rho^2}}{\rho^2}, \qquad \rho^2 = x_1^2 + x_2^2, \; \nu = 1, 2.$$

A straight-forward calculation yields

$$(\partial_1 P)\left(\alpha_\nu, -\frac{\lambda + x_1\alpha_\nu}{x_2}, 1\right) \cdot x_2 - (\partial_2 P)\left(\alpha_\nu, -\frac{\lambda + x_1\alpha_\nu}{x_2}, 1\right) \cdot x_1$$

$$= 2i\,(-1)^\nu(\lambda_1 - \lambda_2)\,\mathrm{sign}\,x_2\sqrt{\lambda^2 + \lambda_\nu \rho^2}.$$

Hence

$$F = \frac{1}{4\pi(\lambda_1 - \lambda_2)} \int_{-\infty}^{x_3} (x_3 - \lambda)\left[\frac{1}{\sqrt{\lambda^2 + \lambda_1\rho^2}} - \frac{1}{\sqrt{\lambda^2 + \lambda_2\rho^2}}\right] d\lambda$$

$$= \frac{1}{4\pi(\lambda_1 - \lambda_2)}\left[\sqrt{x_3^2 + \lambda_2\rho^2} - \sqrt{x_3^2 + \lambda_1\rho^2} + x_3\log\left(\frac{x_3 + \sqrt{x_3^2 + \lambda_1\rho^2}}{x_3 + \sqrt{x_3^2 + \lambda_2\rho^2}}\right)\right].$$

Note that F differs from the unique homogeneous and *even* fundamental solution E in (3.2.23) just by the linear term $\dfrac{x_3}{8\pi(\lambda_1 - \lambda_2)}\log\dfrac{\lambda_1}{\lambda_2}$. The fundamental solution E is also a limit case of formula (3.1.18). ☐

Example 5.2.9 Let us deduce now from Proposition 5.2.7 the even and homogeneous fundamental solution E of *I. Fredholm's operator* $P(\partial) = \partial_1^4 + \partial_2^4 + \partial_3^4$. Up to a multiplicative constant, E had been calculated in Fredholm [82] using the addition theorem for elliptic integrals. By the method employed here, which avoids the use of this addition theorem with the help of the transformation formula for double integrals in Proposition 5.2.10 below, E and, more generally, the fundamental solutions of the elliptic real-valued operators $\sum_{j=1}^{3}\sum_{k=1}^{3} c_{jk}\partial_j^2\partial_k^2$ were derived in Wagner [292], Prop. 3, p. 1198.

(a) We first notice that E is homogeneous of degree one and hence $E = \sum_{j=1}^{3} x_j\partial_j E$ holds by Euler's identity. By the symmetry of $P(\partial)$ with respect to the three coordinates, it then clearly suffices to find a representation of $\partial_3 E(x_1, 1, x_3)$.

Since $\partial_3 E$ is odd with respect to x_3, we then obtain from formula (5.2.7) in the proof of Proposition 5.2.7 that

$$\partial_3 E(x_1, 1, x_3) = -\frac{1}{4\pi^2}\int_0^{x_3} d\lambda \int_{-\infty}^{\infty} \frac{d\alpha}{P(\alpha, -\lambda - x_1\alpha, 1)}$$

$$= -\frac{1}{4\pi^2}\int_0^{x_3} d\lambda \int_{-\infty}^{\infty} \frac{d\alpha}{\alpha^4 + (\lambda + x_1\alpha)^4 + 1}.$$

Let us assume that $x_3 > 0$ and substitute $(\alpha, \lambda) = \mu^{-1/4}(s, 1)$. This yields

$$\partial_3 E(x_1, 1, x_3) = -\frac{1}{16\pi^2} \int_{x_3^{-4}}^{\infty} \frac{d\mu}{\sqrt{\mu}} \int_{-\infty}^{\infty} \frac{ds}{s^4 + (1 + x_1 s)^4 + \mu}.$$

Abbreviating $A = 1 + x_1^4$ and using the shift $t = s + (x_1^3/A)$ furnishes

$$\partial_3 E(x_1, 1, x_3) = -\frac{1}{16\pi^2 A} \int_{x_3^{-4}}^{\infty} \frac{d\mu}{\sqrt{\mu}} \int_{-\infty}^{\infty} \frac{dt}{t^4 + pt^2 + qt + r}, \qquad (5.2.8)$$

where

$$p = \frac{6x_1^2}{A^2}, \quad q = \frac{4x_1(1 - x_1^4)}{A^3}, \quad r = \frac{x_1^8 - x_1^4 + 1}{A^4} + \frac{\mu}{A}.$$

(b) For the evaluation of the double integral in (5.2.8), we make use of formula (5.2.10) in Proposition 5.2.10 below. In our case, $r(\mu) = c\mu + d$, $c = A^{-1}$, $d = (x_1^8 - x_1^4 + 1)A^{-4}$ and hence

$$\mu(z) = \frac{1}{4c}\left(z^2 - 2pz + p^2 - 4d + \frac{q^2}{z}\right)$$

$$= \frac{A}{4}\left(z^2 - \frac{12x_1^2}{A^2}z - \frac{4x_1^8 - 40x_1^4 + 4}{A^4} + \frac{16x_1^2(1 - x_1^4)^2}{A^6 z}\right).$$

Therefore, by Eq. (5.2.10) below,

$$\partial_3 E(x_1, 1, x_3) = -\frac{1}{16\pi^2 A} \cdot A\pi \int_{z_1}^{z_2} \frac{1}{\sqrt{\mu(z)}} \frac{dz}{\sqrt{z}}$$

$$= -\frac{1}{8\pi\sqrt{A}} \int_{z_1}^{z_2} \frac{dz}{\sqrt{z^3 - 12x_1^2 A^{-2} z^2 - (4x_1^8 - 40x_1^4 + 4)A^{-4}z + 16x_1^2(1 - x_1^4)^2 A^{-6}}}.$$

The limits $z_{1,2}$ of the integration are the largest real roots of $\mu(z) = \infty$ and of $\mu(z) = x_3^{-4}$, respectively, i.e., $z_2 = \infty$ and $\mu(z_1) = x_3^{-4}$. The translation $y = z - 4x_1^2 A^{-2}$ and the substitution $y = 4u/A$ result in

$$\partial_3 E(x_1, 1, x_3) = -\frac{1}{8\pi\sqrt{A}} \int_{y_1}^{\infty} \frac{dy}{\sqrt{y^3 - 4A^{-2}y}} = -\frac{1}{8\pi} \int_{u_1}^{\infty} \frac{du}{\sqrt{4u^3 - u}}, \qquad x_3 > 0,$$

where $\mu(z_1) = x_3^{-4}$, i.e.,

$$y_1^3 - 4A^{-2}y_1 = \frac{4}{A} x_3^{-4}(y_1 + 4x_1^2 A^{-2}).$$

and thus

$$8x_3^6u_1^3 - P(x_1, 1, x_3) \cdot 2x_3^2u_1 - 2x_1^2x_3^2 = 0.$$

Setting $\zeta = 2x_3^2u_1$ and employing that E is even in each of the variables x_1, x_2, x_3 and, moreover, homogeneous of degree one, we finally obtain

$$E = -\frac{1}{8\pi} \sum_{j=1}^{3} |x_j| \int_{\zeta/(2x_j^2)}^{\infty} \frac{du}{\sqrt{4u^3 - u}} \tag{5.2.9}$$

$$= -\frac{1}{8\pi} \sum_{j=1}^{3} x_j F\left(\arcsin\left(\frac{\sqrt{2}x_j}{\sqrt{\zeta + x_j^2}} \right), \frac{1}{\sqrt{2}} \right),$$

ζ being the largest real root of $\zeta^3 - (x_1^4 + x_2^4 + x_3^4)\zeta - 2x_1^2x_2^2x_3^2$ and F denoting the elliptic integral of the first kind, i.e.,

$$F(\varphi, k) = \int_0^{\varphi} \frac{d\alpha}{\sqrt{1 - k^2 \sin^2 \alpha}}, \qquad \varphi \in \mathbf{R}, \ 0 \le k < 1.$$

Up to the factor $-\frac{1}{8\pi}$, formula (5.2.9) coincides with the formula in Fredholm [82], p. 6. $\qquad\square$

The following proposition from Wagner [292], Prop. 2, p. 1196, expresses a double integral containing a polynomial of fourth degree in the denominator by a simple integral parameterized by a root of the cubic resolvent of the polynomial.

Proposition 5.2.10 *Let* $\mu_1, \mu_2 \in \mathbf{R}$ *with* $\mu_1 < \mu_2$ *and let* p, q, r *be once continuously differentiable real-valued functions on* $[\mu_1, \mu_2]$ *such that*

$$Q(\mu, t) = t^4 + p(\mu)t^2 + q(\mu)t + r(\mu)$$

does not vanish for $\mu \in [\mu_1, \mu_2]$ *and* $t \in \mathbf{R}$. *Define*

$$R(\mu, z) = z^3 - 2p(\mu)z^2 + (p(\mu)^2 - 4r(\mu))z + q(\mu)^2$$

and denote by $z_1(\mu)$ *the largest (necessarily simple and positive) root of the three real roots of* $R(\mu, z) = 0$. *Let us furthermore assume that*

$$\forall \mu \in [\mu_1, \mu_2] : (\partial_\mu R)(\mu, z_1(\mu)) \neq 0.$$

Let $a = z_1(\mu_1)$, $b = z_1(\mu_2)$, *and define* $\mu(z)$ *for* z *between* a *and* b *as the continuous function which satisfies* $R(\mu(z), z) = 0$ *and* $\mu(a) = \mu_1$, $\mu(b) = \mu_2$.

Then, for each $f \in L^1([\mu_1, \mu_2])$, *the following equation holds:*

$$\int_{\mu_1}^{\mu_2} f(\mu)\,d\mu \int_{-\infty}^{\infty} \frac{dt}{t^4 + p(\mu)t^2 + q(\mu)t + r(\mu)} = -4\pi \int_a^b f(\mu(z))\, \frac{\sqrt{z}}{(\partial_\mu R)(\mu(z), z)}\,dz.$$

In particular, if p, q *are constant and* r *is a linear function of* μ, $r(\mu) = c\mu + d$, $c \neq 0$, *we have*

$$\int_{\mu_1}^{\mu_2} f(\mu)\,d\mu \int_{-\infty}^{\infty} \frac{dt}{t^4 + pt^2 + qt + c\mu + d} = \frac{\pi}{c} \int_a^b f(\mu(z))\, \frac{dz}{\sqrt{z}}, \qquad (5.2.10)$$

where

$$\mu(z) = \frac{1}{4c}\left(z^2 - 2pz + p^2 - 4d + \frac{q^2}{z}\right)$$

and $\{^a_b$ *denote the largest real roots of* $\mu(z) = \{^{\mu_1}_{\mu_2}$, *respectively.*

For the **proof** we refer to Wagner [292], p. 1196.

Example 5.2.11 Let us now exemplify Proposition 5.2.10 in a more complicated context and calculate the (uniquely determined) even and homogeneous fundamental solution E of the homogeneous quartic operator

$$P(\partial) = \partial_1^4 + \partial_2^4 + \partial_3^4 + 2a\partial_1^2\partial_2^2,$$

which is elliptic for $a > -1$ and which contains Fredholm's operator as the special case $a = 0$. For $a = 1$, we obtain the decomposable operator

$$\Delta_2^2 + \partial_3^4 = (\Delta_2 + i\partial_3^2)(\Delta_2 - i\partial_3^2);$$

its fundamental solution can be deduced from Example 3.2.7 (c) by analytic continuation, see also Example 5.2.8.

The calculation of E runs along similar lines as that in Example 5.2.9, and we shall mainly point out the new features.

(a) As in Example 5.2.9, we assume that $x_3 > 0$ and obtain

$$\partial_3 E(x_1, 1, x_3) = -\frac{1}{4\pi^2} \int_0^{x_3} d\lambda \int_{-\infty}^{\infty} \frac{d\alpha}{\alpha^4 + (\lambda + x_1\alpha)^4 + 2a\alpha^2(\lambda + x_1\alpha)^2 + 1}$$

$$= -\frac{1}{16\pi^2} \int_{x_3^{-4}}^{\infty} \frac{d\mu}{\sqrt{\mu}} \int_{-\infty}^{\infty} \frac{ds}{s^4 + (1 + x_1s)^4 + 2as^2(1 + x_1s)^2 + \mu}.$$

Now we set $A = 1 + x_1^4 + 2ax_1^2$ and $t = s + (x_1^3 + ax_1)/A$ and this yields

$$\partial_3 E(x_1, 1, x_3) = -\frac{1}{16\pi^2 A} \int_{x_3^{-4}}^{\infty} \frac{d\mu}{\sqrt{\mu}} \int_{-\infty}^{\infty} \frac{dt}{t^4 + pt^2 + qt + r}$$

where

$$p = \frac{6x_1^2 + 2a(1 - ax_1^2 + x_1^4)}{A^2}, \qquad q = \frac{4x_1(1 - x_1^4)(1 - a^2)}{A^3},$$

$$r = \frac{x_1^8 - x_1^4 + 1 + a[2x_1^2(1 + x_1^4)(1 + a^2) + x_1^4 a^2(a^2 + 6)]}{A^4} + \frac{\mu}{A}.$$

Setting $r(\mu) = c\mu + d$ (with $c = A^{-1}$) and employing formula (5.2.10) in Proposition 5.2.10, i.e., the substitution

$$\mu(z) = \frac{1}{4c}\left(z^2 - 2pz + p^2 - 4d + \frac{q^2}{z}\right)$$

$$= \frac{A}{4}\left(z^2 - 2pz - \frac{4}{A^4}(1 - a^2)[x_1^8 - 4ax_1^6 - 10x_1^4 - 4ax_1^2 + 1] + \frac{q^2}{z}\right),$$

this furnishes

$$\partial_3 E(x_1, 1, x_3) =$$

$$= -\frac{1}{8\pi\sqrt{A}} \int_{z_1}^{\infty} \frac{dz}{z^3 - 2pz^2 - 4A^{-4}(1 - a^2)[x_1^8 - 4ax_1^6 - 10x_1^4 - 4ax_1^2 + 1]z + q^2}.$$

Herein z_1 denotes the largest real root of $\mu(z) = x_3^{-4}$.

The translation $y = z - 4x_1^2 A^{-2}(1 - a^2)$ and the substitution $y = 4u/A$ result in

$$\partial_3 E(x_1, 1, x_3) = -\frac{1}{8\pi\sqrt{A}} \int_{y_1}^{\infty} \frac{dy}{\sqrt{y[y^2 - 4aA^{-1}y + 4(a^2 - 1)A^{-2}]}}$$

$$= -\frac{1}{16\pi} \int_{u_1}^{\infty} \frac{du}{\sqrt{u[u^2 - au - \frac{1}{4}(1 - a^2)]}}$$

where $\mu(z_1) = x_3^{-4}$ and thus

$$y_1[y_1^2 - 4aA^{-1}y_1 + 4(a^2 - 1)A^{-2}] = \frac{4(y_1 + 4x_1^2 A^{-2}(1 - a^2))}{Ax_3^4},$$

i.e.,

$$u_1\left[4x_3^4u_1^2 - 4ax_3^4u_1 + a^2x_3^4 - P(x_1, 1, x_3)\right] = x_1^2(1 - a^2).$$

Setting $\zeta = 2x_3^2 u_1$ we finally obtain

$$\partial_3 E = -\frac{\text{sign}\, x_3}{16\pi} \int_{\zeta/(2x_3^2)}^{\infty} \frac{du}{\sqrt{u[u^2 - au - \frac{1}{4}(1 - a^2)]}}$$

where ζ is the largest real root of

$$\zeta^3 - 2ax_3^2\zeta^2 - \left(P(x) - a^2x_3^4\right)\zeta - 2(1 - a^2)x_1^2x_2^2x_3^2. \tag{5.2.11}$$

(b) Due to the symmetry of P with respect to x_1, x_2, it remains to calculate one of the derivatives $\partial_1 E$ and $\partial_2 E$. Let us assume now $x_1 > 0$ and set $x_2 = 1$. Then

$$\partial_1 E(x_1, 1, x_3) = -\frac{1}{4\pi^2} \int_0^{x_1} d\lambda \int_{-\infty}^{\infty} \frac{d\alpha}{\alpha^4 + (\lambda + x_3\alpha)^4 + 2a(\lambda + x_3\alpha)^2 + 1}.$$

We substitute successively $\alpha = \lambda s$ and $\mu = \lambda^{-2}$ and obtain

$$\partial_1 E(x_1, 1, x_3) = -\frac{1}{8\pi^2} \int_{x_1^{-2}}^{\infty} d\mu \int_{-\infty}^{\infty} \frac{ds}{s^4 + (1 + x_3 s)^4 + 2a\mu(1 + x_3 s)^2 + \mu^2}.$$

Upon setting $A = 1 + x_3^4$ and $t = s + x_3^3/A$, this yields

$$\partial_1 E(x_1, 1, x_3) = -\frac{1}{8\pi^2 A} \int_{x_1^{-2}}^{\infty} d\mu \int_{-\infty}^{\infty} \frac{dt}{t^4 + pt^2 + qt + r}$$

where

$$p = \frac{2x_3^2(3 + aA\mu)}{A^2}, \quad q = \frac{4x_3(1 - x_3^4 + aA\mu)}{A^3}, \quad r = \frac{1 - x_3^4 + x_3^8 + 2aA\mu + A^3\mu^2}{A^4}.$$

In this case, the cubic resolvent R in Proposition 5.2.10, i.e.,

$$R(\mu, z) = z^3 - 2p(\mu)z^2 + \left(p(\mu)^2 - 4r(\mu)\right)z + q(\mu)^2$$

is a quadratic polynomial $\alpha\mu^2 + \beta\mu + \gamma$ in μ. Its derivative $\partial R/\partial\mu$ with respect to μ evaluated in the zeros $\mu_{1,2}(z)$ of $R(\mu, z) = 0$ yields the discriminant of R with respect to μ, i.e.,

$$\frac{\partial R}{\partial \mu}(\mu_{1,2}(z), z) = \pm\sqrt{\beta^2 - 4\alpha\gamma}.$$

For $z = 0$, this discriminant vanishes since it is the discriminant of $q(\mu)^2$. Therefore,

$$\frac{\sqrt{z}}{\sqrt{\beta^2 - 4\alpha\gamma}} = \frac{A^{7/2}}{4\sqrt{A^2 z - 4x_3^2}\sqrt{A^4 z^2 - 8x_3^2 A^2 z - 4(1 - x_3^4)^2 + 4a^2 A^2}}$$

and hence

$$\partial_1 E(x_1, 1, x_3) = -\frac{A^{5/2}}{8\pi} \int_{z_1}^{\infty} \frac{dz}{\sqrt{A^2 z - 4x_3^2}\sqrt{A^4 z^2 - 8x_3^2 A^2 z - 4(1 - x_3^4)^2 + 4a^2 A^2}}$$

where z_1 is the largest real root of $R(x_1^{-2}, z)$. Substituting $A^2 z = 4uA + 4x_3^2$ we obtain

$$\partial_1 E(x_1, 1, x_3) = -\frac{1}{16\pi} \int_{u_1}^{\infty} \frac{du}{\sqrt{u[u^2 - \frac{1}{4}(1 - a^2)]}}$$

where u_1 is the largest real root of $R(x_1^{-2}, 4A^{-1}u + 4x_3^2 A^{-2})$. This implies $u_1 = \zeta/(2x_1^2)$ where ζ is the largest real root of the polynomial in (5.2.11) for $x_2 = 1$. Hence finally,

$$E = -\frac{1}{16\pi} \left[\sum_{j=1}^{2} |x_j| \int_{\zeta/(2x_j^2)}^{\infty} \frac{du}{\sqrt{u[u^2 - \frac{1}{4}(1 - a^2)]}} \right.$$
$$\left. + |x_3| \int_{\zeta/(2x_3^2)}^{\infty} \frac{du}{\sqrt{u[u^2 - au - \frac{1}{4}(1 - a^2)]}} \right], \qquad (5.2.12)$$

ζ being the largest real root of the polynomial in (5.2.11).

More generally, if

$$P(\partial) = (\partial_1^2, \partial_2^2, \partial_3^2) C \begin{pmatrix} \partial_1^2 \\ \partial_2^2 \\ \partial_3^2 \end{pmatrix}$$

with a real-valued symmetric matrix C fulfilling $c_{jj} > 0$ and $c_{jk}\sqrt{c_{ii}} + c_{ji}\sqrt{c_{kk}} \geq 0$ for each permutation ijk of 123, then the (uniquely determined) even and homogeneous fundamental solution E of $P(\partial)$ is given by

$$E(x) = -\frac{1}{16\pi} \sum_{j=1}^{3} |x_j| \int_{\zeta/(2x_j^2)}^{\infty} \frac{du}{\sqrt{u[c_{jj}u^2 + C_{ik}^{ad}u - \frac{1}{4}\det C]}} \qquad (5.2.13)$$

where $\{i, j, k\} = \{1, 2, 3\}$ and ζ is the largest of the three real roots of the polynomial

$$\zeta^3 - 2(c_{23}x_1^2 + c_{13}x_2^2 + c_{12}x_3^2)\zeta^2 - (C_{11}^{\mathrm{ad}}x_1^4 + C_{22}^{\mathrm{ad}}x_2^4 + C_{33}^{\mathrm{ad}}x_3^4$$
$$- 2C_{12}^{\mathrm{ad}}x_1^2x_2^2 - 2C_{13}^{\mathrm{ad}}x_1^2x_3^2 - 2C_{23}^{\mathrm{ad}}x_2^2x_3^2)\zeta - 2(\det C)x_1^2x_2^2x_3^2,$$

see Wagner [292], Prop. 3, p. 1198.

Note that the operator $P(\partial) = \partial_1^4 + \partial_2^4 + \partial_3^4 + 2a\partial_1^2\partial_2^2$ corresponds to the particular case of the matrix

$$C = \begin{pmatrix} 1 & a & 0 \\ a & 1 & 0 \\ 0 & 0 & 1 \end{pmatrix}, \text{ with } C^{\mathrm{ad}} = \begin{pmatrix} 1 & -a & 0 \\ -a & 1 & 0 \\ 0 & 0 & 1-a^2 \end{pmatrix} \text{ and } \det C = 1 - a^2.$$

As in Example 5.2.9, the definite integrals in (5.2.13) are elliptic integrals of the first kind.

Another particular case is the operator

$$P(\partial) = \partial_1^4 + \partial_2^4 + \partial_3^4 + 2c(\partial_1^2\partial_2^2 + \partial_1^2\partial_3^2 + \partial_2^2\partial_3^2), \quad c \in \mathbf{R},$$

which corresponds to the matrix $C = \begin{pmatrix} 1 & c & c \\ c & 1 & c \\ c & c & 1 \end{pmatrix}$. This operator is elliptic iff $c >$

$-\frac{1}{2}$, and (5.2.13) yields the following representation for the even and homogeneous fundamental solution E, see Wagner [292], pp. 1202, 1203:

$$E = \begin{cases} -\dfrac{1}{8\pi\sqrt{c^2-1}} \displaystyle\sum_{j=1}^{3} x_j F\left(\arcsin\left(\dfrac{\sqrt{2(c^2-1)}\,x_j}{\sqrt{\zeta + x_j^2(c-1)(2c+1)}}\right), \sqrt{\dfrac{c+(1/2)}{c+1}}\right), \text{ if } c > 1, \\[20pt] -\dfrac{1}{8\pi\sqrt{1-c^2}} \displaystyle\sum_{j=1}^{3} x_j F\left(\arcsin\left(\dfrac{\sqrt{2(1-c^2)}\,x_j}{\sqrt{\zeta + (1-c)x_j^2}}\right), \dfrac{1}{\sqrt{2(1+c)}}\right), \qquad \text{ if } c \in [0, 1), \\[20pt] -\dfrac{1}{8\pi\sqrt{1-c^2}} \displaystyle\sum_{j=1}^{3} |x_j| \left[2Y(-2cx_j^2 - x_i^2 - x_k^2)\mathbf{K}\left(\dfrac{1}{\sqrt{2(1+c)}}\right) \right. \\[15pt] \left. + \operatorname{sign}(2cx_j^2 + x_i^2 + x_k^2)F\left(\arcsin\left(\dfrac{\sqrt{2(1-c^2)}\,|x_j|}{\sqrt{\zeta + (1-c)x_j^2}}\right), \dfrac{1}{\sqrt{2(1+c)}}\right) \right], \quad \text{ if } c \in (-\frac{1}{2}, 0], \end{cases}$$

$$(5.2.14)$$

where

$$F(\varphi, k) = \int_0^\varphi \frac{d\alpha}{\sqrt{1 - k^2 \sin^2 \alpha}}, \qquad \varphi \in \mathbf{R}, \ 0 \le k < 1, \ \mathbf{K}(k) = F(\tfrac{\pi}{2}, k)$$

and ζ denotes the largest of the three real roots of

$$\zeta^3 - 2c|x|^2\zeta^2 + (c-1)(c|x|^4 + x_1^4 + x_2^4 + x_3^4)\zeta - 4(c-1)^2(c + \tfrac{1}{2})x_1^2 x_2^2 x_3^2.$$

(c) For reasons of control, let us yet derive some special cases and limit cases from formulas (5.2.12–5.2.14). E.g., the even and homogeneous fundamental solution E of the decomposable operator

$$\partial_1^4 + \partial_2^4 + \partial_3^4 + 2\partial_1^2\partial_2^2 = (\Delta_2 + i\partial_3^2)(\Delta_2 - i\partial_3^2)$$

arises by setting $a = 1$ in (5.2.12). This immediately yields

$$E = -\frac{\rho^2}{4\pi\sqrt{2\zeta}} - \frac{x_3}{8\pi}\arctan\left(\frac{x_3\sqrt{2\zeta}}{\rho^2}\right), \tag{5.2.15}$$

where $\rho^2 = x_1^2 + x_2^2$ and $\zeta = x_3^2 + \sqrt{x_3^4 + \rho^4}$, cf. Wagner [292], Ex. 1, p. 1204. Furthermore, formula (5.2.15) can also be deduced from (3.2.23) or Example 5.2.8 by analytic continuation with respect to λ_1, λ_2.

More generally, we can infer from formula (5.2.13) the fundamental solutions of products of anisotropic Laplaceans in $\mathbf{R}^3$, i.e., of the operators

$$P(\partial) = (a_1\partial_1^2 + a_2\partial_2^2 + a_3\partial_3^2)(b_1\partial_1^2 + b_2\partial_2^2 + b_3\partial_3^2), \qquad 0 < a_i, b_i, \ i = 1, 2, 3.$$

This leads to the expression in (3.1.19), which was originally found in Herglotz [125] and rederived in Garnir [100], Bureau [35], cf. also Wagner [292], Prop. 4, p. 1203 and Ex. 2, p. 1204.

Also note that the limit $c \to 1$ in formula (5.2.14) yields correctly the fundamental solution $E = -|x|/(8\pi)$ of $P(\partial) = \Delta_3^2$. Of course, according to Proposition 5.2.3, the even and homogeneous fundamental solution E depends analytically on the coefficients of the homogeneous elliptic polynomial P.

Another limit case of (5.2.14) is the non-elliptic operator $P_0(\partial) = \partial_1^2\partial_2^2 + \partial_1^2\partial_3^2 + \partial_2^2\partial_3^2$. A generalization of it, namely

$$1 + \sum_{1 \le j, k \le n} \partial_j^2\partial_k^2,$$

is considered in Hörmander [134] for the study of regular and temperate fundamental solutions.

Indeed, $P_0(\partial) = \lim_{\epsilon \searrow 0} P_\epsilon(\partial)$, where

$$P_\epsilon(\partial) = \epsilon(\partial_1^4 + \partial_2^4 + \partial_3^4) + P_0(\partial) = \epsilon[\partial_1^4 + \partial_2^4 + \partial_3^4 + 2cP_0(\partial)], \qquad c = \frac{1}{2\epsilon},$$

and hence we apply formula (5.2.14) in the case $c > 1$ to obtain the fundamental solution E_ϵ of $P_\epsilon(\partial)$ for $0 < \epsilon < \frac{1}{2}$:

$$E_\epsilon = -\frac{1}{4\pi\sqrt{1-4\epsilon^2}} \sum_{j=1}^{3} |x_j| F\left(\arcsin\left(\frac{\sqrt{1-4\epsilon^2}\sqrt{2}\,|x_j|}{\sqrt{4\epsilon^2\zeta + 2x_j^2(1-2\epsilon)(1+\epsilon)}}\right), \sqrt{\frac{1+\epsilon}{1+2\epsilon}}\right).$$

Here ζ satisfies the equation

$$\zeta^3 - \frac{1}{\epsilon}|x|^2\zeta^2 + \left(\frac{1}{2\epsilon}-1\right)\left(\frac{1}{2\epsilon}|x|^4 + x_1^4 + x_2^4 + x_3^4\right)\zeta - 4\left(\frac{1}{2\epsilon}-1\right)^2\left(\frac{1}{2\epsilon}+\frac{1}{2}\right)x_1^2 x_2^2 x_3^2 = 0,$$

and hence $2\epsilon\zeta$ converges to μ for $\epsilon \searrow 0$ where μ denotes the largest of the three real roots of

$$Q(\mu) = \mu(\mu - |x|^2)^2 - 4x_1^2 x_2^2 x_3^2. \qquad (5.2.16)$$

Development in series yields

$$\frac{\sqrt{1-4\epsilon^2}\sqrt{2}\,|x_j|}{\sqrt{4\epsilon^2\zeta + 2x_j^2(1-2\epsilon)(1+\epsilon)}} = 1 - \frac{\epsilon(\mu - x_j^2)}{2x_j^2} + O(\epsilon^2), \quad \arcsin(1-\delta) = \frac{\pi}{2} - \sqrt{2\delta} + O(\delta)$$

for $\epsilon \searrow 0$, $\delta \searrow 0$, and hence

$$\arcsin\left(\frac{\sqrt{1-4\epsilon^2}\sqrt{2}\,|x_j|}{\sqrt{4\epsilon^2\zeta + 2x_j^2(1-2\epsilon)(1+\epsilon)}}\right) = \frac{\pi}{2} - a\sqrt{\epsilon} + O(\epsilon), \qquad a = \sqrt{\frac{\mu}{x_j^2} - 1}.$$

On the other hand,

$$\sqrt{\frac{1+\epsilon}{1+2\epsilon}} = 1 - \frac{\epsilon}{2} + O(\epsilon^2) \quad \text{for } \epsilon \searrow 0,$$

and we must therefore investigate $F(\frac{\pi}{2} - a\sqrt{\epsilon}, 1 - \frac{\epsilon}{2})$ for $\epsilon \searrow 0$.
Elementary estimates furnish

$$F\left(\frac{\pi}{2} - a\sqrt{\epsilon}, 1 - \frac{\epsilon}{2}\right) = \log 4 - \frac{1}{2}\log\epsilon - \log(a + \sqrt{1+a^2}) + O(\sqrt{\epsilon}) \text{ for } \epsilon \searrow 0$$

and hence

$$E_\epsilon = O(\sqrt{\epsilon}) - \frac{1}{4\pi} \sum_{j=1}^{3} |x_j| \left[\log 4 - \frac{1}{2} \log \epsilon - \log\left(\frac{\sqrt{\mu - x_j^2} + \sqrt{\mu}}{|x_j|} \right) \right].$$

Since $P_0(\partial)T = 0$ if $T \in \mathcal{D}'(\mathbf{R}^3)$ depends on only one coordinate, we obtain the following homogeneous fundamental solution E of $P_0(\partial)$:

$$E = \frac{1}{4\pi} \sum_{j=1}^{3} |x_j| \log\left(\frac{\sqrt{\mu - x_j^2} + \sqrt{\mu}}{|x_j|} \right)$$

where μ is the largest real root of the polynomial Q given in (5.2.16).   □

Appendix: Table of Operators/Systems with References to Fundamental Solutions/Matrices

A.1 Ordinary Differential Operators

No.	Operator (name)	References
A.1.1	$\frac{\mathrm{d}}{\mathrm{d}x}$	Example 1.3.6
A.1.2	$\frac{\mathrm{d}}{\mathrm{d}x} - \lambda,\ \lambda \in \mathbf{C}$	Example 1.3.6
A.1.3	$(\frac{\mathrm{d}}{\mathrm{d}x} - \lambda)^{r+1},\ \lambda \in \mathbf{C}, r \in \mathbf{N}_0$	Example 1.3.8 (b)
A.1.4	$(\frac{\mathrm{d}}{\mathrm{d}x} - \lambda)^{r+1}(\frac{\mathrm{d}}{\mathrm{d}x} - \mu)^{s+1}$, $\lambda \neq \mu \in \mathbf{C}, r, s \in \mathbf{N}_0$	Example 1.3.8 (b)
A.1.5	$\frac{\mathrm{d}^2}{\mathrm{d}x^2} + \omega^2,\ \omega \in \mathbf{C} \setminus \{0\}$ (vibrating string, Helmholtz)	Examples 1.3.8 (a), 2.4.6
A.1.6	$(\frac{\mathrm{d}^2}{\mathrm{d}x^2} - \lambda^2)^{r+1},\ \lambda \in \mathbf{C} \setminus \{0\}, r \in \mathbf{N}_0$ (iterated metaharmonic)	Example 1.3.8 (b)
A.1.7	$\frac{\mathrm{d}^4}{\mathrm{d}x^4} + \lambda^4,\ \lambda \in \mathbf{C}$ (static elastically supported bar)	Garnir [98], p. 183
A.1.8	$(\frac{\mathrm{d}^2}{\mathrm{d}x^2} - \lambda^2)^2 + p^2,\ p, \lambda \in \mathbf{C}$ (static prestressed elastically supported bar)	Oberhettinger [196], p. 5
A.1.9	$\prod_{j=1}^{m}(\frac{\mathrm{d}}{\mathrm{d}x} - \lambda_j)^{\alpha_j + 1},\ \alpha \in \mathbf{N}_0^m$, $\lambda_j \in \mathbf{C}$ pairwise different	Propositions 1.3.7, 2.4.5 Petersen [228], Ex. 12.18, p. 141 Komech [154], Lemma 2.1, p. 147 Vo-Khac Khoan [283], Thm., p. 109

© Springer International Publishing Switzerland 2015
N. Ortner, P. Wagner, *Fundamental Solutions of Linear Partial Differential Operators*, DOI 10.1007/978-3-319-20140-5

A.2   Systems of Ordinary Differential Operators

No.	System	References
A.2.1	$\begin{pmatrix} \frac{d}{dx} & 1 \\ 0 & 1 \end{pmatrix}$	Example 2.1.2
A.2.2	$A\frac{d}{dt} + B,$ $A, B \in \mathbf{C}^{n \times n}, \det A \neq 0$	Gel'fand and Shilov [105], Ch. II, § 4, p. 58 Treves [272], pp. 27, 28 $E = Y(t)A^{-1}\exp(-BA^{-1}t)$
A.2.3	$A\frac{d^2}{dt^2} + B,$ $A, B \in \mathbf{C}^{n \times n}, \det A \neq 0$	$E = Y(t)A^{-1}\dfrac{\sin(\sqrt{BA^{-1}}t)}{\sqrt{BA^{-1}}}$

A.3   Elliptic Operators

No.	Operator (name)	References
A.3.1	$\partial_1 + i\partial_2 = 2\partial_{\bar{z}}$ (Cauchy–Riemann)	Examples 1.3.14 (b), 1.3.16, 2.3.4 Rudin [239], ex. 8, p. 205
A.3.2	$\partial_1 + \lambda\partial_2 + \mu, \ \lambda, \mu \in \mathbf{C},$ $\operatorname{Im}\lambda \neq 0$	Ortner [200], Op. 2, p. 156
A.3.3	$(\partial_1 + \lambda\partial_2 + \mu)^l, \ \lambda, \mu \in \mathbf{C},$ $\operatorname{Im}\lambda \neq 0, l \in \mathbf{N}$	Example 2.5.2
A.3.4	$\displaystyle\prod_{j=1}^{l}(\partial_1 - \lambda_j\partial_2)^{\alpha_j+1}, \lambda_j \in \mathbf{C} \setminus \mathbf{R},$ λ_j pairwise different, $\alpha \in \mathbf{N}_0^l$	Proposition 3.3.2
A.3.5	Δ_2 (Laplace)	Example 1.3.14 (a) Ortner [200], Op. 8, p. 157
A.3.6	$\Delta_2^k, \ k \in \mathbf{N}$ (iterated Laplace)	Ortner [200], Op. 9, p. 157
A.3.7	Δ_n (Laplace)	Examples 1.3.14 (a), 1.3.16, 1.4.10
A.3.8	$\Delta_n^k, \ k \in \mathbf{N}$ (iterated Laplace)	Examples 1.6.11, 2.3.2 (c), 2.6.2, 5.2.2
A.3.9	$\Delta_n + \lambda, \ \lambda > 0$ (Helmholtz)	Example 1.3.14 (c)
A.3.10	$(\Delta_n + \lambda)^k, \ \lambda > 0, k \in \mathbf{N}$ (iterated Helmholtz)	Example 1.4.3

No.	Operator (name)	References
A.3.11	$\Delta_n - \lambda$, $n \geq 2, \lambda > 0$ (metaharmonic)	Examples 1.4.11, 1.6.11, 2.4.2
A.3.12	$(\Delta_n - \lambda)^k$, $n \geq 2, \lambda > 0, k \in \mathbf{N}$ (iterated metaharmonic)	Example 1.6.11 / Ortner [200], Op. 13, p. 158
A.3.13	$(\partial_1 - \lambda_1 \partial_2)(\partial_1 - \lambda_2 \partial_2)$, $\lambda_1, \lambda_2 \in \mathbf{C}$, $\operatorname{Im}\lambda_1 \cdot \operatorname{Im}\lambda_2 < 0$	Example 2.4.9
A.3.14	$\partial_1^2 + \partial_2^2 + 2i\partial_1\partial_2 + \partial_3^2$	Example 2.4.9
A.3.15	$\partial_1^2 + i(\partial_2^2 + \partial_3^2)$	Example 1.4.12 (a)
A.3.16	$\nabla^T A \nabla$, $A = A^T \in \mathbf{C}^{2\times 2}$ (anisotropic Laplace)	Examples 1.4.12, 2.4.9 (a) / Ortner [200], Op. 17, p. 159
A.3.17	$(\nabla^T A \nabla)^k$, $A = A^T \in \mathbf{C}^{n\times n}, k \in \mathbf{N}$ (iterated anisotropic Laplace)	Example 2.4.9 (c)
A.3.18	$(\nabla^T A \nabla)(\nabla^T B \nabla)$, $A, B \in \mathbf{R}^{n\times n}$, $A = A^T, B = B^T$ positive definite (product of anisotropic Laplaceans)	Example 3.1.8 (a) / Ortner [200], Op. 27, p. 160
A.3.19	$\prod_{j=1}^{l}(\nabla^T A_j \nabla)^{\alpha_j+1}$, $A_j \in \mathbf{R}^{n\times n}$, $A_j = A_j^T$ positive definite, $\alpha \in \mathbf{N}_0^l$ (product of anisotropic Laplaceans)	Example 3.1.8 (a)
A.3.20	$\partial_1^4 + 2(1 - 2\epsilon^2)\partial_1^2\partial_2^2 + \partial_2^4$, $0 < \epsilon < 1$ (static orthotropic plate)	Examples 3.1.8 (b), 3.3.3, 5.2.4
A.3.21	$(\partial_1^4 + 2(1 - 2\epsilon^2)\partial_1^2\partial_2^2 + \partial_2^4)^2$, $0 < \epsilon < 1$ (iterated orthotropic plate)	Galler [86], § 14
A.3.22	$(\partial_1^4 + 2(1 - 2\epsilon^2)\partial_1^2\partial_2^2 + \partial_2^4)^l$, $0 < \epsilon < 1$, $l \in \mathbf{N}$ (iterated orthotropic plate)	Example 5.2.4 / Galler [86], p. 55
A.3.23	$(a_1\partial_1^2 + a_2\partial_2^2 + a_3\partial_3^2)(b_1\partial_1^2 + b_2\partial_2^2 + b_3\partial_3^2)$ $a_j > 0, b_j > 0$	Example 3.1.8 (c)
A.3.24	$\Delta_2^2 + \lambda^4$, $\lambda > 0$ (static elastically supported plate)	Garnir [98], p. 184 / Ortner [200], Op. 21, p. 159
A.3.25	$(-1)^m \Delta_2^m + \lambda^{2m}$, $\lambda \in \mathbf{C} \setminus \{0\}, m \in \mathbf{N}$	Ortner [200], Op. 20, p. 159

No.	Operator (name)	References
A.3.26	$(-\Delta_n + \lambda)^m \Delta_n$, $\lambda \in \mathbf{C} \setminus \{0\}$, $m \in \mathbf{N}, n = 2, 3$	Ortner [200], Ops. 22, 23, pp. 159, 160
A.3.27	$\prod_{j=1}^{l}(\Delta_n + \lambda_j)^{\alpha_j+1}$, $\alpha \in \mathbf{N}_0^l$, $\lambda_j > 0$ pairwise different (product of Helmholtz operators)	Example 1.4.5
A.3.28	$(\nabla^T A \nabla + b^T \nabla - \lambda)^m$, $A = A^T \in \mathbf{R}^{n \times n}$ positive definite, $b \in \mathbf{C}^n, \lambda \in \mathbf{C}, m \in \mathbf{N}$ (iterated anisotropic metaharmonic)	Example 2.5.4
A.3.29	$\prod_{j=1}^{l}(\nabla^T A_j \nabla + b_j^T \nabla - d_j)^{\alpha_j+1}$, $A_j = A_j^T \in \mathbf{R}^{n \times n}$ positive definite, $b_j \in \mathbf{C}^n, d_j \in \mathbf{C}, \alpha \in \mathbf{N}_0^l$ (product of anisotropic metaharmonic)	Example 3.1.6 $\alpha = 0, l = 2$: Ortner [200], Op. 30, p. 161
A.3.30	$\Delta_2^2 + \partial_3^4$	Example 5.2.11 (c)
A.3.31	$(\Delta_2 + \lambda_1 \partial_3^2)(\Delta_2 + \lambda_2 \partial_3^2)$, $\lambda_1, \lambda_2 > 0$, $\lambda_1 \neq \lambda_2$	Examples 3.2.7 (c), 5.2.8 Wagner [285], Bsp. 4, p. 45
A.3.32	$\prod_{j=1}^{l}(\Delta_{n-1} + b_j^2 \partial_n^2)^{\alpha_j+1}$, $b_j > 0$ pairwise different, $\alpha \in \mathbf{N}_0^l$ (product of anisotropic Laplaceans)	Examples 3.2.7, 5.2.5
A.3.33	$\Delta_2^2 - 4c^2 \partial_1^2$, $c \in \mathbf{C} \setminus \{0\}$ (static one-sided stretched plate)	Ortner [200], Bsp. 3.3, p. 143, Op. 31, p. 161 Wagner [284] Dundurs and Jahanshahi [66]
A.3.34	$\partial_1^{2l} + \partial_2^{2l}$, $l \in \mathbf{N}$	Galler [86], § 12, p. 49
A.3.35	$\prod_{j=1}^{l}(\partial_1^2 + b_j^2 \partial_2^2)$, $b_j > 0$ pairwise different	Example 3.3.3

No.	Operator (name)	References
A.3.36	$\partial_1^4 + \partial_2^4 + \partial_3^4$	Example 5.2.9
A.3.37	$\partial_1^4 + \partial_2^4 + \partial_3^4 + 2a\partial_1^2\partial_2^2,\ a > -1$	Example 5.2.11
A.3.38	$\partial_1^4 + \partial_2^4 + \partial_3^4 + 2c(\partial_1^2\partial_2^2 + \partial_1^2\partial_3^2 + \partial_2^2\partial_3^2),\ c > -\frac{1}{2}$	Example 5.2.11
A.3.39	$(\partial_1^2, \partial_2^2, \partial_3^2)C(\partial_1^2, \partial_2^2, \partial_3^2)^T,$ $C = C^T \in \mathbf{R}^{3\times3}$ positive definite	Example 5.2.11
A.3.40	$\Delta_2^4 + 16\lambda^4\partial_1^4,\ \lambda \in \mathbf{C} \setminus \{0\}$ (static circular cylindrical shell)	Jahanshahi [150] Wagner [284]

A.4 Elliptic Systems

No.	System (name)	References
A.4.1	$(\rho\tau^2 + \mu\Delta_3)I_3 + (\lambda + \mu)\nabla\cdot\nabla^T,\ \lambda, \mu, \rho, \tau > 0$ (time-harmonic Lamé system)	Example 2.4.3
A.4.2	$\begin{pmatrix} a_1\partial_1^2 + a_4\partial_2^2 + a_5\partial_3^2 & (a_1 - a_4)\partial_1\partial_2 & a_3\partial_1\partial_3 \\ (a_1 - a_4)\partial_1\partial_2 & a_4\partial_1^2 + a_1\partial_2^2 + a_5\partial_3^2 & a_3\partial_2\partial_3 \\ a_3\partial_1\partial_3 & a_3\partial_2\partial_3 & a_5\Delta_2 + a_2\partial_3^2 \end{pmatrix},$ $a_1 > 0,\ a_2 > 0,\ a_4 > 0,\ a_5 > 0$ and $|a_3| < a_5 + \sqrt{a_1 a_2}$ (hexagonal elastostatics)	Example 5.2.6 (b)

A.5 Hyperbolic Operators

No.	Operator (name)	References
A.5.1	$\partial_1 \cdots \partial_l,\ 1 \le l \le n$	Ortner [200], Op. 42, p. 164
A.5.2	$\partial^\alpha = \partial_1^{\alpha_1} \cdots \partial_n^{\alpha_n},\ \alpha \in \mathbf{N}_0^n$	Example 1.5.5 (a)
A.5.3	$\partial_1^m\partial_2^m,\ m \in \mathbf{N}$	Example 1.5.5 (b)
A.5.4	$\partial_1 + \lambda\partial_2 + \mu,\ \lambda \in \mathbf{R}, \mu \in \mathbf{C}$ (transport operator)	Ortner [200], Op. 44, p. 164
A.5.5	$(a^T\nabla + \lambda)^m,\ a \in \mathbf{R}^n \setminus \{0\}, \lambda \in \mathbf{C}, m \in \mathbf{N}$ (iterated transport operator)	Example 2.5.2 (a)

No.	Operator (name)	References
A.5.6	$\prod_{j=1}^{l}(a_j^T\nabla + d_j)^{\alpha_j+1}$, $a_j \in \mathbf{R}^n \setminus \{0\}$, $d_j \in \mathbf{C}, \alpha \in \mathbf{N}_0^l$ (product of transport operators)	Examples 3.1.3, 3.4.5
A.5.7	$\prod_{j=1}^{l}(a_j^T\nabla)$, $a_j \in \mathbf{R}^n \setminus \{0\}$ (product of homogeneous transport operators)	Example 3.4.6
A.5.8	$\prod_{\epsilon \in \{\pm 1\}^2}(\partial_t + \epsilon_1\partial_x + \epsilon_2\partial_y)$ $= \partial_t^4 - 2\partial_t^2(\partial_x^2 + \partial_y^2) + (\partial_x^2 - \partial_y^2)^2$	Example 3.4.6
A.5.9	$(\partial_t^2 - c^2\partial_x)^m$, $c > 0, m \in \mathbf{N}$ (iterated wave operator)	Example 1.5.5 (b)
A.5.10	$\partial_t^2 - \Delta_2$ (wave operator in two space dimensions)	Example 1.4.12 (b)
A.5.11	$\partial_t^2 - \Delta_3$ (wave operator in three space dimensions)	Example 1.4.12 (b) Ortner [200], Op. 51, p. 165
A.5.12	$(\partial_t - \alpha\partial_3)^2 - \beta\Delta_2$, $\alpha \in \mathbf{R}, \beta > 0$ (wave operator in two space dimensions)	Example 4.3.9
A.5.13	$(\partial_t - \alpha\partial_1 - \beta\partial_3)^2 - 4\alpha^2\Delta_2$, $\alpha > 0, \beta > 0$ (wave operator in two space dimensions)	Example 4.4.8
A.5.14	$(\partial_t^2 - \Delta_3)^k$, $k \geq 2$ (iterated wave operator)	Examples 4.4.4, 2.3.6 Ortner [200], Op. 53, p. 165
A.5.15	$\partial_t^2 - \Delta_n$ (wave operator in n space dimensions)	Examples 1.6.17, 2.3.6
A.5.16	$(\partial_t^2 - \Delta_n)^k$, $k \in \mathbf{N}$ (iterated wave operator)	Example 2.3.6
A.5.17	$\partial_t^{n-2k}(\partial_t^2 - \Delta_n)^k$, $k \in \mathbf{N}, n \geq 2k$	Lemma 3.3.5
A.5.18	$\prod_{j=1}^{l}(\partial_t^2 - \lambda_j\Delta_n)^{\alpha_j+1}$, $\alpha \in \mathbf{N}_0^l$, $\lambda_j > 0$ pairwise different (product of wave operators)	Examples 3.2.2, 3.2.3, 3.2.4, 3.2.5, 3.3.4

No.	Operator (name)	References
A.5.19	$(\partial_t^2 - \Delta_3)(\partial_t^2 - a\Delta_2 - b\partial_3^2),\ a > 0, b > 0$ (product of anisotropic wave operators)	Example 4.2.7
A.5.20	$\prod_{j=1}^{l}(\partial_t^2 - \Delta_2 - a_j^2\partial_3^2),$ $a_j > 0$ pairwise different (product of anisotropic wave operators)	Wagner [285], Bsp. 5, p. 27
A.5.21	$\prod_{j=1}^{l}(\partial_t^2 - a_j^2\Delta_2 - \partial_3^2),$ $a_j > 0$ pairwise different (product of anisotropic wave operators)	Wagner [285], Bsp. 8, p. 30
A.5.22	$(\partial_t + \alpha\partial_3)\big((\partial_t + \alpha\partial_3)^2 - \beta\Delta_2\big),$ $\alpha \in \mathbf{R}, \beta > 0$ (product of transport and wave operators)	Example 4.3.10
A.5.23	$\partial_t^2 - \partial_x^2 + m^2,\ m \in \mathbf{C}$ (Klein–Gordon operator)	Examples 2.3.7, 3.5.12 Ortner [200], Op. 56, p. 166
A.5.24	$(\partial_t^2 - \Delta_n + m^2)^k,\ m \in \mathbf{C}, k \in \mathbf{N}$ (iterated Klein–Gordon operator)	Examples 1.6.18, 2.3.7, 2.6.6, 3.5.12
A.5.25	$(\partial_t^2 + \beta\partial_t - \nabla^T A\nabla + b^T\nabla - \lambda)^m,$ $A = A^T \in \mathbf{R}^{n\times n}$ positive definite, $b \in \mathbf{C}^n, \beta, \lambda \in \mathbf{C}, m \in \mathbf{N}$ (anisotropic Klein–Gordon operator)	Example 2.5.6
A.5.26	$\prod_{j=1}^{l}(\partial_t^2 - \Delta_n - c_j)^{\alpha_j+1},$ $c_j \in \mathbf{C}$ pairwise different, $\alpha \in \mathbf{N}_0^l$ (product of Klein–Gordon operators)	Proposition 1.4.4 Ortner [200], Op. 62, p. 167
A.5.27	$\prod_{j=1}^{l}(\partial_t^2 - a_j\Delta_n - d_j)^{\alpha_j+1},$ $a_j > 0, d_j \in \mathbf{C}, \alpha \in \mathbf{N}_0^l, n \leq 4$ (product of Klein–Gordon operators)	Example 3.4.8
A.5.28	$(\partial_t^2 - a\partial_x^2 + b)^2 - (c\partial_x^2 - d)^2 - e^2,$ $a > 0, c > 0, b, d, e \in \mathbf{C}$ (Timoshenko beam operator)	Examples 3.5.4, 4.1.6

No.	Operator (name)	References
A.5.29	$(a_0\partial_t^2 - b_0\Delta_3 + c_0)(a_1\partial_t^2 - b_1\Delta_3 + c_1) - d^2,$ $a_0, b_0, a_1, b_1 > 0, c_0, c_1, d \in \mathbf{C}$ (generalized Timoshenko operator)	Example 3.5.5
A.5.30	$(\partial_t^2 - \Delta_2)(\partial_t^2 - \alpha\Delta_2) + \beta\partial_t^2,\ \alpha, \beta > 0$ (Uflyand–Mindlin plate operator)	Example 3.5.4
A.5.31	$(\partial_t^2 - \Delta_2)^2 + 4c^2\partial_t^2,\ c \in \mathbf{C} \setminus \{0\}$	Ortner and Wagner [210], p. 192
A.5.32	$(\partial_1^2, \partial_2^2, \partial_3^2)C(\partial_1^2, \partial_2^2, \partial_3^2)^T,$ $C = C^T \in \mathbf{R}^{3\times3}, c_{33} > 0, c_{11}, c_{22} \geq 0,$ $c_{12} \geq -\sqrt{c_{11}c_{22}}, c_{13} \leq -\sqrt{c_{11}c_{33}},$ $c_{23} \leq -\sqrt{c_{22}c_{33}}, C_{12}^{ad} \geq -\sqrt{C_{11}^{ad}C_{22}^{ad}}$	Wagner [294], Prop. 3, p. 150
A.5.33	$\partial_1^3 + \partial_2^3 + \partial_3^3 + 3a\partial_1\partial_2\partial_3,\ a < -1$	Wagner [290], Thm., p. 286

A.6 Hyperbolic Systems

No.	System (name)	References
A.6.1	$(\rho\partial_t^2 - \mu\Delta_3)I_3 - (\lambda + \mu)\nabla\nabla^T,\ \lambda, \mu, \rho > 0$ (Lamé system, isotropic elastodynamics)	Example 2.1.3
A.6.2	$(\rho\partial_t^2 - \mu\Delta_2)I_2 - (\lambda + \mu)\nabla\nabla^T,\ \lambda, \mu, \rho > 0$ (two-dimensional Lamé system)	Ortner and Wagner [217], p. 329 Eringen and Şuhubi [69], p. 412
A.6.3	$\begin{pmatrix} P_1(\partial) & (a_4 - a_1)\partial_1\partial_2 & -a_3\partial_1\partial_3 \\ (a_4 - a_1)\partial_1\partial_2 & P_2(\partial) & -a_3\partial_2\partial_3 \\ -a_3\partial_1\partial_3 & -a_3\partial_2\partial_3 & P_3(\partial) \end{pmatrix},$ $P_1(\partial) = \rho\partial_t^2 - a_1\partial_1^2 - a_4\partial_2^2 - a_5\partial_3^2,$ $P_2(\partial) = \rho\partial_t^2 - a_4\partial_1^2 - a_1\partial_2^2 - a_5\partial_3^2,$ $P_3(\partial) = \rho\partial_t^2 - a_5\Delta_2 - a_2\partial_3^2$ (elastodynamics in hexagonal media)	Examples 2.1.4 (d), 4.3.10, 4.4.5

No.	System (name)	References
A.6.4	$(\rho\partial_t^2 - c\Delta_3)I_3 - b\nabla\nabla^T +$ $+ (b-a)\begin{pmatrix} \partial_1^2 & 0 & 0 \\ 0 & \partial_2^2 & 0 \\ 0 & 0 & \partial_3^2 \end{pmatrix}$, (elastodynamics in cubic media)	Examples 2.1.4 (c), 4.3.9
A.6.5	$I_3\partial_t^2 - \begin{pmatrix} d_3\partial_2^2 + d_2\partial_3^2 & -d_3\partial_1\partial_2 & -d_2\partial_1\partial_3 \\ -d_3\partial_1\partial_2 & d_3\partial_1^2 + d_1\partial_3^2 & -d_1\partial_2\partial_3 \\ -d_2\partial_1\partial_3 & -d_1\partial_2\partial_3 & d_1\partial_2^2 + d_2\partial_1^2 \end{pmatrix}$, $0 < d_1 < d_2 < d_3$ (crystal optics)	Example 4.4.8
A.6.6	$\begin{pmatrix} \partial_t^2 - d\partial_2^2 - \partial_3^2 & d\partial_1\partial_2 & \partial_1\partial_3 \\ d\partial_1\partial_2 & \partial_t^2 - d\partial_1^2 - \partial_3^2 & \partial_2\partial_3 \\ \partial_1\partial_3 & \partial_2\partial_3 & \partial_t^2 - \Delta_2 \end{pmatrix}$, $0 < d \neq 1$ (uniaxial crystal optics)	Example 4.4.8 (d)
A.6.7	$\sum_{\mu=0}^{3} i\gamma^\mu \partial_\mu - mI_4 = i\nabla\!\!\!\!/ - mI_4, m > 0$ (Dirac system)	Vladimirov [279], §§ 2.8, 11.12 Ortner [200], * Op. 75, p. 169
A.6.8	$\begin{pmatrix} \sigma\partial_t & \partial_1 & \partial_2 & \partial_3 \\ \partial_1 & \partial_t & \eta & 0 \\ \partial_2 & -\eta & \partial_t & 0 \\ \partial_3 & 0 & 0 & \partial_t \end{pmatrix}$, $\sigma, \eta > 0$ (rotating liquid)	Vladimirov, Drozzinov and Zavialov [281], p. 224

A.7 Non-elliptic Non-hyperbolic Operators

No.	Operator (name)	References
A.7.1	$\partial_t - \Delta_n$ (heat operator)	Examples 1.3.14 (d), 1.3.16, 1.6.16, 2.6.3
A.7.2	$(\partial_t - \Delta_n)^k,\ k \in \mathbf{N}$ (iterated heat operator)	Example 2.3.8

No.	Operator (name)	References
A.7.3	$\partial_t - i\Delta_n$ (Schrödinger operator)	Example 1.4.13
A.7.4	$\partial_t - \nabla^T A \nabla$, $A = A^T \in \mathbf{R}^{n \times n}$, A positive definite (anisotropic heat operator)	Example 1.4.13
A.7.5	$\partial_t - \nabla^T A \nabla$, $A = A^T \in \mathbf{C}^{n \times n}$, Re A positive semi-definite (anisotropic Schrödinger operator)	Example 1.4.13
A.7.6	$(\partial_t - i\Delta_n)^k$, $k \in \mathbf{N}$ (iterated Schrödinger operator)	Example 2.3.8
A.7.7	$(\partial_t - \nabla^T A \nabla + b^T \nabla - \lambda)^m$, $b \in \mathbf{C}^n, \lambda \in \mathbf{C}$, $A = A^T \in \mathbf{C}^{n \times n}, \det A \neq 0$, Re A positive semi-definite (anisotropic Schrödinger / heat operator)	Example 2.5.5
A.7.8	$\partial_1^m + \partial_2^m$, m odd	Galler [86], Satz 12.2, p. 51
A.7.9	$\partial_1^m - \partial_2^m$	Galler [86], Satz 13.1, p. 52
A.7.10	$\partial_t - i\partial_1\partial_2$	Example 2.6.5
A.7.11	$\partial_t - \partial_1\partial_2\partial_3$ (Sobolev operator)	Examples 2.3.8, 2.6.4
A.7.12	$\partial_t - \partial_1\partial_2\partial_3\partial_4\partial_5$	Examples 2.6.5
A.7.13	$\partial_1^2 + \cdots + \partial_p^2 - \partial_{p+1}^2 - \cdots - \partial_{p+q}^2$ (ultrahyperbolic operator)	Gel'fand and Shilov [104] Ch. III, § 2.5, p. 279 Ortner and Wagner [219], Ex. 2.7.5, p. 74
A.7.14	$\prod_{j=1}^{l}\left(\partial_t - \nabla^T A_j \nabla - d_j + \sum_{k=1}^{n} b_{jk}\partial_k\right)^{\alpha_j+1}$, $B = (b_{jk}) \in \mathbf{C}^{l \times n}, d = (d_j) \in \mathbf{C}^l, \alpha \in \mathbf{N}_0^l$, $A_j = A_j^T \in \mathbf{C}^{n \times n}, \det A_j \neq 0$, Re A_j positive semi-definite (product of anisotropic Schrödinger / heat operators)	Example 3.1.4

No.	Operator (name)	References
A.7.15	$\prod\limits_{j=1}^{l}(\partial_t - a_j\Delta_n - d_j)^{\alpha_j+1}$, $a_j > 0$ pairwise different, $d_j \in \mathbf{C}$, $\alpha \in \mathbf{N}_0^l$ (product of heat operators)	Example 3.4.7
A.7.16	$\prod\limits_{j=1}^{l}(\partial_t - a_j\Delta_n - d)$, $a_j > 0$ pairwise different, $d \in \mathbf{C}$ (product of heat operators)	Example 3.4.7
A.7.17	$\prod\limits_{j=1}^{l}(\partial_t \pm ia_j^2\Delta_n)$, $a_j > 0$ pairwise different (product of Schrödinger operators)	Galler [86], Sätze 18.1/2, p. 67
A.7.18	$\partial_t^2 + (\Delta_n + a)^2 + c^2$, $a, c \in \mathbf{C}$ (prestressed, elastically supported beam / plate)	Examples 3.5.8, 3.5.10
A.7.19	$\partial_t^2 + \partial_x^4$ (Euler–Bernoulli beam)	Ortner [200], Op. 38, p. 163
A.7.20	$\partial_t^2 + \Delta_2^2$ (Lagrange–Germain's plate operator)	Ortner [200], Op. 38, p. 163
A.7.21	$(\partial_t - \Delta_3)(\partial_t^2 - \Delta_3) - \epsilon\partial_t\Delta_3$, $\epsilon \geq 0$ (thermoelastic operator)	Examples 4.1.11, 4.2.11
A.7.22	$\partial_t^2 - \partial_t\partial_x^2 - \partial_x^2$ (Stokes' operator)	Example 4.1.10
A.7.23	$\partial_t - \partial_t\Delta_3 - \Delta_3$	Ortner and Wagner [207], Rem. 3, p. 451
A.7.24	$(\partial_t - a\Delta_n - b)^2 - (c\Delta_n + 2\omega^T\nabla + d)^2 - h^2$, $a, b, c, d, h \in \mathbf{C}$, $\omega \in \mathbf{C}^n$, $|\mathrm{Re}\,c| \leq \mathrm{Re}\,a$	Ortner and Wagner [207], Prop. 4, p. 450
A.7.25	$\partial_t^2 - \alpha\partial_t^2\Delta_n - \delta\Delta_n$, $\alpha > 0$, $\delta \in \mathbf{C}$ (Boussinesq operator)	Examples 3.5.5 (b), 3.5.6 Ortner [203], p. 552
A.7.26	$\partial_t^2 - \alpha\partial_t^2\Delta_n + \beta\Delta_n^2$, $\alpha > 0$, $\beta \geq 0$ (Rayleigh operator)	Examples 3.5.5 (b), 4.1.9

No.	Operator (name)	References
A.7.27	$(\partial_t^2 - \lambda \Delta_3)\Delta_3,\ \lambda > 0$	Ortner [200], Op. 65, p. 167
A.7.28	$\Delta_2(\partial_1^2 - \epsilon^2),\ \epsilon > 0$	Ortner [200], Op. 68, p. 168
A.7.29	$\partial_1^3 + \partial_2^3 + \partial_3^3$ (Zeilon's operator)	Wagner [291]
A.7.30	$\partial_1^3 + \partial_2^3 + \partial_3^3 + 3a\partial_1\partial_2\partial_3,\ a > -1$	Wagner [290]

A.8 Non-elliptic Non-hyperbolic Systems

No.	System (name)	References
A.8.1	$\begin{pmatrix} \alpha^2\partial_t^2 - \partial_x^2 & \partial_x \\ \beta^2\partial_x & \partial_x^2 - \beta^2 \end{pmatrix},\ \alpha, \beta > 0$ (Rayleigh's system)	Examples 2.4.14, 4.1.9
A.8.2	$\begin{pmatrix} (\rho\partial_t^2 - \mu\Delta_3)I_3 - (\lambda + \mu)\nabla \cdot \nabla^T & \beta\nabla \\ \eta\partial_t\nabla^T & \partial_t - \kappa\Delta_3 \end{pmatrix},$ $\rho, \lambda, \mu, \beta, \kappa, \eta > 0$ (dynamic linear thermoelasticity)	Example 4.1.11

References

1. Abramowitz, M., Stegun, I.A.: Handbook of Mathematical Functions, 9th edn. Dover, New York (1970)
2. Achenbach, J.D.: Wave Propagation in Elastic Solids. North-Holland, Amsterdam (1973)
3. Achenbach, J.D., Wang, C.Y.: A new method to obtain 3–D Green's functions for anisotropic solids. Wave Motion **18**, 273–289 (1993)
4. Agranovich, M.S.: Partial differential equations with constant coefficients. Russ. Math. Surv. **16**, 23–90 (1961); Transl. from М.С. Агранович: Об уравненях в частных производнях с постоянными коэффициентами. Успехи Мат. Наук **16**(2), 27–93 (1961)
5. Atiyah, M.F., Bott, R., Gårding, L.: Lacunas for hyperbolic differential operators with constant coefficients I. Acta Math. **124**, 109–189 (1970)
6. Atiyah, M.F., Bott, R., Gårding, L.: Lacunas for hyperbolic differential operators with constant coefficients II. Acta Math. **131**, 145–206 (1973)
7. Badii, L., Oberhettinger, F.: Tables of Laplace Transforms. Springer, Berlin (1973)
8. Barros-Neto, J.: An Introduction to the Theory of Distributions. Marcel Dekker, New York (1973)
9. Bass, J.: Cours de mathématiques, vol. I. Masson, Paris (1968)
10. Becker, R., Sauter, F.: Theorie der Elektrizität. Band I, 18th edn. Teubner, Stuttgart (1964)
11. Bernstein, I.N.: Modules over a ring of differential operators. Study of fundamental solutions of equations with constant coefficients. Funct. Anal. Appl. **5**, 89–101 (1971); Transl. from И.Н. Бернштейн: Модули над кольцом дифференциальных опера-тороб. Иследование фундаментальных решений уравнений с постоянными коэфициентами. Функц. Анализ и Прил. **5**(2), 1–16 (1971)
12. Bernstein, I.N.: The analytic continuation of generalized functions with respect to a parameter. Funct. Anal. Appl. **6**, 273–285 (1972)
13. Bernstein, I.N., Gel'fand, S.I.: Meromorphy of the function P^λ. Funct. Anal. Appl. **3**, 68–69 (1969)
14. Biot, M.A.: Thermoelasticity and irreversible thermodynamics. J. Appl. Phys. **27**, 240–253 (1956)
15. Björk, J.: Rings of Differential Operators. North-Holland, Amsterdam (1979)
16. Blanchard, P., Brüning, E.: Mathematical Methods in Physics. Birkhäuser, Berlin (2003)
17. Bochner, S.: Lectures on Fourier integrals. Annals of Mathematical Studies, vol. 42. Princeton University Press, Princeton, NJ (1959)
18. Bogolubov, N.N., Logunov, A.A., Todorov, I.T.: Axiomatic Quantum Field Theory. Benjamin, Reading, MA (1975); Transl. from Н.Н. Боголюбов, А.А. Логунов и И.Т. Тодоров:

Основы аксиоматического подхода в квантовой теории поля. Наука, Москва (1969)

19. Boley, B.A., Chao, C.C.: Some solutions of the Timoshenko beam equation. J. Appl. Mech. Trans ASME **77**, 579–586 (1955)
20. Boley, B.A., Weiner, J.H.: Theory of Thermal Stresses. Wiley, New York (1960)
21. Borovikov, V.A.: Fundamental solutions of linear partial differential equations with constant coefficients. Amer. Math. Soc. Transl. Ser. 2, **25**, 11–66 (1963); Transl. from В.А. Боровиков: Фундаментальные решения линейных уравнений в частных производных с постоянными коэфициентами. Труды Моск. Матем. Об-ва **8**, 199–257 (1959)
22. Brédimas, A.: La différentiation d'ordre complexe, le produit de convolution généralisé et le produit canonique pour les distributions. CRAS Paris **282**, 37–40 (1976)
23. Bresse, M.: Cours de mécanique appliquée. Mallet-Bachelier, Paris (1859)
24. Bresters, D.W.: On distributions connected with quadratic forms. SIAM J. Appl. Math. **16**, 563–581 (1968)
25. Bresters, D.W.: Initial value problems for iterated wave operators. Thesis, Twente Technological University, Groningen (1969)
26. Bresters, D.W.: The initial value problem for operators which can be factorized into a product of Klein–Gordon operators. In: Conference on the Theory of Ordinary and Partial Differential Equations, University of Dundee, Dundee, 1972. Lecture Notes in Mathematics, vol. 280, pp. 227–231. Springer, Berlin (1972)
27. Brillouin, M.: Sources électromagnétiques dans les milieux uniaxes. Bull. Sci. Math. **53**, 13–36 (1918)
28. Brychkov, Yu.A., Prudnikov, A.P.: Integral Transforms of Generalized Functions. Gordon & Breach, New York (1989)
29. Brychkov, Yu.A., Marichev, O.I., Prudnikov, A.P.: Integrals and Series, vol. 1. Elementary Functions. Gordon & Breach, New York (1986); Transl. from А.П. Прудников, Ю.А. Брычков и О.И. Маричев: Интегралы и ряды. Наука, Москва (1983)
30. Brychkov, Yu.A., Marichev, O.I., Prudnikov, A.P.: Integrals and series, vol. 3. More Special Functions. Gordon & Breach, New York (1990); Transl. from А.П. Прудников, Ю.А. Брычков и О.И. Маричев: Интегралы и ряды, Дополнительные главы, Наука, Москва (1986)
31. Buchwald, V.T.: Elastic waves in anisotropic media. Proc. R. Soc. Lond. A **253**, 563–583 (1959)
32. Bureau, F.: Les solutions élémentaires des équations linéaires aux dérivées partielles, pp. 1–37. Marcel Hayez, Bruxelles (1936)
33. Bureau, F.: Essai sur l'intégration des équations linéaires aux dérivées partielles, Mém. Cl. Sci. Acad. Roy. Belgique Sér. 2 **15**(1491), 1–115 (1936)
34. Bureau, F.: Le problème de Cauchy pour une équation linéaire aux dérivées partielles, totalement hyperbolique d'ordre quatre et à quatre variables indépendantes. Bull. Cl. Sci. Acad. Roy. Belgique Sér. 5 **33**, 379–402 (1947)
35. Bureau, F.: Sur la solution élémentaire d'une équation linéaire aux dérivées partielles d'ordre quatre et à trois variables indépendantes, Bull. Cl. Sci. Acad. Roy. Belgique Sér. 5 **33**, 473–484 (1947)
36. Bureau, F.: Sur l'intégration d'une équation linéaire aux dérivées partielles totalement hyperbolique d'ordre quatre et à trois variables indépendantes. Mém. Cl. Sci. Acad. Roy. Belgique Sér. 2 **21**(1570), 1–64 (1948)
37. Bureau, F.: Quelques questions de géométrie suggérées par la théorie des équations aux dérivées partielles totalement hyperboliques. In: Colloque de Géométrie Algébrique tenu à Liège, 155–176 (1949)
38. Bureau, F.: Divergent integrals and partial differential equations. Comm. Pure Appl. Math. **8**, 143–202 (1955)
39. Burridge, R.: The singularity on the plane lids of the wave surface of elastic media with cubic symmetry. Quart. J. Mech. Appl. Math. **20**, 41–56 (1967)

40. Burridge, R., Qian, J.: The fundamental solution of the time-dependent system of crystal optics. Euro. J. Appl. Math. **17**, 63–94 (2006)
41. Calderón, A.P.: Singular integrals. Bull. Am. Math. Soc. **72**, 427–465 (1966)
42. Carleman, T.: L'intégrale de Fourier et questions qui s'y rattachent. Almqvist & Wiksels, Uppsala (1967)
43. Carlson, D.E.: Linear thermoelasticity. In: Truesdell, C.A. (ed.) Handbuch der Physik VI a/2, pp. 297–346 (1972)
44. Chadwick, P.: Thermoelasticity. The dynamical theory. In: Sneddon, I.N., Hill, R. (eds.) Progress in Solid Mechanics I, pp. 263–328. North-Holland, Amsterdam (1964)
45. Chadwick, P., Norris, A.N.: Conditions under which the slowness surface of an anisotropic elastic material is the union of aligned ellipsoids. Quart. J. Mech. Appl. Math. **43**, 589–603 (1990)
46. Chadwick, P., Smith, G.D.: Surface waves in cubic elastic materials. In: Hopkins, H.G., Sewell, M.J. (eds.) Mechanics of Solids, pp. 47–100. Pergamon, Oxford (1982)
47. Chadwick, P., Sneddon, I.N.: Plane waves in an elastic solid conducting heat. J. Mech. Phys. Solids **6**, 223–230 (1958)
48. Chazarain, J., Piriou, A.: Introduction à la théorie des équations aux dérivées partielles linéaires. Gauthiers-Villars, Paris (1981)
49. Cheng, A.H.-D., Antes, H., Ortner, N.: Fundamental solutions of products of Helmholtz and polyharmonic operators. Eng. Anal. Boundary Elem. **14**, 187–191 (1994)
50. Chou, T.W., Pan, Y.C.: Point force solution for an infinite transversely isotropic elastic solid. J. Appl. Mech. **43**, 608–612 (1976)
51. Courant, R., Hilbert, D.: Methods of Mathematical Physics, vol. I. Interscience, New York (1953)
52. Courant, R., Hilbert, D.: Methods of Mathematical Physics, vol. II. Interscience, New York (1962)
53. Dautray, R., Lions, J.L.: Mathematical analysis and numerical methods for science and technology. Functional and Variational Methods, vol. 2. Springer, Berlin (1988)
54. Dautray, R., Lions, J.L.: Mathematical analysis and numerical methods for science and technology. Evolution Problems. I, vol. 5. Springer, Berlin (1992)
55. Dederichs, P.H., Leibfried, G.: Elastic Green's function for anisotropic cubic crystals. Phys. Rev. **188**, 1175–1183 (1969)
56. Delache, S., Leray, J.: Calcul de la solution élémentaire de l'opérateur d'Euler-Poisson-Darboux et de l'opérateur de Tricomi-Clairaut, hyperbolique, d'ordre 2. Bull. Soc. Math. France **99**, 313–336 (1971)
57. Deresiewicz, H., Mindlin, R.D.: Timoshenko's shear coefficient for flexural vibrations of beams. Proceedings of 2nd U.S. National Congress of Applied Mechanics, pp. 175–178. 1954 (1955); Collected papers I: 317–320
58. Dierolf, P., Voigt, J.: Convolution and S'-convolution of distributions. Collect. Math. **29**, 185–196 (1978)
59. Dieudonné, J.: Éléments d'analyse. Tome III. Gauthier-Villars, Paris (1970)
60. Dieudonné, J.: History of Functional Analysis. North-Holland, Amsterdam (1981)
61. Donoghue, W.F., Jr.: Distributions and Fourier Transforms. Academic Press, New York (1969)
62. Duff, G.F.D.: The Cauchy problem for elastic waves in an anisotropic medium. Philos. Trans. Roy. Soc. London, Ser. A, **252**, 249–273 (1960)
63. Duff, G.F.D.: Positive elementary solutions and completely monotonic functions. J. Math. Anal. Appl. **27**, 469–494 (1969)
64. Duff, G.F.D.: Hyperbolic differential equations and waves. In: Garnir, H.G. (ed.) Boundary Value Problems for Linear Evolution Partial Differential Equations, pp. 27–155. Reidel, Dordrecht (1977)
65. Duistermaat, J.J., Kolk, J.A.C.: Distributions. Birkhäuser, Berlin (2010)
66. Dundurs, J., Jahanshahi, A.: Concentrated forces on unidirectionally stretched plates. Quart. J. Mech. Appl. Math. **18**, 129–139 (1965)
67. Duoandikoetxea, J.: Fourier Analysis. American Mathematical Society, Providence, RI (2001)

68. Ehrenpreis, L.: Solution of some problems of division. Part I. Division by a polynomial of derivation. Am. J. Math. **76**, 883–903 (1954)
69. Eringen, A.C., Şuhubi, E.S.: Elastodynamics II. Academic Press, New York (1975)
70. Esser, P.: Second analytic wave front set in crystal optics. Applicable Anal. **24**, 189–213 (1987)
71. Estrada, R., Kanwal, R.P.: Asymptotic Analysis: A Distributional Approach. Birkhäuser, Boston (1994)
72. Fedorov, F.I.: Theory of Elastic Waves in Crystals. Plenum, New York (1968)
73. Feynman, R.P.: Space-time approach to quantum electrodynamics. Phys. Rev. **76**, 769–789 (1949)
74. Feynman, R.P., Leighton, R.B., Sands, M.: The Feynman Lectures on Physics, vol. II, 6th edn. Addison-Wesley, Reading, MA (1977)
75. Fladt, K., Baur, A.: Analytische Geometrie spezieller Flächen und Raumkurven. Vieweg, Braunschweig (1975)
76. Folland, G.B.: Lectures on Partial Differential Equations. Springer, Berlin (1983)
77. Folland, G.B.: Introduction to Partial Differential Equations, 2nd edn. Princeton University Press, Princeton, NJ (1995)
78. Folland, G.B.: Quantum Field Theory. American Mathematical Society, Providence, RI (2008)
79. Flügge, W.: Die Ausbreitung von Biegungswellen in Stäben. Z. Angew. Math. Mech. **22**, 312–318 (1942)
80. Frahm, C.P.: Some novel delta-function identities. Am. J. Phys. **51**, 826–829 (1983)
81. Fredholm, I.: Sur les équations de l'équilibre d'un corps solide élastique. Acta Math. **23**, 1–42 (1900). Œuvres complètes: 17–58
82. Fredholm, I.: Sur l'intégrale fondamentale d'une équation différentielle elliptique à coefficients constants. Rend. Circ. Mat. Palermo **25**, 346–351 (1908). Œuvres complètes: 117–122
83. Friedlander, F.G.: The Wave Equation on a Curved Space-Time. Cambridge University Press, Cambridge (1975)
84. Friedlander, F.G., Joshi, G.M.: Introduction to the Theory of Distributions, 2nd edn., Cambridge University Press, Cambridge (1998)
85. Friedman, A.: Partial Differential Equations of Parabolic Type. Prentice Hall, Englewood Cliffs, NJ (1964)
86. Galler, M.: Fundamentallösungen von homogenen Differentialoperatoren. Diss. Math. **283** (1989)
87. Gal'pern, S.A., Kondrashov, V.E.: The Cauchy problem for differential operators which decompose into wave factors. Trans. Moscow Math. Soc. **16**, 117–145 (1967); Transl. from С.А. Гальперн и В.Е. Кондрашов: Садача Коши для дифференциальныьх операторов распадающыхся на волнобые множители. Труды Моск. Мат. Общ. **5**, 109–136 (1967)
88. Gårding, L.: Linear hyperbolic partial differential equations with constant coefficients. Acta Math. **85**, 1–62 (1951)
89. Gårding, L.: Transformation de Fourier des distributions homogènes. Bull. Soc. Math. France **89**, 381–428 (1961)
90. Gårding, L.: Hyperbolic differential operators. In: Perspectives in Mathematics. Anniversary of Oberwolfach, vol. 1984, pp. 215–247. Birkhäuser, Basel (1984)
91. Gårding, L.: Singularities in linear wave propagation. Lecture Notes in Mathematics, vol. 1555, Springer, Berlin (1987)
92. Gårding, L.: History of the mathematics of double refraction. Arch. History Exact Sci. **40**, 355–385 (1989)
93. Gårding, L.: Some Points of Analysis and their history. American Mathematical Society, Providence, RI (1997)
94. Gårding, L.: Hyperbolic equations in the twentieth century. In: Sémin. Congr., vol. 3, pp. 37–68. Société Mathématique de France, Paris (1998)

95. Gårding, L.: Mathematics and Mathematicians: Mathematics in Sweden before 1950. American Mathematical Society, Providence, RI (1998)
96. Gårding, L., Lions, J.-L.: Functional analysis. Nuovo Cimento, 14(10), 9–66 (1959)
97. Garnir, H.G.: Détermination de la distribution résolvante de certains opérateurs d'évolution décomposables. Bull. Soc. Roy. Sci. Liège 20, 96–99 (1951)
98. Garnir, H.G.: Sur les distributions résolvantes des opérateurs de la physique mathématique, 1ère partie, Bull. Soc. Roy. Sci. Liège 20, 174–185 (1951)
99. Garnir, H.G.: Sur les distributions résolvantes des opérateurs de la physique mathématique, 3ème partie. Bull. Soc. Roy. Sci. Liège 20, 271–287 (1951)
100. Garnir, H.G.: Sur la solution élémentaire pour l'espace indéfini d'un opérateur elliptique décomposable du quatrième ordre. Bull. Cl. Sci. Acad. Roy. Belgique Sér. 5 38, 1129–1141 (1952)
101. Gawinecki, J., Kirchner, G., Łazuka, J.: The fundamental matrix of the system of linear micropolar elasticity. Z. Anal. Anw. 23, 825–831 (2004)
102. Gel'fand, I.M.: Some aspects of functional analysis and algebra. In: Proceedings of International Congress on Mathematicians, vol. 1, pp. 253–276. North-Holland, Amsterdam (1954)
103. Gel'fand, I.M., Shapiro, Z.Ya.: Homogeneous functions and their extensions. Am. Math. Soc. Transl. 8(2), 21–86 (1958); Transl. from И.М. Гельфанд и З.Я. Шапиро: Успехи Мат. Наук (Н.С.) 10(3), 3–70 (1955)
104. Gel'fand, I.M., Shilov, G.E.: Generalized functions. Properties and Operations, vol. I. Academic Press, New York (1964); Transl. from И.М. Гельфанд и Г.Е. Шилов: Обобщённые функции, Вып. 1, Физматгиз, Москва (1958)
105. Gel'fand, I.M., Shilov, G.E.: Generalized functions. Theory of Differential Equations, vol. III. Academic Press, New York (1967); Transl. from И.М. Гельфанд и Г.Е. Шилов: Обобщённые функции. Вып. 3, Физматгиз, Москва (1958)
106. Gel'fand, I.M., Shilov, G.E.: Generalized functions. Spaces of Fundamental and Generalized Functions, vol. II. Academic Press, New York (1968); Transl. from И.М. Гельфанд и Г.Е. Шилов: Обобщённые функции, Вып. 2, Физматгиз, Москва (1958)
107. Gindikin, S.G.: Tube domains and the Cauchy problem. Translations of Mathematical Monographs, vol. 11. American Mathematical Society, Providence, RI (1992)
108. Gindikin, S.G., Volevich, L.R.: The Cauchy problem. In: Partial differential equations, vol. III. In: Egorov, Yu.V., Shubin, M.A. (eds.) Encyclopaedia of Mathematical Sciences, vol. 32, pp. 1–86. Springer, Berlin (1991)
109. Gindikin, S.G., Volevich, L.R.: Distributions and Convolution Equations. Gordon & Breach, Philadelphia (1992); Transl. from Л.Р. Волевич и С.Г. Гиндикин: Обобщённые функции и уравнения в свёртках. Наука, Москва (1994)
110. Glasser, M.L.: The evaluation of lattice sums. III. Phase modulated sums. J. Math. Phys. 15, 188–189 (1974)
111. Goldberg, R.R.: Fourier Transforms. Cambridge University Press, Cambridge (1970)
112. González-Vieli, F.J.: Inversion de Fourier ponctuelle des distributions à support compact. Arch. Math. 75, 290–298 (2000)
113. Gradshteyn, I.S., Ryzhik, I.M.: Table of Integrals, Series, and Products, 5th edn. Academic Press, New York (1972); Transl. from И.С. Градштейн и И.М. Рыжик: Таблицы интегралов, сумм, рядов и произведений. Наука, Москва (1971)
114. Graff, K.F.: Wave Motion in Elastic Solids. Clarendon, Oxford (1975)
115. Gröbner, W., Hofreiter, N.: Integraltafel, 2. Teil: Bestimmte Integrale, 5th edn. Springer, Wien (1973)
116. Gröbner, W., Hofreiter, N.: Integraltafel, 1. Teil: Unbestimmte Integrale, 5th edn. Springer, Wien (1975)
117. Grothendieck, A.: Sur certains espaces de fonctions holomorphes I. J. reine angewandte Math. 192, 35–64 (1953)
118. Grubb, G.: Distributions and Operators. Springer, New York (2009)
119. von Grudzinski, O.: Quasihomogeneous Distributions. Elsevier, Amsterdam (1991)
120. Grünwald, J.: Über die Ausbreitung der Wellenbewegungen in optisch-zweiachsigen elastischen Medien. Festschrift für Ludwig Boltzmann, J.A. Barth, Leipzig, 1904, pp. 518–527

121. Gurarii, V.P.: Group methods in commutative harmonic analysis. Commutative Harmonic Analysis, vol. II. In: Havin, V.P., Nikolski, N.K. (eds.) Encylopaedia of Mathematical Sciences, vol. 25. Springer, Berlin (1998)

122. Guzmán-Partida, M., Ortner, N., Wagner, P.: M. Riesz' kernels as boundary values of conjugate Poisson kernels. Bull. Sci. Math. **135**, 291–302 (2011)

123. Hadamard, J.: Le problème de Cauchy et les équations aux dérivées partielles linéaires hyperboliques. Nouv. éd., Hermann, Paris (1932)

124. Heald, M.A., Marion, J.B.: Classical Electromagnetic Radiation, 3rd edn. Harcourt Brace, New York (1995)

125. Herglotz, G.: Über die Integration linearer partieller Differentialgleichungen mit konstanten Koeffizienten I–III. Ber. Sächs. Akademie Wiss. **78**, 93–126, 287–318 (1926); **80**, 69–114 (1928) Ges. Schriften: 496–607

126. Herglotz, G.: Gesammelte Schriften. Vandenhoek&Ruprecht, Göttingen (1979)

127. Herglotz, G.: Vorlesungen über die Mechanik der Kontinua. Teubner, Leipzig (1985)

128. Hetnarski, R.B.: Solution of the coupled problem of thermoelasticity in the form of series of functions. Arch. Mech. Stos. **16**, 919–941 (1964)

129. Hervé, M.: Transformation de Fourier et distributions. Presses university, France (1986)

130. Hewitt, E., Stromberg, K.: Real and Abstract Analysis. Springer, Berlin (1965)

131. Hirsch, F., Lacombe, G.: Elements of Functional Analysis. Graduate Texts in Mathematics, vol. 192. Springer, Berlin (1999)

132. de Hoop, A.T.: A modification of Cagniard's method for solving seismic pulse problems. Appl. Sci. Res. B **8**, 349–356 (1959/60)

133. Hörmander, L.: On the theory of general partial differential operators. Acta Math. **94**, 161–248 (1955)

134. Hörmander, L.: Local and global properties of fundamental solutions. Math. Scand. **5**, 27–39 (1957)

135. Hörmander, L.: On the division of distributions by polynomials. Ark. Mat. **3**, 555–568 (1958)

136. Hörmander, L.: Linear Partial Differential Operators. Grundlehren Math. Wiss., vol. 116, Springer, Berlin (1963)

137. Hörmander, L.: Distribution Theory and Fourier Analysis. Lectures 1972/73, Mimeographed Notes, Lund (1972)

138. Hörmander, L.: The analysis of linear partial differential operators, vol. II. Differential operators with constant coefficients. Grundlehren Math. Wiss., vol. 257, Springer, Berlin (1983)

139. Hörmander, L.: The analysis of linear partial differential operators, vol. I. Distribution theory and Fourier analysis. In: Grundlehren Math. Wiss., vol. 256, 2nd edn. Springer, Berlin (1990)

140. Hörmander, L.: Lectures on Harmonic Analysis. Lund (1995)

141. Horváth, J.: Topological Vector Spaces and Distributions, vol. I. Addison-Wesley, Reading, MA (1966)

142. Horváth, J.: An introduction to distributions. Amer. Math. Monthly **77**, 227–240 (1970)

143. Horváth, J.: Distribuciones definidas por prolongación analítica. Rev. Colombiana Mat. **8**, 47–95 (1974)

144. Horváth, J.: Sur la convolution des distributions. Bull. Sci. Math. **98**(2), 183–192 (1974)

145. Horváth, J.: Distributionen und Pseudodifferentialoperatoren. Vorlesung, Innsbruck, pp. 1–68 (1977)

146. Horváth, J.: Composition of hypersingular integral operators. Appl. Anal. **7**, 171–190 (1978)

147. Horváth, J., Ortner, N., Wagner, P.: Analytic continuation and convolution of hypersingular higher Hilbert–Riesz kernels. J. Math. Anal. Appl. **123**, 429–447 (1987)

148. Ibragimov, N.H., Mamontov, E.V.: On the Cauchy problem for the equation $u_{tt} - u_{xx} - \sum_{i,j=1}^{n} a_{ij}(x, t)u_{y_i y_j} = 0$. Mat. Sb. (N.S.) **102(144)**, 391–409 (1977)

149. de Jager, E.M.: Theory of distributions. In: Roubine, E. (ed.) Mathematics Applied to Physics, pp. 52–110. Springer, Berlin (1970)

150. Jahanshahi, A.: Some notes on singular solutions and the Green's functions in the theory of plates and shells. J. Appl. Mech. **31**, 441–447 (1964)

151. John, F.: Plane Waves and Spherical Means Applied to Partial Differential Equations. Interscience, New York (1955)
152. Jones, D.S.: The Theory of Generalised Functions, 2nd edn. Cambridge University Press, Cambridge (1982)
153. Jörgens, K.: Linear Integral Operators. Pitman, Boston (1982)
154. Komech, A.I.: Linear partial differential equations with constant coefficients. Partial differential equations, vol. II. In: Egorov, Yu.V., Shubin, M.A. (ed.) Encylopaedia of Mathematical Sciences, vol. 31, pp. 121–255. Springer, Berlin (1994)
155. König, H.: An explicit formula for fundamental solutions of linear partial differential equations with constant coefficients. Proc. Am. Math. Soc. **120**, 1315–1318 (1994)
156. Krantz, S.G., Parks, H.R.: A Primer of Real Analytic Functions, 2nd edn. Birkhäuser, Boston (2002)
157. Krée, P.: Distributions quasihomogènes et intégrales singulières. Bull. Soc. Math. France, Mém. 20, pp. 47 (1969)
158. Kröner, E.: Das Fundamentalintegral der anisotropen elastischen Differentialgleichungen. Zeitschrift f. Physik **136**, 402–410 (1953)
159. Larsen, R.: Functional Analysis. Dekker, New York (1973)
160. Lavoine, J.: Transformation de Fourier. CNRS, Paris (1963)
161. Lemoine, C.: Fourier transforms of homogeneous distributions. Ann. Scuola Normale Pisa **26**(3), 117–149 (1972)
162. Léonard, P.: Solution élémentaire de l'opérateur des ondes. Bull. Soc. Roy. Sci. Liège **50**, 36–46 (1981)
163. Leray, J.: Hyperbolic Differential Equations. Institute of Advanced Study, Princeton (1952)
164. Levi, E.E.: Sulle equazioni lineari totalmente ellitiche alle derivate parziali. Rend. Circ. Mat. Palermo **24**, 275–317 (1907)
165. Liess, O.: Conical refraction and higher microlocalization. Lecture Notes in Mathematics, vol. 1555. Springer, Berlin (1993)
166. Linés, E.: Resolución en forma finita del problema de Cauchy sobre una superficie cualquiera en la ecuación de ondas con cualquier número de variables, y en otras notables de tipo hiperbólico con coeficientes constantes. Collect. Math. **2**, 1–86 (1949)
167. Lions, J.L., Magenes, E.: Non-homogeneous boundary value problems and applications, vol. I. Grundlehren Math. Wiss., vol. 181. Springer, Berlin (1972); Transl. from: Problèmes aux limites non homogènes, vol. I. Dunod, Paris (1968)
168. Łojasiewicz, S.: Sur le problème de la division. Studia Math. **18**, 87–136 (1959)
169. Łojasiewicz, S.: Sur le problème de la division. Rozprawy Matematyczne **22**, 1–57 (1961)
170. Lorenzi, A.: On elliptic equations with piecewise constant coefficients. II. Ann. Scuola Norm. Sup. Pisa **26**(3), 839–870 (1972)
171. Love, A.E.H.: The Mathematical Theory of Elasticity, 4th edn. Dover, New York (1944)
172. Ludwig, D.: Conical refraction in crystal optics and hydromagnetics, Comm. Pure Appl. Math. **14**, 113–124 (1961)
173. Lützen, J.: The Prehistory of the Theory of Distributions. Springer, Berlin (1981)
174. Malgrange, B.: Existence et approximation des solutions des équations aux dérivées partielles et des équations de convolution. Ann. Inst. Fourier **6**, 271–355 (1955/56)
175. Malgrange, B.: L. Schwartz et la théorie des distributions. Astérisque **131/132**, 25–34 (1985)
176. Méthée, P.-D.: Sur les distributions invariantes dans le groupe des rotations de Lorentz. Comment. Math. Helv. **28**, 225–269 (1954)
177. Méthée, P.-D.: Transformées de Fourier de distributions invariantes liées à la résolution de l'équation des ondes. In: Colloque int. du CNRS "La théorie des équations aux dérivées partielles", Nancy, pp. 145–163 (1956)
178. Metzner, W., Neumayr, A.: Reduction formula for Fermion loops and density correlations of the 1D Fermi gas. J. Stat. Phys. **96**, 613–626 (1999)
179. Meyer, Y., Coifman, R.: Wavelets. Calderón–Zygmund and Multilinear Operators. Cambridge University Press, Cambridge (1997)

180. Mindlin, R.D.: Influence of rotatory inertia and shear on flexural motions of isotropic elastic plates. J. Appl. Mech. **18**, 31–38 (1951)

181. Mindlin, R.D.: In: Deresiewicz, H. Bienick, M.P., DiMaggio, F.L. (eds.) Collected Papers, vol. I. Springer, New York (1983)

182. Mizohata, S.: The Theory of Partial Differential Equations. Cambridge University Press, Cambridge (1973)

183. Morrison, J.A.: Wave propagation with three-parameter models. Quart. J. Appl. Math. **14**, 153–169 (1956)

184. Muhlisov, F.G.: Construction of the fundamental solution for certain equations of elliptic type. Funct. Anal. Theory Funct. (Izdat. Kazan. University) **8**, 134–141 (1971). original Russ.: Ф.Г. Мухлисов: Построение фундамнетального решения для некоторых уравнений эллиптического типа, Функциональныи Анализ и Теория Функции. Вып. 8

185. Mura, T.: Micromechanics of Defects in Solids, 2nd edn. Kluwer, Dordrecht (1987)

186. Musgrave, M.J.P.: Crystal Acoustics. Holden-Day, San Francisco (1970)

187. Musgrave, M.J.P.: On the propagation of elliptically polarized plane elastic waves along acoustic (or optical) axes. Quart. J. Mech. Appl. Math. **39**, 567–580 (1986)

188. Nardini, R.: Sul comportamento asintotico della soluzione di un problema al contorno della magnetoidrodinamica I, II. Atti Ac. Naz. Lincei Rend. **16**(8), 225–231, 341–248 (1954)

189. Newman, D.J.: Fourier uniqueness via complex variables. Am. Math. Monthly **81**, 379–380 (1974)

190. Norris, A.N.: Dynamic Green's functions in anisotropic piezoelectric, thermoelastic and poroelastic solids. Proc. Roy. Soc. London A **447**, 175–188 (1994)

191. Nowacki, W.: Thermoelasticity. Pergamon, Oxford (1962)

192. Nowacki, W.: Coupled fields in elasticity. In: Fichera, G. (ed.) Trends in Applications of Pure Mathematics to Mechanics, pp. 263–280. Pitman, London (1976)

193. Nowacki, W.: Distortion problems of elasticity. In: Sneddon, I.N. (ed.) Application of Integral Transforms in the Theory of Elasticity, pp. 171–240. Springer, Wien (1976)

194. Oberhettinger, F.: Tables of Bessel Transforms. Springer, Berlin (1972)

195. Oberhettinger, F.: Fourier Expansions. Academic Press, New York (1973)

196. Oberhettinger, F.: Tables of Fourier Transforms and Fourier Transforms of Distributions. Springer, Berlin (1990)

197. Ortner, N.: Fundamentallösungen und Existenz von schwachen Lösungen linearer, partieller Differentialgleichungen mit konstanten Koeffizienten. Ann. Acad. Sci. Fenn. Ser. A1 **4**, 3–30 (1978/79)

198. Ortner, N.: Convolution des distributions et des noyaux euclidiens. In: Choquet, G. et al. (eds.) Sém. Initiation à l'Analyse, exp. no. 12 (1979/80)

199. Ortner, N.: Faltung hypersingulärer Integraloperatoren. Math. Ann. **248**, 19–46 (1980)

200. Ortner, N.: Regularisierte Faltung von Distributionen, I, II. Z. Angew. Math. Physik **31**, 133–173 (1980)

201. Ortner, N.: A fundamental solution of the product of Laplace operators. Publ. Math. Debrecen **33**, 180–182 (1986)

202. Ortner, N.: Methods of construction of fundamental solutions of decomposable linear differential operators. In: Brebbia, C.A. (ed.) Boundary Elements IX, vol. 1, pp. 79–97. Computational Mechanics Publications, Southampton (1987)

203. Ortner, N.: Die Fundamentallösung des Timoshenko- und des Boussinesq-Operators. Z. Angew. Math. Mech. **68**, 547–553 (1988)

204. Ortner, N.: On convolvability conditions for distributions. Monatsh. Math. **160**, 313–335 (2010)

205. Ortner, N.: Fourier transforms in the "classical sense", Schur spaces and a new formula for the Fourier transforms of slowly increasing $O(p, q)$—invariant functions. Indag. Math. **24**, 142–160 (2013)

206. Ortner, N., Wagner, P.: Applications of weighted $\mathcal{D}'_{L^p}$-spaces to the convolution of distributions. Bull. Polish Acad. Sci. Math. **37**, 579–595 (1989)

207. Ortner, N., Wagner, P.: Some new fundamental solutions. Math. Meth. Appl. Sci. **12**, 439–461 (1990)
208. Ortner, N., Wagner, P.: On the fundamental solutions of the operators of S. Timoshenko and R.D. Mindlin. Math. Meth. Appl. Sci. **15**, 525–535 (1992)
209. Ortner, N., Wagner, P.: On the fundamental solution of the operator of dynamic linear thermoelasticity. J. Math. Anal. Appl. **170**, 524–550 (1992)
210. Ortner, N., Wagner, P.: Fundamental solutions of hyperbolic differential operators and the Poisson summation formula. Integral Transforms Spec. Funct. **1**, 183–196 (1993)
211. Ortner, N., Wagner, P.: Feynman integral formulae and fundamental solutions of decomposable evolution operators. Proc. Steklov Inst. **203**, 305–322 (1995)
212. Ortner, N., Wagner, P.: On the evaluation of one-loop Feynman amplitudes in Euclidean quantum field theory. Ann. Inst. H. Poincaré Phys. Théor. **62**, 81–110 (1995)
213. Ortner, N., Wagner, P.: A short proof of the Malgrange–Ehrenpreis theorem. In: Dierolf, S., Dineen, S., Domański, P. (eds.) Functional Analysis. Proceedings of the 1st International Workshop in Trier, Germany, 1994, pp. 343–352. de Gruyter, Berlin (1996)
214. Ortner, N., Wagner, P.: Solution of the initial-boundary value problem for the simply supported semi-infinite Timoshenko beam. J. Elast. **42**, 217–241 (1996)
215. Ortner, N., Wagner, P.: A survey on explicit representation formulae for fundamental solutions of linear partial differential operators. Acta Appl. Math. **47**, 101–124 (1997)
216. Ortner, N., Wagner, P.: Deduction of L. Hörmander's extension of Ásgeirsson's mean value theorem. Bull. Sci. Math. **127**, 835–843 (2003)
217. Ortner, N., Wagner, P.: Fundamental matrices of homogeneous hyperbolic systems. Applications to crystal optics, elastodynamics and piezoelectromagnetism. Z. Angew. Math. Mech. **84**, 314–346 (2004)
218. Ortner, N., Wagner, P.: Convolution groups for quasihyperbolic systems of differential operators. Note di Matematica **25**(2), 139–157 (2005/2006)
219. Ortner, N., Wagner, P.: Distribution-valued analytic functions. Theory and applications, Max-Planck-Institut, Leipzig (2008); http://www.mis.mpg.de/preprints/ln/lecturenote-3708.pdf; published in tredition, Hamburg (2013)
220. Ortner, N., Wagner, P.: The fundamental matrix of the system of linear elastodynamics in hexagonal media. Solution to the problem of conical refraction. IMA J. Appl. Math. **73**, 412–447 (2008)
221. Ortner, N., Wagner, P.: On conical refraction in hexagonal and cubic media. SIAM J. Appl. Math. **70**, 1239–1259 (2009)
222. Ortner, N., Wagner, P.: On the static term for the electric field in crystal optics. Quart. J. Mech. Appl. Math. **62**, 311–319 (2009)
223. Ortner, N., Wagner, P.: On the Fourier transform of Lorentz invariant distributions. Functiones et Approximatio **44**, 133–151 (2011). Volume dedicated to the memory of S. Dierolf
224. Palamodov, V.P.: Distributions and harmonic analysis. In: Nikol'skij, N.K. (ed.) Commutative harmonic analysis, vol. III. Encylopaedia of Mathematical Sciences, vol. 72, pp. 1–127. Springer, Berlin (1995)
225. Paneyah, B.P.: On the existence and uniqueness of the solution of the n-metaharmonic equation in unbounded space. Vestnik Mosk. University **5**, 123–135 (1959); original Russ.: Б.П. Панеях: О существовании и единственности решения n-метагармонического уравнения в неограниченном пространстве, Вестник Моск. Университета
226. Payton, R.G.: Elastic Wave Propagation in Transversely Isotropic Media. Kluwer, The Hague (1983)
227. Payton, R.G.: Wave propagation in a restricted transversely isotropic elastic solid whose slowness surface contains conical points. Quart. J. Mech. Appl. Math. **45**, 183–197 (1992)
228. Petersen, B.E.: Introduction to the Fourier Transform and Pseudo-differential Operators. Pitman, Boston (1983)
229. Piskorek, A.: Radon-Transformation und hyperbolische Differentialgleichungen der mathematischen Physik. In: Brosowski, B., Martensen, E. (eds.) Methoden und Verfahren der mathematischen Physik, Bd. 10, pp. 85–97. Bible Institute, Mannheim (1973)

230. Poincaré, H.: Leçons sur la théorie mathématique de la lumière, Paris (1889)
231. Poruchikov, V.B.: Methods of the Classical Theory of Elastodynamics. Springer, Berlin (1993)
232. Rauch, J.: Partial differential equations. Graduate Texts in Mathematics, vol. 128. Springer, New York (1991)
233. Ricci, F., Stein, E.M.: Harmonic analysis on nilpotent groups and singular integrals. I. Oscillatory integrals. J. Funct. Anal. **73**, 179–194 (1987)
234. Riesz, M.: L'intégrale de Riemann-Liouville et le problème de Cauchy pour l'équation des ondes. In: Soc. Math de France. Conf. à la Réunion Int. des Math., Paris, Juillet, 153–170 (1937); Collected papers: 527–544, Springer, Berlin (1988)
235. Riesz, M.: L'intégrale de Riemann-Liouville et le problème de Cauchy. Acta Math. **81**, 1–223 (1949). Collected papers: 571–793, Springer, Berlin (1988)
236. Robertson, A.P., Robertson, W.I.: Topological Vector Spaces, 2nd edn. Cambridge University Press, Cambridge (1973)
237. Roider, B.: Sur la convolution des distributions. Bull. Sci. Math. **100**(2), 193–199 (1976)
238. Rudin, W.: Lectures on the edge-of-the-wedge theorem. Conference Board of the Mathematical Sciences Regional Conference Conference Board of the Series in Mathematics, vol. 6. American Mathematical Society, Providence, RI (1971). Extended reprint 1997
239. Rudin, W.: Functional Analysis. McGraw-Hill, New York (1973)
240. Rudin, W.: Real and Complex Analysis, 3rd edn. McGraw-Hill, New York (1987)
241. Schwartz, L.: Les équations de convolution liées au produit de composition. Ann. Inst. Fourier **2**, 19–49 (1951)
242. Schwartz, L.: Séminaire Schwartz. Année 1953/54. Produits tensoriels topologiques d'espaces vectoriels topologiques. Espaces vectoriels topologiques nucléaires. Applications, Fac. Sci. Paris (1954)
243. Schwartz, L.: Séminaire Schwartz. Année 1954/55. Équations aux dérivées partielles. Fac. Sci. Paris (1955)
244. Schwartz, L.: Matemática y física cuántica, Notas, Universidad de Buenos Aires (1958)
245. Schwartz, L.: Méthodes mathématiques pour les sciences physiques. Hermann, Paris (1965)
246. Schwartz, L.: Théorie des distributions. Nouv. éd., Hermann, Paris (1966)
247. Schwartz, L.: Application of Distributions to the Theory of Elementary Particles in Quantum Mechanics. Gordon & Breach, New York (1968)
248. Schwartz, L.: Notice sur les travaux scientifiques de L. Schwartz. In: Nachbin, L. (ed.) Mathematical Analysis and Applications, Part A, pp. 1–25. Academic Press, New York (1981)
249. Seeley, R.: Distributions on surfaces, Report TW 78, Mathematical Centre, Amsterdam (1962)
250. Shilov, G.E.: Generalized Functions and Partial Differential Equations. Gordon & Breach, New York (1968)
251. Shimakura, N: Partial Differential Operators of Elliptic Type. American Mathematical Society, Providence, RI (1992)
252. Shiraishi, R.: On the definition of convolution for distributions. J. Sci. Hiroshima University Ser. A **23**, 19–32 (1959)
253. Shreves, R.W., Stadler, W.: The transient and steady-state response of the infinite Bernoulli-Euler beam with damping and an elastic foundation. Quart. J. Mech. Appl. Math. **23**, 197–208 (1970)
254. Sneddon, I.N.: The Linear Theory of Thermoelasticity. Springer, Wien (1974)
255. Sobolev, S.L.: The Fundamental Solution of Cauchy's Problem for the Equation $\frac{\partial^3 u}{\partial x \partial y \partial z} - \frac{1}{4}\frac{\partial u}{\partial t} = F(x, y, z, t)$. Dokl. Akad. Nauk SSSR **129**, 1246–1249 (1959). original Russ.: С.Л. Соболев: Фундаментальное решение задачу Коши для уравнения $\frac{\partial^3 u}{\partial x \partial y \partial z} - \frac{1}{4}\frac{\partial u}{\partial t} = F(x, y, z, t)$, Докл. Акад. Наук СССР
256. Sokhotski, Yu.V.: On Definite Integrals and Functions Used in Series Expansions. M. Stalyusevich, St. Petersburg, Russia (1873). original Russ.: Ю.В. Сохоцкий: Об определённых интегралах и функциях, употребляемых при разложении в ряды, С.-Петерсбург (1873)

257. Sokolnikoff, I.S.: Mathematical Theory of Elasticity, 2nd edn. McGraw-Hill, New York (1956)
258. Somigliana, C.: Sui sistemi simmetrici di equazioni a derivate parziali. Ann. Mat. Pura Appl. **22**(2), 143–156 (1894)
259. Sommerfeld, A.: Vorlesungen über theoretische Physik II (Mechanik der deformierbaren Medien), 5. Aufl. Geest&Portig, Leipzig (1964)
260. Stampfer, F., Wagner, P.: A mathematically rigorous formulation of the pseudopotential method. J. Math. Anal. Appl. **342**, 202–212 (2008)
261. Stampfer, F., Wagner, P.: Mathematically rigorous formulation of the Fermi pseudopotential for higher-partial-wave scattering in arbitrary dimension. Phys. Rev. A **81**, 052710 (2010)
262. Stein, P.: Die Anwendung der Singularitätenmethode zur Berechnung orthogonal anisotroper Rechteckplatten, einschließlich Trägerrosten. Stahlbau-Verlag, Köln (1959)
263. Stein, E.M.: Singular integrals and differentiability properties of functions. Princeton University Press, Princeton, NJ (1970)
264. Stein, E.M., Weiss, G.: Fourier Analysis on Euclidean Spaces. Princeton University Press, Princeton, NJ (1971)
265. Stokes, G.G.: On the dynamical theory of diffraction. Trans. Cambridge Phil. Soc. **9**, 1–62 (1849)
266. Strichartz, R.S.: A Guide to Distribution Theory and Fourier Transforms. World Scientific, Singapore (1994)
267. Szmydt, Z., Ziemian, B.: Explicit invariant solutions for invariant linear differential equations. Proc. Roy. Soc. Edinburgh **98**, 149–166 (1984)
268. Szmydt, Z., Ziemian, B.: An invariance method for constructing fundamental solutions for $P(\square_{mn})$. Ann. Polon. Math. **46**, 333–360 (1985)
269. Timoshenko, S., Young, D.H.: Vibration problems in engineering, 3rd edn. Van Nostrand, New York (1955)
270. Treves, F.: Solution élémentaire d'équations aux dérivées partielles dépendant d'un paramètre. C. R. Acad. Sci. Paris **242**, 1250–1252 (1956)
271. Treves, F.: Lectures on linear partial differential equations with constant coefficients. Notas de Matemática, N. 27, Rio de Janeiro (1961)
272. Treves, F.: Linear partial differential equations with constant coefficients. Gordon & Breach, New York (1967)
273. Treves, F.: Topological Vector Spaces, Distributions and Kernels. Academic Press, New York (1967)
274. Treves, F.: Basic Linear Partial Differential Equations. Academic Press, New York (1975)
275. Treves, F., Zerner, M.: Zones d'analyticité des solutions élémentaires. Bull. Soc. Math. France **95**, 155–192 (1967)
276. Treves, F., Pisier, G., Yor, M.: Laurent Schwartz (1915–2002). Not. Am. Math. Soc. **50**, 1072–1078 (2003)
277. Trimèche, K.: Transmutation Operators and Mean-Periodic Functions Associated with Differential Operators. Harwood, Chur (1988)
278. Uflyand, Ya.S.: Wave propagation with transverse vibrations of bars and plates. Prikl. Mat. Mekh. **12**, 287–300 (1948). original Russ.: Я. С. Уфлянд: Распространение волн при поперечных колебаниях стержней и пластин, Прикл. Мат. Мех
279. Vladimirov, V.S.: Equations of Mathematical Physics. Dekker, New York (1971)
280. Vladimirov, V.S.: Generalized Functions in Mathematical Physics, 2nd edn. Mir, Moscow, 1979; Transl. from В.С. Владимиров: Обобщённые функции в математической физике, Наука, Москва (1976)
281. Vladimirov, V.S., Drozzinov, Yu.N., Zavialov, B.I.: Tauberian Theorems for Generalized Functions. Kluwer, Dordrecht (1988); Transl. from В.С. Владимиров, Ю.Н. Дрожжинов, Б.И. Завьялов: Многомерные тауберовы теоремы для обобщённых функций, Наука, Москва (1986)
282. Vo-Khac, K.: Distributions, analyse de Fourier, opérateurs aux dérivées partielles. Tome 1, Vuibert, Paris (1972)

283. Vo-Khac, K.: Distributions, analyse de Fourier, opérateurs aux dérivées partielles. Tome 2, Vuibert, Paris (1972)
284. Wagner, P.: Die Singularitätenfunktionen der gespannten Platte und der Kreiszylinderschale, Z. Angew. Math. Physik **35**, 723–727 (1984)
285. Wagner, P.: Parameterintegration zur Berechnung von Fundamentallösungen, Diss. Math. **230**, 1–50 (1984)
286. Wagner, P.: Zur Faltung von Distributionen. Math. Annalen **276**, 467–485 (1987)
287. Wagner, P.: Bernstein-Sato-Polynome und Faltungsgruppen zu Differentialoperatoren. Z. Anal. Anw. **8**, 407–423 (1989)
288. Wagner, P.: Fundamental matrix of the system of dynamic linear thermoelasticity. J. Therm. Stresses **17**, 549–565 (1994)
289. Wagner, P.: Evaluation of non-relativistic one-loop Feynman integrals by distributional methods. J. Math. Phys. **39**, 2428–2436 (1998)
290. Wagner, P.: Fundamental solutions of real homogeneous cubic operators of principal type in three dimensions. Acta Math. **182**, 283–300 (1999)
291. Wagner, P.: A fundamental solution of N. Zeilon's operator. Math. Scand. **86**, 273–287 (2000)
292. Wagner, P.: On the fundamental solutions of a class of elliptic quartic operators in dimension 3. J. Math. Pures Appl. **81**, 1191–1206 (2002)
293. Wagner, P.: On the explicit calculation of fundamental solutions. J. Math. Anal. Appl. **297**, 404–418 (2004)
294. Wagner, P.: On the fundamental solutions of a class of hyperbolic quartic operators in dimension 3. Ann. Mat. Pura Appl. **184**, 139–159 (2005)
295. Wagner, P.: The Herglotz formula and fundamental solutions of hyperbolic cubic operators in $\mathbf{R}^4$. Integral Transforms Spec. Funct. **17**, 307–314 (2006)
296. Wagner, P.: A new constructive proof of the Malgrange–Ehrenpreis theorem. Amer. Math. Monthly **116**, 457–462 (2009)
297. Wagner, P.: Distributions supported by hypersurfaces. Applicable Anal. **89**, 1183–1199 (2010)
298. Wagner, P.: The singular terms in the fundamental matrix of crystal optics. Proc. Roy. Soc. A **467**, 2663–2689 (2011)
299. Wagner, P.: On convolution in weighted $\mathcal{D}'_{L^p}$-spaces. Math. Nachrichten **287**, 472–477 (2014)
300. Weierstrass, K.: Zur Integration der linearen partiellen Differentialgleichungen mit constanten Coeffizienten. In: Mathematische Werke, 1. Band, Abhandlungen I, 275–295, Mayer & Müller, Berlin (1894)
301. Whitham, G.B.: Linear and Nonlinear Waves. Wiley, New York (1974)
302. Willis, J.R.: Dislocations and Inclusions. J. Mech. Phys. Solids **13**, 377–395 (1965)
303. Willis, J.R.: The elastic interaction energy of dislocation loops in anisotropic media. Quart. J. Mech. Appl. Math. **18**, 419–433 (1965)
304. Yano, T.: On the theory of b-functions. Publ. Res. Inst. Math. Sci., Kyoto University **14**, 111–202 (1978)
305. Zeidler, E.: Quantum Field Theory I: Basics in Mathematics and Physics, corrected 2nd printing. Springer, Berlin (2009)
306. Zeilon, N.: Das Fundamentalintegral der Allgemeinen Partiellen Linearen Differentialgleichung mit konstanten Koeffizienten. Arkiv f. Mat. Astr. o. Fys. **6**, 1–32 (1911)
307. Zemanian, A.H.: Generalized Integral Transforms. Interscience, New York (1968)
308. Zorich, V.A.: Mathematical Analysis, vol. II. Springer, Berlin (2004)
309. Zuily, C.: Problems in Distributions and Partial Differential Equations. Elsevier, Paris (1988); Transl. from C. Zuily: Problèmes de distributions, Hermann, Paris (1978)

Index

© Springer International Publishing Switzerland 2015
N. Ortner, P. Wagner, *Fundamental Solutions of Linear Partial Differential*
Operators, DOI 10.1007/978-3-319-20140-5

Printed in the United States
By Bookmasters